PLANT ANATOMY AND EMBRYOLOGY

By the same author

A TEXTBOOK OF BOTANY VOL. 1
(Algae, Fungi, Bacteria, Mycoplasmas, Viruses, Lichens and Elementary Plant Pathology)

A TEXTBOOK OF BOTANY VOL. 2
(Bryophyta, Pteridophyta, Gymnosperms and Palaeobotany)

A TEXTBOOK OF BOTANY VOL. 3.
(Plant Anatomy and Economic Botany)

PLANT PHYSIOLOGY

A TEXTBOOK OF ALGAE

A TEXTBOOK OF PTERIDOPHYTA

ECONOMIC BOTANY

PLANT ANATOMY AND EMBRYOLOGY

S.N. PANDEY
Department of Botany
D.A.V. College
University of Kanpur

AJANTA CHADHA
Department of Botany
S.N. Sen P.G. College
University of Kanpur

VIKAS® PUBLISHING HOUSE PVT LTD

VIKAS® PUBLISHING HOUSE PRIVATE LIMITED
E-28, Sector-8, Noida-201301 (UP), India
Phone: +91-120-4078900 • Fax: +91-120-4078999
Registered Office: A-27, 2nd Floor, Mohan Co-operative Industrial Estate, New Delhi-110044
E-mail: helpline@vikaspublishing.com • Website: www.vikaspublishing.com

SALES AND MARKETING OFFICES OF VIKAS® AND S CHAND AND COMPANY LIMITED

Chennai : Ph: 044-2363 2120, chennai@schandpublishing.com
Guwahati : Ph: 0361-2738 811, 2735 640, guwahati@schandpublishing.com
Hyderabad : Ph: 040-4018 6018, hyderabad@schandpublishing.com
Jalandhar : Ph: 0181-4645 630, jalandhar@schandpublishing.com
Kolkata : Ph: 033-2335 7458, 2335 3914, kolkata@schandpublishing.com
Lucknow : Ph: 0522-4003 633, lucknow@schandpublishing.com
Mumbai : Ph: 022-2500 0297, mumbai@schandpublishing.com
Patna : Ph: 0612-2260 011, patna@schandpublishing.com

Plant Anatomy and Embryology
ISBN: 978-81259-0240-9

Product Code: V6PAE68BOTN10ENAA96O

First Published 1996
Reprinted several times between 1996 and 2022, 2023
Reprint 2024

PRINTED IN INDIA
By Vikas Publishing House Private Limited, Plot 20/4, Site-IV, Industrial Area Sahibabad, Ghaziabad-201010 and Published by Vikas Publishing House Private Limited, E-28, Sector-8, Noida-201301

Preface

Plant Anatomy and Embryology is a much awaited requel to the already popular and appreciated books by the authors. The present book is being humbly offered to satisfy the demand for a better, exhaustive and authoritative coverage on various aspects of plant anatomy, embryology and palynology. These branches of angiosperms usually form a part of the same paper in most of the Universities. Such a combination would certainly augment its value to the students.

Simplicity coupled with originality of presentation will make the text most lucid. The vast fund of literature has been compiled in a manner of suitable understanding for students. The book comprises two parts.

Part 1 deals with Plant Anatomy covering all topics related to the histological aspects of the various plant parts. Topics such as *Nodal Anatomy*, *Anatomy of Floral parts*, *Fruit wall and Seed coat*, *Ecological Anatomy* and *Systematic Plant Anatomy* will be of great utility to the students of various levels.

Part 2 deals with Plant Embryology And Palynology and covers all topics starting from microsporogenesis and megasporogenesis to Embryogenesis. Of special interest are chapters on *Sexual compatibility*, *Apomixis*, *Polyembryony*, *Embryology in relation to Taxonomy*, *Experimental Embryology*, *Pollen allergy*, etc. The exhaustive and recent literature on these related topics will be of considerable help to the competitive and regular students.

Of late, in certain Indian Universities, Palynology has been introduced both at under-graduate and post-graduate levels. In view of this step, it has been included in separate chapters in its Part 2. Special emphasis has been laid on *Pollen terminology*, *Method of Studying Pollens*, *Applied aspects of Palynology*, *Pollen Allergies*, etc. Inclusion of Palynology, covering its essential aspects, will be the first approach in text form at this level of studies. Hope, such students will certainly be benifitted by this coverage.

The aesthetic value of the book has been greatly enhanced by the lucid and true-to-life diagrams for which sincere thanks are expressed to the artist Mr. S.N. Dey.

We have sincerely endeavoured to present a complete book which aims at satisfying academicions and students equally. Suggestions, comments and valued criticisms are invited for the improvement of the book.

AUTHORS

Contents

PART ONE
PLANT ANATOMY

PART TWO

EMBRYOLOGY AND PALYNOLOGY

PART ONE

PLANT ANATOMY

Plant Body
Tissue Systems
Anatomy of Stem, Root and Leaves
Secondary Growth
Abscission and Healing of Wounds
Nodal Anatomy
Anatomy of Floral parts, Fruit Wall and Seed Coat
Ecological Anatomy
Systematic Plant Anatomy

CHAPTER 1

The Plant Body

Plant kingdom shows great diversity in their forms and organisations, ranging from unicellular microscopic to giant red wood trees. The most evolved plants are seed plants which constitute the conspicuous and dominant part of earth's vegetation. These show considerable division of labour accompanied with differentiation of specialised organs.

The plant body of a typical seed plant basically consists of two parts, the *root* and the *shoot*. The shoot grows above the ground, upward into the air and comprises of stem, bearing leaves at the nodes. The root is subterranean forming the lower part and grows downward into the soil.

The characteristic feature of stem is the presence of leaves on them. Leaves are not found on roots. Leaves are regarded as lateral expansions of stem. The root, stem and the leaves are thus three main organs of a plant. The flowers are the reproductive organs which are formed after the plants have attained maturity. Goethe proposed that flowers are modified shoot and each part of it is homologous to a leaf.

The primary functions of the stem are: (i) to bear foliage leaves, (ii) the conduction of water and mineral substances from the root to the leaves, (iii) conduction of food material from leaves to various parts of the plant, and (iv) to produce reproductive structures, the flowers. The growth of shoot is restricted to its terminal part, commonly known as apex. In the apex, the apical meristem is found which produces leaves, nodes and internodes. Leaves are organs of manufacturing food due to presence of green pigment containing cells in them. Leaves actively participate in transpiration.

The main function of root is anchorage and absorption of water and minerals from the soil. Water absorption is facilitated by the presence of epidermal tubular elongations called root hairs commonly found near the apical region. The apex of root remains protected by the root cap which is derived from the apical meristem along with other tissues of the root.

The primary meristem found in the apical region contributes to the three main types of tissue system which form the primary plant body. The three main tissue systems are, the **epidermis**, the **primary vascular tissues**, the **cortex**, and the **pith**. The epidermis is derived from **protoderm**, the primary vascular tissues from **procambium** and cortex and pith from **ground meristem**.

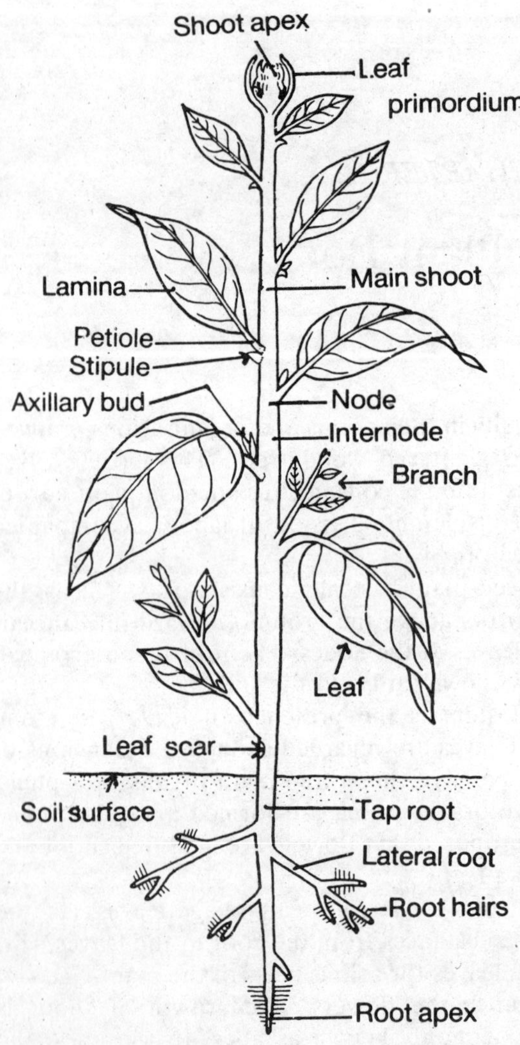

Fig. 1.1. Diagrammatic representation of the plant body.

The epidermis is the outermost layer which is usually uniseriate and protective in nature. Internally, next to epidermis is the multilayered cortex composed of different types of tissues performing functions like support, storage etc. The innermost layer of cortex is called **endodermis** which functions like a jacket or water dam. The central part of the axis is called **stele**. Stele comprises of **vascular bundles,** which consists of a set of xylem, phloem and other tissues. Vascular bundles are associated with two main functions, conduction and mechanical support. Xylem conducts water and phloem conducts food material. The tissues which provide the mechanical support are usually developed within xylem and phloem. However, in stems of dicotyledons and gymnosperms, a lateral meristem, called cambium, may be present between xylem and phloem. The cambium is responsible for growth in thickness by the formation of secondary vascular tissues.

All the tissues derived from apical meristem are commonly called **primary tissues** of which the primary body of the plant is made up of. During primary growth, the axis grows in length, appendages develop and branch systems are laid down. The tissues formed by the activity of vascular cambium are called **secondary tissues**. The vascular cambium forms secondary phloem towards the outerside and secondary xylem towards inner side of the axis. The growth in girth, due to increase in secondary tissues, causes rupture in cortex and epidermis, exposing the inner living cells. To protect these cells, a new meristem, called **phellogen** or cork cambium arises in the peripheral portions of the organs. The phellogen cuts off **phellem** or cork cells with suberised cell walls on the outerside and parenchymatous cells called **phelloderm** or secondary cortex on the inner side. The phellem, phellogen and phelloderm together form the **periderm**.

Thus, the plant body consists of root, stem and leaves, each with distinct morphological structures and specific functions to perform. They have their specific tissue system to carry on restricted and allotted functions. The tissues, in turn, are groups of cells developed in response to a basic division of labour. So, the unit of tissue or the unit of organisation of different plant organs is the cell. Cell, thus, constitutes the foremost chapter of Plant anatomy.

CHAPTER 2

The Cell

Life on earth may be visualized as a dynamic equilibrium of hundreds and thousands and millions of chemical and biochemical reactions going on between non-living substances at ionic and molecular levels. All these various reactions are mediated through the single, universally occurring unit-*cell*. Cell is the most enigmatic mystery of nature and the structural and functional unit of the living plant and animal body. The word cell was coined by Robert Hooke (1665), who with the help of compound microscope observed the texture of a thin section of cork as composed of small spaces surrounded by a firm boundary wall and the whole structure resembling a honey comb. He published his work in *Micrographia,* London. Cell is a Latin word derived from *cellula* meaning small apartments. M.J. Schleiden (1809-1881) observed the nucleoli and said that cells are the units of structure in plants and formulated his famous *cell theory.* T. Schwann (1810-1882) extended this view of cellular concept to animals also and stated that all organisms are composed of cells and cell products.

The cell theory in present state may be briefly summed up as follows:

(i) All living things are composed of cells and their products.
(ii) All cells arise from pre-existing cells.
(iii) All cells are basically alike in chemical composition and metabolic activities.
(iv) The function of an organism as a whole is the outcome of the activities and interactions of the constituent cells.

During recent past, the cell theory has been challenged and the **organismal theory** has been proposed according to which, an organism is regarded as a protoplasmic unit which is incompletely divided into small centres–the cells for the performance of various biological ativities.

Cell size. The size of cells varies widely. The smallest living cells are bacteria (0.2-5.0μ) and the largest cell is the ostrich egg which measures about six inches. The largest unicellular plant is *Acetabularia.* Table 2.1 reveals some interesting average measurements.

Cell shape. The shape of cells also varies depending upon their function, needs of the organisms and environmental conditions. It can be specially observed when the various shapes of unicellular organisms such as *Amoeba,*

Table 2.1. Average measurements of some living cells and molecules.

Name of the cell	*Measurement*	*Name*	*Measurement*
Ostrich egg	170 × 135 mm.	Typhoid bacillus	2.4 × 0.5 micron
Hen egg	60 × 45 mm.	Influenza bacillus	0.5 × 0.2 micron
Humming bird egg	13 × 8 mm.	Pneumococcus bacterium	500 × 200 mm.
Human egg	0.1 mm		
Amoeba	100 microns (μ)	Bacterial virus	80 m μ
Sea urchin egg	70 microns (μ)		
Liver cell	20 microns (μ)	Protein molecule	10 m μ
Red blood cells	7 microns (μ)	Haemoglobin	7 m μ

Paramoecium, Chlamydomonas, Yeast, Bacteria etc. are considered. Some bacteria are rods, spirals and even comma-like, while *Acetabularia* consists of a stalk and a cap, upto 10 cm in height. The shape of the cell in multicellular organisms shows some regularity where it is partly controlled by the mutual pressure of the cells. However, in these cases, some specialised cells modified for various functions show typical shapes, e.g. cells of glandular hairs of leaf, guard cells and root hairs.

Cell number. The cell number varies to a large extent. Some organisms are small and unicellular while others are giants and multicellular. Among unicellular forms *Chlamydomonas* and *Chlorella* are of common occurrence while those of multicellular forms are *Ulothrix*, mosses, ferns and phanerogams. An average man possesses about 100 trillion cells and if all of them are extended in one line, they will take 100 turns around the earth. However, some forms have definite number of cells in their organisation such as *Gonium* (4 or 8 cells), *Pandorina* (4, 8 or 16 cells), *Eudorina* (8, 16, 32 or 64 cells) etc.

Organisation of cells. A cell has three distinct regions (i) cell wall, (ii) protoplasm and (iii) vacuole. Cell wall and vacuoles have been considered as non-living components. The protoplasm consists of two important components, the cytoplasm and the nucleus. It remains externally bounded by cell membrane or plasma membrane. The cytoplasm contains several cell organelles such as plastids, mitochondria, endoplasmic reticulum, ribosomes, lysosomes, cilia, flagella etc. A plant cell is distinguished from animal cell in presence of (i) cell wall, (ii) plastids and (iii) vacuoles. An animal cell has centrosome which is not found in plants, except in few microrganisms.

CELL WALL

Cell wall is the non-living component of the cell and is secreted and maintained by the living portion of the cell, called protoplasm. The synthesis of cell wall is controlled by Golgi bodies. A typical cell wall is composed of three different regions–(i) middle lamella, (ii) primary wall (1-3 μ thick and elastic) and (iii) secondary wall (5-10 μ thick and rigid).

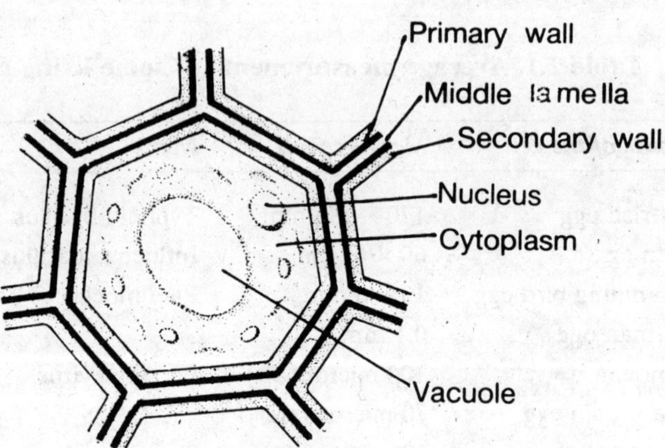

Fig. 2.1. A diagram of a typical cell showing position of cell wall and different components.

Middle Lamella

The formation of this layer starts in the telophase stage of cell division and is formed by deposition. It works as a cementing layer between the two daughter cells and is called middle lamella. It consists of Ca and Mg pectates (composed of chains of hexuronic acids). Pectic substances are a mixture of polygalacturon (D-galacturonic acid) and polysaccharides (neutral sugars). This common layer between the two cells may be dissolved with the help of pectolytic enzymes and the two cells separated. This is evident during the loosening of mature ripe fruits, which is the result of the dissolution of middle lamella.

Primary Wall

Bordering middle lamella, the primary wall is the first layer of individual cell, synthesised by protoplasm. In young enlarging cells, the primary wall remains thin and elastic becoming thick and rigid with the approach of cell maturity. The dry primary wall contains hemicellulose upto 50%, cellulose upto 25% and smaller amounts of pectic substances, non-cellulosic polysaccharides, fats and proteins. Hemicellulose is composed of complex polysaccharides (xylans, mannans, galactans, arabinogalactons, glucomannans) and forms the matrix of wall in which cellulose microfibrils are embedded. The flexibility of primary cell wall during elongation is due to presence of these pectic substances. In primary wall, the microfibrils are roughly transverse to the cell axis and do not show a longitudinal orientation at the cell corners. Primary wall layer is the principle wall layer of meristematic cells, chlorophyllous cells and true parenchyma.

Secondary Wall

The formation of the secondary wall gives structural independence to the cell and provides rigidity to the wall. It is formed next to primary wall towards protoplasm. The secondary wall mainly consists of cellulosic or varying mixture of cellulose and non-cellulosic material.

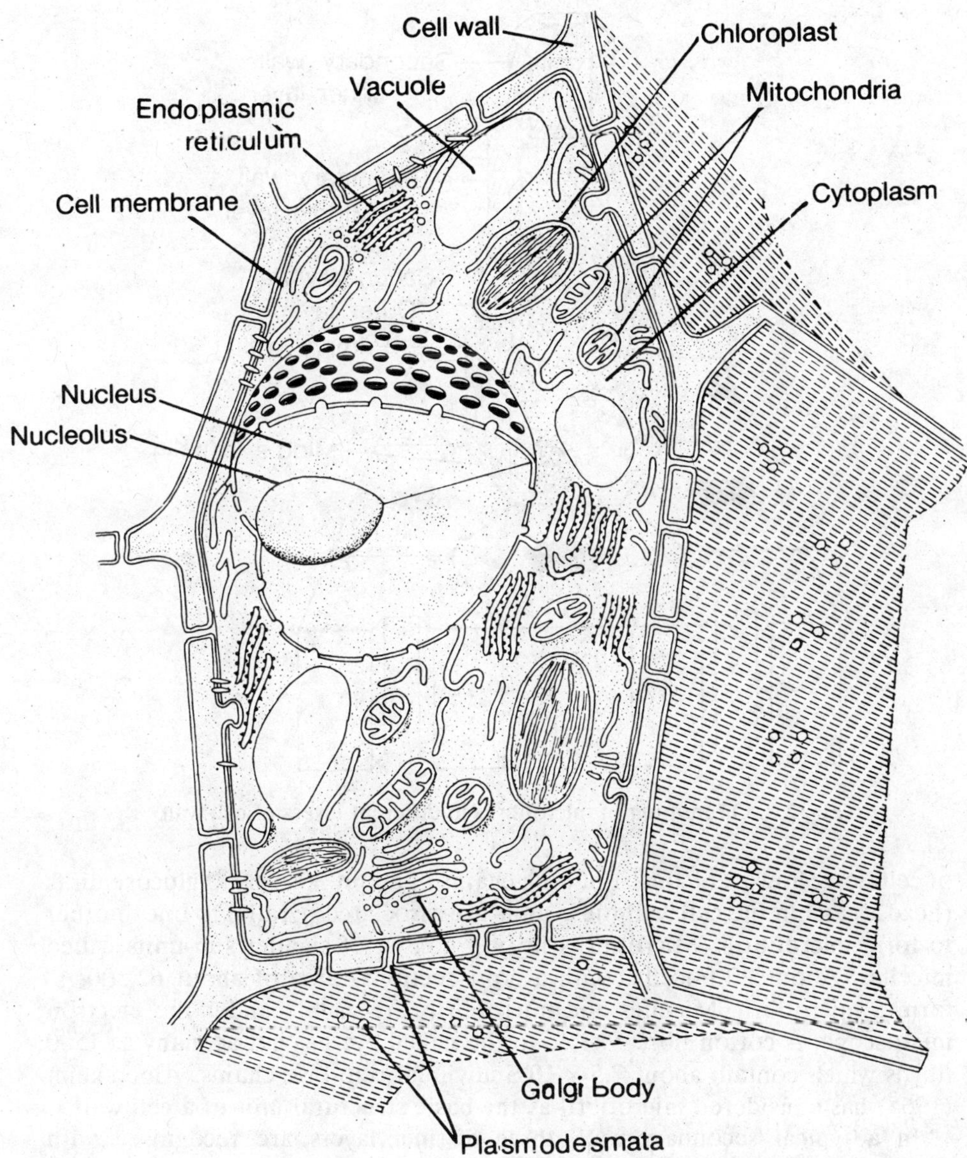

Fig. 2.2. A generalised diagram of plant cell based on electron microscopic studies. The perforations on the wall indicate the sites of plasmodesmata.

Structurally, the secondary wall may be thought of as a fine interwoven network of cellulose strands of varying complexity and size. According to Siegel (1962), the elementary fibrils or micelles are the smallest structural units of cell wall. These are composed of approximately 100 individual cellulose chains with a cross-sectional area of about 3000 A°. The molecules

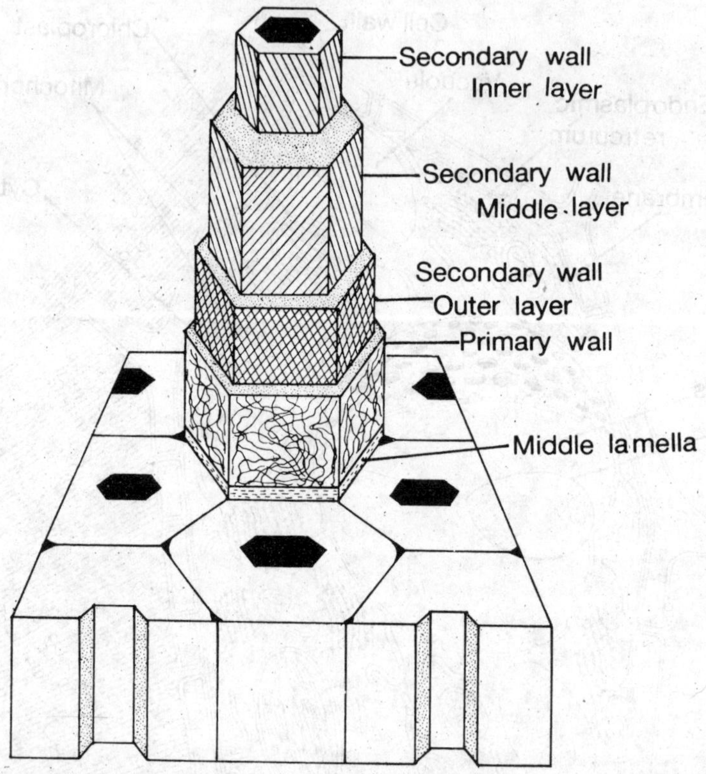

Fig. 2.3. Orientation of microfibrils in different layers of cell wall.

of cellulose, however, consist of long chains of about 3000-6000 glucose units (hexose) and in the cell wall these units are associated parallel to one another to form microfibrils. Microfibrils in turn consist of smaller sub-units called micelles. About 20 micelles with a cross-sectional area of about 62,500 A° form a microfibril. Micelles and microfibrils are visible under the electron microscope. A cotton fibre, visible to the naked eye has got as many as 1500 fibrils which contain about 7.5×10^8 individual cellulose chains. Albersheim (1965) has considered microfibril as the basic structural unit of a cell wall.

In a typical secondary wall, three distinct layers are recognised with different arrangement of microfibrils. In the outer layer of the secondary wall two spirals of microfibrils may be seen forming a large angle with the cell axis. In the middle section, a steep spiral in addition to concentric rings of spirals can be observed and inner layer is thought to have a helical arrangement.

At places in the cell wall the deposition of material does not take place. These places are known as plasmodesmata through which contents of neighbouring cells remain in communication with each other. (Fig. 2.4)

Chemical changes in the cell wall. As the cell matures, various depositions on the cell wall take place, which bring about a change in its chemical nature. These chemical changes bring about a corresponding change in the structure and function of the cell. The various depositions are–

1. Lignification. The deposition of lignin on the cell wall or the conversion of cellulose into lignin is called lignification. As a result of lignification, the cells become hard, thick walled and dead. The cell wall retains its permeability to some extent but mainly serves mechanical purpose, e.g. tracheids, vessels. Chemically, lignin is a polyalcohol containing coumaryl, coniferyl and sinapyl. In sclerenchyma, the lignification is uniform all around the cell, but in tracheids and vessels the thickening is localized to specific areas. Usually, after the thickenings of the cell wall, the protoplasm of the cell diminishes in size and the cell becomes dead and rigid. These secondary depositions of lignin may form various patterns on the cell wall. These are–

(i) ***Annular or ring like***–Lignin is deposited in the form of rings, one above the other along the interior of the secondary cell wall.

(ii) ***Scalariform or ladder-like***–Lignin is deposited in the form of the steps of a ladder along the inner side of the cell wall.

(iii) ***Reticulate or net-like***–Lignin is deposited irregularly in the form of a network.

(iv) ***Spiral***–The thickenings of lignin are deposited in the form of spiral bands.

(v) ***Pitted***–The entire inner surface of the cell wall is thickened due to depositions, with small unthickened areas here and there. These unthickened areas lying in pairs are called pits. The two pits are separated from one another by a piece of the original cell wall called, closing membrane or pit membrane. The pits may be simple or of bordered type.

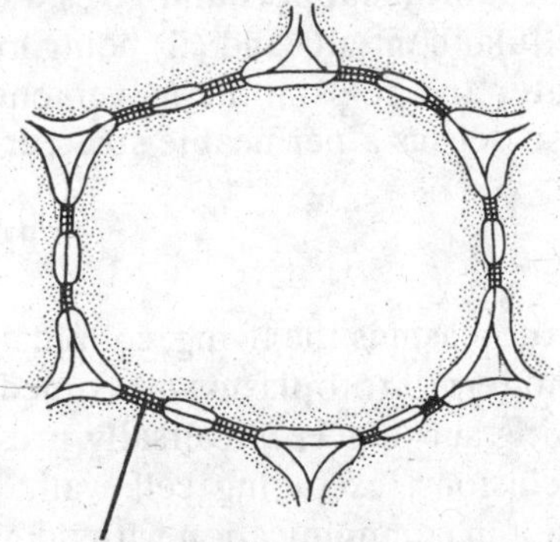

Fig. 2.4. A cell showing plasmodesmata.

2. Suberization. The cell wall of cork cells and Casparian strips of endodermis get deposited with a layer of suberin by a process known as suberization. Suberin is light, impervious to water, hence a piece of cork floats in water and does not get wet. Chemically it is a polymerised form of suberic acid (a lipid derivative).

3. Cutinization. In some cells, in the outer layers of the cell wall, the cellulose gets converted to cutin by the process of cutinization. This forms a definite, impermeable membrane on the cell wall known as cuticle. Cutinization of cell wall helps in checking evaporation of water. Chemically cutin consists of fatty acids. Cutin is commonly found deposited on the outer surface of exposed green parts of the plant, e.g. epidermal cells of green stem and leaves.

4. Mucilaginous changes. Sometimes the cellulose is changed to mucilage which has the property of absorbing and retaining water. This forms a viscous, mucilaginous coating on the cell walls and helps to tide over dry conditions. Chemically mucilages are polysaccharides derived from galactose and mannose, e.g. the leaves and flowers of *Hibiscus* (China rose), fruits of Lady's finger, seeds of isabgol *(Plantago),* linseed *(Linum)* etc. Many sea weeds yield mucilaginous substances such as agar, alginic acid, and carageenan of great commercial value.

5. Mineralization. The deposition of minerals on the cell walls by the process of infiltration or by deposition of inorganic salts is known as mineralization. The minerals usually deposited are silica, carbonates and oxalates of calcium. The phenomena of silica and calcium deposition are known as silicification and calcification, respectively.

Functions of Cell Wall

The cell wall in general performs two important functions–(i) it provides mechanical support and gives a definite shape and protection to the cell and cellular contents and (ii) being hydrophilic in nature, it is capable of imbibing water and helps in the movement of water and solutes towards protoplasm. It also acts as a permeable structure during absorption of minerals and solutes.

PROTOPLASM

Protoplasm is the living, colloidal semifluid substance of changing consistency. The term protoplasm was coined by the Bohemian biologist Purkinje for the living substance, universally present in all cells. Protoplasm with non-living inclusions, excluding cell wall, is called *protoplast.* Each protoplast keeps itself in communication with neighbouring protoplasts through small openings in the cell wall known as plasmodesmata. Protoplasm is characterized by certain physical and chemical properties.

Physical properties. The protoplasm appears as a colourless, elastic, colloidal semifluid containing numerous tiny granules and droplets. The tiny particles show erratic and random movements within protoplasm. The phenomenon is referred to as *Brownian movement.* In plant cells, vacuoles are also present, and the protoplast shows a state of movement around these vacuoles, referred to as cyclosis. The cyclosis movement can be easily seen in the cells of *Hydrilla, Eleocharis, Nitella, Chara,* staminal hairs of *Tradescantia* and *Rhoeo dis-color* etc. The protoplasm exhibits sol-gel properties and frequently undergoes such transformations by taking in or losing water. Optically, the protoplast is transparent and permits to pass through it the wavelengths of visible spectrum. The *tyndall effect* is also observed. The viscosity of protoplasm is affected by mechanical injury, electric shocks, temperature and pH changes. To the best, the protoplasm is considered as *a complex polyphasic colloidal system.*

Chemical nature. Water is the chief constituent of an active protoplast and normally constitutes 90% of the system. In woody parts of the plants, water is about 50%, but in soft parts 75%, in succulent parts 85-95%, in hydrophytes 95-98% and in dry seeds only 10% water is found. The remaining part of the protoplast contains two types of constituents i.e. organic and inorganic materials.

The common organic compounds are carbohydrates, proteins, nucleic acids, lipids, fats, phosphatides etc. Protein is one of the main constituents which constitutes about 40-60% of the total dry weight of the protoplasm. Carbohydrates form 12-14% of the dry matter. Nucleic acids exist in both free and combined form with proteins, in the form of nucleoproteins, forming about 1.5% of the protoplasm. About 12-14% of the protoplasm (dry matter) is lipid, which occurs in the form of natural fats, fatty acids, fatty oils, phosphatides, waxes and steroids. The average chemical composition of plant body is summarised below:

Elements	*% wt.*	*Elements*	*% wt.*
Oxygen	62	Hydrogen	10
Carbon	20	Nitrogen	3
Phosphorus	1.14	Sulphur	0.14
Chlorine	0.16	Potassium	0.11
Iodine	0.014	Magnesium	0.07
Calcium	2.50	Iron	0.01
Other trace elements	0.756	Sodium	0.1

Thirteen elements are constantly found in plant ash. Among metals are potassium, calcium, magnesium, iron and sodium and non-metals are carbon, hydrogen, oxygen, nitrogen, sulphur, phosphorus, chlorine and silicon. The other important elements found in traces are boron, manganese, zinc, copper, molybdenum, aluminium etc. Being found in extremely minute quantities these are called trace elements.

Protoplasm consists of cytoplasm and nucleus and is externally bounded by the cell membrane or plasmalemma.

CELL MEMBRANE

All cells remain enclosed by a thin, film like pliable membrane called plasma membrane or plasmalemma. In addition, the eukaryotic cell also possesses intracellular membranes that surround the vacuole and organelles. The plasma membrane and the subcellular membrane are collectively called biological membrane.

Composition of membrane

Cell membrane mainly consists of proteins and lipids but in certain cases, polysaccharides, sialic acid, RNA and DNA have also been found.

Proteins. Three different categories of membrane proteins have been found present, (i) structural protein, (ii) enzymatic protein and (iii) carrier protein.

(i) Structural proteins–form the backbone of the membrane. Lenard and Singer (1966) suggested that these are strongly lipophilic in nature in which from one-fourth to one-third of the amino acid residues show α-helix secondary conformation (arrangement) and the remaining portion forms the random coils. There may be several types of structural proteins in a membrane.

(ii) Enzymatic proteins–vary considerably in structure and amount. Such proteins specify the physiological function of the membrane. For example, the marker enzymes of endoplasmic reticulum include glucose 6-phosphatase and for mitochondria cytochrome oxidases.

(iii) Carrier proteins–of membrane are not well defined. But these have definitely been reported in *Salmonella typhimarium* and *Escherichia coli* where these act as sulphate carrier and glucose binding factor, respectively. The structural and carrier proteins are of low molecular weight and interestingly, in certain cases, both these may have dual functions.

Lipids. The amount of lipids found in the membrane varies from 30 per cent (in mitochondria) to 70 per cent of the total dry matter of cell membrane. The lipids are mostly in the form of phospholipids of which five different types have been identified. One is simple phosphatidic acid (diglyceride with the phosphate molecule attached to the glycerol) and the other four are complex in nature. These have either choline, inositol, ethanolamine or serine in addition. The choline containing lipid is called lecithin which forms more than half of the total phospholipid content of the membrane.

The ratio of lipid and protein shows variation depending upon physico-chemical and functional differences. For instance, the proportion of lipids to proteins may vary from 1:0.8 to 1:4.

Other constituents. Some polysaccharides, sialic acid, RNA and DNA have also been found in certain cases. The amount of polysaccharide found in the membrane is extremely low. Bal (1966) reported the presence of RNA in the membrane of nucleus, endoplasmic reticulum and plasma membrane of *Allium cepa.* DNA has been detected in the membranes of mitochondria, chloroplast and plasma membrane but in most cases, this DNA is easily separable.

Other components playing vital roles in membrane physiology have also been found, such as certain metals (Ca, Mg, Zn), coenzymes, porphyrins etc. Electron microscopic studies revealed the finer structure of plasmalemma to be 6 nm to 10 nm (10 nm = 100 A°; 1 nm = 10^{-6} nm) on surface of all cells. Plasma membranes of two cells remain separated usually by a space of 11-15 nm.

Structure of Membrane

Several models have been proposed to explain the organisation of cell membrane. A few of them are Danielli-Davson model, Unit membrane hypothesis, Kavanau's lipid pillars, Hydrophobic binding models, Benson's model, Lenard and Singer's model, Mosaic membrane concept, Composite model, Fluid mosaic model etc. Only a few are explained here.

Unit membrane concept. Robertson (1962) studied the structure of membrane with the help of electron microscope and found that all biological membranes show three layers, viz., two outer layers (electron dense) separated by a lighter middle layer. These correspond to two layers of proteins and a middle layer of phospholipids. Together these form a unit membrane. His findings confirmed the observations of Danielli and Davson (1930) who

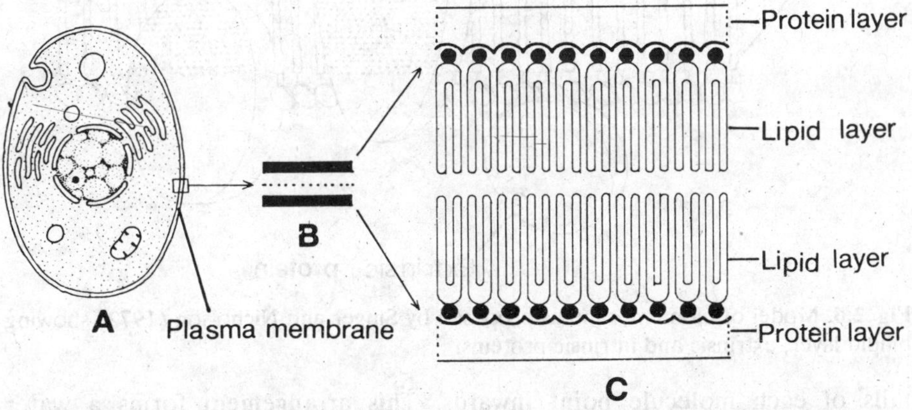

Fig. 2.5. Structural details of plasma membrane. A–as seen under ordinary magnification of electron microscope, B–a portion more magnified and C–a detailed structure showing sandwich arrangement.

proposed that the plasma membrane is made up of lipid layer, sandwiched between two continuous layers of proteins. The protein and lipid molecule layers are electrostatically held together, the former being positively charged and the latter negatively charged. In this sandwich, lipid molecules show orderly arrangement and remain set at right angles to the surface. The lipophilic tails of fatty acids face one another along the double row and their hydrophilic heads face outwards. Each protein layer is found to be 20Å thick and lipid layers 35Å thick, thus making the sandwich 75-100Å thick. At places, the lipid layers are penetrated by pores of variable sizes. The pores may form or reform.

This model cannot be applied to all membranes, as it reveals the definite proportion of lipids and proteins, which has not been found true. Moreover, recent studies reveal that proteins are found in the form of particles and not as a continuous layer.

Fluid-mosaic model. Singer and Nicholson (1972) proposed fluid-mosaic model of cell membrane. They proposed that a central bilipid layer composed of phospholipids, with their spherical, polar head groups on the outer surfaces, form the basic structure of membrane (Fig. 2.6). The two non-polar

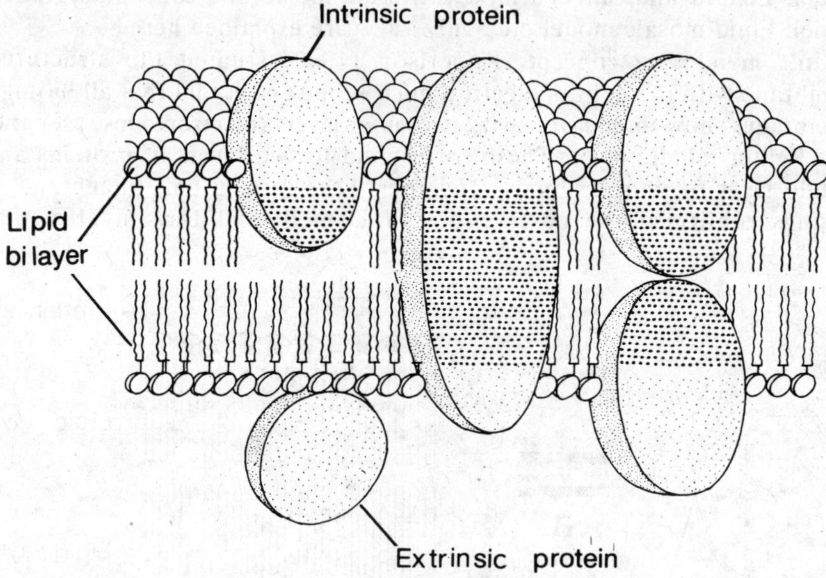

Fig. 2.6. Model of biomembrane as proposed by Singer and Nicholson (1972) showing bilipid layer, extrinsic and intrinsic proteins.

tails of each molecule point inwards. This arrangement forms a water resistant barrier through which lipid soluble substances can pass through. Lipid molecules show various kinds of motions such as

(i) internal motion within each lipid molecule,
(ii) lipid molecules might diffuse laterally,
(iii) lipid molecules may rotate rapidly as a whole, about their long axes, and
(iv) transfer of lipid molecules from one side of the bilayer to the other. It is referred to as *flip flop movement.*

As regards proteins, they described the model as *protein icebergs in a sea of lipids.* In fact, proteins are found in the form of descrete particles which may be either embedded in the lipid layer (integral proteins or *intrinsic* proteins) or superficially attached (peripheral proteins or *extrinsic* proteins). The peripheral proteins can be easily dislodged from the membrane, while for the removal of integral proteins, the membrane has to be disrupted. Some large globular integral proteins are believed to have channels, through which water soluble materials can pass.

In certain cases, some oligosaccharides are found attached to lipids and integral proteins, which project into the extracellular fluid, from the outer surface of the plasma membrane. These influence the manner in which cells interact with one another. Cells recognise one another on the basis of these glycoproteins and glycolipids.

There are specific proteins in the membrane, called *membrane receptors* which play an important role in the flow of materials and information into the cell, by binding with specific molecules reaching the cell's surface.

Some membrane proteins function as enzymes, for example, mitochondrial membrane carries electron transport enzymes which participate in respiration.

Functions of Cell Membrane

Some of the important functions of the cell membrane are as follows:

(i) It acts as a permeability barrier which controls and co-ordinates the rate of substrate transfer and diffusion. Compounds with polar groups such as –OH, –COOH, NH_2, –CHO and inorganic salts enter the cell slowly. Non-polar compounds, alcohol, chloroform etc. penetrate rapidly. Cell membrane is impermeable to polysaccharides, phospholipids and proteins.

(ii) It acts as a cytoskeleton providing mechanical frames on which enzymes can be specifically oriented.

(iii) It acts as a vehicle for transport of substances from one organelle to another, from inside to the outside of the cell and from outside to the inside of the cell.

(iv) It acts as elements, capable of synthesis of various macro-molecules.

CYTOPLASM

The living part of the protoplast surrounding the nucleus is known as cytoplasm. So protoplasm consists of cytoplasm and nucleus. Cytoplasm is bounded externally by the cell membrane and internally by the tonoplast or vacuolar membrane. It is semipermeable in nature and maintains connections with neighbouring cells through plasmodesmata. The cytoplasm is distinguishable into ectoplasm, mesoplasm and endoplasm. The cytoplasm functions as a medium for transport of the various cellular products; intracellular movements and provides a matrix in which the cytoplasmic organelles remain floating.

Cytoplasmic inclusions. Embedded in the cytoplasm are a number of structures of immense physiological importance. These may be of 2 types–(i) Living cytoplasmic organelles; (ii) Non-living inclusions.

CYTOPLASMIC ORGANELLES

The living cytoplasmic organelles are the site of various important metabolic activities such as photosynthesis, respiration, protein synthesis etc. Some of these important organelles are briefly discussed here.

Plastids

Plastids are discoid organelles about 5μ in diameter and 3μ in thickness, found lying free in the cytoplasm of all plant cells. The term plastid is derived from the Greek word *Plastikas,* and was first used by Schimper (1885). Three different types of plastids are recognised on the basis of their types of pigments or colouration.

(i) *Chromoplasts*–These are coloured plastids containing variously coloured pigments other than green photosynthetic ones. They may arise from chloroplasts, as in petals and in pericarp of fruits of Solanaceae, Rutaceae etc. or from leucoplasts, as in carrot roots. Their coloured pigments, belonging to carotenoid groups (carotenes, xanthophyll etc.) are mostly present in vacuolar sap and provide resistance to flowers against thermal fluctuations. Anthocyanin is characteristically found in vacuolar sap of petals and it is never found in chloroplast.

(ii) *Leucoplasts*–These are colourless plastids, originating from proplastids and may be rod-like, spherical or oval in shape. They are usually involved in the storage of various kinds of reserve food materials and are named accordingly, as amyloplasts (storage of starch), aleuroplasts (storage of proteins) and elaioplasts or lipoplast (storage of oil).

(iii) *Chloroplasts*–They were first observed by Antony Von Leewenhoek in 1679 and from the time Sachs (1862) observed starch grains inside chloroplasts, these are believed to be the site of photosynthesis. Chloroplasts are small green bodies present in the cytoplasm of higher plants and green algae. The chloroplasts of higher plants are mainly found in palisade and spongy tissue of leaves. They are small, green, discoid or ellipsoidal, 4-10μ in diameter and 1-3μ thick. Their number may vary from 20 (spongy parenchyma) to 30 (palisade parenchyma) or more.

Structure of chloroplast. The important organisational components of a typical chloroplast are:

(i) Chloroplast membrane,
(ii) Stroma or matrix,
(iii) Lamellae-distinguishable into grana lamellae and stroma lamellae,
(iv) Osmophilic granules,
(v) Ribosomes, and
(vi) DNA and RNA.

Stroma or matrix is proteinaceous in nature that provides medium for freely lying 70S types of ribosomes, osmophilic granules, DNA and RNA molecules. The lamellae also remain dispersed within the stroma. Stroma is the principle site of dark reaction of photosynthesis as well as other activities.

Each chloroplast is surrounded by two unit membrane of lipoproteinaceous nature. Each unit membrane in turn is composed of two lipid molecules packed in between two protein molecules. The external surface of outer membrane is smooth, whereas inner membrane is thrown into lamellated

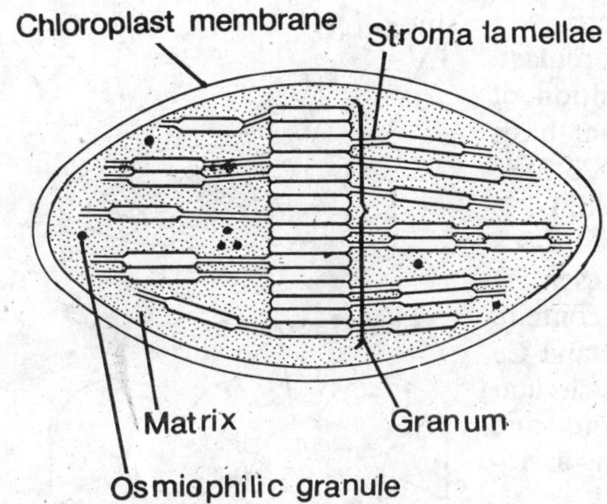

Fig. 2.7. A cross-section of chloroplast based on electron microscopic studies (diag.).

structures showing two distinct regions:

(a) ***Stroma lamellae***–The lamellae in this region are colourless, loosely arranged and referred to as stroma lamellae or intergrana lamallae.

(b) ***Grana lamellae***–This is the most important and complicated part of the chloroplast. In this region, the lamelle are compactly arranged one above the other just like stacks of coins and referred to as grana lamellae. A pile of grana lamellae is termed as *granum*. Electron microscopic studies have shown that grana lamellae form sac-like structures, to which Menke gave the name thylakoids. Park and Beggins (1964, 67) studied thylakoid structure by freeze etch technique and observed small projections known as thylakoid bodies. On these thylakoid bodies are found small rounded photosynthetic units, known as quantasomes, arranged in a monolayer fashion.

Structure of quantasome. The photosynthetic unit or quantasome may be defined as the smallest group of collaborating pigment molecules necessary to affect a photochemical act, that is the absorption and migration of a light quantum to a trapping centre where it promotes the release of an electron. The size of a quantasome is about 175A° with a molecular weight of about two million. Internally, quantasome can be differentiated into two clear regions–the peripheral region enriched with pigment molecules and the middle region or energy centre. Each quantasome contains about 230 chlorophyll mols (160 chl *a* and 70 chl *b*), 48 carotenoids, 46 quinone compounds, 116 phospholipids, 144 diagalactosyldiglyceride, 346 monogalactosyldiglyceride, 48 sulfolipid, some

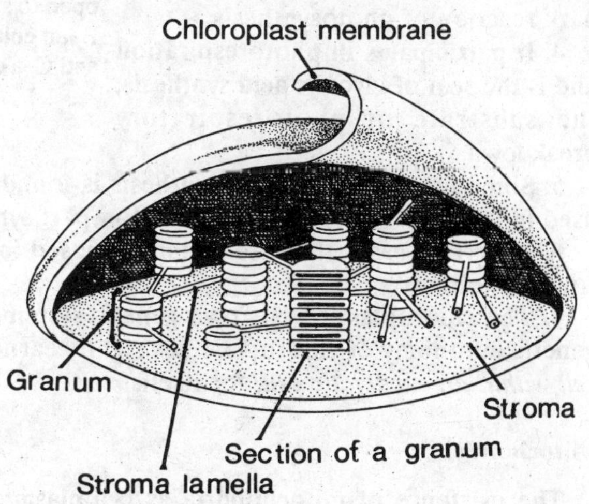

Fig. 2.8. Three dimensional representation of a typical chloroplast of higher plants (diag. based on electron microscopic studies).

sterols and unidentified lipids etc.

Chemical composition of chloroplast. The average chemical composition of the chloroplast on a dry weight basis consists of 40-50% protein, 23-25% phospholipids, 5-10% chlorophyll, 1-2% carotenoids, 5% RNA and about 4-8 × 10^8 daltons of DNA. Mg is present in about 2-3% of the total ash content. The other elements such as Fe and Cu are found in traces. The DNA is circular and of non-histone type and one chloroplast may contain as many as 15-30 molecules of DNA.

Functions of chloroplast. The functions performed by the chloroplast are of such magnitude that some scientists have named it as *a cell within a cell.* Some of the important functions are–

1. The all important processes of photosythesis, i.e. light and dark reaction occur within the chloroplast.
2. The granum is the site of NADP reduction forming NADPH + H^+ and photophosphorylation, i.e. formation of ATP in presence of light.
3. The stroma is the main site for the dark reaction of photosynthesis.
4. It participates in photorespiration and is the seat of glycolic acid synthesis, the substrate of photorespiratory breakdown.
5. Some amount of protein synthesis is found in the chloroplast which is used in its structural organisation and in part synthesis of rubisco enzyme.
6. In certain cases they are metamorphosed into chromoplasts, such as in fruits of tomato.
7. The chloroplast has its own genetic system and is self-replicating. The genetic autonomy of chloroplast has rightly earned the nickname *procaryotic cell within an eucaryotic cell*, for itself.

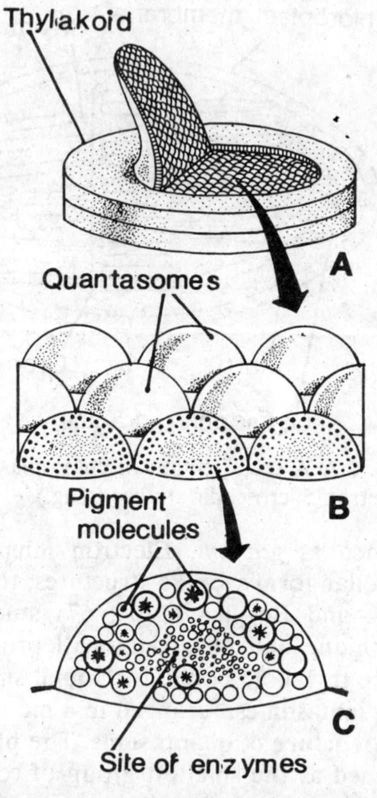

Fig. 2.9. Schematic representation of a granum-thylakoid. A–a thylakoid cut open to show position of quantasomes, B–an enlarged portion of quantasomes and C–a cross-section of a quantasome.

Mitochondria

The existence of mitochondria as cytoplasmic particles was observed by Kalliker, Fleming and Altman, but the name mitochondria was given by C. Benda (1897). Mitochondria, is a Greek word derived from *mitos* = filament and *chondros* = granules. They have been variously called bioplasts, chondriosomes, chondriomes, chondrioconts etc. They are of universal occurrence in every living cell except prokaryotic cell.

Structure. A typical mitochondrion is a sausage shaped body, 1.5 μm long and 5 μm in diameter. Electron microscopy reveals that it is made up of two unit lipoproteinaceous membrane lying parallel to one another. The outer membrane appears to be smooth or finally sculptured but the inner membrane forms involutions inside the lumen, which are called cristae or cristae mitochondriales. These cristae are delicate and finger-shaped and may be branched or unbranched. On the outer surface of outer membrane and inner surface of inner membrane are found numerous minute structures called oxysomes or F_1 particles. The oxysomes on the outer surface are simple, globular, but those on the inner surface have a stalk and bacteriophage-like head. The stalk is hollow, 30-35Å in width and 45-50Å long, whereas the head is 75-80 A° in diameter. Oxysomes are proteinaceous in nature and have been variously named as elementary particles, inner lamellae spheres, F_1 particles etc. The cavity of mitochondria is occupied by homogenous and finely granular mitochondrial matrix.

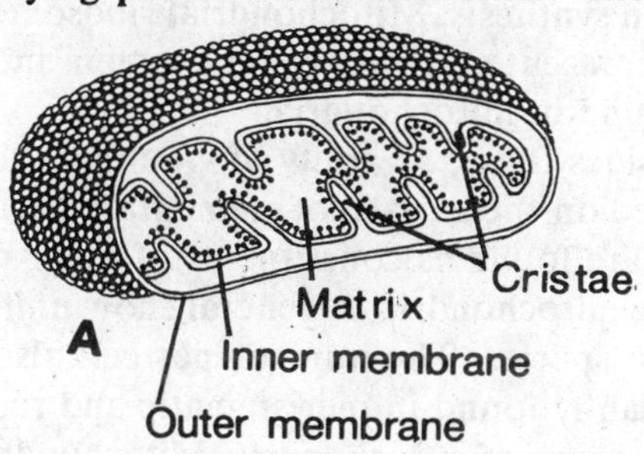

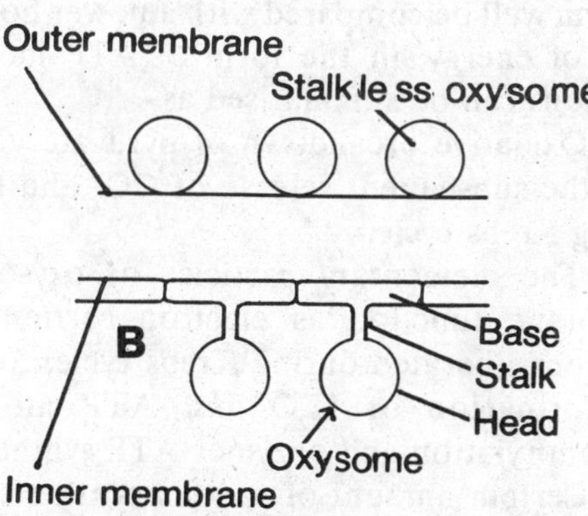

Fig. 2.10. A three dimensional and a cut open portion of mitochondrion (A) and an enlarged portion of the same showing oxysomes (B).

Chemically, mitochondria consist mainly of proteins (65-70%), lipids (25-30%), RNA (0.5%) and DNA. The outer membrane fraction has 40% lipid content, while inner membrane has only 20% lipid content. Mitochondria are rich in manganese. The presence of

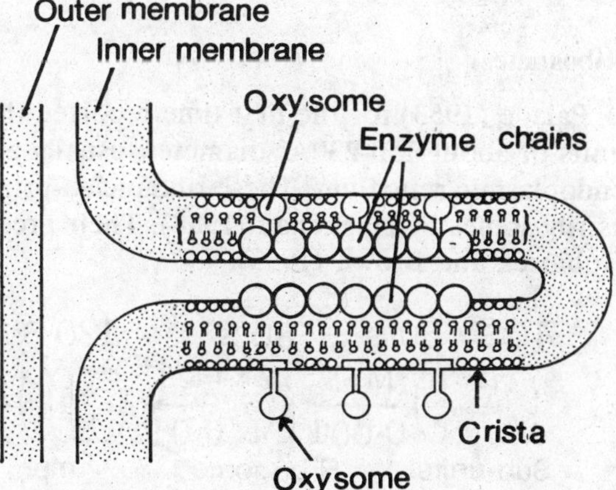

Fig. 2.11. The molecular and internal organisation of a portion of crista.

circular and naked DNA(non-histone type) is a well established fact. There may be one or more DNA molecules in a single mitochondrion, which is sufficient to code for about 25 polypeptide chains and which codes for some structural proteins only. Ribosomes present in mitochondria participate in protein synthesis. Mitochondrial ribosomes are formed in mitochondria itself and represent a second genetic system in a cell. Only 70S types of ribosomes are found in mitochondria.

Besides these, a variety of vitamins and enzymes of Krebs cycle are found situated on the surface of elementary particles or in stroma. Enzymes of ETS chain are found as constituents of inner membrane.

The mitochondria in general show uniform distribution, but are characteristic to species. The number per cell also varies considerably, but these are abundantly found in meristematic and reproductive cells.

Functions of mitochondria. Mitochondria are the seat of cellular respiration and can well be compared with a power house, as they supply readily available form of energy, in the form of ATP molecules, to the plant. The various functions can be summarised as–

1. Oxidative breakdown of pyruvate via four oxidation steps occur in it, with the subsequent release of CO_2 and formation of $NADH_2$, and $FADH_2$ during Krebs cycle.

2. The elementary particles or oxysomes lining the cristae and inner membrane function as electron carriers or transporters of hydrogen or electrons liberated during Krebs cycle, to the final acceptor O_2, resulting in the formation of H_2O and ATP molecules. These are the sites of phosphorylation and enzymes ATP synthetases are found in these particles.

3. Certain amount of protein synthesis is found in mitochondria which is used for its structural organisation.

4. Mitochondrial DNA codes for about 25 polypeptide chains, which are used for some structural proteins only.

5. The evolution of light induced CO_2 in photorespiration occurs in the mitochondria.

Ribosomes

Palade (1953) for the first time reported the presence of some particulate units of about 150-230A° diameter, usually attached to the outer surface of endoplasmic reticulum. These ribonucleo-protein bodies (RNP) were named as ribosomes by Haguenau (1958). Their presence in plants was reported by Robinsen and Brown (1953).

Fig. 2.12. Ribosomes with the formation of dimers and polysomes.

Structure. Ribosomes are composed of two sub-units of unequal sizes. The cytoplasmic ribosomes are corn-shaped, with one larger sub-unit (60S) and other smaller cap-like sub-unit (40S). Both together form 80S type of ribosome. Both sub-units are lobed in such a manner that a continuous hole is formed between them, through which the *m*RNA strand passes. The ribosomes of mitochondria, chloroplast and prokaryotic cells have sub-units of 30S and 50S. The composition and the differences between the two ribosomes are tabulated in Table 2.2.

Table 2.2. Composition and differeences betweeen 70S and 80S ribosomes

Character	*70S ribosomes*	*80S ribosomes*
Occurrence	In prokaryotes, chloroplast and mitochondria.	In cytoplasm and nucleus of eukaryotes.
Sedimentation coefficient	Average 70S	Average 80S
Molecular wt.	3 million	4-5 million
Sub-units	Small 30S and large 50S	Small 40S and large 60S
RNAs	3 molecules of RNA-16S RNA in 30S sub-unit and 5S and 23S RNAs in 50S sub-unit.	4 molecules of RNA-18S in 40S sub-unit and 5S, 5.8S, and 28S RNAs in 60S sub-unit.
Mol. wt. of RNAs	5S RNA-40,000 16S RNA-550,000 23S RNA-1,100,000	5S RNA-39,000 5.8S RNA-51,000 18S RNA-700,000 28S RNA-1,700,000
No. of proteins	In small sub-unit 21 (S1-S21) and in large sub-unit 34 (L1-L34). Total 50-60 proteins.	In small sub-unit 33 (S1-S33) and in large sub-unit 49 (L1-L49). Total 70-80 proteins.
Average mol.wt. of protein.	118,00	21,000
RNA-protein ratio	2:1	1:1
First critical Mg^{++} level	0.5 mole Mg^{++} per mole of phosphorus.	0.3 - 0.1 mole of Mg^{++} per mole of phosphorus.

The two sub-units of a ribosome are stapled by magnesium ions. The larger sub-unit of each ribosome has two cavities-P (polypeptidyl site) and A (aminoacyl site). A group of ribosomes found connected with a single *m*RNA strand is known as polysome, polyribosome, ergosome or ribosomal cluster. The two 70S ribosomes (dimers) have S value of 100 while two 80S have the S value of 120, i.e. 70S + 70S = 100S and 80S+80S=120S. The two or more ribosomes become connected by a single *m*RNA and then may be called polyribosome. Such condition is found only during protein synthesis.

Function. The major function of smaller sub-unit of ribosome is to provide proper site for binding of *m*RNA and its translation. The larger sub-unit of ribosome supports translation and translocation processes coupled with polypeptide synthesis (protein synthesis).

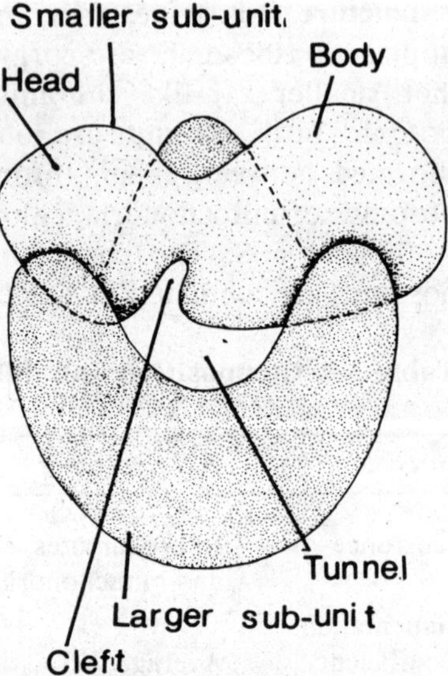

Fig. 2.13. The structure of a ribosome (ultrastructure based on studies of Stoffler and Wittmann, 1977).

Endoplasic Reticulum

One of the most fascinating discoveries of electron microscopy is the endoplasmic reticulum (Porter 1955), present in all living cells of plants and animals. The endoplasmic reticulum (ER) arises from the outer membrane of the nucleus, forming an intermediate meshed network of branching and anastomosing tubules 400 to 700 A° in dia and extending upto the cell membrane. The flat, saccular expansions of ER are termed as cisternae, which are usually aggregated to form parallel lying lamellae system, interconnected through partitions and resulting in compartmentalisation of ER. Besides tubules and cisternae, vesicles are also present.

Endoplasmic reticulum may be of two types–(i) the granular or rough surfaced form, in which the outer surface of ER is studded with ribosomes, and (ii) the agranular or smooth surfaced form in which ribosomes are not attached. The ribosomes present on the ER occur in groups and are called

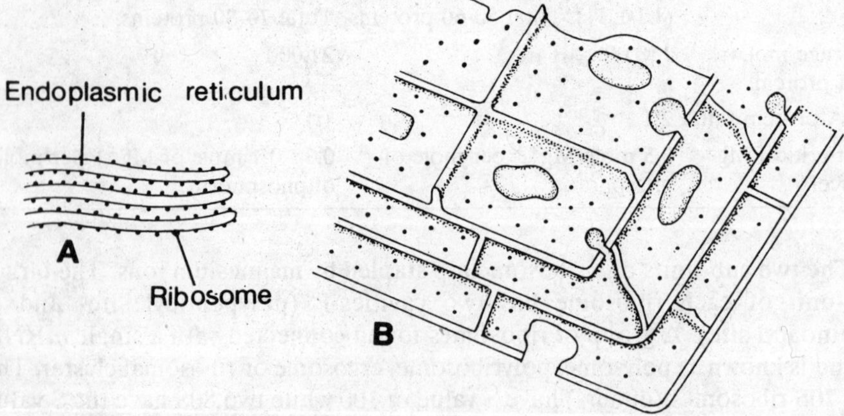

Fig. 2.14. Endoplasmic reticulum. A–a cross-section of the same showing parallel and tubular appearance and B–a three dimensional figure of ER showing flat structure with compartmentalisation.

polysomes. The polysomes and ER together are called ergastoplasm. The ER measures about 6 nm wide. At places in the region of plasmodesmata, ER occasionally passes through it and connects with the branches of ER of the neighbouring cell. It develops strongly in meristematic cells or in metabolically active cells.

Chemically, ER consists of proteins (40-60%), lipids (30-50%) and some RNAs. The lipids are mostly phospholipids (50-60%), whereas protein is partly of structural and partly of enzymatic types.

Functions. ER participates in a number of biochemical acitivities of the cell. Some of these are:

1. It forms the skeleton framework of the cell.
2. It helps in intracellular transport of various metabolites.
3. The compartmentalisation helps in keeping certain enzymes and metabolites separate.
4. Some enzymes, synthesized on the ribosomes move through the supporting membrane of the ER into the enclosed cavity and accumulate in the form of granules.
5. It provides surface for certain enzymes of carbohydrate metabolism.
6. The rough ER are indirecly involved in protein synthesis as they provide a larger surface for the attachment of ribosomes.
7. The smooth ER participate in the biosynthesis of fatty acids, phospholipids as well as in the oxidation and reduction reactions of steroids.
8. The smooth ER are in some way or the other involved in electron transport (ET) and other related processes, emphasized by the presence of certain elements of ETS, such as NADH, NADPH, active ATPase complex etc.

Golgi Complex

It was discovered by Camillo Golgi (1898) in the neural tissue, in the form of extensive network of dark strands. It is variously called Golgi bodies, lipochondria, Golgi complex, Golgiosome etc. The name Golgi bodies was given by his students, Raman and Cajal.

Structure. Golgi complex is an assemblage of flat or curved saccules or cisternae lying one above the other in close parallel array. Each cisternal unit represents an individual lamellae enclosing a thin cavity of about 150A°. About 4-8 cisternal units are piled one upon the other forming multilayered cisternae. On the outer, convex surface of cisternae are present several vesicles, 440-800A° in dia. and termed as *Golgi vesicles*. The vesicles may be of two types: (i) Smooth surfaced, flattened vesicles and (ii) rough surfaced, spherical vesicles. The cisternae are separated from each other by an intercisternal region, which is occupied by fibrous proteinaceous elements known as *intercisternal elements*. Several small vesicles and vacuoles are enclosed within the cisternae. In such a system, the peripheral lamellar system is referred to as *Golgi externum* and the vacuolar elements in the centre as *Golgi internum*.

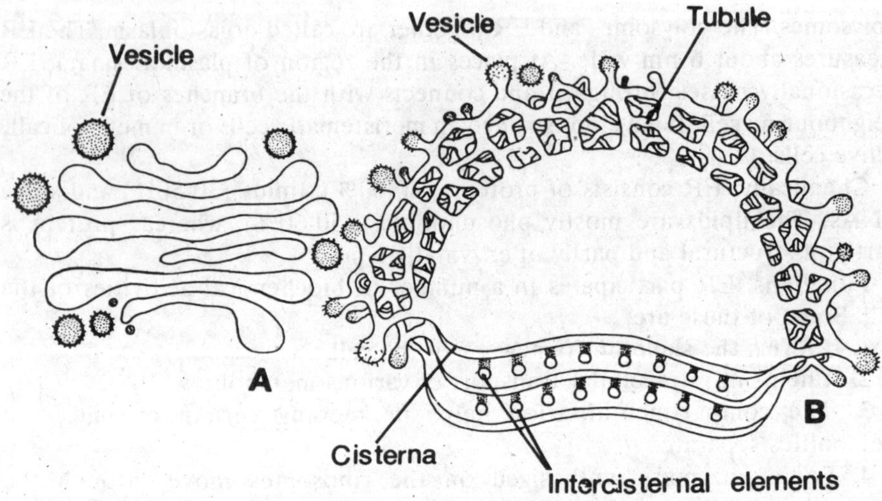

Fig. 2.15. Golgi bodies. A–as appear in cross-section and B–a three dimensional figure showing details.

Some botanists use the term dictyosome to refer to a pile of lamellae or cisternae with their attached tubules and vesicles. Farquhar (1985) stated that Golgi complex exhibit polarity i.e. the side facing ER is the *cis*, whereas the side facing the centrioles is the *trans* end. By *cis* end it receives vesicles of ER containing immature enzymes, which are processed step by step in different cisternal elements, and ultimately mature enzymes enclosed in a vesicle are released from *trans* end.

Functions. The various important functions are–

1. Their main function is to process the immature enzymes into functional enzymes and to release them in a closed vesicle.
2. Golgi complex participates in cell wall formation i.e. cell plate, primary and secondary wall formation.
3. They are involved in secretion e.g. slime secretion from leaf glands of certain insectivorous plants like *Drosera* and *Drosophyllum*.
4. Dictyosomes are concerned with intra-and intercellular transport.
5. They are also concerned with giving rise to primary lysosome.
6. Golgi complex pinches off vesicles which later fuse to form plasma membrane.
7. They are believed to play a role in regulation of fluid balance.
8. Golgi complex is supposed to play a role in phospholipid synthesis, lipid absorption etc.

Lysosomes

These are rounded dense bodies observed in the liver tissues and renamed as lysosomes by Christian de Duve (1955). These contain hydrolytic enzymes and as a matter of fact these are the lytic bodies functioning as suicidal bags.

These are found both in plants and animals.

Structure. These are submicroscopic bodies ranging in size from 0.4 to 0.8 micron. A lysosome has a single limiting membrane composed of lipoprotein. Chemically, it is filled with more than a dozen hydrolytic enzymes such as Acid phosphatase, Acid ribonuclease, Acid deoxyribonuclease etc.

Functions. Some important functions of lysosomes are enumerated below:

1. *Intracellular digestion*–Lysosomes are responsible for the digestion of particles that are taken into the cell by phagocytosis.
2. *Autophagy*–During adverse conditions, lysosomes begin to digest its own cell inclusions such as mitochondria and ER. This is called cellular autophagy.
3. *Ageing*–Lysosomes are considered as agents of ageing, for they produce autolytic enzymes that slowly disrupt the intracellular machinery.
4. *Autolysis*–They perform the function of removing the dead or degenerating cells and this process is known as autolysis.
5. *Cell division*–Lysosomes may be involved in repressing mitotic division of a cell.

Lomasomes

These structures, referred to as border bodies, were discovered by Moore and Mc Alear (1961). Structurally, these are membraneous or vesicular structures present between the cell wall and cell membrane and are formed by evaginations delimited by cell membrane. Functionally, lomasomes play a role in–(i) secretions, (ii) increasing the surface area for the diffusion of substances involved in cell wall formation or breakdown, (iii) in endocytosis and (iv) in membrane proliferation.

Microtubules

The term microtubule was proposed by Slautterback (1963) for the subcellular filamentous components of eukaryotic cells. Structurally, these are straight cylinders, of dia. 240 A° with a hollow core of 150A°. A single microtubule consists of 13 longitudinal protofilaments, each composed of alpha and beta tubulin in linear chains. Microtubules are self-assembling polymers and are specially observed in the mitotic apparatus, cortex of meristematic plant cells, elongating cells etc.

Microtubules have multiferous functions, such as (i) chromosomal movement, (ii) intracellular transport of materials, (iii) cellular motility, (iv) ciliary and flagellar movement, (v) development and maintenance of cell form, (vi) movement of organelles within cell and (vii) they establish asymmetries, polarities and changes of shape in higher plants.

Microbodies

Rhodin (1954) used the term microbody to include uni-membrane bounded organelles, ranging 0.2 to 1.0 μm in size. Lazarow and Fujiki (1985) identified four types of microbodies, (i) peroxisome, (ii) glyoxysome, (iii) glycosome and (iv) hydrogenosome. Some of these microbodies are briefly discussed here.

1. **Peroxisomes.** Peroxisomes were discovered by Christian de Duve (1965). Salisbury and Ross (1986) reported that they are found exclusively in photosynthetic tissue closely appressed with chloroplast, but their presence in animals and fungal cells is also reported.

Peroxisomes are found to be interconnected with each other through narrow branched and tortuous tubules called *peroxisomal reticulum*. These organelles are bounded by one unit membrane, 6.5-7 nm thick. Peroxisomes are rich in catalase and occur abundantly during later stages of life cycle. The most important function is its participation in photo-respiration in C_3 plants.

2. **Glyoxysomes.** These microbodies were discovered by Breidenbach and Beevers (1967) from the extracts of endosperm of germinating castor beans. These are 0.5 to 1.0 μm in size and made up of a single layered bounding membrane enclosing a granular stroma. They are considered to be analogous particles to peroxisomes, and contain RNA and the important enzymes of the glyoxylate cycle, isocitrate ligase, malate synthetase, glycolate oxidase and catalase activities.

Both peroxisome and glyoxysome are considered as separate organelles due to differences in their enzyme systems. Peroxisomes are rich in catalase while glyoxysomes are rich in isocitrase. Lazarow and Fujiki (1986) proposed that these two organelles are same. Immuno-electron microscopy reveals that both types of enzymes are present in the same organelle during transition, and transition occurs by a change in the complement of enzymes. Thus, these two organelles are interconvertible. In young stages, particularly during seed germination, glyoxysomes are found in a large number within a cell, while during later stages of life cycle, peroxisomes are found in abundance in the cytoplasm. These are bounded by a single lipoproteinaceous unit membrane. They are believed to be formed by the pinching off small bits of endoplasmic reticulum and contain mostly phospholipids, reserve oil droplets and hydrolytic enzymes.

Cilia and Flagella

These are delicate, protoplasmic extensions of variable sizes and number. Flagella are longer, cylindrical, fewer in number showing independent movement, while cilia are shorter, flat, more in number and exhibiting combined action during movement.

A cilium or flagellum arises from a basal body, the blepharoplast granule and consists of a central axial filament, *the axoneme*.

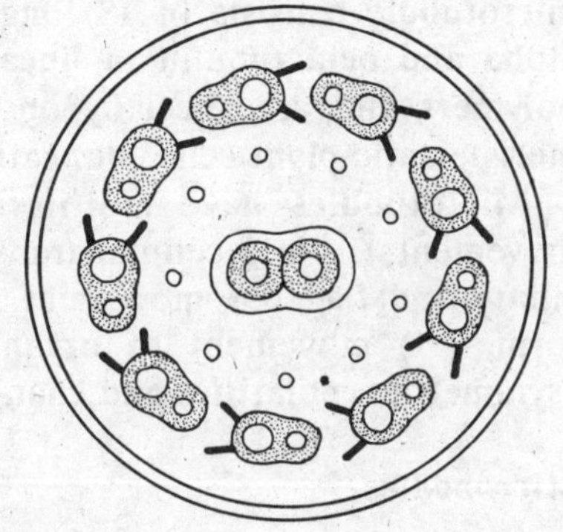

Fig. 2.16. C.S. of a flagellum showing 9+2 arrangement of fibrils.

The axoneme consists of nine peripheral

duplets of tubulin protein and two protein strands in centre, i.e., 9+2 arrangement of tubulin fibrils.

NUCLEUS

The nucleus as a prominent body within a cell was first discovered by Robert Brown in 1831, in an orchid cell.

Structure and composition. The nucleus is more or less spherical, lying in the cytoplasm and occupying about two-thirds of the cell space. Normally a cell contains a single nucleus but multinucleate cells have been reported in *Cladophora, Pithophora,* laticiferous cells in angiosperms etc. A typical nucleus is composed of a few prominent structures–

(i) **Nuclear membrane**–The nucleus remains bounded by two unit membranes of lipoproteinaceous nature, which run parallel to one another with a gap of about 200 to 400Å. This gap is called perinuclear space and is an electron transport area. Electron microscopy shows that the nuclear envelope has a complex double membrane structure. The outer membrane at places gives out tubular structure, referred to as endoplasmic reticulum. Small tubules known as microtubules remain attached on the surface of nucleus. The nuclear envelope possesses openings of 300-600 A° size, called *annuli*, through which nucleoplasm communicates with the cytoplasm. Yoo and Bayley (1967) observed eight spheroid protein particles arranged around the circumference of each annulus, which regulate the size of the annulus.

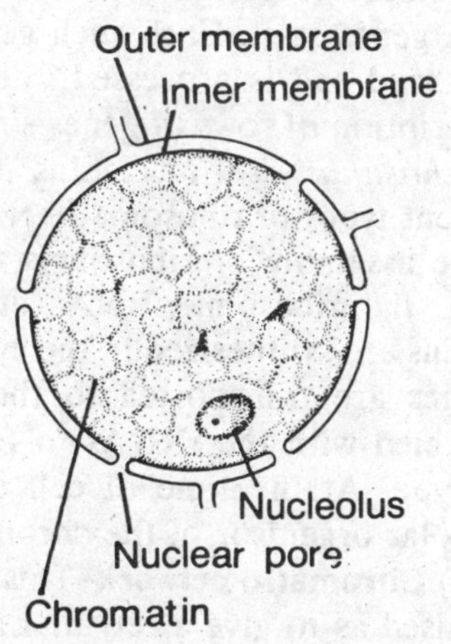

Fig. 2.17. A schematic figure of nucleus.

Next to inner membrane, an internal dense lamellae region of 200-800 A° size has been recognised.

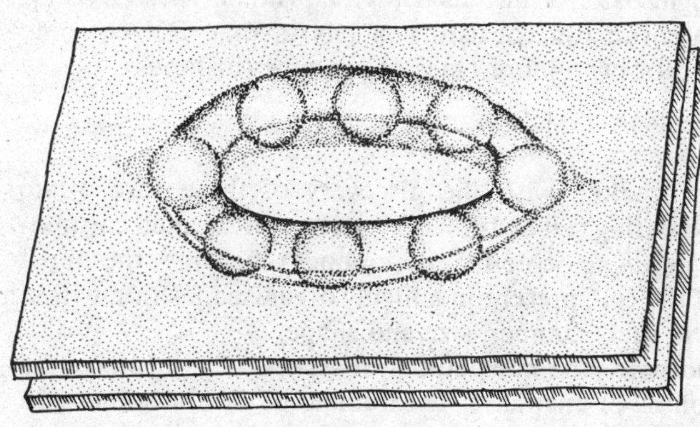

Fig. 2.18. Schematic representation of an annulus of nuclear membrane surrounded by eight spheroids.

(ii) **Nuclear sap**–Encircled by the nuclear envelope is found the nuclear sap (nuclear matrix) variously termed as *nucleoplasm or karyolymph*. It is a semi-fluid substance and consists of proteins which also contains nucleic acids, nucleoproteins (both histones and protamines) etc.

(iii) **Nucleolus**–It is found lying in the nucleoplasm closely associated to the nucleolar organisor of one specific chromosome. The position of the nucleolus is a *species specific trait*. The nucleolus is a spherical body made up largely of proteins followed by *r*RNAs and *r*DNAs. Electron microscopy shows that a typical nucleolus exhibits two distinct structural regions–the *core* (central region composed of compact fibrous elements) and the *cortex* (the surrounding granular area).

Recent studies have shown that DNA of nucleolus is circular and called *r*DNA (ribosomal DNA) which actively transcribes *r*RNA. Nucleolar DNA is highly redundant (highly repeatitious type) and is an example of a giant pallindrome (inverted repeat sequence). There are thousands of copies of *r*RNA genes in it. Each such gene transcribes to form 45S *r*RNA, which is short-lived and is processed into 5.8S, 18S and 28S *r*RNAs. The systematic transcription of rows of genes simultaneously, gives to *r*DNA the appearance of a *christmas tree*. Nucleolus is thus actively involved in the synthesis of different types of ribosomal RNAs.

The assembly of ribosomes takes place at the surface of nucleolus, of which, the ribosomal RNAs are formed within nucleolus and ribosomal proteins are synthesized in the cytoplasm. From the cytoplasm, the ribosomal proteins are transported to the surface of the nucleolus where they are assembled with the *r*RNAs to form the ribosomes. These ribosomes are of 80S type. At the time of cell division, the nucleolus is withdrawn within nucleolar organisor of the chromosome.

(iv) **Chromatin network**–This forms the skeleton of the nucleus and are so organised as to give a reticular appearance. Chromatin threads are the sites of main genetic material which controls all the activities of the cell, metabolism as well as heredity. During cell division, the chromatids become tightly coiled, get enclosed in a definite proteinaceous matrix and are identified as chromosomes.

Chemically on dry weight basis, a nucleus has 70% protein, 10% DNA, 3-5% phospholipids and a chromatin consisting of DNA, some RNAs, some RNA polymerases and 3-5 or even more types of proteins (histones).

Chromosomes

Strasburger (1875) is credited with the discovery of these thread-like structures which appeared during cell division. These structures called chromosomes (Greek word: *chroma* = colour; *soma*=body) became the most significant component of the cell, present in a highly organised fashion within the nucleus and specific in their number, size and shape.

Morphology of chromosome. The morphological aspects of a chromosome are studied during metaphase or anaphase. The number of chromosomes in a nucleus varies to a great extent and is a *species-specific trait*. In *Haplopappus gracilis*, a diploid nucleus contains only 4 chromosomes, while in *Ophioglossum regalis* (a pteridophyte) there are 1262 chromosomes in a diploid nucleus.

Haploid set of chromosomes which is inherited is called **genome** and the whole collection of chromosomes in a nucleus is referred to as chromosome complement. The genome in *Pisum sativum* is 7, *Allium cepa* is 8 etc. The size of chromosome shows a variation from 0.25 μ in fungi to 3.0 μ in some plants.

The shape of the chromosome is decided by the position of the centromere. Accordingly these may be of four types:

(i) *Telocentric*–The centromere is at the tip of the chromosome.

(ii) *Acrocentric*–The centromere is sub-terminal and chromosomes appear rod-like, having one small arm and the other very long.

(iii) *Sub-metacentric*–The centromere is close to the centre and the chromosome has unequal arms resembling L-shape.

(iv) *Metacentric*–Centromere is situated in the centre or middle part of the chromosome and has two equal arms. It appears V-shaped.

Structure. Chromosome is the most wonderful gift of nature. Structurally, they appear as a rod-shaped structure, bounded by a proteinaceous pellicle which encloses a jelly-like substance called matrix. The presence of matrix has been clearly shown in *Luzula campestris* (woodrush).

During metaphase, a chromosome is seen to consist of two symmetrical structures called chromatids, which become intertwined in the matrix. These chromatids are attached to each other by the centromere. During prophase and sometimes during metaphase, the chromosomal material becomes visible as very thin filaments called *chromonemata*. Several chromonemata coiled together and enclosed in a separate common matrix form a chromatid. The chromonema may be composed of 2, 4 or more fibrils which are coiled with each other. If the coiling is easily separable, it is termed as paranemic coiling and if closely intertwined and inseparable, it is termed as plectonemic coiling. The chromonema of one chromosome is identical in nature with the chromonemata of other chromosomes in the same cell.

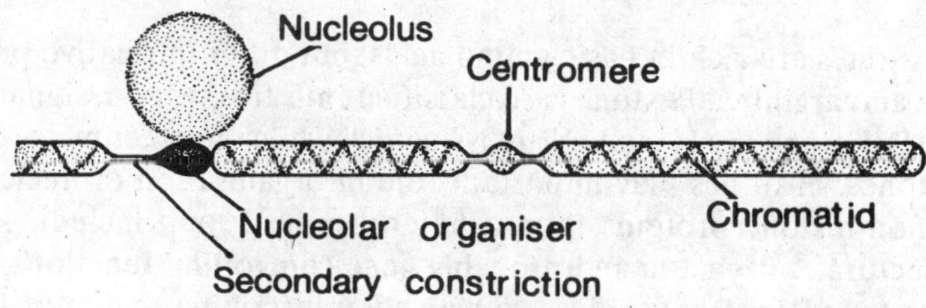

Fig. 2.19. Schematic figure of a chromosome.

The group of chromonemata at places show superimposed coiling, forming a dark band-like structure called *chromomere*, whereas the regions in between two chromomeres are designated as *interchromomeres*. The chromomeres were earlier believed to be the sites of genes, but these dark band structures found on chromosomes are now called heterochromatin segments and are the sites of gentically inactive genes. Euchromatin segments are light in colour and are genetically active.

The chromosome is not uniform throughout, but has a constriction called *centromere*. These have a basic structure consisting of a dense, dark-staining region called spherule or kinosome surrounded by inter-chromomeral fibrillae. Centromere has two important functions, viz.

(i) It holds the two sister chromatids together until the end of metaphase.

(ii) It is the site for the attachment of spindle fibres in order to transport the sister chromatids to opposite poles. This is mediated by the kinetochore. Chromosomes are usually monocentric (one centromere) but some are dicentric (two centromeres) or polycentric (many centromeres).

The chromosome in addition to primary constriction (centromere) possess secondary constriction at any point of the chromosome. It is associated with the nucleolus and may take part in the reorganization of the nucleolus at the end of cell division, as such it may be called nucleolus-organising region. The part of the chromosome, which is present beyond the secondary constriction is called *satellite body*, and the chromosome, bearing satellite is termed as *SAT-chromosome*.

Ultrastructure of chromosome. In view of recent studies related with ultrastructure of chromosome, following points are considered here.

1. ***Strandedness of chromosome:*** The eukaryotic cell contains an enormous amount of DNA which remains greatly compressed and packaged inside a chromosome. A question is raised at this stage, is there one DNA duplex in a chromatid (unineme or single strandedness) or more than one (multineme or multistrandedness)? This question remains much debated. However, it is now generally accepted that the chromosome is unineme, containing a single DNA duplex per chromatid.

2. ***Chromosomal proteins:*** Different types of protein found in a chromosome are divided into two broad categories, the *histones* (or basic proteins) and the *hertones* or *non-histones* (or less basic proteins). Histones represent the largest proportion of chromosomal proteins and usually exceed DNA by weight.

All histones are rich in basic amino acids, but differ in relative proportion of lysine and arginine. Histones are classified into five types designated as H_1, H_2A, H_2B, H_3, and H_4. H_1 and H_2 are lysine rich while the other two are arginine rich histones. Histones play important role in organisation of nucleosomes.

The non-histone proteins form a heterogenous group including proteins with structural, enzymatic and probably gene controlling functions. Some of these are found aggregated close to actively transcribing region of DNA and have been found associated with *hn*RNA.

3. ***The nucleosome:*** The inactive chromatin shows a *beads-on-a-string* appearance. Olins and Olins (1974) called these knob-like structures, the **nu bodies**. The term **nucleosome** was given to these beads by Oudet, Goss-Bellard and Chambon (1975). Nucleosomes are present in all the eukaryotic cells, both in plants and animals.

The bead and the connecting string forms one repeat unit. The bead is called **core particle** or **nucleosome** and the connecting string is called **linker region** or **spacer region** or internucleosomal region. Structurally, the repeat

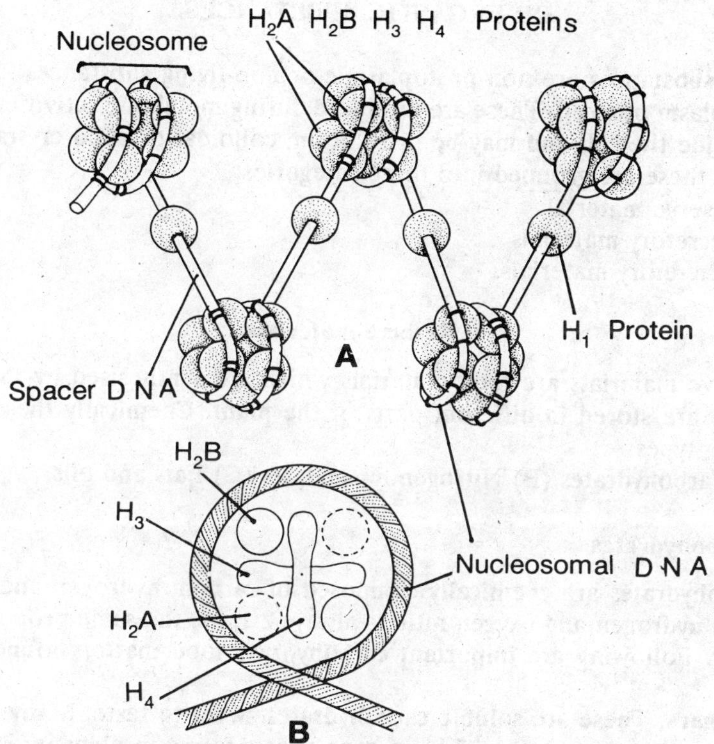

Fig. 2.20. A–An ultrastructure of part of chromatin thread showing nucleosomes, and B–figure showing arrangement of histone proteins in a nucleosome.

unit consists of 200 base pairs (bp) of DNA and 9 molecules of 5 types of histone proteins. The core particle or nucleosome contains 140 bp of DNA wound in two complete (or 1.75 turns) superhelical turns around eight molecules (octomer unit) of histone proteins, two each of H_2A, H_2B, H_3 and H_4. About 60 bp of DNA are found in the linker or spacer region with which histone H_1 remains associated. The nucleosome should not be considered as a gene as it is much much smaller than that.

4. ***The supra-mucleosomal structure*** (or ***Solenoid model***): The supra-nucleosomal structure explains that how the nucleosomes are packed into the 200-300A° thick nucleofilament of chromatin. Finch and Klug (1976) found a close packing of nucleosomes to produce a nucleofilament, a fibre 100 A° in dia. The nucleofilament is further coiled up to form the solenoid with a dia. of 300-350A°. There are about six nucleosomes per turn of the solenoid coils. Kornberg and Klug (1981) suggested that H_1 leads to folding of nucleofilament to form the solenoid.

Chemical composition of chromosome. The chief constituents of a chromosome are DNA, RNA and proteins. Proteins are both histone and non-histone types. Calcium is also present.

NON-LIVING CYTOPLASMIC INCLUSIONS OR ERGASTIC SUBSTANCES

Ergastic substances are non-protoplasmic or non-living substances found in the cytoplasm of a cell. These are produced during metabolic activities taking place inside the cell and may be in solution, colloidal or solid crystal form. Basically these are grouped into three categories:

(i) Reserve materials

(ii) Secretory materials

(iii) Excretory materials.

I. Reserve materials

Reserve materials are food materials which are synthesised by the living cells and are stored in different parts of the plant. Chemically these are of following types:

(A) Carbohydrates (B) Nitrogenous matters (C) Fats and oils.

(A) Carbohydrates

Carbohydrates are chemically composed of carbon, hydrogen and oxygen in which hydrogen and oxygen ratio is always 2:1, i.e. the same proportion as in water. Following are important carbohydrate food matters found in the plants.

1. **Sugars**. These are soluble carbohydrates sweet to taste. Many of them are strong reducing agents. The common sugars found in plants are *hexoses*, having formula $C_6H_{12}O_6$. Pentosans are less common. Sugars having one basic group is called monosaccharides such as glucose, fructose, mannose and galactose. **Glucose** ($C_6H_{12}O_6$) is the basic sugar formed during photosynthesis and is therefore found in all green plants. **Fructose**, also called *fruit sugar* has the same formula, $C_6H_{12}O_6$, but it differs in the structure of molecule. It is commonly found in fruits. Mannose and galactose are less common.

Sucrose is a disaccharide as it consists of two monosaccharides. It is commonly called *cane sugar* and it is our table sugar. It is widely distributed in plants but commercially it is extracted from stem of sugarcane and root of beets. On hydrolysis it yields one molecule of glucose and one molecule of fructose. Another important disaccharide is **maltose,** which consists of two molecules of glucose. It is commonly found in germinating seeds during digestion of starch.

Raffinose is the most widely spread trisaccharide found among plants but it is always found in small quantities. Chemically, it consists of one molecule each of glucose, fructose and galactose.

2. **Inulin.** Inulin is a less common polysaccharide found among plants. It occurs in colloidal condition in the cell sap of roots of *Dahlia* and also in artichoke or other members of Compositae (Asteraceae). In alcohol or glycerine preserved root sections when seen under microscope, inulin depositions appear in the form of beautiful fan-like crystals such as in *Dahlia*.

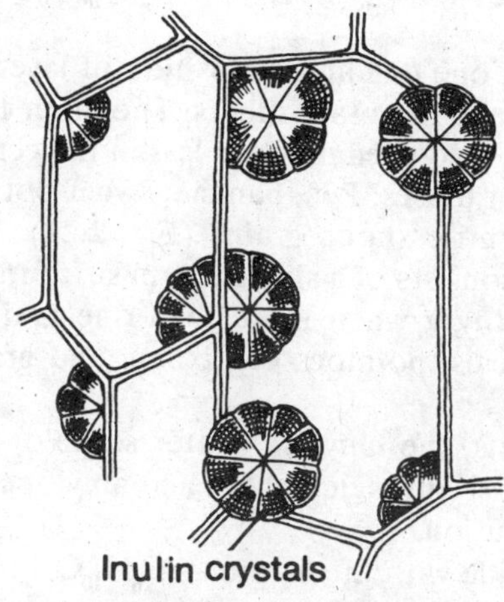

Fig. 2.21. Figure showing inulin crystals in roots of *Dahlia.*

Inulin crystals turn yellowish brown when treated with acid phloroglucin (Fig. 2.21)

3. Starch grains. These are made up of insoluble carbohydrates and are the most common reserve materials found among green plants. Starch is a complex polysaccharide made up of a large number of glucose units, the chemical formula being $(C_6H_{10}O_5)_n$. Starch is commonly accumulated in the form of different types of starch grains, which are stored by amyloplasts (leucoplasts). Each starch grain has a shiny refractive point, called *hilum*, which is the starting point or area of active growth. Around the hilum, the layers of starchy materials are deposited and these layers are called lines of stratification.

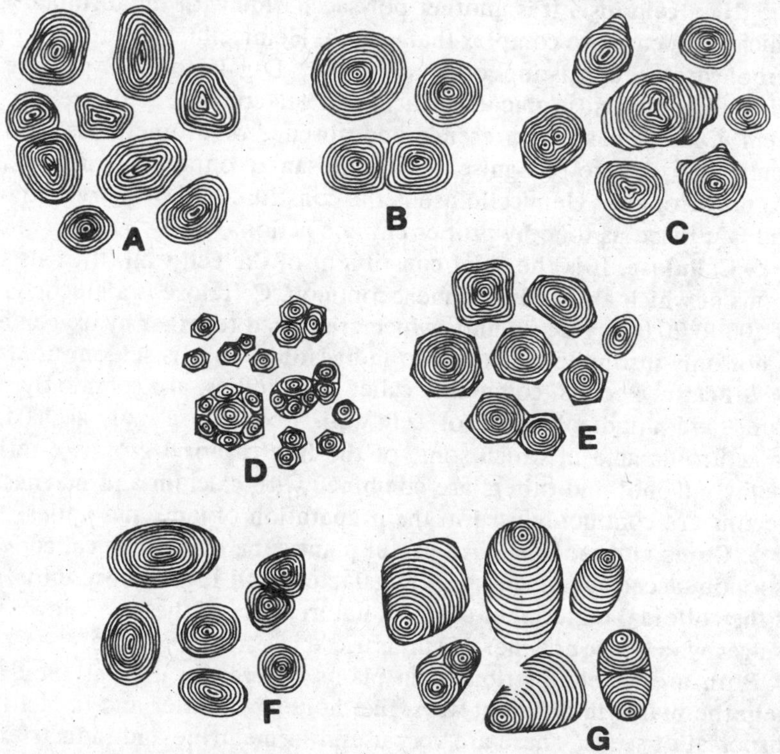

Fig. 2.22. Starch grains of different plants. A–from cotyledon of *Pisum* seed, B–from flesh of *Musa* (banana), C–from tuberous root of *Ipomoea batatas* (sweet potato), D–from endosperm of rice grain (*Oryza sativa*), E–from endosperm of maize grain (*Zea mays*), F–from cotyledon of gram seed (*Cicer arietinum*) and G-from tuber of potato (*Solanum tuberosum*).

Starch grains are thus sphaero-crystals consisting of layers of $C_6H_{10}O_5$ units arranged in a regular space-lattice.

In a starch grain, if hilum is located at one end and deposition of layers is unequal, it is called **eccentric grain,** as found in potato tubers. The other type of grain is **concentric type** in which hilum is located more or less in the centre due to uniform deposition of layers, as in pulses. Pea, banana, sweet potato, rice, maize, wheat and bean have concentric starch grains (Fig. 2.22). The starch grain is called *simple type* when it consists of a single hilum surrounded by layers of stratification and *compound type* when more than one grain is closely appressed. In rice and other cereals, potatoes etc. compound grains are of common occurrence.

Each starch grain has two components, the *amylose* (water soluble) and the *amylopectin* (water insoluble). Amylose turns deep blue and amylopectin turns light-blue to voilet with iodine solution.

4. **Glycogen.** It is another polysaccharide with the formula $(C_6H_{10}O_5)_n$. It is commonly called animal starch due to its common presence in muscles and livers of the animals. However, it is also found in blue green algae, slime moulds, fungi and bacteria.

5. **Hemicellulose.** It is another polysaccharide with the formula $(C_6H_{10}O_5)_n$ which is much more complex than starch. Hemicelluloses are water insoluble carbohydrates consisting of L-arabinose, D-galactose, D-glucose and D-xylose. Uronic acids (glucuronic acid and galacturonic acid) are also found in them. On the basis of presence and absence of uronic acids, two types of hemicelluloses are recognised, (i) pentosans (contains no uronic acid) and (ii) polyuronides. Hemicellulose is the constituent of primary wall of the cell and is utilised as food by proper enzyme action.

6. **Cellulose.** It is the chief constituent of the cell wall. It exists in several forms of which alpha form is most common. Cellulose is a glucogen made up of 500 to 20,000 glucose units which are linked together by oxygen bridges. It is not only insoluble but it is also indigestible, so it is seldom used as food.

7. **Pectin.** Pectins, commonly called fruit jellies, are commonly present in fruits and middle lamella of cell wall. Pectin is a polysaccharide of D-galacturonic acid in which some of the free carboxyl groups esterified with methyl alcohol and others are combined with calcium and magnesium ions. Pectins are commonly used in the preparation of jams and jellies.

8. **Gums and mucilage.** An useful plant gum, commonly called *gum arabic* is a polysaccharide made up of D-galactose and L-arabinose units. Mucilage is the colloidal material of complex nature. One of the best known mucilages is agar which is a polymer of sulphuric acid ester of galactose.

Both are widely distributed in plants. These are used as food as well as help the plants in different ways, like holding of water and in facilitating the dispersal of seeds. These are very useful in medicine and industries.

(B) Nitrogenous matters

The most common nitrogenous matter present in the plants is protein. Protein may be a food matter or a constituent of some organelles, enzymes etc. In general, the nitrogenous matter of non-protoplasmic nature may be present in two forms, (i) simple soluble ones, such as amino acids, and (ii) complex insoluble ones such as in the form of grains called protein grains. Sometime, these grains are very complicated and have very large protein molecules. For example, gliadin of wheat is $C_{635}H_{1068}N_{196}O_{211}S_5$ and zein of corn is $C_{736}H_{1161}N_{174}O_{208}S_3$. In storage regions, protein grains are called aleurone grains, which may be found in crystalline or amorphous forms.

(C) Fats and oils

These are important reserve materials of plants which provide a high amount of energy on digestion. That is why these are commonly preferred in reserve food material of seeds. As compared to carbohydrates, they are poor in oxygen, e.g. the common fats, olein is $C_{57}H_{104}O_6$ and palmitin is $C_{51}H_{98}O_6$. Fats which are liquid at room temperature are called *oils*. Their storage as reserve food materials is associated with elaioplast (leucoplast). Fats usually occur in plants as oils, which are non-volatile in nature, hence may be called fixed oils. Coconut oil, mustard oil and ground nut oil are some of common examples of commercial oils obtained from plants.

II. Secretory materials

Secretory materials are commonly found within special sacs or glands in plants. These do not serve as food and hence are not related with growth of plants. The common secretory matters found in plants are pigments (photosynthetic and non-photosynthetic types), enzymes and nectar. Nectar is a sweet substance usually secreted by the floral parts of some plants.

III. Excretory materials

Since these are by-products of metabolic activities and are of no use to plants themselves, these are called excretory materials or waste products. But these are of high commercial value to the man. These are commonly accumulated in leaves, fruits and bark of the plants. Some of the important excretory materials are as follows:

1. Tannins

These are bitter and astringent substances related to glucosides. Tannins are present in the vacuolar sap of cells in soluble forms in barks, unripe fruits, leaves, seeds and woody parts of some plants. Bark of *Acacia* and fruits of *Myrobalans* are good sources of commercial tannins. Tannins are frequently used in leather industries and in the manufacture of ink.

2. Resins

Resins are oxidation products of essential oils and are complex in composition. These are found in special glands or ducts either alone or in association with gums and essential oils. There are specialised canals or ducts in plants like *Pinus*. Resins may be of different types. Hard resins like lacquer are used in paint and varnish industries. Oleo-resins are resins associated with essential oils and are liquid in nature, e.g. turpentine and canada balsam. Gum resins are mixtures of gum and resins, such as common asafoetida.

3. Alkaloids

Alkaloids are decomposition products of proteins and are mostly found in root, seed, bark and leaf of the plants. Alkaloids are bitter in taste and some of them are deadly poisonous. A few of them are medicinally very important, e.g.

Quinine (obtained from the bark of *Cinchona*)
Reserpine (from roots of *Rauwolfia*)
Strychnine (from seeds of nux vomica, *Strychnos*)
Coccaine (from the leaves of coco, *Erythroxylon*)
Caffeine (from seeds of *Coffea*)
Ephedrine (from leaf and stem of *Ephedra*)
Nicotine (from the leaves of *Nicotiana*)
Atropine (from roots of *Atropa*)

4. Glucosides

These are similar in several properties to alkaloids but are the decomposition products of carbohydrates. Some common examples are *amygdalin* (obtained from almond) and *digitoxin* (obtained from digitalis). These are used in medicines. Digitoxin is a valuable heart stimulant.

5. Latex

These are milky-white, water soluble substances found in leaves, stems and roots of some plants. It is a kind of emulsion which contains substances like proteins, gums, resins, alkaloids etc. suspended in water. Latex are variously used by plant themselves and by human beings. In plants, these are helpful in healing of wounds. For men, latex are valuable commercial source of rubber. Rubber yielding plants commonly belong to families Moraceae, Euphorbiaceae and Apocynaceae. Latex from *Hevea brasiliensis* and *Ficus elastica* are most commonly used in the manufacture of rubber. The latex of the famous cow tree of Venezuela (*Brosimum galactodendron*, Moraceae) is taken by natives as a substitute for milk.

6. Gums

Gums are found in large number of plants and it is a decomposition product of cellulose cell wall. Gums are water soluble and swell up on getting water. These are usually released in plants through wounds or cuts by the process referred to as *gummosis*. Some gums are of considerable economic value, e.g. gum arabic (obtained as an exudate of *Acacia senegal*).

7. Essential oils

These are volatile oils found in specialised glands which exist in the leaves (*Citrus* and *Eucalyptus* species), pericarp (*Citrus* species), flowers (*Jasminum*) etc. These are used for the manufacture of soap, oil, scent and medicines.

8. Organic acids

Organic acids are most commonly found accumulated in the leaves and fruits of certain plants. Fruits of tamarind and unripe grapes have tartaric acid, lemon, tomatoes and oranges are rich in citric acid and apples are rich in malic acid. These acids are water soluble and may make the cell sap toxic when present in high concentration.

9. Mineral crystals

Crystals of minerals like calcium oxalate, calcium carbonate and silica are found as waste products in many plants.

Calcium oxalate crystals. These are most common and widely distributed type of crystals which are found in storage organs, pith and cortical tissues. These are of different shapes-rod-like, needle-like, prismatic or pyramidal

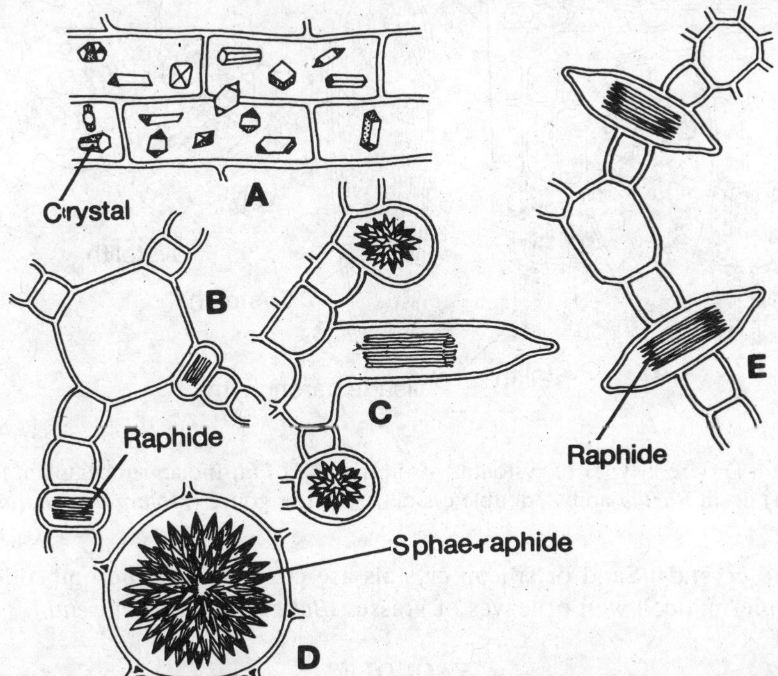

Fig. 2.23. Different types of oxalate crystals. A–isolated crystals in the tunic of onion (*Allium cepa*), B–raphides found in the petiole of arum (*Colocasia* spp.), C–raphide and sphaeraphides in the leaf cells of water cabbage (*Pistia stratiotes*), D–sphaeraphide in the petiole of papaya (*Carica papaya*) and E–raphides in the petiole cells of water-hyacinth (*Eichhornia crassipes*).

etc. Solitary crystals are found in the tunic of onion (*Allium cepa*) (Fig. 2.23). Needle-like or acicular crystals found in the form of bundles are called raphides, as in *Colocasia, Pistia* and *Eichhornia* (water hyacinth). Raphides are usually found in special mucilage containing cells called idioblasts. Some raphides are found in spherical groups and hence called sphae-raphides as in case of petiole of *Carica* and *Pistia* (Fig. 2.23). Group of raphides are also called **druses.**

Calcium carbonate crystals. These are less common. These crystals occur in the form of crystalline mass in the epidermal cells, by impregnating a portion of cell wall. These are characteristically found in the upper epidermis of leaves of Indian rubber (*Ficus elastica*) and banyan (*F. bengalensis*) where these appear in the form of a bunch of grapes (Fig. 2.24). These calcium carbonate crystal aggregates are commonly called **cystoliths**. These are also found in the leaves of some members of Cucurbitaceae (e.g. *Momordica*) and Acanthaceae (e.g. *Ruellia*). In *Momordica*, double cystoliths and in *Ruellia*, elongated cystoliths are found (Fig. 2.24).

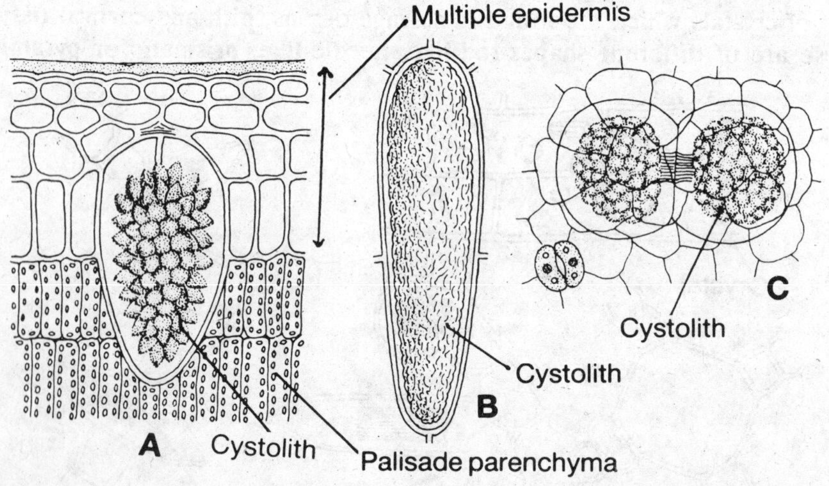

Fig. 2.24. Different types of cystoliths. A–in the leaf of an Indian rubber plant (*Ficus elastica*), B–in *Ruellia* and C–double cystolith in bitter gourd (*Momordica charantia*)

Sand crystals. Sand or silicon crystals are commonly found embedded in the epidermal cell well of leaves of grasses, *Belladona* and *Equisetum*.

VACUOLES

Vacuoles are not visible in young stage, but as the cell grows in size and enlarges, they begin to make their appearance like small droplets which go on coalescing with each other to form a bigger vacuole at maturity. At maturity the size of vacuole may enlarge to such an extent that cytoplasm is left only in the form of a thin, peripheral cytoplasmic layer, enclosing the nucleus. Surrounded by the tonoplast, a vacuole contains cell sap with dissolved atmospheric gases, minerals, sugars, organic acids (such as acetic acid, malic acid and tartaric acid) and at times anthocyanin pigments. The various colours of petals are due to the presence of different anthocyanin pigments present in vacuoles. Some solid bodies of aleurone grains and soluble proteins are also present. Several types of enzymes, including hydrolases are also present. These play an important role in cell metabolism and may act as storage compartments and also as lysosomes. Vacuoles help to regulate turgor and detoxification.

PROKARYOTIC CELLS

A prokaryotic cell is found in case of bacteria and blue green algae and differs in certain respects from eucaryotic cell found among plants and animals. In prokaryotic cells, the protoplasm can be differentiated into cytoplasm and incipient nucleus. The cytoplasm contains photosynthetic lamellae or chromatophores which are the sites of photosynthesis. Certain other lamellar structures or infoldings of cell membrane called *mesosomes* are found to carry on respiration. The ribosomes lie free in the cytoplasm and are the site

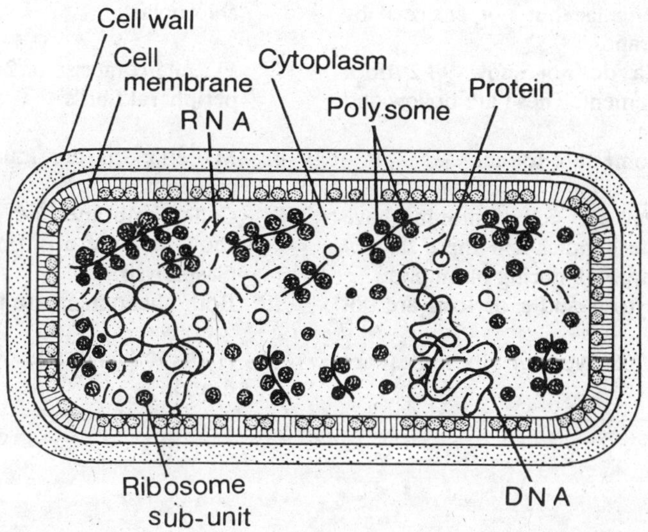

Fig. 2.25. Diagram of a typical prokaryotic cell (*E. coli*) showing different structures.

of protein synthesis. Plastids, mitochondria, endoplasmic reticulum, Golgi apparatus, peroxisomes, lysosomes etc. are not found. The nucleus is not true, i.e. nuclear membrane is absent. The chromatin material lies free in the centre of the cytoplasm in the form of a tangled mass. DNA of prokaryotic cells is non-histone type and circular in nature. The prokaryotic cell differs

from a eukaryotic cell by different features as summarised in table 2.3.

Table 2.3. Diference between Prokaryotic and Eukaryotic cell.

Prokaryotic cell	*Eukaryotic cell*
1. Nucleus is incipient type. It lacks nuclear membrane, nucleolus and chromatin reticulum (chromosome).	1. It has a true nucleus in which nuclear membrane, nucleolus and chromatin reticulum are present.
2. DNA is circular and non-histone type, i.e. it lacks protein association and lies in a tangled mass called nucleoid.	2. DNA is associated with protein to form chromosomes or chromatids.
3. All membrane-bounded structures such as chloroplast, mitochondria, ER, Golgi complex, vacuoles etc. are absent.	3. Membrane bounded structures are present.
4. Cell wall contains aminosugars and muramic acids.	4. These substances are absent in the cell wall.
5. Cytoplasmic streaming (cyclosis) is not observed.	5. It can be observed.
6. In autotrophic forms, photosynthetic pigments are found associated with lamellae but not enclosed by membranes.	6. Lamellae are enclosed in two unit membrane envelope and called chloroplast.
7. Flagella do not show 9+2 fibril arrangement. These are hollow and tubular.	7. Flagella comprise of 2 central and 9 peripheral fibrils.
8. Mesosomes are found.	8. Mitochondria are found instead of mesosomes.
9. Cell wall and cell membrane have intricate contact and these do not fully separate during plasmolysis.	9. Cell membranes lie separate and show complete shrinkage during plasmolysis.
10. Endocytosis and exocytosis are not found.	10. Both phenomena are found.
11. Only 70S type of ribosomes are found.	11. Both 70S and 80S ribosomes are found.

GENERAL QUESTIONS

1. Justify the statement "Cell is the chemical and biochemical basis of life."
2. Write notes on
   (i) Cell wall
   (ii) Ultrastructure of cell membrane
   (iii) Plastids
   (iv) Ultrastructure of chromosome.
3. Give the structure and function of :
   (i) Ribosomes
   (ii) Nucleolus
   (iii) Mitochondria.
4. Write a note on differences between eukaryotic and prokaryotic cells.
5. Write short notes on
   (i) Nucleosome
   (ii) Solenoid model
   (iii) Golgi bodies
   (iv) Peroxisomes
   (v) Lysosomes.

CHAPTER 3

Cell Cycle and Cell Division

CELL CYCLE

The term cell cycle is commonly used to describe the history of a cell from the time of its formation as a product of mitosis to the completion of its own division into two daughter cells. A typical cell cycle consists of four phases of which first three (G_1, S and G_2) form interphase.

(i) **G_1 phase** or gap 1 phase: It is characterised by the presence of transcription and absence of DNA replication. During this phase, the different metabolic activities are found in a cell but there is no synthesis of DNA. Active synthesis of RNA and proteins takes place at this stage.

(ii) **S phase:** During this phase, replication of DNA takes place and a cell begin preparation for its division. Each chromosome now carries a duplicate set of genes. It is now composed of two sister chromatids held together by a common centromere.

(iii) **G_2 phase:** Again in this stage, replication of DNA is not found. However, transcription process and protein synthesis are found. This protein is used in structural make up of chromosomes.

(iv) **M phase:** It refers to mitosis or nuclear division.

Such a cell cycle is of common occurrence in eukaryotic cells but it is not found among prokaryotic cells. In prokaryotes, the replication or synthesis of DNA is almost a continuous process. Why this kind of definite time of DNA replication is found in the organisms having definite nuclear membrane is not clearly known. Cell cycle may show variation as regards to timing pattern of different phases. The relative length of different phases shows variation in different species, sometimes in different tissues or even in the same cell depending upon different conditions.

The period of G_1 and G_2 varies remarkably between different species. G_1 may be virtually absent in some cases. In the conventional cell cycle, G_1 phase is commonly of 8 to 10 hours, the S-phase 6 to 8 hours, the G_2 phase 2 to 4 hours and M-phase about 1 hour. The total duration of cell cycle in *Vicia faba* and *Tradescantia paludosa* is 19.3 h and 17 h, respectively.

CELL DIVISION

The cell is the basic unit of life and division of the cell is the characteristic feature of every organism. It was first pointed out by Nageli (1864) that the new cells are always formed by the division of the pre-existing ones. Schleicher (1878) proposed that the main dividing structure of a cell is the nucleus for which he used the term karyokinesis.

In each cell division, two important phases are recognised.

1. Karyokinesis or division of nucleus.
2. Cytokinesis or division of cytoplasm.

KARYOKINESIS

The term is used to refer division of nucleus. Depending upon pattern of division, three fundamental types of nuclear division are recognised.

(i) Amitosis, (ii) Mitosis and (iii) Meiosis.

Amitosis

It is the simplest type of division and may also be called direct nuclear division. Amitosis is commonly found among unicellular organisms such as yeast, bacteria and cyanobacteria. During the process, the nucleus elongates and becomes dumb-bell shaped. A constriction in the middle of it is produced which deepens and ultimately a nucleus is broken into two daughter nuclei. The different stages like prophase, metaphase, anaphase etc. are not found in this kind of karyokinesis.

Mitosis

Mitosis was first discovered by Walter Flemming (1878-1882) but its details were worked out by Schneider (1883). The term mitosis (*mitos* = thread) is a Greek word which basically refers the thread-like appearance of chromosome in early cell division. The growth of an individual from zygote to adult stage takes place through repeated mitotic cell divisions. So the principal feature of mitosis is that the new nuclei receive identical chromosomes and genes to those of parent cell. While it's a characteristic to somatogenetic reproduction it maintains the same number of chromosomes after the cell has divided.

Before mitosis, a nucleus is present in a resting stage called **interphase.** This may however be referred as a critical period when the cell prepares itself for the division. The nucleus appears then as a dark staining body, the nucleolus being pronounced and the chromatids while showing minimum degree of condensation or coiling remain dispersed in the nuclear sap. In fact, interphase is the most active stage for the nucleus when it controls the synthesis of different types of enzymes to regulate metabolic processes and also it undergoes with the process of nucleic material *(DNA)* duplication.

Three stages in the interphase nucleus have been recognised. They are G_1, S and G_2, each separately described under heading 'cell cycle'.

Besides interphase, four different stages–Prophase, Metaphase, Anaphase and Telophase can be recognised in mitosis.

1. **Prophase.** Being the longest stage of mitosis, it may last long from few minutes to several hours. In onion root it is completed in 7.1 minutes and in pea endosperm in about 40 minutes. At the beginning of prophase, the chromosomes appear as a thin, thread-like, uncoiled structure that becomes coiled, shortened and more distinct as the mitosis progresses through the prophase.

Another important feature of prophase is the *longitudinal splitting of each chromosome into two sister chromatids.* The two sister chromatids are attached only at the position of centromere. The coiling occurs in between these two sister chromatids which may be of two levels–larger one called *major coils* and smaller ones *minor coils.* In most cases, the pattern of coiling is **plectonemic type** in which chromatids are twisted about each other relationally so much that it is not easy to separate them. In a few cases, **paranemic coiling** may also be seen.

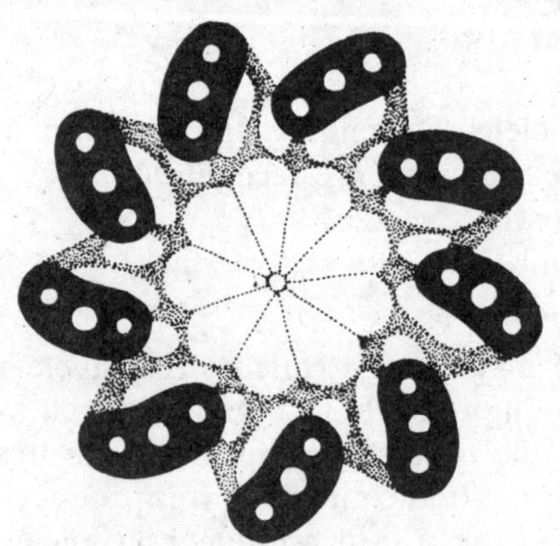

Fig. 3.1. T.S. of centriole as seen with the help of electron microscope.

Towards the end of prophase *spindle formation occurs.* In case of animal cells, the centriole, a special structure close to nucleus, also divides into two daughter centrioles. Ultrastructure studies reveal that a centriole is built up of nine obilquely oriented peripheral fibrils, each composed of three sub-fibrils. Structurally they are quite similar to basal body of flagellum. The division of centriole may be of two types. In one case, the divided centrioles separate and each shows an aster or astral rays and in between the two asters, a bundle of delicate spindle filaments appear. The centrioles along with aster continue to migrate towards their respective poles. While in the second case, the centrioles are already polarised prior to nuclear division. Such mitoses are called amphiastral types being common to animals and a few lower plants (*Spirogyra* sp.). In contrast, plants show anastral type of mitoses.

Finally, at the end of prophase, the nuclear membrane begins to disappear into the cytoplasm. Similarly nucleolus also disappears before the cell enters the metaphase.

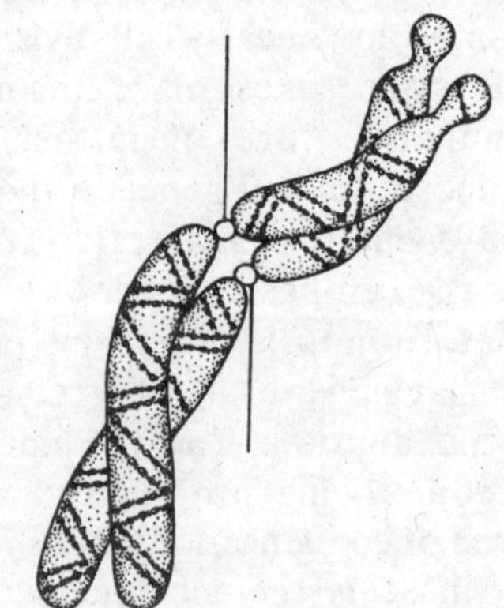

Fig. 3.2. Condition of chromosome as seen during metaphase.

2. Metaphase. As described earlier that towards the end of prophase the spindle tubules begin to appear. Of these some are attached to chromosomes called chromosomic fibres and others are continuous from one pole to other called continuous fibres. The attachment of spindle tubules to the chromosomes is instrumental in directing the chromosomes to the cell equator. Simultaneously the centromeres of the chromosomes divide so that each sister chromatid has its own centromere.

Thus, in metaphase, the *duplicated chromosomes* come to lie in the equatorial plate of cell. The chromosomes are best studied at this stage.

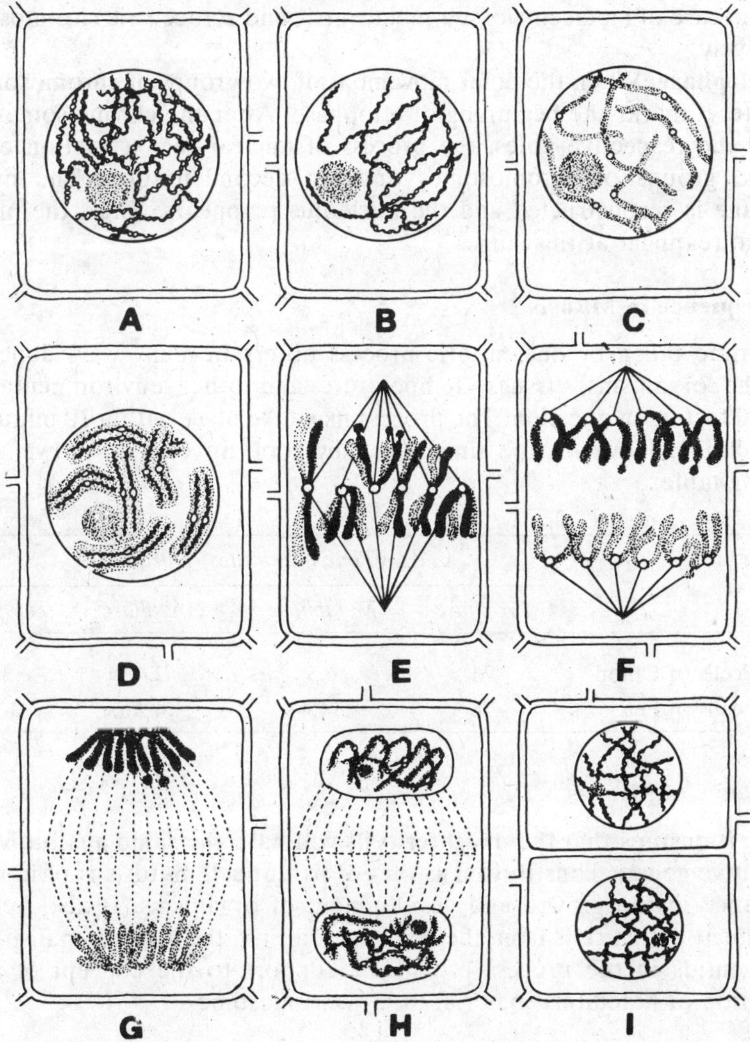

Fig. 3.3. Schematic representation of mitosis in plant cell, A to D–different stages of prophase; E–metaphase, F–anaphase, G and H–telophase and I–newly formed daughter cells.

3. **Anaphase.** Anaphase begins with the movement of chromosomes towards their respective poles. At this stage as soon as the centromere becomes functionally double (divided), sister chromatids separate from each other due to a kind of repulsive force produced between the two centromeres. Each separated chromatid with its own centromere is now called chromosome. The *movement of chromosomes towards opposite poles* of spindle is the characteristic feature of anaphase and is believed to be supported by contraction of spindle fibres.

In anaphase, the chromosomes assume typical shapes depending upon the position of their centromeres. They may look like V-shaped, L-shaped or J-shaped in case of metacentric, sub-metacentric and acrocentric chromosomes, respectively.

4. **Telophase.** When the polar movement of two groups of chromosomes is complete, it marks the beginning of telophase. After the chromosomes have reached the respective poles, the process of nuclear reconstruction occurs. The two groups of chromosomes undergo decondensation. The nuclear membrane is reconstructed and the nucleolus reappears. Thus, the nucleus begins to reappear at this stage.

Time Sequence in Mitosis

The time taken by one mitotic process differs in many ways depending upon the organisms, tissues, temperature and other environmental and chemical factors in question. The process may take place within 10 minutes to several hours. However the time taken by root tip cells is shown in the following table.

Examples	*Time in minutes (Duration)*			
	Prophase	*Metaphase*	*Anaphase*	*Telophase*
Root tip cells of Onion	71	6.5	2.4	3.8
Root tip cells of Pea	78	14.4	4.2	13.2

Significance of Mitosis

Since it ensures that the daughter cells will have the same qualitative and quantitative genetic constitution as that of parent cell, mitosis is of immense significance in the growth and development of organisms. It also helps in replacement of lost cells from the body. In addition, the longitudinal splitting of chromatids in the process provides a support to the concept of linear distribution of hereditary material in the chromosome.

Meiosis

Meiosis comprises of two nuclear divisions--(i) heterotypic division or first meiotic division or reductional division and (ii) homotypic division or second meiotic division or equational division. However, there is only one chromosomal division in the process.

(A) First Meiotic Division

First meiotic division consists of most complicated but very important stages of cell division since the process of reduction in chromosome number occurs within its various steps. Basically it is divided into 4 stages–Prophase I, Metaphase I, Anaphase I and Telophase I but its prophase I needs special mention.

Prophase I. Longest among all other stages prophase I differs from mitotic prophase in several respects. Owing to its complication the stage is further divided into five sub-stages as follows.

(i) **Leptotene**–Leptotene is not much distinct from the mitotic prophase. In this stage, the nucleus enlarges in size and the chromosomes begin to appear as a long, uncoiled, thread-like, loosely interwoven structure. The arrangement of chromosomes is often irregular. Occasionally, in certain plants, particularly in *Lilium,* a phenomenon called **syninzesis** is seen where all chromosomes are clumped to one side of nucleus and there being no chromatin material seen in the remaining part of the nucleus. Progressive condensation and coiling of chromatin fibres occur at this stage. The chromosomes appear at first, as fine single threads, which later contract and become thick. The nucleolus and nuclear membrane remain distinct.

(ii) **Zygotene**–The zygotene stage is characterised by the presence of pairing or synapsis of homologous chromosomes. The two homologous chromosomes pair lengthwise with each other in a zipper-like fashion, starting from centromere or at the chromosome ends. These form syneptinemal complex which contains RNA, proteins and DNA. It holds the homologous chromosomes together. The pattern of pairing is characteristic to species in question. Of the two synaptic chromosomes, one is paternal and other maternal, collectively these are called **bivalents.** In eukaryotes, synapsis is the major DNA–DNA recognition process.

However, coiling in chromosomes continues in this stage so as to result chromosomes shorter and thicker in comparison to those of leptotene.

(iii) **Pachytene**–Pachytene begins with the end of chromosomal synapsis. The chromosomes are further shortened and coiled. At the height of pachytene, the paired chromosomes may twist about one another as relation coils and further the two chromatids of each chromosome may also coil around each other. Each chromatid undergoes a longitudinal splitting, if no splitting has occurred. At this stage, two sister chromatids of each homologue become visible and the bivalent thus becomes a **tetrad** with four chromatids. At this stage the crossing over or the exchange of segments of chromatids is brought about. When chromosomes have longitudinally splitted, transverse breaks occur in two of the adjacent homologous chromatids. Soon, both segments of chromatids interchange and fuse together with those of homologous chromosomes. So, genetic material is exchanged between non-sister chromatids (of the homologues) by breakage and rejoining. The nucleolus still persists.

(iv) **Diplotene**–Diplotene is characterised by the separation of homologous chromosomes from one another which begins from the centromere to the ends. This process is called **terminalisation** (Darlington, 1930). The name diplotene is characteristic to this stage as the dual nature of bivalent becomes distinct due to such separation. The separation is not complete as the two homologous chromosomes remain held together at certain point along the length. Such points of contact between the two homologous chromosomes are called **chiasmata** which represent the place of actual crossing over. With the process of terminalisation, these chiasmata move towards the end of the chromosomes.

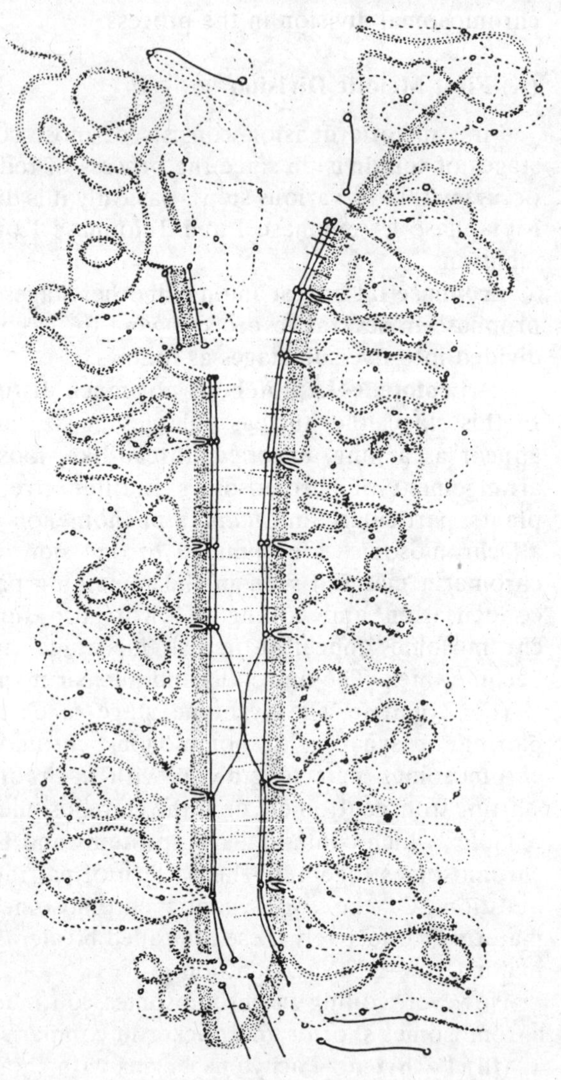

Fig. 3.4. Diagram to show the probable assembly of a syneptinemal complex from two axial cores. The central element may represent *DNA* spun out on either sides of two homologues. (Redrawn from Moens, 1968). (State–starting zygotene to pachytene).

The nucleolus and nuclear membrane at this stage begin to disorganise. Nucleolus may still persist but nuclear membrane cannot be seen. The process of contraction of chromosome still continues.

(v) **Diakinesis**–The diakinesis can be distinguished from diplotene by the presence of more contracted state of chromosomes. Due to further contraction and terminalisation these appear as rounded bodies evenly scattered throughout the cell or have the tendency to be at the periphery near the nuclear envelope. The nucleolus also disorganises and is no longer visible at the end of this stage. The spindle begins to be formed at the end of this stage.

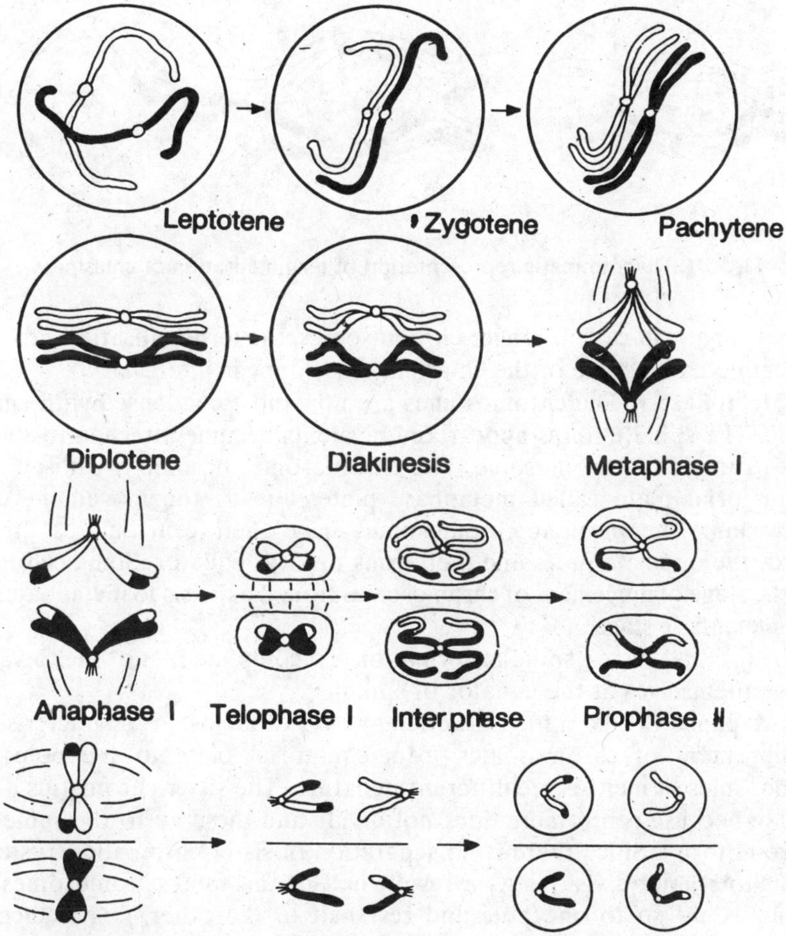

Leptotene
Zygotene
Pachytene
Diplotene
Diakinesis
Metaphase I
Anaphase I
Telophase I
Interphase
Prophase II
Metaphase II
Anaphase II
Telophase II

Fig. 3.5. Schematic representation of various stages of meiosis.

Zygotene
Pachytene
Diplotene

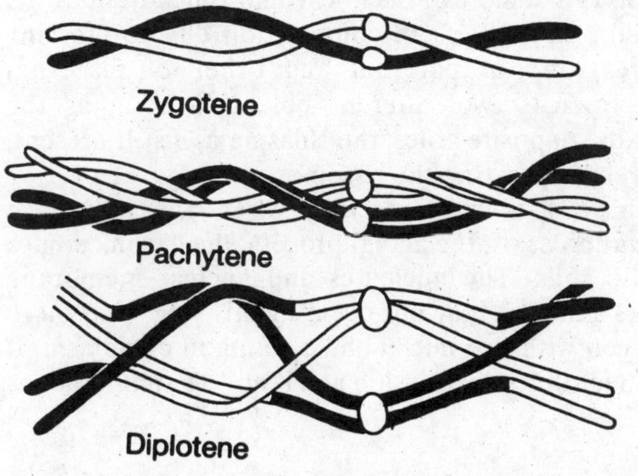

Fig. 3.6. Diagrammatic illustration of behaviour of homologous chromosomes during zygotene, pachytene and diplotene.

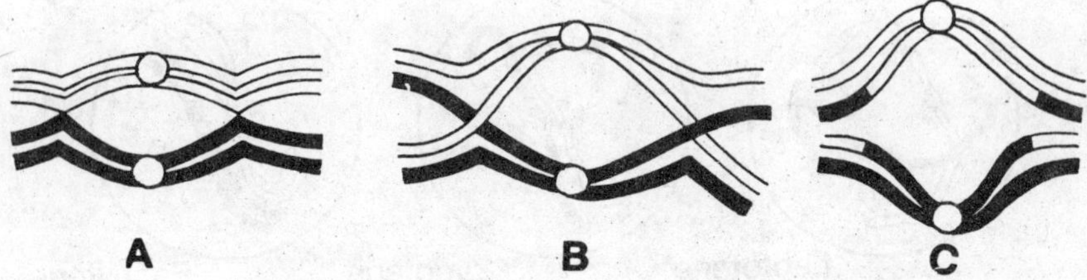

Fig. 3.7. Diagrammatic representation of terminalisation of chiasmata.

Occasionally in case of longer chromosomes, the terminalisation process is not complete and some of the chiasmata may appear intercalary.

2. **Metaphase I.** The chromosomes are now most condensed with smooth outline. The spindle fibres appear and bivalents become attached to them at the centromere of chromosome. The chromosomes become organised along an equatorial plate called metaphase plate, due to the movement called **congression.** The bivalent chromosomes have their centromeres directed towards the opposite poles and their arms towards equator. Some cytologists refer the stage of movement of chromosomes from diakinesis to metaphase plate as prometaphase stage.

During this stage, spindle formation is completed and the bivalents arrange themselves at the equator of spindle.

3. **Anaphase I.** Like mitosis, the anaphase I of meiosis is characterised by the movement of chromosomes from equatorial plate to the poles, but chromosomes of meiosis are different in nature. The sister chromatids do not separate because centromere does not divide and these go to the same pole (unlike mitosis). Since there is no separation of sister chromatids, it's called **non-disjunctional** division. As a result, half of the total chromosomes of a nucleus would go to one pole and rest half to the other, i.e., reductional division. Now each group contains haploid number of chromosomes.

In the process, the two chromatids of the chromosome widen apart except at the centromere and assume a V-shaped structure. The shapes however vary and depend on position of centromeres. Each chromosome, besides its double number of chromatids, also differs in composition due to crossing over. The two centromeres, one of paternal and other of maternal chromosomes are directed towards two different poles. As soon as the chromosomes move towards the opposite poles, the chiasmata, if still present, loosen, and finally free, thereby separating the chromosomes.

4. **Telophase I.** Telophase I is characterised when the anaphase chromosomes conglomerate at poles. In the actual process, the chromosomes elongate by loosening of the coils. The nucleolus and nuclear membrane reappear. However, cytokinesis may or may not occur at this stage and there may be two cells or a single cell with two nuclei and a common cytoplasm. If the cell wall is formed, it's called successive division and if not it's simultaneous division.

Interkinesis. Two such formed haploid daughter nuclei undergo a short resting stage which occurs in between telophase I and prophase II. In certain cases, such as in *Trillium* sp., the chromosomes while retaining coiling at telophase pass on to prophase II. However, the interkinesis, also called interphase, is either very short or may be entirely absent. DNA synthesis does not take place during this stage.

(B) Second Meiotic Division

It is a simple mitotic division and therefore may be called meiotic mitosis. The characteristic features of different stages of mitosis have been discussed earlier. The stages may be called–prophase II, metaphase II, anaphase II and telophase II.

The chromosomes of prophase II are already double, i.e., each one has two sister chromatids with a single functional centromere. Shortly these chromosomes arrange on metaphase plate where their centromeres split. Each separated part is called chromosome (daughter chromosome). With the help of spindle fibres, these chromosomes soon begin to move towards the two poles (anaphase II). Since the chromosomes were doubled, each group will contain similar number of chromosomes. The division therefore may be called an **equational** or a **homotypic** division. Simultaneously, telophase II sets in which is soon followed by cytokinesis, Cytokinesis whether successive or simultaneous type, will give rise to four haploid cells.

Significance of Meoisis

In preparation of sexual reproduction, haploid male and female gametes are formed by meiosis. During fertilization, these two gametes fuse and thus restores the original diploid chromosome number in the zygote. This mechanism ensures the maintenance of the same chromosome number, characteristic of species, from generation to generation.

Secondly, due to crossing over, meiosis leads to the production of four daughter cells, differing slightly from one another, and also from the parent cell in their genetic constitution. Thus, meiosis results in variations of genetic character in the offsprings.

Comparison Between Mitosis and Meiosis

Mitosis	*Meiosis*
1. Resting stage preceding mitosis is considerably long.	1. Resting stage is shorter.
2. Prophase is short stage and without substages.	2. Prophase is long and composed of leptotene, zygotene, pachytene, diplotene and diakinesis.
3. During the preceding resting stage, the chromosome longitudinally splits into two sister chromatids. Therefore, from the early stage, chromosomes appear as double threads.	3. During resting stage preceding prophase, chromosomes do not appear as a double thread rather as a single thread (leptotene).

Mitosis	*Meiosis*
4. The homologous chromosomes are not attracted to each other.	4. The homologous chromosomes are attracted to each other and form bivalents. The chromosomes then split into sister chromatids and at four-strand stage crossing over takes place. During crossing over, there is exchange of segments of chromatids between the maternal and paternal chromosomes. Chromosomes later on separate and chiasmata appear. Depending upon the number and size of chiasmata, the chromosomes have characteristic appearance.
5. During metaphase, the centromere of each chromosome divides and the sister chromatids move towards the opposite poles. Thus the number of chromosomes remains maintained.	5. Without the division of centromere, homologous chromosomes move towards opposite poles. Therefore, the number of chromosomes at any place is reduced to half, i.e., reduction in chromosome number occurs at this stage. It is called metaphase I.
6. The chromosomes which begin to move towards opposite poles are of both maternal and paternal types (due to longitudinal division). Therefore chromosomes at both the poles would be identical.	6. Chromosomes at metaphase I are so oriented that either paternal or maternal chromosome only, goes to one pole. Therefore, chromosomes of both the poles are never identical.
7. The movement of chromosomes towards the poles is due to repulsion between the parts of divided centromere. The two divided centromeres belong to same chromosome.	7. The repulsion occurs between the centromeres of homologous chromosomes. The two centromeres are of different chromosomes.
8. Anaphase mechanism is similar.	8. Anaphase I mechanism is similar to that of mitosis.
9. The two daughter nuclei formed at telophase have same number of chromosomes as that of parental cell.	9. The chromosome number is reduced to half.
10. As a result, by the end of division, neither chromosome number nor quality of chromosomes has changed.	10. The number and quality (genetic constitution) of chromosomes are changed.
1. No further division occurs and therefore only two daughter cells are formed.	11. This division is followed by mitotic division (similar to mitosis) and therefore 4 daughter cells, each with reduced number of chromosomes, are formed.
12. It is found in both haploid and diploid cells.	12. It is found only in diploid cells.

CYTOKINESIS

The nuclear division is followed by cytoplasmic division which begins as early as the late anaphase. This division of cytoplasm is called cytokinesis. The pattern of cytokinesis differs in animals and plants.

(i) **In animals**–An animal cell is devoid of cell wall, and surrounded only by cell membrane, and therefore more flexible. During cytokinesis, a circular constriction appears at the equator which deepens on all sides and ultimately separating two daughter cells. Since the constriction appears as a furrow, the process may be called cytokinesis forming cell furrow.

(ii) **In plants**–In plants, the cell wall is rigid and during cytokinesis a cell plate is laid down in the centre which is completed towards the periphery. The formation of cell plate occurs with the deposition of certain cytoplasmic droplets or microspheres which is soon followed by the deposition of materials of primary and secondary walls. The process is also called cytokinesis by cell plate formation. The cell wall synthesis is controlled by vesicles released from the Golgi bodies.

PHYSIOLOGY OF CELL DIVISION

Cell division is of considerable interest and is something wonderful within a cell which can be seen with the help of most easily available compound microscope. The universal occurrence of such stages in eukaryotes adds further interest and invites keen cyto-geneticists and inquisitive students to maintain some prudent concept of this magic fact. Undoubtedly, the available concerned literature offers conclusive explanation to simple thinkers, yet if seen the various steps with deep concentration, each one creates a new problem. Some of them being possible to discuss within the scope of the book have been taken up here. Of course, these would serve as prenote only.

The fact that cell division is a classical name for chromosome reproduction initiates a question in the mind to explore the temporal controls in chromosome reproduction. To ask *what is the molecular basis of chromosome reproduction?* might appear a precise and clean question. Actually, it is more undefined and confusing. Had chromosome reproduction been a single event, it would have been the most ideal question. The fact is that it is not. Hubert Stern (1972) while replying to this question said that "the degree to which a molecular mechanism elucidates a particular cellular event can be no better than the degree to which that event has been defined." He suggested that to the various facets of chromosome reproduction molecular analyses could be applied to more restricted targets.

What regulates the cell division is probably present in both cytoplasm as well as in the nucleus. A cytoplasm from the dividing cell is capable to induce chromosome replication when in contact with an otherwise non-replicating nucleus. It's not the sole factor but is subject to other determinants that supply a temporal ordering of regional replications both within and between chromosomes. Specific sites that control the initiation of chromosomal replication are also known present within the chromosomes and it is visualised that a kind of message must exist between different regions of a chromosome and between different chromosomes in order to account for the fine aspects of the pre-determined temporal pattern of chromosome reproduction.

Problems when visualised at the cellular level offer several specific opportunities for molecular explanations. At the very first stage, the capacity of cytoplasm to induce chromosome replication only at the time of nuclear division provides tentative conclusion that the onset of *S* phase is due to some cytoplasmic transformations. This fact was explained to be the involvement of provision of essential enzymes for *DNA* replication. However, studies made on the enzyme changes in association with *DNA* replication, do not support this interpretation and the problem remains a total mystery at the molecular level.

Meiosis is considered to be a system for achieving a reduction in chromosome number. Undoubtedly it is a common method but by no means the sole pattern, specially in insects, where in many species, unusual is usual. However, to the cytologists more important is to investigate and comprehend the centromere behaviour in meiosis. Two points need clear explanations–(i) the timing of effective centromere division and (ii) its physical association with spindle fibres. An explanation to the first point, based on the fact that the centromere division is a result of *DNA* replication in the centromere region, and that the timing of this replication is affected by mechanism similar to those in other regions of chromosome, has not been substantiated by any kind of chemical evidence. Moreover, cytological studies indicate that the centromere is a compound body formed, prior to metaphase. The second point provides even less scope for speculation. Stern (1970) holds the opinion, that chromosome fibre running to the pole of the spindle formed at the centromere is reasonably well established and that this characteristic property is retained by the centromere even in the fragmented chromosomes. Since the chromosome is considered to be a linearly differentiated *DNA* templete, it is difficult to think of the chromosome as to have some special site for the formation of such special structures. If not, then whether centromeres have a specific mechanism for synthesising some special protein which might serve either as monomers for fibre formation or the centromeres are the site of polymerisation of pre-existing monomers into a fibril? These questions are extremely difficult to answer.

Geneticists have other problems of major interests, such as phenomena of pairing and crossing over. At the same time it should come to the mind that the occurrence of pairing and crossing over is not unique to meiosis. Somatic pairing occurs regularly in certain insects and is also known in the salivary gland chromosomes of *Drosophila*, and crossing over, in fact, between the sister chromatids is common in mitosis, where it is called sister strand exchange. Crossing over occurs most frequently in viruses than higher organisms. However, crossing over occurs in meiosis when homologous chromosomes are brought into close proximity. Should it be considered as a mere coincidence of pairing or the synchronised occurrence of chromosomes?

When this phenomenon is compared with mitosis, an explanation comes out itself. The prophase in mitosis extends for about an hour while that in meiosis extends for days or even weeks. The extended time interval must be associated with a sequence of distinctive changes in the chromosomes. Obviously, it's a presumption, just as crossing over is presumed to occur during pachytene, because pairing is completed at the end of zygotene and persists no later than the beginning of diplotene. This argument is quite logical but without any direct evidence.

Similarly the problems like–What initiates meiosis? What causes synpasis? What causes diplotene opening? Does coiling have any role in meiosis? What is the exact mechanism of crossing over? How are the chromosomes brought and organised along a metaphase plate? etc. are various delicate questions with insufficient replies.

Let us look, some of the problems as explained on the basis of circumstantial evidence produced during establishing temporal correlations between macromolecular syntheses and cytological events. Chromosomes in general consist of nucleic acid (*DNA*) and proteins.

DNA: If synthesis of *DNA* is inhibited at the beginning of zygotene, the cells show no further development. These cells fail to form synaptinemal complex. It means that synthesis of *DNA* is essential to the chromosome pairing. However, if *DNA* synthesis is inhibited at different intervals of zygotene, in each case further progress stopped. Thus *DNA* synthesis is not only essential to the initiation of pairing but also to the continuation of the pairing. When considered the absolute amount of *DNA* synthesised during meiosis, it has been found that 99.6 per cent is synthesised at *S* phase while the rest at different stages of cell division. Stern (1970) states that amount formed during zygotene and pachytene are very small, even though they have major functional significance to meiosis. Besides, the formation of syneptinemal complex, DNA synthesis during zygotene is essential to the structural integrity of each of the chromosomes.

Protein: In fact, the total cell division is the result of various drama of proteins. If protein synthesis is inhibited at any stage of meiosis, the effects are always adverse and the result is either partial or complete meiotic arrest. Whether a few or all the proteins synthesised during cell division are essential to maintain the various acts of meiosis has also been unravelled. It has been found that among many proteins produced, only one or two are essential and that these alone are responsible for meiotic arrest. It has been found that when protein synthesis was inhibited for 2 or 3 days at the early stage zygotene, the inhibitor acted as though it stopped all the processes, and when it is inhibited in late zygotene or early pachytene for same period, the chromosomes proceed through pachytene, but when they reach diplotene no chiasmata were formed. These findings permitted to visualise that proteins responsible for chromosomal pairing are synthesised at different time whereas proteins essential to chiasmata formation at some other time. The absence of chiasmata could be explained by the failure of chromosome pairs to hold together long enough to reach the crossing over stage.

The study based on *DNA* and protein syntheses reveals that *DNA* synthesis is essential to the occurrence of pairing and protein synthesis to its stabilisation. Furthermore stabilisation appears to be a necessary condition for crossing over, and that crossing over between homologues is not an accidental consequence of replication in two juxtaposed DNA strands, but a deliberately, regulated event following the completion of pairing.

GENERAL QUESTIONS

1. What is cell cycle? What are G_1 and G_2 phases? Discuss briefly the various types of cell cycles met in plants.
2. Write an essay on physiology of cell division.
3. "Mitosis constitutes an example of fundamental cell division." Comment upon the statement.
4. "Meiosis is lengthy and complicated but most important type of cell division." Discuss critically and also mention its significance.
5. Give a comparative account of mitosis and meiosis.
6. Write short notes on (i) Pachytene chromosomes, (ii) Chiasmata, (iii) Bivalents, (iv) Syneptinemal complex.

CHAPTER 4

Meristematic Tissues

The term meristem has been derived from *meristos* which means *divisible* and thus a meristem can be defined as an immature, not well differentiated cell which has the capacity of division. True meristem cells (eumeristems) have following important features.

1. They are normally isodiametric, spherical, oval or polygonal in shape.
2. They are compactly arranged and lack intercellular spaces.
3. They have thin, homogenous and cellulosic cell wall. Secondary wall is normally absent.
4. Each cell has dense cytoplasm with a large distinct and prominent nucleus. Vacuoles and ergastic substances are generally absent.
5. Plastids are indistinct or may be present in the form of proplastids.
6. The most important character of these cells is their **capacity to divide.** In fact, meristems are formative regions from where new cells are added to the plant body.

CLASSIFICATION OF MERISTEMS

Meristems are normally classified on basis of their origin, plane of division, functions and position in the plant body.

I. Classification based on origin and development

On the basis of origin and development of initiating cells, three categories are recognised.

1. **Promeristem or primordial meristem.** The term promeristem is used to refer to a group of earliest and youngest meristematic cells of a growing organ. It is the early embryonic meristem from which other advanced meristems are derived. In a plant, it occupies a small area at the tip of stem and root. Promeristem further divides and forms primary meristem.

2. **Primary meristem.** These are derived from promeristem and found at the tips of root, stem and appendages. The cells divide in all possible planes and form the fundamental part or primary structure of a plant. The cambium of dicot stem is also primary meristem but its cells divide in only one plane.

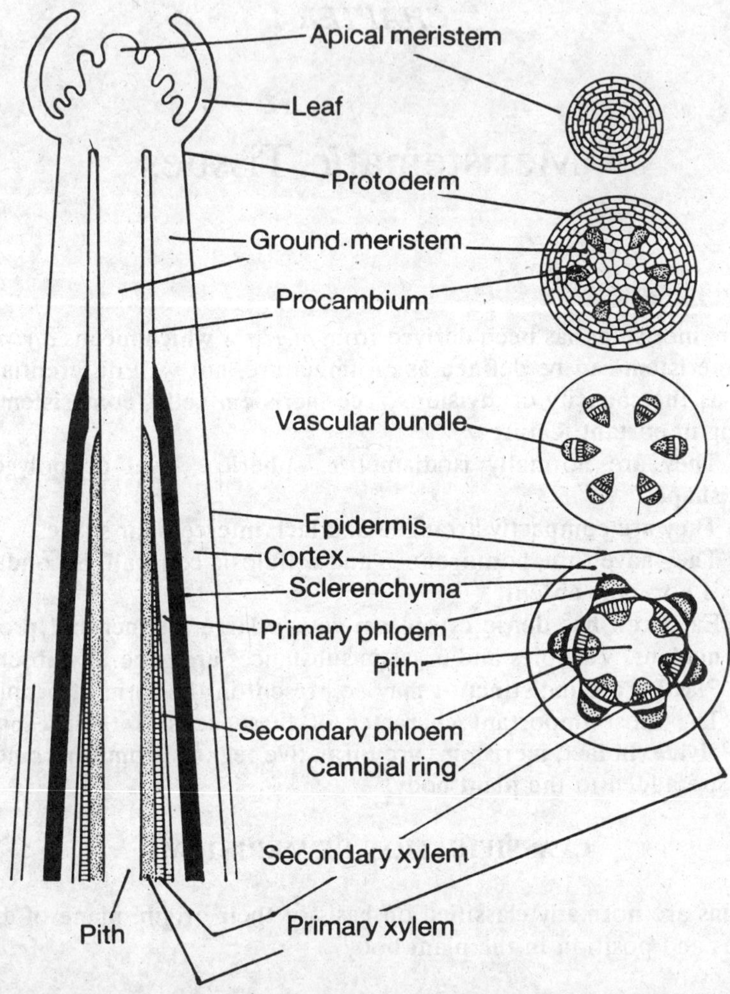

Fig. 4.1. Longitudinal section and transverse sections of various regions of stem apex showing different stages of tissue differentiation.

3. **Secondary meristem.** It is derived from primary permanent tissues which have the capacity of division. These are always lateral in position and develop either at the time of emergency to heal up the wounded portion or to effect secondary growth or to form cork.

II. Classification on basis of position

On the basis of their position in the plant body, three categories are recognised.

1. **Apical meristem.** These are found at the apices or growing points of root and shoot and bring about increase in length. It includes both promeristem as well as primary meristem. The apical meristematic tissues divide and gradually

attain the shape of permanent tissues.

2. **Intercalary meristem.** It lies between the regions of permanent tissues and is considered as a part of primary meristem which has become detached due to formation of intermediate permanent tissues. It is found at various places, e.g. at the base of leaf in *Pinus*, at the base of internode in grasses and *Equisetum* and below the node in *Mentha*.

3. **Lateral meristem.** These are aranged parallel to the sides of organs and normally divide periclinally or radially and give rise to secondary permanent tissues. These increase the thickness of the plant part. Vascular cambium and cork cambium are common examples of such meristem of which former is a primary meristem and latter is a secondary meristem.

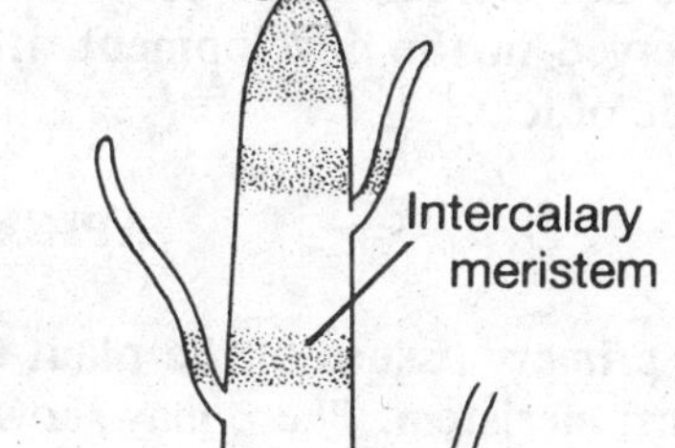

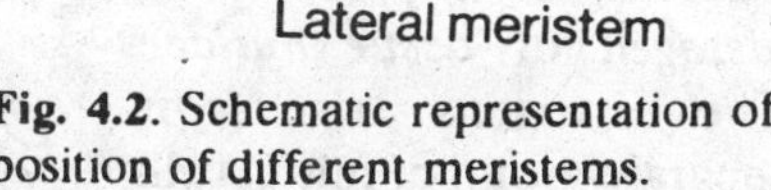

Fig. 4.2. Schematic representation of position of different meristems.

III. Classification on basis of function

Haberlandt considering the nature of function of meristem recognised three important categories.

1. **Protoderm meristem.** It is the outermost layer of the young growing region which develops to form epidermal tissue system.

2. **Procambium meristem.** It is composed of narrow, elongated, prosenchymatous, meristematic cells that gives rise to the vascular tissue system.

3. **Ground meristem.** It is composed of large, thin-walled cells which develop to form ground tissue system, i.e. hypodermis, cortex and pith.

IV. Classification on basis of plane of divisions

The growth, pattern and plane of division of meristematic tissues is of considerable importance to govern the mode of growth. In general, three important types have been recognised.

1. **Mass meristem.** In such meristems, cell divisions occur in all planes so that plant body shows isodiametric or irregular shape. This type of growth takes place in endosperm and young embryo and also in the formation of spores and sperms.

2. **Rib or file meristem.** In such type, the plane of division is restricted at right angles or anticlinally to the longitudinal axis of the cell row, i.e. cells divide only anticlinally. This gives rise to longitudinal files or ribs. Such

growth pattern is found in the development of the cortex and pith.

3. **Plate meristem.** It consists of parallel layers of cells which divide anticlinally in two planes so that new cells are formed but number of layers does not increase and thus a plate-like structure is formed. This pattern is observed in the development of uniseriate epidermis and multiseriate flat blade of leaf.

APICAL MERISTEMS

All primary tissues of the plant body are derived either from shoot or root apical meristem. These may show variation in their organisation.

SHOOT APEX ORGANISATION

Several theories have been put forward from time to time to explain the mode of growth found in shoot apical meristem.

1. **Apical cell theory.** The theory was first proposed by Hofmeister (1857) and advanced by Nageli (1878). According to the theory, a single apical cell is the structural and functional unit of apical meristem which governs the entire process of apical growth. However, such organisation has been found only in cryptogams and not in phanerogams.

2. **Histogen theory.** Hanstein (1868) postulated this theory revealing that three distinct meristematic zones (layers) are found in apical region. Each zone consists of a variable number of layers called *histogen* or *tissue builder*. The histogens arise from separate group of initial cells and have different mode of development. The three histogens are, (i) **dermatogen**, the outermost uniseriate layer, (ii) **periblem**, the middle region composed of isodiametric cells and (iii) **plerome**, the central or inner mass. Each histogen has definite function. The **dermatogen** cells divide anticlinally and develop into unilayered epidermis, the **periblem** forms the cortex and the **plerome** forms the stele (vascular tissues, pericycle, pith, rays etc.). Further studies revealed that these layers have no morphological significance as they are not specific in their function. Moreover, it is not possible to distinguish these three histogen layers in gymnosperms and angiosperms.

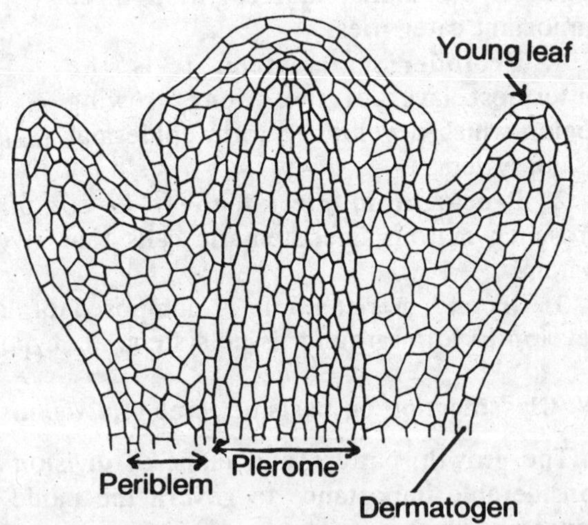

Fig. 4.3. L.S. of growing apex of stem showing different histogen layers.

Hence this theory was rejected for shoot apical meristem.

3. Tunica-corpus theory. Schmidt (1924) proposed tunica-corpus theory to explain shoot apex organisation. According to this theory, there are two

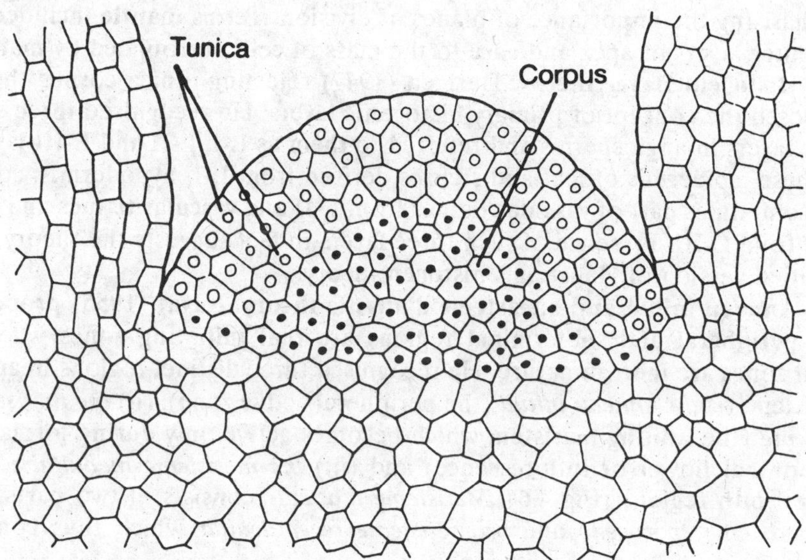

Fig. 4.4. L.S. through stem apex of *Vinca* species showing differentiation of Tunica and carpus.

zones in apical meristem, (i) **tunica**, the outer zone consisting of one or more peripheral layers of cells, forming the outer region by anticlinal divisions, and (ii) **corpus**, the central undifferentiated multilayered mass of cells surrounded by tunica which forms the central part of shoot by irregular divisions (Fig. 4.4).

Tunica and corpus have separate initials which are contiguous with one another at the tip of apex. The tunica cells are smaller and corpus cells are larger. The number of tunica layers may vary even in same species due to influence of seasonal growth changes. According to Schmidt (1924) and Reeve (1942), these variations

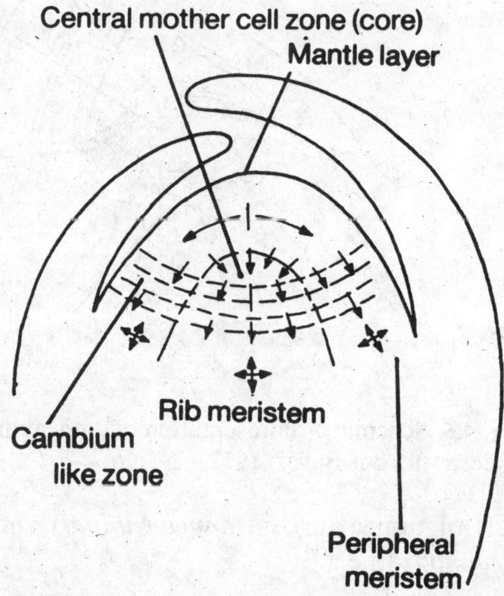

Fig. 4.5. Schematic representation of Mantle and core concept with marked arrows to indicate the direction of growth.

are due to plastochron periodicity.

Tunica term is somewhat a restricted term as regards to cell division which is always anticlinal. Some workers suggested to make it more flexible. Popham and Chan (1950) proposed **mantle-core hypothesis** without emphasising the importance of planes of division. Terms **mantle** included all the outer layers of apex and **core** to the mass of cells surrounded by mantle.

4. Histogenic layer theory. Dermen (1947) rejecting tunica corpus theory proposed the concept of primary histogenic layers. He recognised three such basic layers in angiosperms and referred to them as L-I, L-II and L-III layers. Of these, epidermis of leaf and stem is derived from L-I, hypodermis, cortex and sometimes part of vascular tissues from L-II and vascular tissues and pith arise from L-III. Presence of such three layers in fact suggests the theory as a modified version of Hanstein's histogen theory.

5. Anneau initial and meristeme d'attente theory. Buvat (1955) proposed that peripheral and subterminal regions are real initiating zones whereas distal zones are inert in nature. He recognised three distinct regions in apical meristem, (i) *the anneau initial* (the peripheral active zone), (ii) *the meristeme d'attente* (the waiting meristem which becomes active only during formation of terminal flower or inflorescence) and (iii) *the meristeme medullaire* (the central pith region) (Fig 4.6). *Meristeme d'attente* consists of two parts, the *promeristeme sporogene* and *promeristeme receptaculair* which thus contain

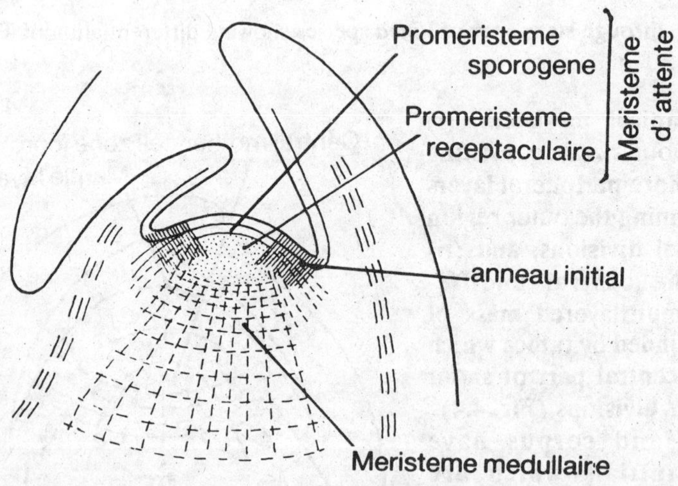

Fig. 4.6. Schematic representation of Anneau initial and meresteme d' attente concept (Redrawn from Buvat, 1952).

part of tunica (except *anneau initial*) and part of corpus (except *meristeme medullaire*) i.e.,

Tunica	anneau initial	
	promeristeme sporogene	meristeme d'attente
Corpus	promeristeme receptaculaire	meristeme d'attente
	promeristeme medullaire	

This theory becomes controversial mainly because it considers meresteme d'attente as waiting meristem which were previously regarded as apical initials. In several cases, this region has been found dividing very actively.

Cytohistological zonations

It was found that in gymnosperms, tunica like layer is not found which led Foster (1939) to divide shoot apex organisation of *Ginkgo biloba* on basis of cytohistological zonations, which react differently to staining. He recognised four inter-related zones, (i) **Apical initial group,** (ii) **Central mother cell zone,** (iii) **Rib meristem** and (iv) **Peripheral meristem** or flank meristem (Fig. 4.7). These zones have also been found to show differences in biochemical activities. Such zonations have been reported in several angiosperms. Popham and Chan (1950) added one more zone, the **cambium like zone** found just below the youngest leaf primordium. However, it has not been found in all cases, particularly in *Chrysanthemum morifolium* (Popham and Chan, 1950) and *Ricinus communis* (Singh and Singh, 1976).

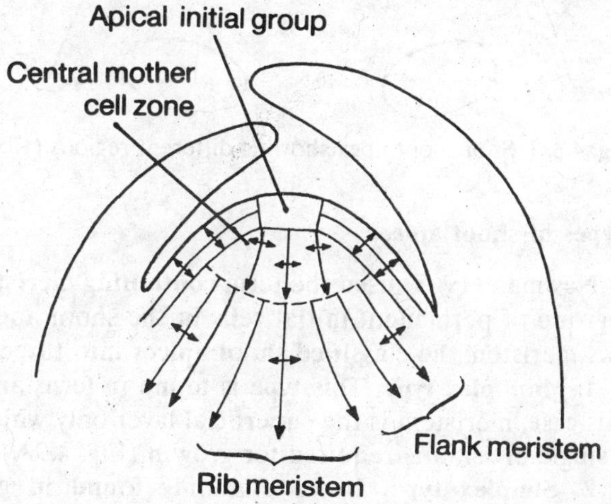

Fig. 4.7. L.S. of shoot apex of *Ginkgo biloba* showing schematic representation of cytohistological zonation (Redrawn from Foster, 1938).

Regions of shoot apex

Wardlaw (1957) recognised five regions in the shoot apex (Fig 4.8).

(i) **Distal region**–a group of initial cells arranged in one or more tiers.

(ii) **Sub-distal region**–superficial layers of meristematic cells where growth centres are located.

(iii) **Organogenic region**–region of leaf initiation and tissue differentiation.

(iv) **Sub-apical region**–region characterised by continued cell division, cell elongation and further internal cellular differentiation.

(v) **Region of maturation**–region of maturation with no division or differentiation of tissue.

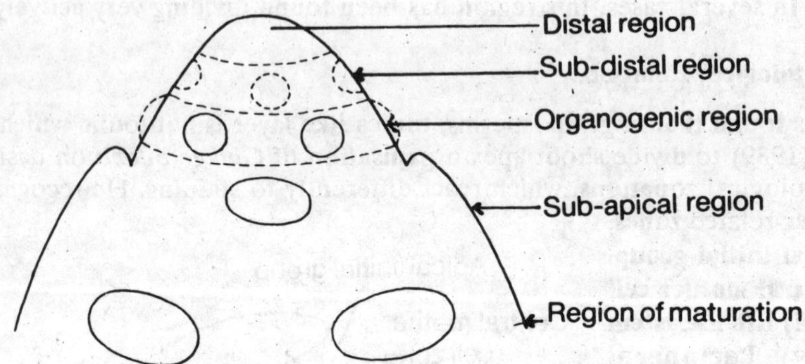

Fig. 4.8. L.S. of shoot apex showing different regions (Redrawn from Wardlaw, 1957).

Types of shoot apices

Newman (1965) used the term **continuing meristematic residue** to refer to a group of permanent initial cells in the shoot apex and on basis of form of this meristem, he classified shoot apices into three main types.

1. **Monoplex type.** This type is found in ferns and their related genera. In this case, meristem is the superficial layer only which is pointed inwardly and a single division is required for growth (Fig. 4.9A).

2. **Simplex type.** It is commonly found in gymnosperms where it is characterised by a parallel sided meristematic residue limited to the superficial layer (Fig 4.9B).

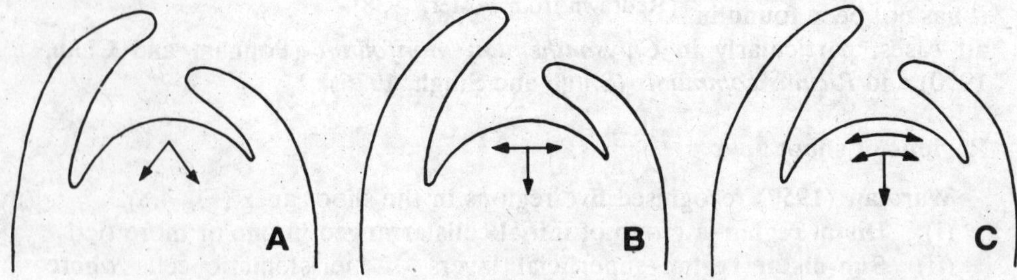

Fig. 4.9. Schematic representation of types of shoot apices; A–monoplex, B–simplex and C–duplex types (Redrawn from Newman, 1965).

3. **Duplex type.** It is commonly found amongst angiosperms where the continuing meristematic residue is parallel-sided but at least in two layers, the *superficial layer(s)* which divides anticlinally and *innermost layer(s)* which divides both anticlinally and periclinally (Fig. 4.9C).

ROOT APEX ORGANISATION

Several theories have been proposed to describe the root apex organisation.

1. **Apical cell theory.** Nageli (1958) proposed that a single tetrahedral apical cell in the root apices brings about growth. It has three cutting faces for giving rise to the tissues of the body of the root. The root cap is derived from its base. Such a proposition was not found true to angiosperms. However, different types of root apices have been found in angiosperms which are schematically shown in figure (Fig 4.10).

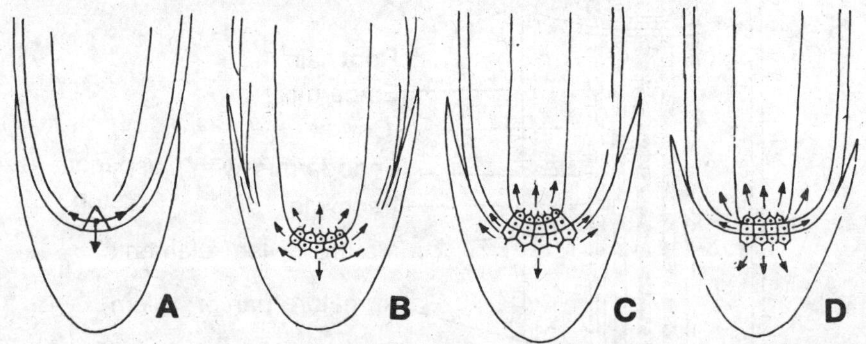

Fig. 4.10. Schematic representation of different root apex types; A–initial solitary (root cap distinct), B–initials in two groups (cap structurally not distinct), C–initials in three groups (cap indistinct) and D–initials in three groups (cap distinct and independent in origin) (Redrawn from Eames and Mac Daniels, 1947).

2. **Histogen theory.** Hanstein (1868) put forth histogen theory proposing that in root apex regions, three distinct histogens are found. The **dermatogen** that gives rise to epidermis, the **periblem** to the cortex and the **plerome** to the vascular cylinder. Haberlandt (1914) proposed the terms *protoderm, ground meristem* and *procambium,* respectively for these histogens.

3. **Körper-Kappe theory.** Schuepp (1917) proposed that the cells at the root apex divide in two planes. The zone with inverted 'T' type of division was referred to as *Körper* (cap) whereas other with straight 'T' as *Kappe* (body). The theory basically recalls the tunica corpus concept of shoot apex and similarly, it also fails to explain the differences in behaviour in different species.

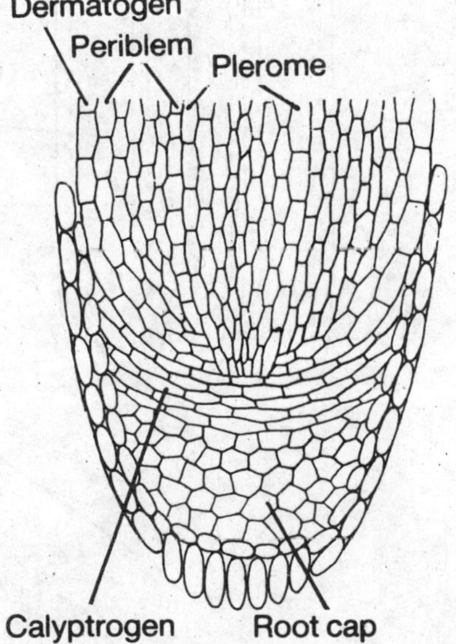

Fig. 4.11. L.S. of root apex showing different histogen layers.

Concept of quiscent zone

Clowes (1958) in his autoradiographic studies of DNA synthesis in the root tip of *Zea mays,* noted a central cup-like region, the cells of which had fewer mitochondria and ER, very small nuclei and low rate of DNA and protein synthesis. He referred this inactive region as **quiscent centre.** Since then such regions have been observed in many plants. It consists of about 500 cells in *Zea mays,* 600 in *Sinapsis alba* and 1100 in *Vicia faba.*

In general, cells of quiscent region do not divide, but may become active

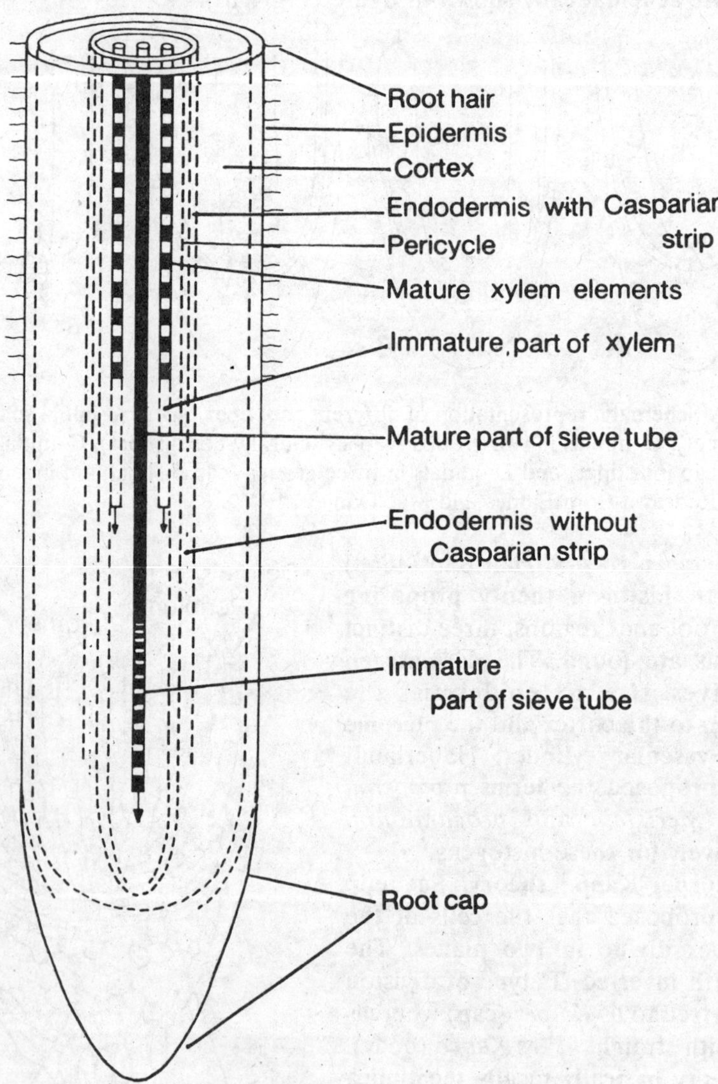

Fig. 4.12. L.S. of root apex showing differentiation of different tissues (Redrawn after Esau, 1953).

when active initials are damaged. Thus, this region acts as reservoir of cells. These undividing cells offer more resistance to damages and irradiations than actively dividing cells and under certain conditions may divide actively. Clowes (1961) proposed that the cells of quiscent centre are inactive because of their position in the apical meristem, and not due to any inherent inability to undergo divisions.

In general, the differentiation of different tissues is schematically shown in figure (Fig 4.12).

GENERAL QUESTIONS

1. Write short notes on meristematic tissues and their organisation.
2. Write a brief account of shoot and root apices organisations.
3. Write short notes on:
   (1) Shoot apex organisation
   (2) Tunica-corpus theory
   (3) Histogen theory
   (4) Quiscent centre.

CHAPTER 5

Permanent Tissues

Meristematic cells gradually divide and get differentiated to form permanent tissues. Permanent tissue is thus composed of cells in which growth has stopped. These may be living or dead and thin or thick-walled. Though living permanent tissue retains the potentiality of growth but it is manifested secondarily and only at the time of real and emergent need. Permanent tissues may be classified into three main groups:

1. Simple, 2. Complex, and 3. Secretory tissues.

I. SIMPLE TISSUE

Simple tissues are homogenous in nature and composed of structurally and functionally similar cells. The main simple permanent tissues of the plants are parenchyma, collenchyma and sclerenchyma.

Parenchyma

Parenchyma is the most common tissue which is morphologically and physiologically simple and unspecialised.

Origin. These basically arise from primary meristematic tissue and form the framework of all plant organs. Parenchyma of cortex, pith, mesophyll of leaf and of floral parts arise from ground meristem and that found in primary vascular tissues arise from the procambium. During secondary growth, they also arise by the activity of cambium and cork cambium.

Distribution. Parenchyma tissue may be found as an independent mass of tissue or associated with other types of cells in complex tissues. These occur in the cortex, pith and pericycle of stems and roots, photosynthetic tissues, mesophyll of leaves, fleshy part of succulent fruits, the endosperm of seeds etc. They are also found in xylem and phloem as vascular rays and vertical files of living cells. Sometimes, certain specific tissues are also parenchymatous in nature like scleroids, laticiferrous tissue, sieve tubes and certain glandular tissues.

Structure. For studying structure of parenchyma some important points are emphasised i.e., shape, arrangement and cell contents.

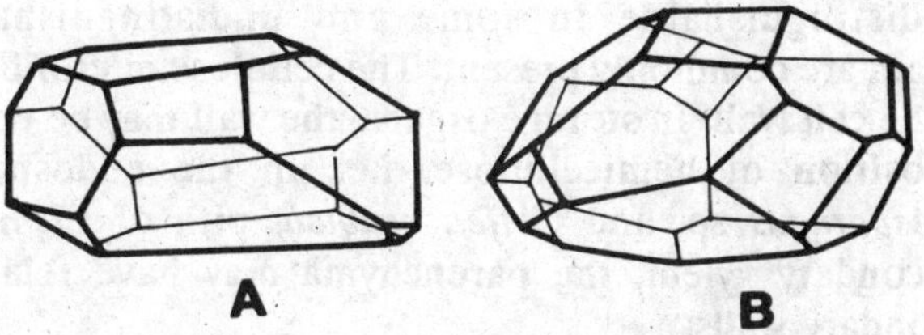

Fig. 5.1. Parenchyma cells, A–with 14 and B–with 17 faces.

(i) **Shape**–Parenchyma cells are usually isodiametric but they vary in form. In several cases, these become elongated and merge with prosenchyma cells in shape and dimensions. (The term prosenchyma was earlier used for elongated cells with tapering ends but in view of the fact that different types of cells may be elongated, the term has gone out of use.) Parenchyma of mesophyll tissues are variously lobed, folded and armed. In general, the parenchyma cells ideally have 14 facets or sides, with 8 hexagonal and 6 quadrilateral faces (Fig. 5.1A). Such polyhedran cells are called *tetrakaidecahedron.* Usually, the number of facets in parenchyma cells is variable. In *Pinus,* the mesophyll parenchyma of leaf are characteristically folded with internal ridges.

(ii) **Arrangement**–Mature parenchyma cells are either packed closely or with small intercellular spaces. The parenchyma of endosperm which serves as storage tissue usually lack intercellular spaces but those of fleshy fruits which are also storage parenchyma have large and numerous intercellular spaces. However, there is always a great variation as regards to arrangement of parenchyma in the plant body.

(iii) **Cell Contents**–(a) *Cell wall*–Parenchyma usually have thin primary

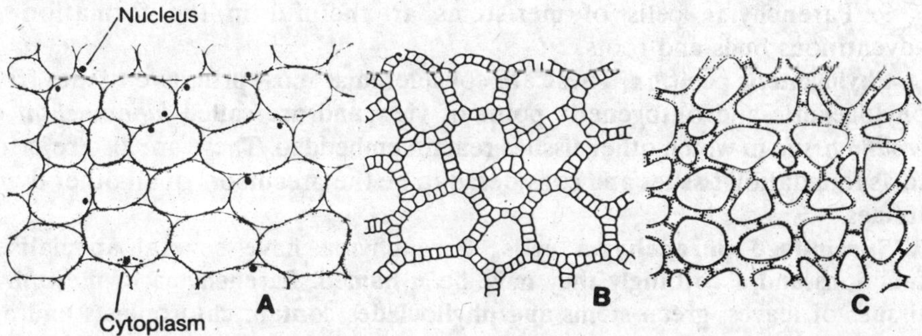

Fig. 5.2. Different types of parenchyma in C.S., A-simple parenchyma as found in stem and root cortex, B–aerenchyma in petiole of *Zantedeschia* and C–aerenchyma in leaf of *Canna.*

wall composed of cellulose, calcium pectate and hemicellulose. The middle lamella is distinguishable in some and undistinguishable in others. Plasmodesmata are commonly present. The cellulose microfibrils are variously oriented in the cell wall. In storage organs, the wall may be considerably thick due to deposition of hemicellulose, i.e., in the endosperm of *Phoenix dactylifera, Asparagus* sp. and *Coffea arabica,* pith of *Clematis* and leaf of *Pinus*. In secondary xylem, the parenchyma may have relatively thick and lignified secondary walls.

(b) *Cell contents*–The cell contents show variation particularly in respect to their activities, i.e., photosynthetic cells contain numerous chloroplasts and starch; non-chlorophyllous cells are highly vacuolated and contain leucoplasts; storage parenchyma contain ergastic substances like starch grains, protein granules, oil globules, sugar etc. The parenchyma found in potato tubers and rhizomes contain amides, proteins and starch. Many parenchyma cells accumulate tannins, alkaloids, mineral substances etc. Parenchyma, in general, possess all living inclusions including nucleus.

Functions. Parenchyma are perhaps the most important type of tissues capable of carrying out following important functions:

1. These are the centres of various physiological activities like photosynthesis, assimilation, respiration, storage, secretion, excretion etc.
2. Parenchyma cells of xylem and phloem in the form of trachea and sieve elements help in upward and downward conduction of water and dissolved food materials.
3. These are helpful in wound healing.
4. These retain the power of cell division.
5. Cutinised parenchyma of leaf epidermis are protective in nature which prevents transpiration and upto some extent mechanical injuries.
6. Parenchyma of cambial cells divide and form secondary tissues.
7. In succulent plants, these help in water storage.
8. In aquatic plants, these store air (aerenchyma), provide buoyancy to float easily in or on the water.
9. Parenchyma cells of meristems are helpful in the formation of adventitious buds and roots.

Phylogenetic position. These are considered as most primitive tissues from phylogenetic and ontogenetic point of view and are called *fundamental* or *ground tissue* in which other tissues remain embedded. These are also referred to as foundation tissues and considered to be the precursors of all other living tissues.

Specialised parenchyma cells. Parenchyma have several specialised functions and accordingly they have been named. Parenchyma of mesophyll tissues of leaves, green stems and phylloclades contain chloroplasts and are called chlorenchyma which are photosynthetic in nature. Parenchyma of aquatic plants such as *Eichhornia, Hydrilla, Nymphaea* etc. have large and abundant intercellular spaces and may assume star-like or stellate shape (Fig. 5.3C). These are called aerenchyma. Aerenchyma are also seen in petiole of *Zantedeschia* and leaf of *Canna* (Fig. 5.2). Specialised parenchymatous cells which produce and store tannins, oils and crystals of calcium oxalate are called **idioblasts**. Idioblastic parenchyma is secretory in function and provides various enzymes that digest food substances.

Collenchyma

Collenchyma are living tissues composed of more or less elongated cells with thick primary non-lignified walls.

Origin. These are derived from apical meristem and cortical collenchyma and procambium originate from a common meristem. The collenchyma that differentiates early in a given organ becomes highly specialised in its morphology whereas that formed later is more parenchymatous in nature. The former has its origin in a procambium-like meristem, while the latter ones in a parenchymatic ground meristem.

Structure. Collenchyma do not show much variation in their structure.

(i) **Cell shape**–The shape of cells shows variation. The shorter collenchyma cells are more or less like parenchyma while the longer ones resemble the fibres with tapering ends often overlapped and interlocked like fibres. Collenchyma are usually polygonal in cross sections.

(ii) **Cell wall**–Cell wall is the most distinctive feature of collenchyma which is characteristically unevenly thickened. The wall thickening is primary in nature. It is composed of cellulose, hemicellulose and pectic materials with high percentage of water. In some plants, cellulose-rich and pectin-rich layers alternate on the wall. Rarely, the wall may undergo sclarification. Simple pits may be found in both thin and thick portions of cell wall. Electron microscopic studies reveal various types of layering in collenchyma walls, in which lamellae with transverse orientation of microfibrils alternate with longitudinal

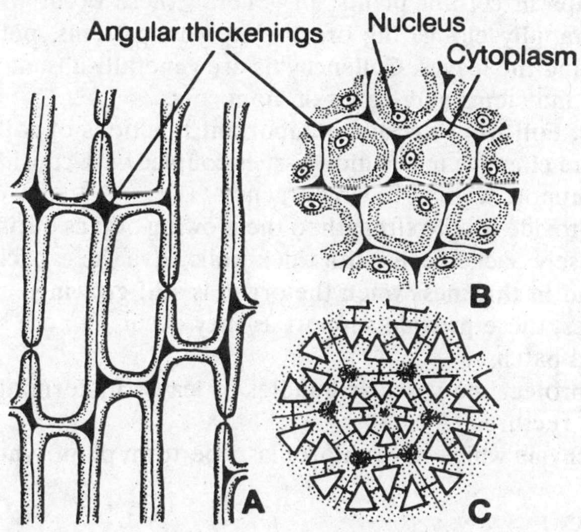

Fig. 5.3. Collenchyma, A–in longitudinal section, B–in transverse section and C–stellate parenchyma.

orientation of microfibrils.

On the basis of thickenings on cell wall and arrangement of cells, three forms of collenchyma are recognised:

(a) *Angular collenchyma*–The thickenings are primarily at the corners or angles of the cells. As a result, this collenchyma becomes a compact tissue consisting of irregularly arranged cells without intercellular spaces, as in stems of *Datura* and *Lycopersicon.* Sometimes, due to continued thickening of cell wall, the lumen may look circular and this type has been called *angular collenchyma* by some workers.

(b) *Lacunar* or *tubular collenchyma*–The thickenings are restricted to the walls of the regions bordering air spaces. These have large intercellular spaces. Such types of collenchyma are found in *Salvia, Malva, Althaea,* aerial roots of *Monstera* etc.

(c) *Plate* or *lamellar collenchyma*–The thickening occurs chiefly on the tangential walls and lesser in the radial walls. As a result, cells appear like plates, bands or lamellae. Such collenchyma are found in the stems of *Raphanus* and *Sambucus.*

(iii) **Cell contents.** These are living tissue with vacuolated protoplast. Each cell is uninucleate which may or may not have chloroplast. Collenchyma are closely related to parenchyma but have thicker walls and longer cells. Collenchyma differ from sclerenchyma in the presence of protoplast at maturity and in wall structure, the thickening being cellulosic in nature.

Distribution. Collenchyma are characteristically found in the hypodermis of herbaceous dicots like *Helianthus, Cucurbita, Datura* etc. These occur either as a homogenous layer as in *Helianthus* stem or in patches as in *Cucurbita* stem. In leaves, collenchyma are found on both sides of vein as in *Mangifera* or several other dicot leaves. Sometimes, these form bundle sheath or bundle caps in certain plants. In general, these occur in the peripheral portions of rapidly elongating organs like young stems, petioles of leaves, floral stalks and the leaves. Collenchyma are generally absent in underground stems, roots and stems and leaves of monocots.

Functions. Following are some important functions of collenchyma.

1. These are effective mechanical tissues coupled with considerable elasticity which give support to the growing organs.
2. These provide tensile strength to the growing organs during development.
3. The closely packed cells with thick walls have the capacity of increasing in surface and in thickness when the organ is still growing.
4. In leaves, these provide support as they occur on both sides of bundles or in isolated patches.
5. These protect the vascular bundles of leaves by forming cap or bundle sheath like structures.
6. Collenchyma which have chloroplasts perform photosynthetic functions.

Sclerenchyma

Sclerenchyma are simple tissues meant specially for mechanical function. The word *sclerous* itself means hard. These are considerably thickwalled and lignified with simple pits in their wall and are characterised by absence of living protoplast. The cells show much variation in form, structure, origin and development and are placed into two groups: *fibres* and *sclereids.*

1. Fibres

Fibres are much elongated and have pointed needle-like ends.

Origin. Fibres arise either from procambium (apical meristem) or from fusiform cambial initials. Some fibres differentiate from ground meristem. Protodermal cells divide periclinally and anticlinally and give rise to fibres. The fibres found in the bundle sheath of monocot stems develop partly from procambium and partly from ground meristem.

Classification. On basis of their position in the plant body, fibres are classified into two large groups, the *xylary fibres* (intra-xylary) and *extra-xylary fibres.*

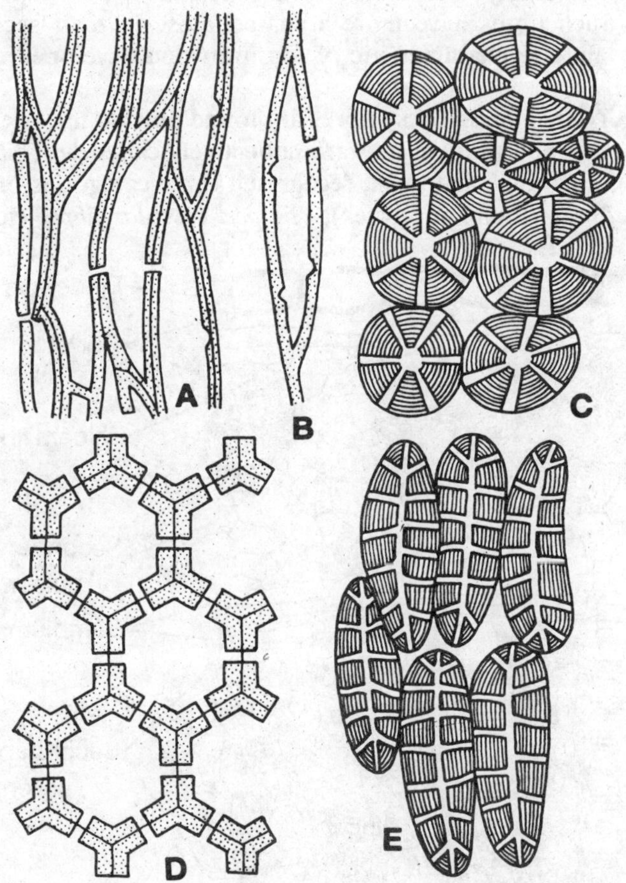

Fig. 5.4. Sclerenchyma. A–L.S. of sclerenchymatous fibres, B–a single fibre, C–T.S. of stone cells, D–T.S. of typical sclerenchymatous cells and E–L.S. of stone cells.

(a) *Xylary Fibres*–These are also called wood fibres for the reason they are found in xylem and develop from the same meristematic tissue which gives rise to other xylem elements. They vary in size, shape, thickness of wall and number of pittings. The pits may be small, round or slit-like in appearance. Sometimes, the secondary wall is so much thickened and lignified that central lumen is almost obliterated.

On the basis of relative thickness of their wall and types of pits, xylary fibres are sub-divided into **libriform fibres and fibre tracheids.** The libriform fibres are characterised by very thick secondary wall and simple pits, while fibre tracheids by relatively thin walls and bordered pits. However, both types of fibres are septate. The septa are true walls and are formed after the deposition of secondary layers. The septate fibres retain their protoplast for longer periods and serve for storage of starch, oils, resins and other reserve materials.

In certain dicotyledonous plants, **gelatinous** or **mucilaginous** fibres are found which are simply xylary fibres having specially a great capacity of water absorption. Such fibres have more alpha-cellulose in their secondary walls than lignins and are characteristic of tension wood developed during stress condition.

(b) *Extra Xylary Fibres*–These fibres are found outside the xylem such as in cortex, pericycle and phloem and are collectively called *bast fibres*. Madam Esau (1953), a famous anatomist, recognised three categories of such fibres: (i) *Cortical fibres* (found in cortex); (ii) *peri-vascular fibres* (found on the

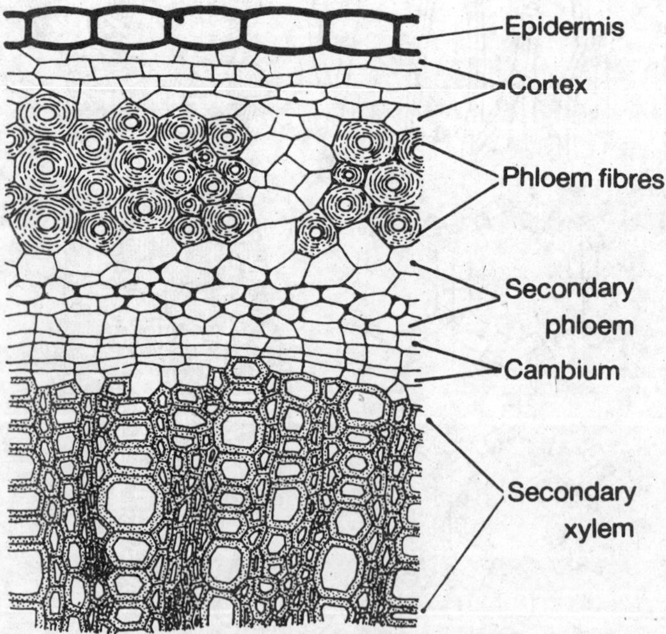

Fig. 5.5. T.S. of stem of *Linum usitatissimum* showing distribution of primary phloem fibres (Redrawn from Esau, 1965).

periphery of vascular cylinder); and (iii) *phloic fibres* or phloem fibres (found in phloem).

Extra-xylary fibres are long spindle-like with acute or acuminate or blunt ends. Their cell wall is quite thick with simple or slightly bordered pits. At maturity, these fibres loose protoplast and become dead. Usually, these remain unbranched but branched fibres may be found both in monocotyledons and dicotyledons. Fibres found in *Cannabis sativa* (hemp) and *Linum usitatissimum* (flax) are very long and have cellulose as major substance in their secondary cell walls. Their wall exhibits lamellations.

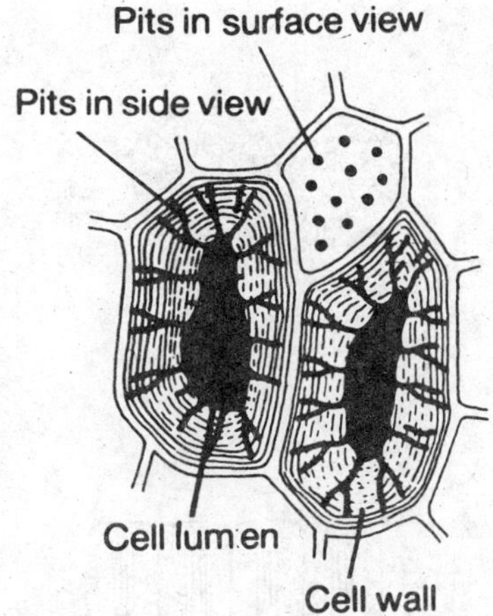

Fig. 5.6. Sclereids. Pits shown in both side and surface views.

Distribution. Fibres occur as groups, as sheets or as cylinders in various parts of plant body. In dicot stems such as in *Linum* and *Nerium,* fibres occur in the outermost part of primary phloem, and in monocot stems such as in *Zea mays* and *Saccharum* species, fibres are found in bundle sheath. In roots, fibres are found in primary and secondary xylem and phloem tissues. In gymnosperms, the primary pholem usually lack fibres while these may be found in secondary phloem.

Functions. Their main function is to provide mechanical strength to the plant organs while maintaining their flexibility and elasticity and to develop resistance against gravitational tension and strains. The xylary fibres also participate in conduction of water along with that of providing support.

Economic fibres. Fibres find many economic and commercial uses to human welfare and plants spread in 44 families are used as the source of commercial fibres. Commercial fibres are separated into hard fibres and soft fibres. The hard fibres are monocot leaf fibres with heavily lignified walls, e.g. *Musa textiles* (manilla hemp), *Yucca, Agave, Sansiviera* (bowstring hemp) etc. The soft fibres are bast fibres obtained from pholem as in *Linum usitatissimum* (flax), *Cannabis sativa* (hemp), *Corchorus capsularis* (jute) etc.

2. Sclereids (sclerotic cells)

These are non-prosenchymatous (without pointed ends) cells of different shapes and may be commonly called stone cells.

Origin. Sclereids arise either through secondary sclerosis or by sclereid primordia or parenchyma cells. In leaves, these may develop from epidermal cells, palisade or spongy parenchyma. In the vascular regions, the sclereids develop from the derivatives of procambial or cambial cells. Stone cells found in the cork arise from phellogen and macro-sclereids found in seed coats arise from protoderm. In early stages, the cell wall remains thin until the wall develops and branching commences and then a secondary wall develops

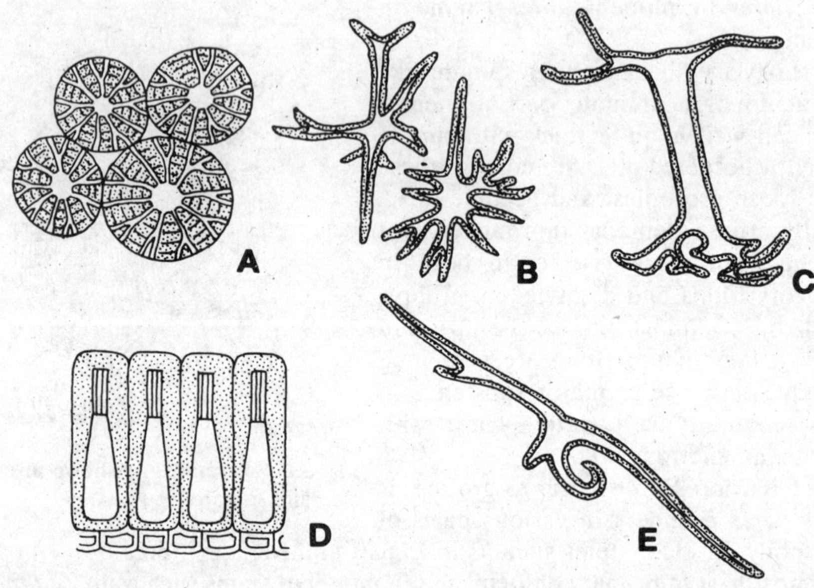

Fig. 5.7. Different types of sclereids. A–brachysclereids, B–astrosclereids, C–osteosclereid, D–macrosclereids and E–trichosclereid.

centripetally. The lignin deposition increases and lumen becomes narrower and narrower.

Classification. Sclereids have been classified on basis of their size, shape, nature of cell wall and mode of deposition of secondary wall materials. Normally, five categories are recognised.

1. *Brachysclereids* (stone cells)–These are short roughly isodiametric sclereids found in cortex, pith and phloem of stems and flesh of fruits (e.g. *Pyrus*) (Fig. 5.7A).

2. *Macrosclereids*–These are elongated columnar sclereids which commonly form palisade like epidermal layer as in seed coats of leguminous seeds (Fig. 5.7D).

3. *Osteosclereids*–These sclereids are bone or barrel-shaped with dilated ends, as commonly found in leaves and seed coats of several monocotyledons (Fig. 5.7C).

4. *Astrosclereids*–These are star-shaped or stellate cells found in the leaves of *Nymphaea* and stem and leaves of *Trochodendron* (Fig. 5.7B).

5. *Trichosclereids*–These are hair-like branched or unbranched sclereids found in the intercellular spaces of leaves and stems of some hydrophytes, aerial roots of *Monstera* and leaves of *Olive* (Fig. 5.7E). These are also called *trichoblasts* or *internal hairs*.

Functions. Some of the important functions of different types of sclereids are:

1. These provide firmness, stiffness and mechanical strength to the part of plant where they are found.

2. Brachysclereids increase the incompressibility of the bark.

3. Sclereids found in seed coat and endocarp of fruits produce a hard texture and thus protect the internal parts.

II. COMPLEX TISSUES

The complex tissues are heterogenous in nature, being composed of different types of cells performing diverse functions. Xylem and phloem are complex tissues which constitute the major part of vascular bundles and hence are also called **vascular tissues.** Vascular bundles form a continuous and interconnected system in different organs of the plant.

Xylem

Structurally, xylem consists of both living and non-living cells. The most important constituents of it are the tracheary elements. However, xylem consists of four elements, (i) tracheids, (ii) vessels, (iii) xylem fibres, and (iv) xylem parenchyma.

1. Tracheids

Origin. Tracheids originate from the procambium of apical meristem or vascular cambium which remain attached end to end in longitudinal series. In fact, **primary xylem** differentiates from procambium while **secondary xylem** from vascular cambium. These two are not histogically different.

Structure. A tracheid is very much elongated or tube-like cell with hard, thick and lignified walls and a large cavity. It is devoid of protoplast and hence dead. Their ends are tapering, blunt or chisel-like. The end walls usually do not uniformly taper in all planes. The tracheids are fundamental and primitive type of cells of xylem, of which, woods of ancient vascular plants were exclusively made of. These are the only type of elements found in the fossils of seed plant.

In transverse section these appear either circular, polygonal or polyhedral in shape.

A tracheid differs from sclerenchymatous fibres in having

(i) Cell wall comparatively less lignified,

(ii) large number of pits,

(iii) bordered pits in secondary wall layers, and

(iv) a very large lumen.

Cell Wall. Cell wall of the tracheid is highly characteristic and specialised type. It is hard, moderately thick and lignified. The secondary wall layers are deposited in different manners which may be annular, spiral, scalariform, reticulate and pitted.

(a) *Annular*–The lignin is deposited in the form of rings all along the walls almost at regular intervals which provides mechanical strength to the wall

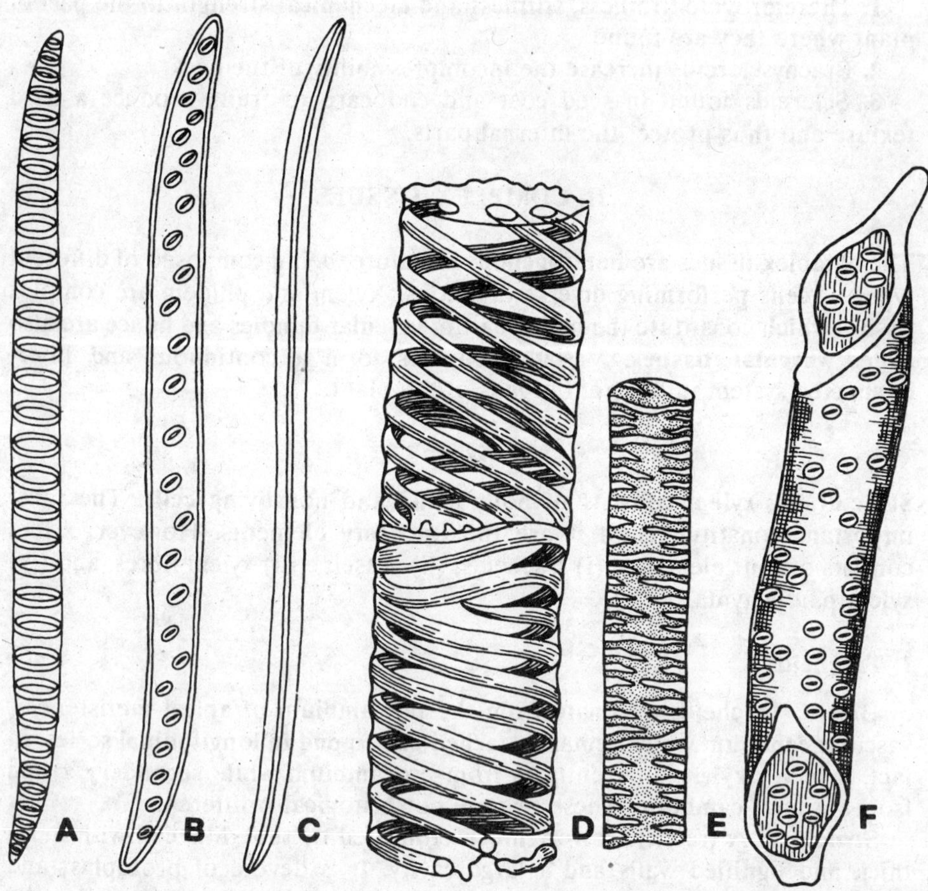

Fig. 5.8. Different types of tracheids. A–annular, B and C–tracheidal fibres, D–Spiral, E–scalariform and F–pitted.

and prevents it from collapsing. It appears polygonal in cross-section (Fig. 5.8A).

(b) *Spiral*–The lignin is deposited in a manner that appears spiral shape in length. It is stronger than angular thickening (Fig. 5.8D).

(c) *Reticulate*–The lignin is deposited in such a manner that it looks a network-like thickening.

(d) *Scalariform*–The lignin deposition is in the form of closely arranged transverse bands which gives an appearance of a ladder (Fig. 5.8E).

(e) *Pitted*–These are the most common types of tracheids. In this case, the entire inner surface of the cell wall is thickened except for certain unthickened circular areas which look-like pits (Fig. 5.8 F). Through these pits, tracheids establish communication with neighbouring cell elements which may be living or non-living type. The nature of pits may be variable. In lower vascular plants, these are elongated, giving them scalariform appearance. In gymnosperms and angiosperms, these are circular, in which two types are recognised, i.e. simple and bordered.

Simple pits are the areas with only primary wall without any secondary thickening. The **bordered pits** are, however, quite complicated as their secondary wall partially outgrows the pit to form a dome-shaped structure with a small perforation in the middle (Fig. 5.9). The primary wall develops a thickening in its central part which is known as **torus.** The torus is well developed in bordered pits of gymnosperms and remains surrounded by microfibrils, known as **margo.** Torus acts as a valve membrane. Pits are so much important that their structure and organisation are taken into account for identifying the different woods of gymnosperms.

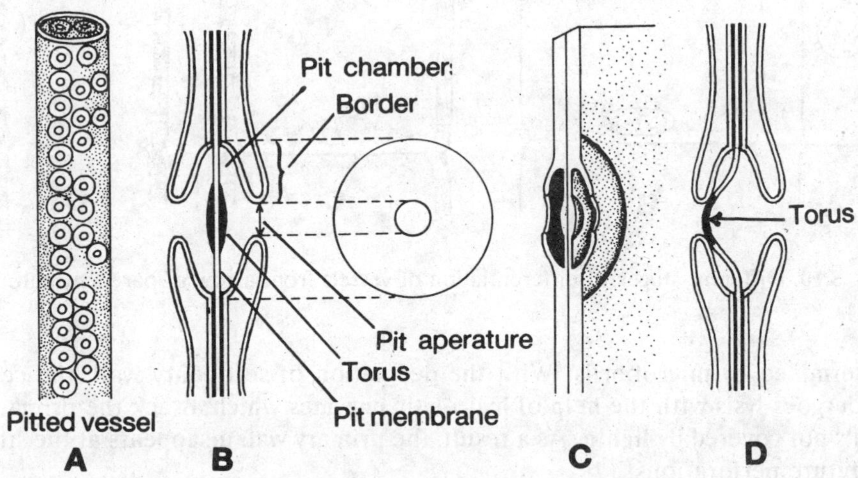

Fig. 5.9. Bordered pits. A–surface view, B–sectional view, C–a diagrammic 3-D appearance of a pit and D–sectional view of a pit with changed position of torus.

Functions. Because of structural specialisation, tracheids are well adapted for transport of water and solutes from root to stem and leaves. In gymnosperms, these are the chief water transporting elements. Besides, these also provide mechanical support.

2. Vessels or Trachea

Origin. The nature of pattern of origin of vessels is much similar to that of tracheids. However, mode of development shows some variation.

Development. Vessels arise from procambial or cambial cells in the primary and secondary xylem, respectively. Vessel elements usually enlarge rapidly prior to deposition of secondary wall material and primary wall remains constant in thickness except at the site of future perforations where it becomes thicker. The pattern of secondary wall deposition is perhaps

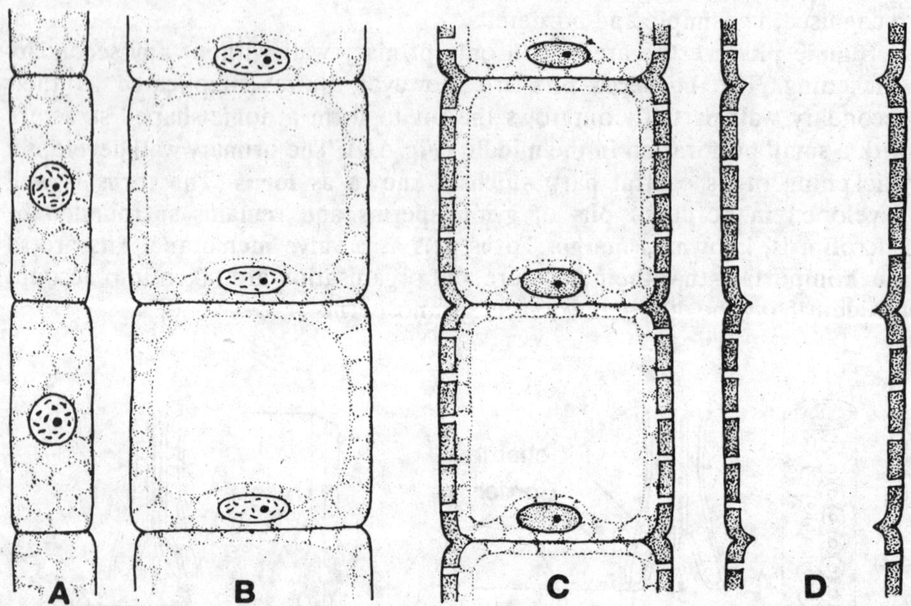

Fig. 5.10. Different stages of differentiation of vessels from a row of parenchymatous cells.

determined by microfibrils. With the deposition of secondary wall, the cell undergoes lysis with the help of hydrolytic enzymes which attack the primary walls not covered by lignin. As a result, the primary wall disappears at the site of future perforations.

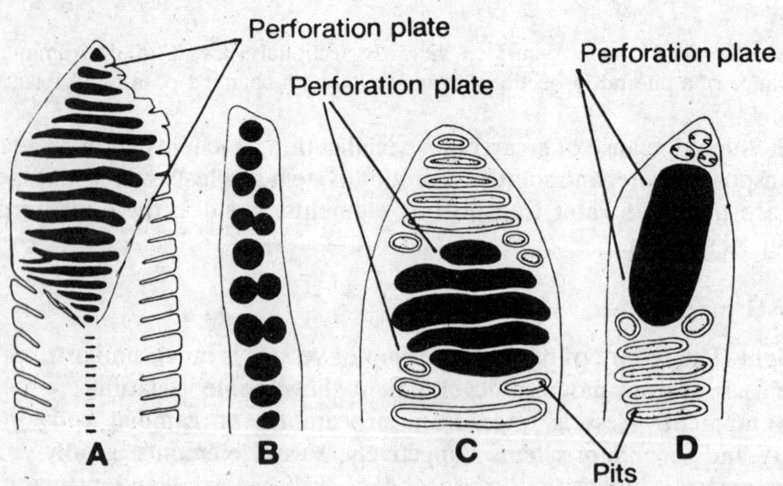

Fig. 5.11. End walls of vessels showing various types of perforation plates. A–scalariform type, B–foraminate type, C–scalariform type of *Vitis* and D–Simple type of *Vitis*.

Structure. A vessel is long, cylindrical, tube-like structure with lignified walls and a wide central lumen. The cells are dead as these do not have protoplast. The cells are arranged in longitudinal series in which the partitioned walls (transverse walls) are perforated, so the entire structure look-like a water pipe. If perforation plate contains one large pore, it is called **simple** perforation plate and if there are several pores in it, it is called **multiple** perforation plate. In multiple perforation plate, the pores have been found arranged in several ways. When the pores are arranged in a reticulate manner, it is called *reticulate type* and if pores are elongated and arranged in parallel series, it is called *scalariform type*. Some times perforations are circular and grouped together, then, the perforation plate is called *foraminate type*.

Besides, the secondary wall layers (radial or longitudinal walls) are deposited in a manner similar to tracheids (i.e. annular, spiral, scalariform, reticulate and pitted).

Distribution. Vessels are found in few pteridophytes (*e.g. Selaginella*), some gymnosperms (*e.g.* members of Gnetopsida) and angiosperms. Vessels are considered more advanced than tracheids.

Functions. Because of structural peculiarities, vessels are more efficient in transport of water and minerals as compared to tracheids. Their efficiency is due to presence of perforation plates. Besides, these also give mechanical strength.

3. Xylem fibres

Xylem fibres are found both in primary and secondary xylem and are basically of two types, the **fibre tracheids** and the **libriform fibres**. Fibre tracheids are intermediate forms between typical fibres and tracheids which possess bordered pits, though borders are less developed. The libriform fibres are narrow with highly thickened secondary wall, obliterated central lumen and simple pits. The latter is more common in woody dicots. Fibres may be septate as found in several dicot families including Anacardiaceae, Meliaceae, Sapindaceae etc.

4. Xylem parenchyma

The parenchymatous cells are found both in primary and secondary xylem but more commonly in primary xylem. The parenchyma found in secondary xylem are classified into two types.

(i) *Axial or wood parenchyma*–These are derived from the fusiform cambial initials together with tracheary elements and fibres. These are somewhat elongated and lie in vertical series.

(ii) *Ray parenchyma*–These orginate from ray initials of fusiform cambium and occur in radial transverse series.

The cell wall of parenchyma found in primary xylem is thin and made up of cellulose but those found in secondary xylem may often have thick, lignified and pitted cell wall.

Xylem parenchyma store food reserves in the form of starch or fat. Sometimes, tannins, crystals and other substances may also be found in them. Ray parenchyma also participates in radial conduction of water.

Some parenchymatous cells are structurally modified for intensive transfer of solutes. These are characterised by wall ingrowths of unlignified secondary wall, to which the plasma membrane remains intimately appressed, increasing its surface to several times. These are referred to as *transfer cells*.

Evolution of tracheary elements. Bailey (1953) proposed that tracheary elements have developed during the evolution of land plants in response to rapid movement of water and inorganic solutes over increasing distances in the axis of land plants. In lower vascular plants, simple tracheids evolved for conduction and support. Both the functions were carried out by the same element. However, with increasing specialisation, woods evolved with conducting elements, the vessels, for more efficient conduction, and fibres, for providing enough mechanical support. Thus, from the primitive tracheids, evolution preceded in two lines, one towards **vessels** and the other towards **fibres.**

The **vessels** have originated independently in five distinct categories, i.e. Selaginellales, Filicales, Gnetales, Monocotyledons and Dicotyledons. The first vessel in these groups developed by modification of tracheids having conspicuous bordered pits. Such primitive vessels have undergone phylogenetic modification through following important changes (adaptations).

1. Shortening of vessels which led to less and less overlapping of ends of adjacent vessel members.
2. Increasing lateral expansion, which resulted in the formation of short and relatively broad vessels with broader ends.
3. Reduction in number and modification in perforations which led to differentiate various types of perforation plates and finally in the formation of a single large oval or circular perforation at each truncated end.
4. Replacement of imperforate scalariform intervessel pitting by imperforate circular bordered pits.

Xylem fibres also underwent specialisation. For more efficient support, there has been considerable increase in thickness of the wall reducing heavily the size of cell lumen, some times completely absent. The pits changed from elongate to circular, the borders reduced to become functionless and ultimately disappeared. Thus, the evolution sequence led from tracheids to libriform fibres through fibre tracheids.

The differences between primary and secondary xylem are summarised in the table 5.1.

Table 5.1. Differences between primary and secondary xylem.

	Primary xylem		*Secondary xylem*
1.	It is formed by the activity of procambium of the apical meristem.	1.	It develops from vascular cambium which is a lateral meristem.
2.	Protoxylem and metaxylem are distinct.	2.	There is no distinction between protoxylem and metaxylem.
3.	Depending upon position of protoxylem, it may be endarch exarch or mesarch.	3.	Such distinction in xylem is not observed.
4.	The tracheids and vessels are relatively narrow and longer.	4.	Both are normally wide and short.
5.	Medullary rays found in this region are formed by the histogen of the apical meristem.	5.	The secondary medullary rays arise by the activity of ray initials of the cambial ring.
6.	Axial and radial systems are not distinct.	6.	The two systems are clearly distinct and well marked.
7.	Tylosis are not found in the vessels.	7.	Vessels show tylosis.
8.	There is no distinction like heart wood and sap wood.	8.	Sap wood and heart wood are clearly demarcated in woody trees.
9.	Annual rings are not found.	9.	Annual rings are clearly distinct in the wood of temperate trees and shrubs.
10.	Medullary rays are always homocellular, i.e. each consists of only one type of cells.	10.	Medullary rays may be homocellular or heterocellular.
11.	Xylem fibres are either absent and if present, these are very few in number.	11.	Xylem fibres are found abundantly.

Phloem

It is another type of complex tissuc found in the vascular system which may also be called bast or leptome. As such phloem consists of (i) Sieve elements (sieve cell and sieve tube), (ii) companion cells, (iii) phloem parenchyma and (iv) phloem fibres. The different elements of phloem originate from the procambium or vascular cambium. The phloem formed from procambium is called primary phloem and that formed from vascular cambium is secondary phloem.

The phloem is typically found outside the vascular cambium. However, in the stem of some dicotyledons, patches of primary phloem are found in the pith, i.e. inner to the xylem. It is called **internal phloem** or **intra-xylary phloem** which is commonly met among members of Apocynaceae, Asclepiadaceae, Convolvulaceae, Myrtaceae and Solanaceae. In other cases, such as among members of Amaranthaceae, Chenopodiaceae and Nyctaginaceae, the phloem patches are found included in the secondary xylem and are called **included phloem** or **interxylary phloem.**

1. Sieve elements

Basically sieve elements include two types of cells, the less specialised sieve cells and more specialised sieve tubes. Esau (1969) proposed following terms to be used in describing sieve elements, (i) Sieve cells, (ii) sieve tube element, (iii) sieve tube, (iv) sieve areas, (v) sieve plate and (vi) lateral sieve areas.

(i) Sieve cells

These are narrow elongated cells without conspicuous sieve areas, and taper at their ends or have greatly inclined walls which overlap in the tissue. These are found among members of lower vascular plants and gymnosperms.

(ii) Sieve tubes

Structure. These are long tube-like slender bodies arranged in longitudinal

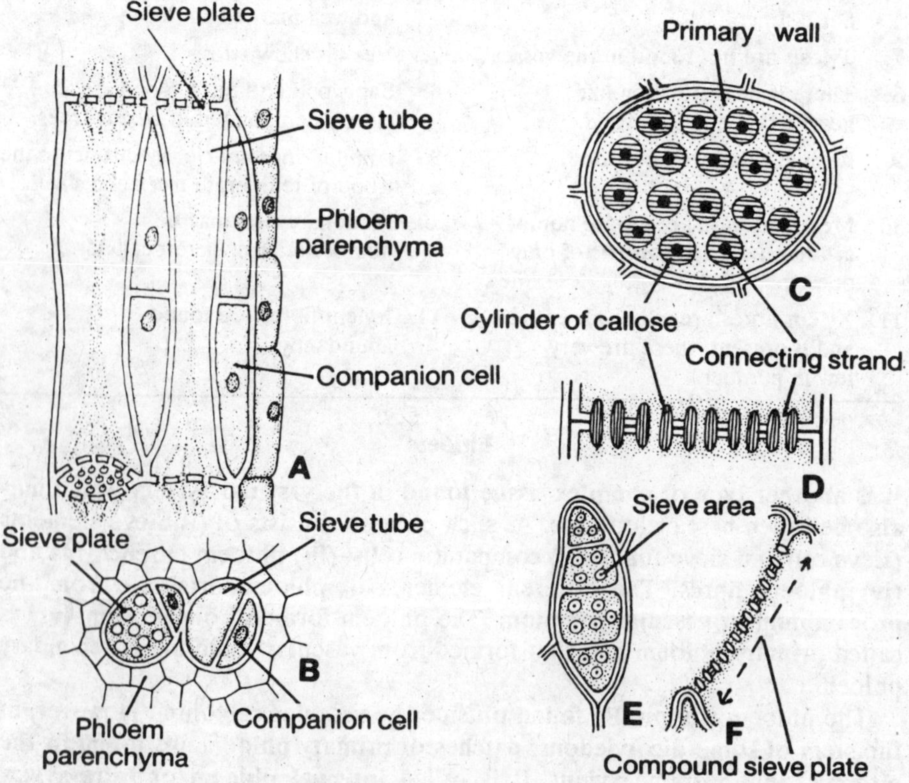

Fig. 5.12. Structure of phloem. A–L.S. of a portion showing different parts, B–T.S. of a portion of phloem, C–surface view of a sieve plate, D–L.S. of a sieve plate, E–surface view of a compound type of sieve plate and F–L.S. or side view of compound sieve plate.

series where the end walls are perforated in a sieve like manner and referred to as sieve plate. The **sieve plate** is the more specialised area found usually in or near an end wall. A sieve plate is called **simple** if it has only one sieve area or compound if it has several sieve areas arranged in scalariform, reticulate or in other manner. A sieve tube differs from vessels on certain points as described in table 5.3.

They usually have primary walls which are composed of cellulose and are usually thicker than those of neighbouring parenchyama cells. These are unique cells in that they are living cells which usually contain no nucleus at maturity. However, Ervin and Evert (1970) have shown normal nuclei in mature sieve tubes of some angiosperms. They stated that the enucleate condition of mature sieve elements might be due to disruption of the fragile nucleus during preparation of material for microscopy. However, Esau and Gill (1972-1973) have conclusively shown various degenerating stages of sieve tube nuclei with the help of electron microscopic studies.

In a mature sieve tube, there is a thin layer of parietal cytoplasm, a large central vacuole and a large number of groups of longitudinally running transcellular strands. Transcellular strands are aggregates of tubules made up of P-protein (Phloem protein). Esau (1971) reported that phloem protein strands promote movement of materials through the cell. However, non-functioning sieve tube in some plants may become filled with tylosoids, i.e. proliferations from contiguous parenchyma cells.

Development of sieve tube. During maturation of sieve tube, cell wall undergoes certain distinctive changes. It develops pores in transverse wall or end walls. Each pore has a single strand of cytoplasm extending through it and connect to the protoplast of adjoining sieve tube. Sometimes, the pores also develop in lateral walls. The border of each pore becomes impregnated at maturity with callose (a polysaccharide) and thus the cytoplasmic strand within the pore remains encased in a cylinder of callose. However, in certain cases, pores without callose lining are also reported.

The protoplast also undergoes remarkable changes during maturation. Nucleus, tonoplast and vacuole undergo disintegration and disappear. Esau (1966) believes that in case of mature sieve tubes, the cytoplasm and vacuole become one system called **mycotoplasm.** Muramanis and Evert (1966) reported that ribosomes and dictyosomes also disappear. Mitochondria, however, remain present but they differ from that of parenchyma in having fewer cristae. A proteinaceous component called P-protein (Phloem protein) makes its appearance in the cytoplasm of young sieve tubes as discrete **slime bodies.** The slime bodies consist of thread-like filaments. These tend to fuse with one another during maturation. Later, the slime spreads out and disperses throughout the sieve tube (Walker and Thaine, 1971). Endoplasmic reticulum aggregate in stacks of parallel cisternae which often occupy parietal position. These modified ER gradually lose their attached ribosomes.

Functions. The main function of sieve tubes is the longitudinal translocation of prepared food materials, from leaves to the storage organs in the downward direction and then to the growing regions in the upward direction. The food moves mainly in the form of sucrose followed by other types of sugars.

2. Companion cells

These are specialised parenchymatous cells found closely associated with sieve elements in origin, position and function. These develop from same meristematic cell that gives rise to the sieve element. Such a meristematic cell divides longitudinally and one of them differentiates into sieve tube and other into companion cell.

Normally, a single companion cell is found associated with a sieve tube in the primary phloem, but in certain plants, such as carrot, more than one cell have been reported. However, solitary long companion cells are common in primary phloem, and in groups, in secondary phloem.

In a cross-section, the companion cell appears rounded or polygonal. These are living cells and have dense cytoplasm, prominent nucleus, several vacuoles, and other cytoplasmic organelles. Starch grains are never found in companion cells.

Companion cells are not found in pteridophytes and gymnosperms, rather found only among members of angiosperms. In gymnosperms and

Table 5.2. Difference between primary and secondary phloem.

	Primary phloem		*Secondary Phloem*
1.	It develops from procambium of apical meristem.	1.	It develops from vascular cambium which is a lateral meristem.
2.	There is distinction between first formed protophloem and later developed metaphloem.	2.	Such distinction is not found.
3.	Phloem fibres are found in the outer part of phloem.	3.	They are found among phloem parenchyma cells.
4.	Sieve tubes are usually longer with narrow lumen.	4.	Sieve tubes are shorter with broad lumen.
5.	Phleom parenchyma is poorly developed.	5.	Phleom parenchyma are abundant and are concerned with storage of food materials or may contain crystals, latex etc.
6.	Sclereids are usually not found.	6.	Sclereids are found in secondary phloem of several plants.
7.	Medullary rays found in this region are developed from the activity of apical meristem.	7.	Secondary medullary rays found in in this region arise by the activity of rays initials of cambium.
8.	Medullary rays are homocellular.	8.	Medullary rays may be homocellular or heterocellular.
9.	Callus formation around sieve pore is either very little or even absent.	9.	Callus formation is abundant.

pteridophytes, some parenchymatous cells remain associated with sieve cells and are called albuminous cells. The sieve tubes and companion cells are connected by pit fields present in their longitudinal walls, which is a common wall for both and with the death of one cell other cell also dies. It is a peculiar but a characteristic feature of companion cells.

Companion cells play an important role in maintenance of a pressure gradient in the sieve tubes. These control the passage of materials and provide energy to sieve tubes in loading and unloading of assimilates, thus help in conduction of food along with sieve tubes. However, there still remains much to work out the exact function of companion cells.

3. Phloem parenchyma

The phloem parenchyma found associated with phloem are elongated, tapering to broadly cylindrical, sub-spherical and polyhedral in shape. These are living cells which have dense cytoplasm and nucleus. The cell wall is composed of cellulose, having primary pit fields, interconnecting, axial parenchyma cells and ray cells, and some times also between companion cell

Table 5.3. Comparison between xylem vessels and sieve tubes.

Vessels	*Sieve Tubes*
A. Anatomical comparisons	
1. Vessels have open ends.	1. Sieve plates are present at the ends of sieve tubes.
2. Vessels are wider.	2. These are very narrow.
3. These have thick, rigid and lignified walls.	3. These have thin, more extensible and cellulosic cell wall.
4. These have various types of thickenings on their wall.	4. No such specialised thickenings are found.
B. Physiological comparisons	
5. They are dead when mature, but functional.	5. They are alive when mature and functionai.
6. These are permeable to both solute and solvent.	6. These are semipermeable in nature.
7. They have low sap concentration.	7. They have high sap concentration (O.P. = 15 to 34 Atm.)
8. They do not have turgor pressure.	8. The turgid cells have high turgor pressure.
9. They are partially collapsed when functioning.	9. They are distended by pressure when functioning.
10. They absorb water or air when cut	10. They exude cell sap when cut.
11. They translocate both solutes and solvent.	11. They translocate only solute.
12. Their translocation speed is upto 75 cm/min.	12. Their translocation speed is upto 5 cm/min, (maximum).

and sieve tube. Lignified cell wall has been found in inactive phloem parenchyma. Phloem parenchyma are characteristically not found in monocots.

The phloem parenchyma are mainly associated with the storage of organic food materials and other susbtances such as resins, tannins, mucilage, latex etc.

4. Phloem fibres or bast fibres

Sclerenchymatous fibres are found both in primary and secondary phloem and are commonly called phløem fibres. Details of fibres has been described under sclerenchyma.

The differences between primary and secondary phloem and those between sieve tube and vessels are given in tables 5.2 and 5.3.

III. SECRETORY TISSUES

Secretory tissues are structurally modified and specially organised for secretory functions. These may occur either as a single cell or in isolated patches or in a mass and likewise may be external or internal in position.

I. External Secretory Structures

External secretory structures may be epidermal in origin or may develop from sub-epidermal layers. These are classified as follows:

1. **Trichomes and glands.** These may also be called glandular trichomes. These are the simplest kind of excretory structure, though well organised and composed of different types of cells. These are of following types:

(i) **Digestive glands:** Special digestive glands are found in the body of insectivorous plants. The leaves of *Pinguicula* (butterwort) are trough-shaped on the surface of which two types of glands are found, (a) digestive glands which have a two celled stalk with eight celled head and (b) mucilage glands which are long with umbrella-shaped head.

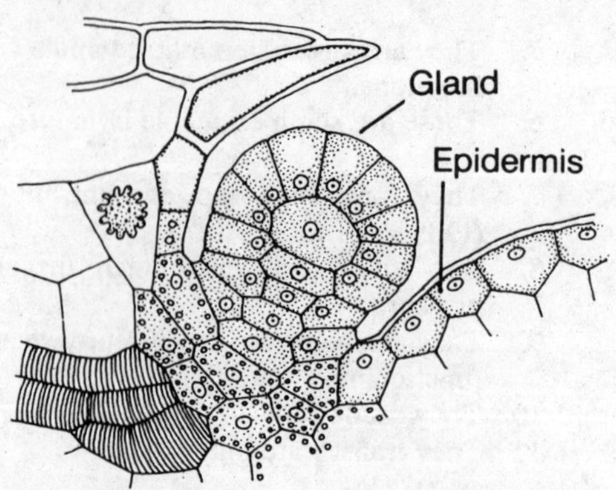

Fig. 5.13. Digestive gland of *Nepenthes* (an insectivorous plant).

Several different kinds of glands are found in the pitchers of *Nepenthes* (hanging pitcher plant), *Sarracenïa* (pitcher plant), *Cephalotus* (Albany pitcher plant), *Heliamphora* (sun pitcher) and *Darlingtonia* (Cobra lily pitcher). In *Nepenthes,* **alluring glands** are found on the undersurface of lid of the pitcher while digestive glands are present on the inner surface of pitcher wall which secrete nepenthesin, acid phosphatase, esterases etc. A large number of glands are found along the margin and adaxial surface of spathulate leaf blades of *Drosera* (sun dew) which act as capturing, digesting and absorbing organs. These glands consist of a multicellular stalk and a head with three or four layers of cells covered by perforated cuticles. In *Utricularia* (bladderwort), the leaf bladder contains glandular hairs surrounding the opening and several stalked or sessile digestive glands on the inner surface of wall. Similar digestive glands are found on the adaxial surface of leaf of *Dionea* (venus fly trap). Thus, secretory glands found in insectivorous plants show highest degree of specialisation in their form as well as function. These secrete digestive enzymes as well as poisonous substances for paralysing the insects.

(ii) **Stinging hairs:** These are found in *Urtica* all along the body, more specifically on the surface of leaves and young stem. Each hair is like a fine capillary tube, calcified at its lower end and silicified at its upper end. The basal part is bladder-like embedded in epidermal tissue and upper end is spherical which ruptures the object on coming in contact. These hairs secrete poisonous and muscle irritating substances.

(iii) **Colleters:** These are commonly found on the bud scales of roses and secrete sticky excretory substances rich in terpenes and mucilage.

(iv) **Nectaries:** Nectaries can be classified into two types on basis of their position. If found on floral parts, these are called **floral nectaries** and if on vegetative parts called **extra-floral nectaries.** The former are very common in occurrence. Cells of nectaries have dense cytoplasm and contain numerous well developed mitochondria, ER etc. These secrete sugary substances called nectar or honey composed of sucrose, glucose and fructose which are derived from the phloem. Fahn (1979) classified floral nectaries into several categories, (*a*) nectaries found at the base of perianth (as in *Ranunculus, Althaea, Hibiscus* etc.), (*b*) found in receptacle between sepals and petals (e.g. *Capparis*), (*c*) found on receptacle between sepals and stamens (*Raphanus, Passiflora*), (*d*) found in the form of a disc between sepals and ovary (*Grevillea*), (*e*) found between the bases of the stamens (*Polygonum*), (*f*) found placed on stamens either on the filament (*Dianthus, Trifolium*), or on the appendix of the connective (*Viola, Asclepias*), (*g*) found between stamens and ovary (*Lathyrus, Phaseolus, Bauhinia, Prunus*), (*h*) found as disc surrounding the base of ovary (*Citrus, Tecoma*), (*i*) found in the septa of ovary also called septal nectary *(Allium, Crocus, Musa)*, (*j*) found on receptacle above the ovary *(Eucalyptus, Cucurbita, Callestemon)*, and (*k*) found placed on the style

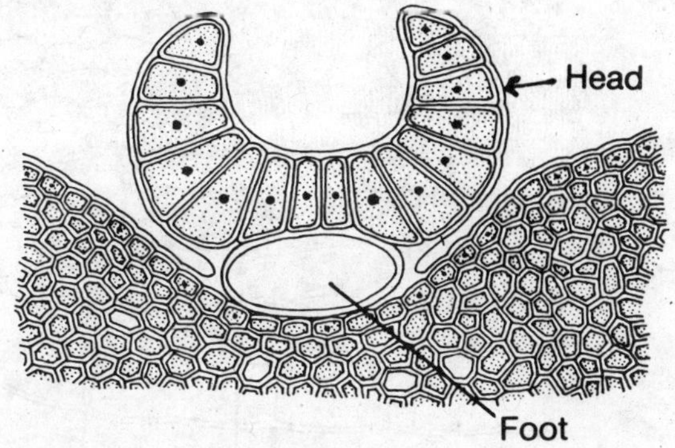

Fig. 5.14. Extra-floral nectary of *Lundia corymbifera* (Redrawn from Seibert, 1948).

(Calendula, Helianthus).

(v) **Osmophores:** These are a kind of oil ducts which secrete some essential oils. Osmophores are in the form of cilia or brushes, present on floral parts which contain secretory tissues responsible for emission of fragrance, as in *Spartium* and *Narcissus.*

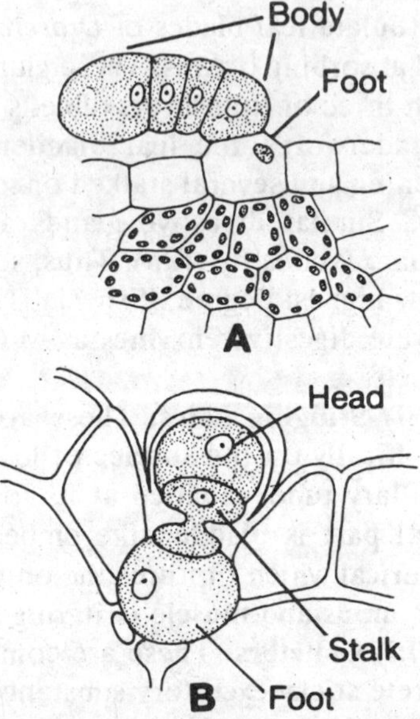

Fig. 5.15. Trichome hydathodes. A–*Phaseolus multiflorus* and B–*Piper nigrum.*

(vi) **Hydathodes:** These are specialised water stomata through which water is released in the form of liquid. The phenomenon is called **guttation,** which is commonly found among plants inhabiting humid tropics. The guttated water contains a mixture of various salts, sugars, amino acids and other organic substances. Guttation is seen when root pressure or exudation pressure increases due to reduced transpiration under humid conditions. Three types of hydathodes are recognised.

(a) *Unicellular hydathodes*–They have small mucilaginous papilla protruding out of the epidermal surface and a narrow longitudinal canal that maintains communication between papillae and cell cavity. These are found both on abaxial and adaxial leaf surfaces of *Anamirta cocculus* and *Gonocaryum pyreforme.*

(b) *Trichome hydathodes*–These look like multicellular trichomes with capitate scale-like structure. These consist of body cells (one or two) with

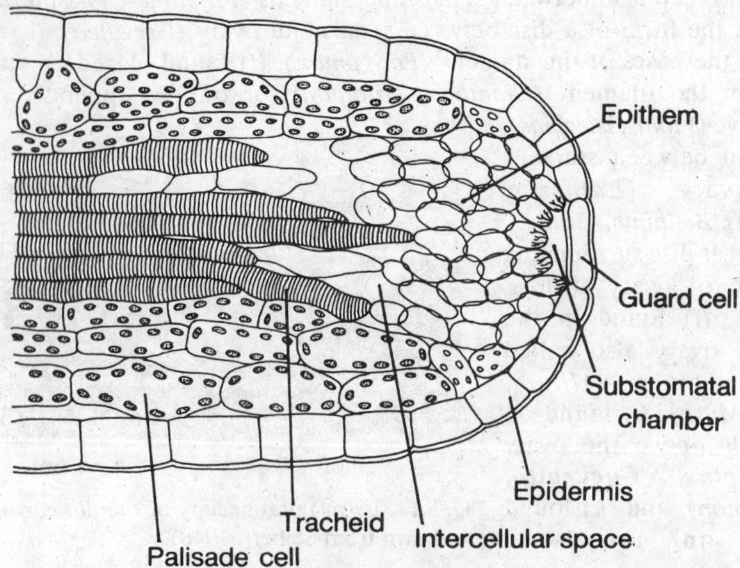

Fig. 5.16. A typical water hydathode (Redrawn from Haberlandt, 1914).

numerous slit-like pit and five to six celled foot embedded in the epidermis. Water is exuded through terminal body cell. Such hydathodes are found in *Machaerium oblongifolium, Phaseolus multiflous, Piper nigrum* etc., and also in some members of Bignoniaceae, Convolvulaceae, Fabaceae and Piperaceae.

(c) *Hydathodes with water conducting elements*–These are found among several angiosperms like *Colocasia, Lycopersicon* (tomato), *Pistia* (water cabbage), *Tropoelum* (garden nasturtium), *Primula* and in some ferns like *Nephrolepis.* Each hydathode has a terminal pore or ostiole followed by loosely arranged thin walled cells called epithem and blindly ending tracheids. *Colocasia antiquorum* is known to guttate heavily and the amount ranges from a few drops to 100 ml or even more water per day. Such hydathodes have been reported in more than 300 plants belonging to 115 families.

II. Internal Secretory Structure

Several types of internal secretory tissues are found among plants, which may consist of a single cell to a complex mass. Some of them are as follows:

(i) **Secretory cells:** These are specialised cells found in ground tissue and are called idioblasts which appear distinct from other neighbouring cells. These have been named on basis of their secretory materials such as oils, resins, tannins, gums, mucilage and crystals of various types. Oil cells are commonly found in the endosperms of *Ricinus,* cotyledons of *Arachis* and mesocarp of *Eleais.* Tanniniferous idioblasts are found among members of Euphorbiaceae, Myrtaceae and Rosaceae.

(ii) **Oil glands:** Many plants have characteristic oil glands composed of small groups of thin walled, densely cytoplasmic cells, secretory in function. Their secretion is collected in a cavity which may develop either *schizogenously* (formed by the separation of cells at the middle lamellae) or *lysigenously* (formed by dissolution of cells). The former are usually lined by intact cells while latter has partly disintegrated cells along periphery. Schizogenous oil glands are found in the leaves and bark of *Eucalyptus* and lysigenous oil glands in the

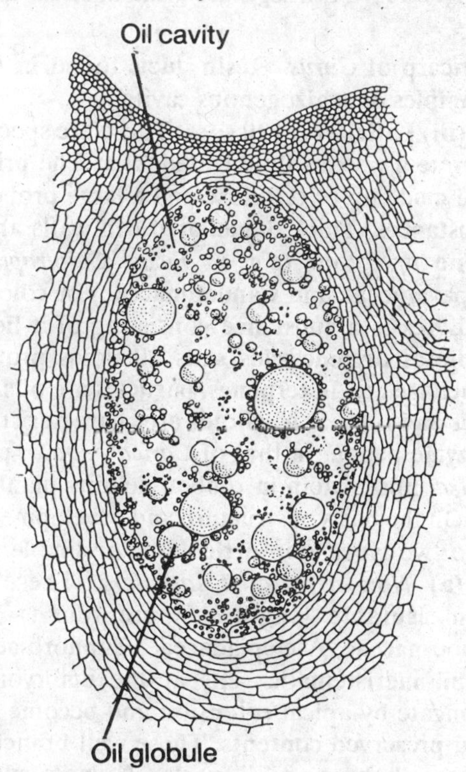

Fig. 5.17. Lysigenous oil gland of *Citrus* (adapted from Brown, 1935).

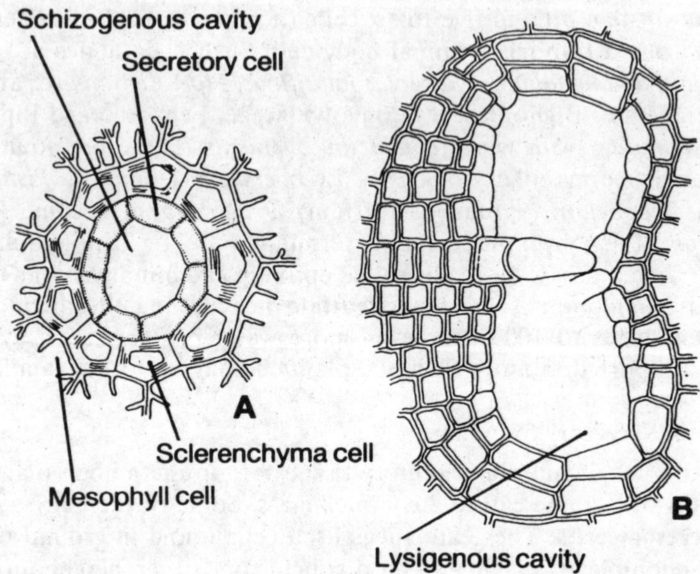

Fig. 5.18. A–Schizogenous gland of *Pinus* and B–Lysigenous gland of *Copaifera*.

pericarp of *Citrus*. Resin ducts found in *Pinus* and other conifers are good examples of schizogenous cavity.

(iii) **Laticiferous tissues:** These are specialised parenchymatous cells which secrete a viscous fluid called *latex*. The primary walls are thin, devoid of pits and made up of cellulose and a large proportion of hemicellulose with pectic substances. Thick, lamellated cell walls are known but rare. Latex is mostly white in colour (e.g. in *Ficus, Calotropis* etc.) but may be yellow (as in Papavaraceae and Cannabinaceae) or colourless (as in Musaceae).

Latex is a substance consisting of a liquid matrix which contains carbohydrates, organic acids, salts, alkaloids, proteins, sterols, fats, tannins, terpenes, mucilage etc. In terpenes, particles are oils, balsam, resins, camphors, carotenoids and rubber. Alkaloids are characteristically found in the latex of *Papaver,* enzyme papain in that of *Carica papaya,* proteins in *Ficus callosa,* tannins in *Musa,* lipoprotein in rubber yielding plant *Hevea* and chickle in *Achras.*

On the basis of structure and development, laticiferous tissues are identified into two types, **non-articulated** or simple laticifers and **articulated.**

(a) *Non-articulated laticifers.* These are multinucleate laticifers of immeasurable length and are found among members of families like Apocynaceae, Asclepiadaceae, Euphorbiaceae and Urticaceae. These arise as small meristematic cells at the embryonic stage. They grow further and elongate by apical extension and become distinct because of their large size and preserved contents. These well branched, elongated tube-like structures enter all the possible parts of plants such as leaves, roots, stem etc. The nuclei of these cells undergo repeated free nuclear divisions (without accompanying wall formation).

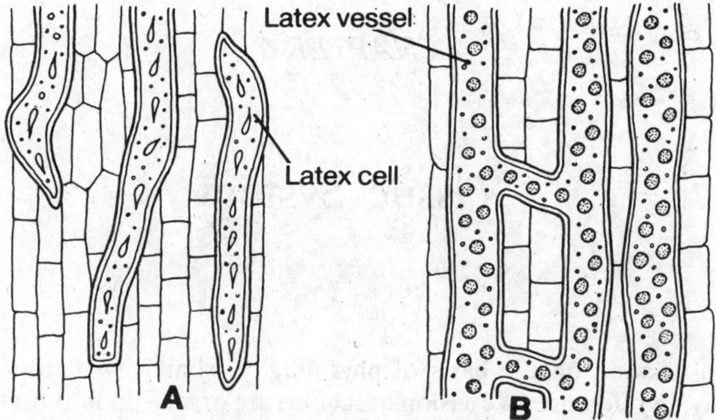

Fig. 5.19. Laticiferous tissue. A–Latex cells (non-articulated latificers) and B–latex vessels (articulated laticifers).

(b) *Articulated laticifers*–These are also called laticiferous ducts which originate from a row of cells in continuous series by partial or total dissolution of end walls. Thus, these arise by fusion of many laticiferous cells becoming coenocytic with numerous nuclei scattered through out the cytoplasm. These may be rightly called living coenocytic laticiferous cells which may be unbranched as in members of Musaceae, Convolvulaceae and Sapotaceae or branched as in Caricaceae and Compositae.

The functions of laticiferous tissues may be summarised as follows:

1. These help in conduction of food substances.
2. These also store food materials or reserve nutrition.
3. These regulate and maintain the water balance in the plant.
4. These work as agency for oxygen transport.
5. The most important function of these tissues is to perform excretory roles.

Latex in some cases is of great economic value, especially as a source of rubber (*Hevea* and *Ficus*), chickel (*Achras*), and papain (*Carica*).

GENERAL QUESTIONS

1. Write a brief account of simple permanent tissues and describe their functions.
2. Describe the structure and functions of complex tissues.
3. What are secretory tissues? Give a brief account of various types of secretory tissues found in plants.
4. Write short notes on: (a) Parenchyma, (b) Collenchyma, (c) Sclerenchyma, (d) Laticiferous tissues, (e) Nectar glands and (f) Sclereids.

CHAPTER 6

Tissue System

Structural tissues, on the basis of physiological similarity or topographical continuity, in reference to division of labour, are organised into tissue system. Sachs (1875) recognised three basic tissue systems found in the primary structure of plants.

I. Epidermal tissue system
II. Ground or fundamental tissue system
III. Vascular tissue system.

The precursors of these tissue systems are meristems which may be outlined as follows.

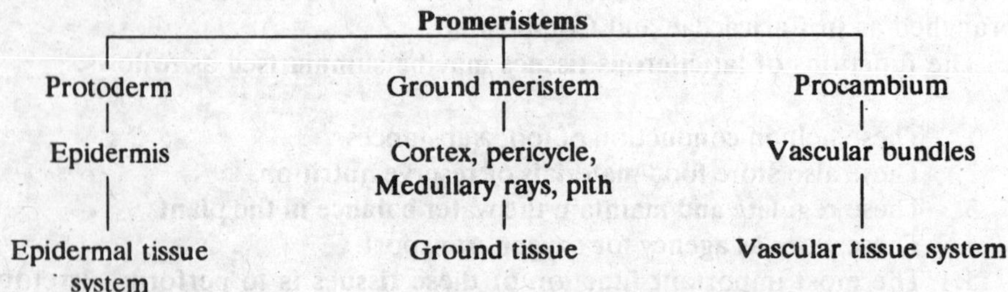

I. THE EPIDERMAL TISSUE SYSTEM

Epidermal tissue system forms the outermost skin layer which is basically a protective layer. It is subjected to a number of adverse factors such as evaporation of water, risk of mechanical injury by rains, wind and biological agencies, exposure to weak and intense light and varied thermal fluctuations. As such several adaptations and variations are found in epidermal tissue system.

Origin

The outermost layer of the apical meristematic cells, called protoderm, divides characteristically anticlinally and by further division gives rise to uniseriate layer called **simple epidermis.** The word epidermis is derived from two Greek words–*epi*-upon; *derma*-skin. It is assumed that protodermal cells divide periclinally in certain cases to form multiseriate or **multiple epidermis.** The multiple epidermis is found in some organs like roots of orchids *(Vanda),* leaves of *Ficus, Nerium, Piperomia,* etc. In *Ficus* leaves, multiple epidermis is found on adaxial (upper) surface only but in oleander *(Nerium),* it is found on both surfaces. In *Piperomia,* the epidermis consists of 14 to 15 layers.

The outermost layer of multiple epidermis is similar to simple epidermis but inner layers may show variation. Some scientists believe, that these inner layers are infact, outer layers of cortex originating from the ground meristem but resemble the epidermis both in structure and function.

Structure of epidermal cells

Epidermal cells are parenchymatous in nature and are compactly arranged without intercellular spaces. The cells appear more or less rectangular in cross section. Some points regarding their structure are as follows:

(a) **Morphology.** These cells show considerable variation as regards their size, shape and arrangement. They are normally tubular, which in surface view appear more or less isodiametric or elongated in shape. These are of irregular shape in the leaves and petals of flowers. In monocot stem and leaves, these are highly elongated, some times may be fibre-like in appearance. In general, elongated epidermal cells are found in stems, petioles, veins of leaves, adjacent to hairs and stomata, etc. The epidermal cells of the petals of *Viola* show ridge-like infoldings of the lateral walls and protruding papillae.

(b) **Arrangement.** Epidermal cells are compactly set, so that a continuous layer without intercellular spaces is formed. But in petals of certain flowers, small intercellular spaces may be present which remain covered with thick outer cuticle. The epidermal cells thus form a continuous external layer of a plant organ except at places where stomata or lenticels are found.

(c) **Cell contents.** The epidermis consists of living cells. Cells have a large central vacuole and thin peripheral uninucleate cytoplasm which contains chromoplasts and leucoplasts, and occasionally chloroplasts. The chloroplasts are found in the guard cells of stomata, aquatic plants and plants growing in humid climates. Mucilage, starch grains, tannins and crystals may be present in certain cases. Silica cells and cork cells are found in sugarcane leaves. The anthocyanin rich chromoplast is found in vacuolar sap of petals and tender, coloured leaves.

(d) **Wall structure.** The epidermal cells show considerable variation as regards to wall thickness. In general, they have unevenly thick walls, the outer and radial walls being much more thick than the inner walls. Sometimes, the thickness is so massive that central lumen of the cell becomes indistinct. Sometimes, plasmodesmata are found.

The walls are strongly cutinised. Cutin is a fatty, wax-like substance which is inert and resistant to oxidation. It is very important for protection against mechanical injuries and prevention of loss of water. It forms a layer on epidermis called *cuticle* which is present all over the plant body except in roots and some submerged aquatic plants. Epidermis of root without cuticle is called **epiblema.** The thickness of cuticle is variable which may be smooth as in sunflower stem or have protrusions and ridges of various shapes. Cuticle often projects into the radial walls of epidermis as peg-like bodies as in case of *Musa* sp.

Sometimes, additional waxy substances are deposited on the cuticle in the form of rods and granules. These crystalline waxy substances may be in the form of rods, often with hooked ends, network of tubes, isolated tufts etc. The glaucous characters of many stems, leaves and fruits is due to these deposits. In certain cases like *Equisetum* and grasses, deposition of silica as silicon oxide is characteristically found. Wax, oil, resin, crystalline salts, lignin and silica are often found present on the epidermal cells of aerial parts.

In some cases, lignified epidermal cells have also been observed as in case of *Cycas* and *Pinus* leaves. In cork cells, suberin is found deposited and the dermal system is referred to as periderm.

Functions. It carries out following important functions.

1. It is a protective layer that protects internal tissues against mechanical injury, high light intensities, injurious thermal changes, attack of microbial parasites, high wind velocity, acid rain effects etc. This is due to deposition of cutin, wax, oils, resin, hairs, lignins etc.

2. It reduces loss of water due to excessive transpiration.

3. The presence of sharp and stiff hairs on epidermis in certain cases protects plants from grazing.

4. Epidermis of some xerophytic plants store water and food materials in the form of mucilage.

5. In Bryophytes, Pteridophytes and certain hydrophytes, epidermal cells contain chloroplasts and thus manufacture the food.

Special epidermal structures

(1) **Lithocytes.** The epidermal cells of plants belonging to families like Urticaceae, Cucurbitaceae, Moraceae and Acanthaceae contain deposition of calcium carbonate crystals, arranged in the form of bunches, like those of grapes. Such depositions are called **cystoliths** and cells containing them are called **lithocytes.** Lithocytes are usually larger than other epidermal cells. (For figures see chapter 2).

(2) **Myrosin cells.** Certain members of family Cruciferae contain special sac-like cells scattered in the epidermis. These remain filled with an enzyme, *myrosin,* and are called myrosin cells. These are idioblastic cells resembling the laticifers in nature.

(3) **Sclerotic cells.** The epidermis of seed coats of *Pisum* and *Phaseolus* of family Leguminosae, and the scales of *Allium sativum* of family Liliaceae, is often made up of a layer of sclereids.

(4) **Bulliform cells.** Some epidermal cells found in groups in the leaves of certain monocots, specially grasses, (maize, sugarcane etc.) are comparatively

larger, thin walled and highly vacuolated. These are called bulliform (meaning, bubble-like) cells. These are commonly found in the upper surface of leaves, sometimes in lower surface also. Bulliform cells are mainly water containing cells, poor in solid and chlorophyll contents. Rarely tannins and crystals may be found. In cross-sections, these appear as a fan-like band because of the median cell being largest in size.

There are three views as regards their function.

(i) These are concerned with the unrolling of developing leaves by undergoing sudden and rapid expansion bringing about unrolling of lamina.

(ii) These are supposed to play an important role in hygroscopic opening and closing movements of mature leaves, due to changes in turgor. They are called *motor cells* by some scientists.

(iii) These are concerned only with water storage and have no other functions.

(5) **Stomata.** Stomata are narrow openings found in the epidermis of green aerial parts, specially leaves of the plant. A pore, the *stoma,* remains surrounded by two specialised epidermal cells, called *guard cells.* The guard cells are kidney-shaped in dicotyledons but dumble-shaped in the members of Graminae (monocotyledon) (Fig. 6.1B and 6.4E). Their wall is unevenly thickened. The wall surrounding the pore is thickened and inelastic due to presence of secondary layer of cellulose but remaining part of the wall remains thin, elastic and permeable.The radial walls have characteristic fan-like micellae, composed of cellulose, due to which guard cells can elongate more than lateral expansion. Each guard cell has cytoplasmic lining and a large central vacuole filled with cell sap. The cytoplasm contains a nucleus and a number of chloroplasts, often poorly developed but capable of photosynthesis. Electron microscopic studies have revealed that this chloroplast has fewer and less well organised lamellae than those of mesophyll cells. The guard cells are also covered with cuticle which extends to the inner wall

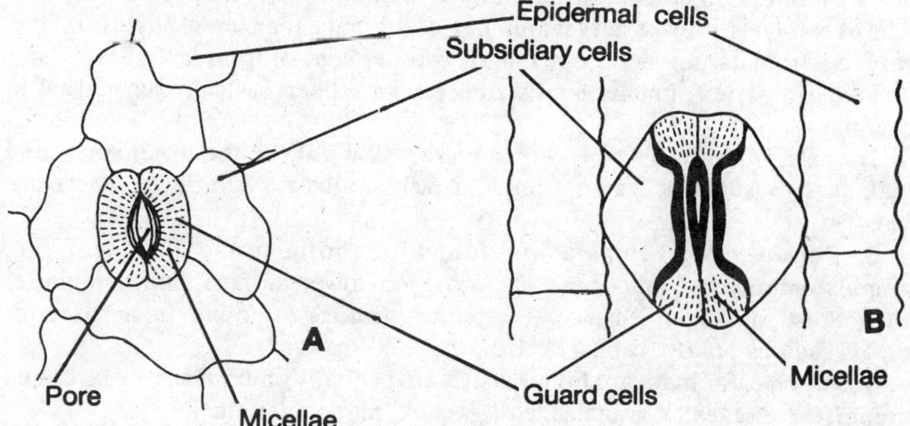

Fig. 6.1. Stomata. A–as found in a dicot leaf and B–in monocot leaf.

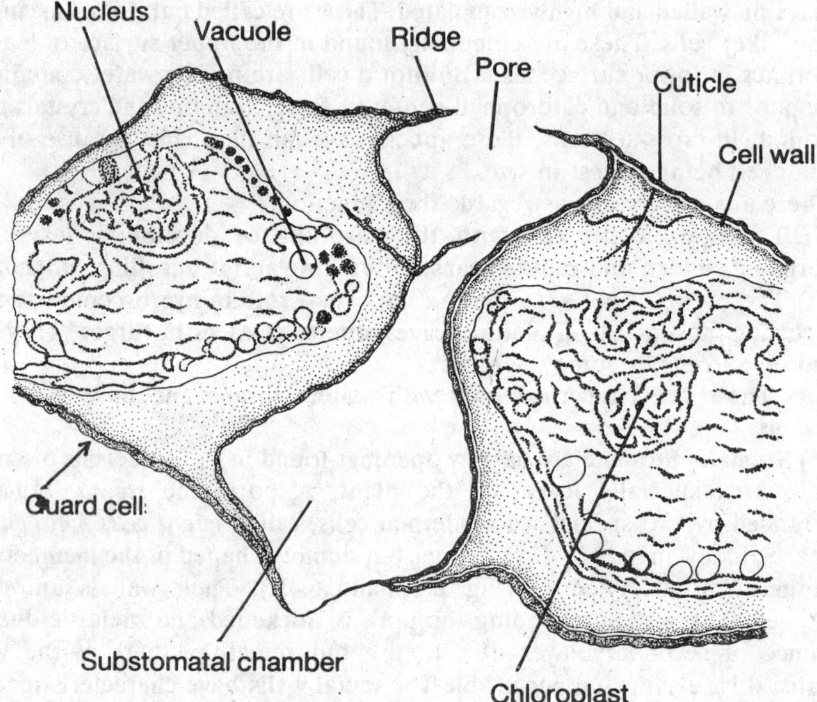

Fig. 6.2. Guard cells of *Vicia faba* as seen under electron microscope.

forming the boundary of the pore and the substomatal chamber. Mitochondria, dictyosomes, ER, ribosomes are found in guard cells.

The epidermal cells surrounding guard cells are specialised and called *subsidiary cells* or *accessory cells* which give support in the movement of guard cells. The size and shape of stomata vary from plant to plant, e.g. in *Phaseolus vulgaris* 7 × 3 μ, in *Avena sativa* 38 × 8 μ and in *Zea mays* 4 × 26 μ.

The number of stomata in a definite area of leaf varies from plant to plant. The xerophytes possess larger number of stomata than mesophytes, in this respect. Stomata nearly occupy one to two per cent of total leaf area.

Five categories of **stomatal distribution** have been basically recognised in the plants.

1. **Apple or mulberry type.** Stomata are found only on the lower surface of leaf, i.e. **hypostomatic leaf,** e.g. apple, peach, mulberry, walnut etc. These are mostly trees.

2. **Potato type.** Stomata are found on both surfaces of leaf, i.e. **amphistomatic leaf,** but these are more on lower surface (**multistomatic**) than those on upper surface (i.e. **paucistomatic**) as found in herbaceous dicots such as potato, cabbage, bean, tomato, pea etc.

3. **Oat type.** Stomata are found distributed equally on both the surfaces i.e., **amphistomatic leaf,** e.g. maize, oat, grasses (monocots) etc.

4. **Wild lily type.** Stomata are found distributed only on the upper surface

Table 6.1. Distribution of stomata.

Name of the plant	*Number of stomata per square cm upon the upper surface*	*Number of stomata per square cm upon the lower surface*
1. *Helianthus annuus*	58	156
2. *Lycopersicon lycopersicum*	12	130
3. *Phaseolus vulgaris*	40	281
4. *Solanum tuberosum*	51	161
5. *Zea mays*	52	68
6. *Avena sativa*	40	43

of the leaf, i.e. **epistomatic leaf** as found in water lily, lotus etc.

5. **Potamogeton type.** Stomata are altogether absent or if present these are vestigeal (non-functional or rudimentary type) as in case of *Potamogeton* and other submerged hydrophytes.

Stomata may occupy **three different positions** with regards to epidermal cells, (*a*) stomata found at the same level as in most mesophytes, (*b*) stomata are sunken, found in depression as in xerophytes, and (*c*) stomata are slightly raised above the surface of epidermis as in *Cucurbita, Prunus, Solanum* etc.

Considering the **behaviour of stomatal movements,** five categories have been recognised.

1. **Photo-active movements.** Light directly or indirectly controls stomatal movements. Such stomata remain open during day time and closed during night.

2. **Skoto-active movements.** Stomata remain closed during day time and open during night. Such examples are found in succulent plants.

3. **Hydro-active movements.** In certain cases, stomata open due to excessive loss of water from epidermal cells and close due to turgid conditions of epidermal cells. This condition is usually observed during mid-day.

4. **Autonomous movements.** In some cases, stomata open and close at a rate of intervals of 10-15 minutes showing diurnal or rhythmic pulsation. Their mechanism is not precisely known.

5. **Passive and active movements.** Opening of stomata is considered as active process while closing is the passive process. This is caused by turgor changes in the guard cells.

Stomata have also been classified by Metcalfe and Chalk (1950) who considering the neighbouring epidermal or subsidiary cells divided them into following four categories.

1. **Anomocytic type.** It is also called **irregular-celled type** where subsidiary cells are indistinguishable from other epidermal cells (Fig. 6.3A). Such cases are found in members of Ranunculaceae, Papavaraceae, Capparidaceae and Nyctaginaceae (Fig. 6.4A).

2. **Anisocytic type.** It is also called **unequal celled type.** This is characterised

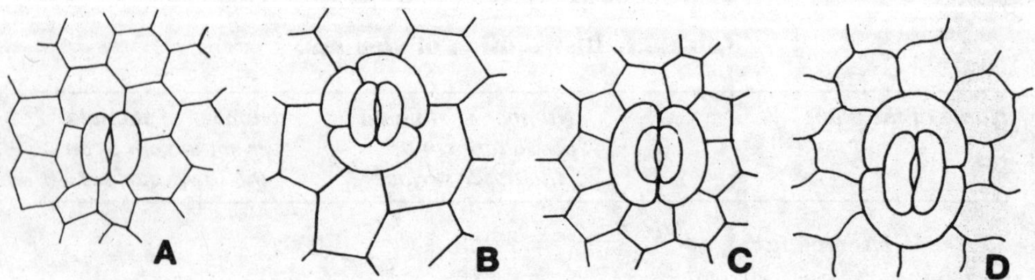

Fig. 6.3. Various types of stomata as suggested by Metcalfe and Chalk (1950). A–anomocytic type, B–anisocytic type, C–paracytic type and D–Diacytic type.

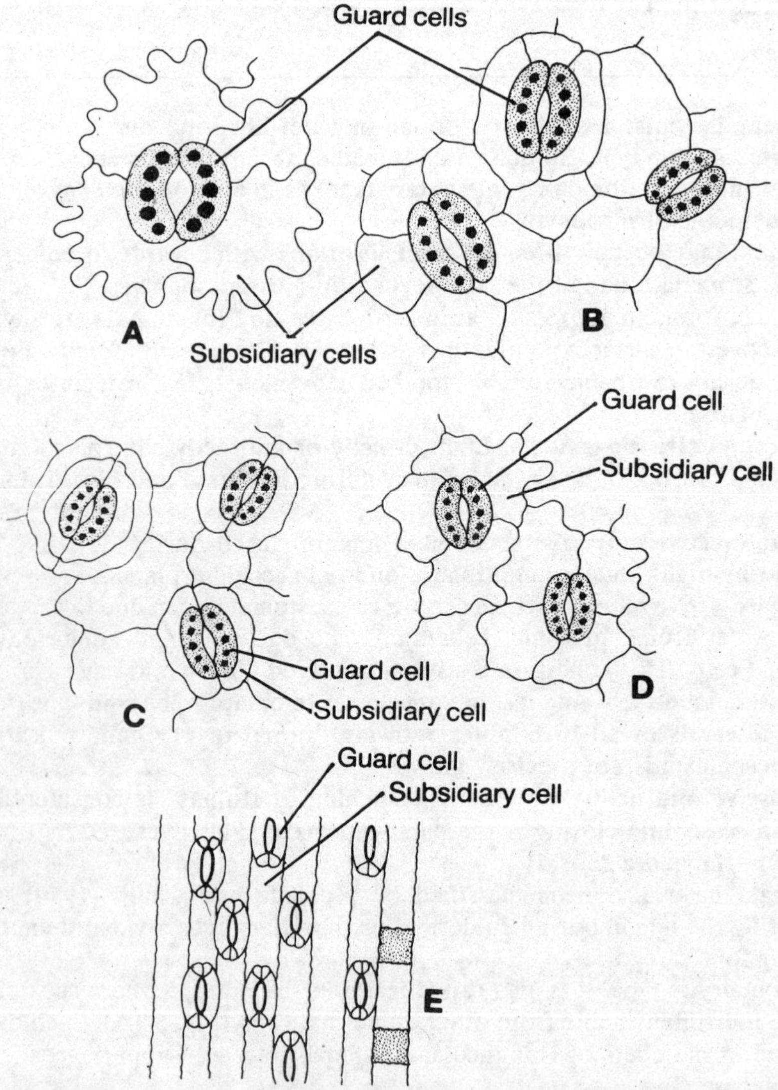

Fig. 6.4. Types of stomata. A–ranunculaceous type, B–cruciferous type, C–rubiaceous type, D–caryophyllaceous type and E–graminaceous type.

by the presence of three subsidiary cells, of which one is small and the other two are relatively larger, (Fig. 6.3B) as found in members of Cruciferae, Solanaceae, Umbelliferae and Convolvulaceae (Fig. 6.4B).

3. Paracytic type. It is also referred to as **parallel celled type.** In this type, stoma is accompanied on either side by one or more subsidiary cells which lie parallel to the long axis of the pore and guard cells (Fig. 6.3C), as found in members of Rubiaceae and Magnoliaceae (Fig. 6.4C).

4. Diacytic type. It is also called **cross-celled type** in which stoma remains enclosed by a pair of subsidiary cells whose common wall is at right angles to the long axis of the guard cells (Fig. 6.3D), as seen among members of Caryophyllaceae, Labiatae and Acanthaceae (Fig. 6.4D).

Stace (1965) included two more categories, i.e. **actinocytic type** (with four or more subsidiary cells elongated radially to the stoma (Fig. 6.5A) and **cyclocytic type** (with four or more subsidiary cells arranged in a closed ring around the stoma (Fig. 6.5B).

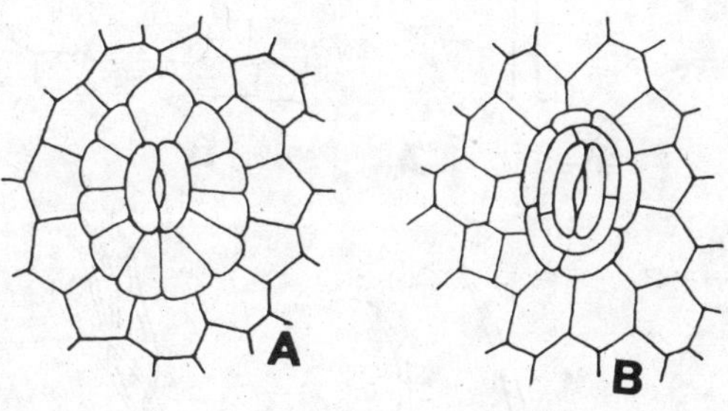

Fig. 6.5. Types of stomata. A–actinocytic type and B–cyclocytic type (after, Stace, 1965).

Florin and his coworkers (1959) recognised two types of stomata based on ontogeny, i.e., (*a*) **haplocheilic type** (when the guard cells arise by a single division of the stomatal initial, and some of the neighbouring cells become modified into subsidiary cells), and (*b*) **Syndetocheilic type** (when the guard cells and subsidiary cells orginate from the same mother cell).

Pant (1965) recognised three categories of stomata on basis of development.

(i) **Mesogenous type**–Both guard and subsidiary cells are derived by consecutive division of same mother cell as found in members of Rubiaceae and Cruciferae.

(ii) **Mesoperigenous type**–In this case, surrounding cells are dual in origin, i.e. some from the mother cell and some from the neighbouring cell, as found in members of Ranunculaceae and Caryophyllaceae.

(iii) **Perigenous type**–In this type, all neighbouring and subsidiary cells have independent origin, as found among members of Cucurbitaceae and Nymphaeaceae.

Functions of stomata. Stomata are very important structures from physiology point of view. Physiological functions like photosynthesis, respiration and transpiration take place with the help of stomata, as it is through them interchange of gases takes place between the intercellular space system of the internal tissues and the outer atmosphere.

(6) Trichomes. The term trichome includes different types of unicellular or multicellular appendages of the epidermis. These are at times so much characteristic, that they have been successfully used in taxonomy, in classification of genera and species. Trichomes are found both on vegetative as well as reproductive parts and may be living or dead, ephimeral or persistent and variable in structure, form and function. Foliar trichomes are important in regulation of moisture exchange with the atmosphere, Covering

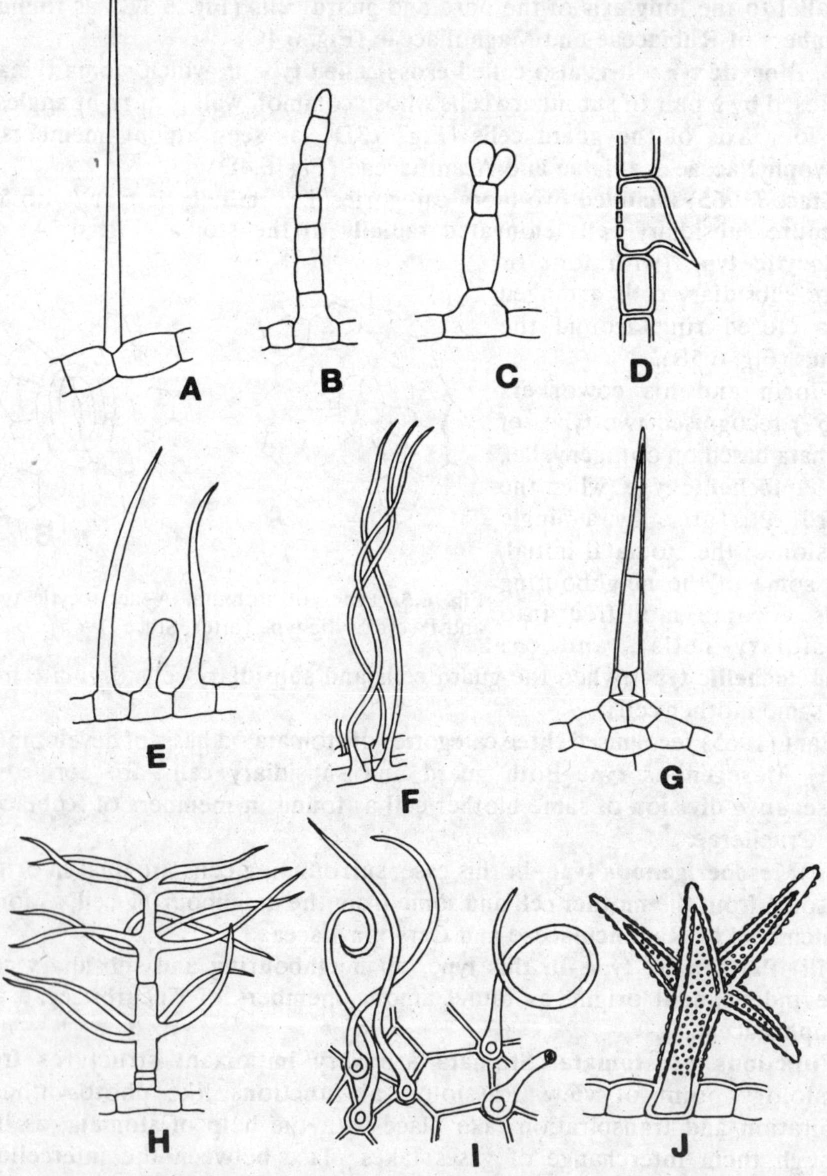

Fig. 6.6. Various types of trichomes as found on A–petal of *Epigaea*, B–leaf of *Coreopsis*, C–petal of *Phryma*, D–leaf of *Avena*, E–sepal of *Heliotropium*, F–stem of *Onoprodum*, G–leaf of *Cucumis*, H–leaf of *Platanus*, I–fruit of *Rubus* and J–stem of *Aubrietia* (Redrawn from Eames and Mac Daniels, 1947).

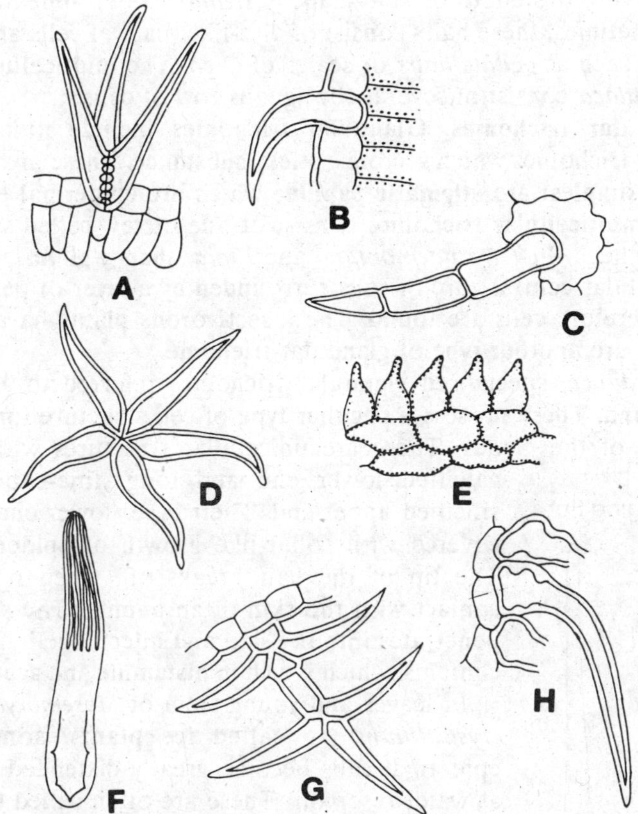

Fig. 6.7. Various types of trichomes as found on A and D–leaf of *Hamamelis*, B–leaf of *Cassia*, C–leaf of *Datura*, E–petal of *Arnica*, F–testa of *Strychnos*, G–leaf of *Verbascum* and H–testa of *Strophanthus*.

hairs for plant defence against phytophagous insects, **Secretory trichomes** for providing chemical defence against insects and **Stinging hairs** for protecting plants from animals.

Uphof (1952) classified trichomes on basis of morphological characters. The important categories are as follows:

I. Non-glandular trichomes. These may be of two types:

(a) *Unicellular trichomes*–These are infact simple epidermal elongations which may be straight or curved, bifurcate at apex, Y-shaped (as in *Buddleja*) or T-shaped (as in *Lobularia* and *Dicrasus*). Unicellular trichomes of *Cannabis* (hemp) are highly silicified with warty protuberances.

(b) *Multicellular trichomes*–These may be branched or unbranched consisting of a single row of cells (uniseriate) or several rows of cells (multiseriate). The uniseriate trichomes may be *vesiculate type* (terminal cell bladder-like as in *Chenopodium*) or *capitate* or *sub-capitate* type.

These hairs often show peculiar branching. Some of them give dendroid or tree-like appearance as in *Mimosa* and *Platanus* or the branches come out in one plane giving it stellate or star-shaped (*stellate hairs*) appearance as in *Althaea.* Sometimes, these hairs consist of disc-like plate of cells arranged on a short staik such as *peltate hairs* or scales of *Olea.* The multicellular shaggy hairs of *Portulaca* consist of several contiguous row of cells.

II. Glandular trichomes. Glandular trichomes include unicellular or multicellular trichomes which secrete a sticky substance. These are of various shapes. The simplest are **stigmatic papillae** which are epidermal extensions. The simple multicellular trichomes consist of one or few celled stalk and a multicellular head. In *Tabernaemontana* and *Viola,* **shaggy glands,** consisting of a multicellular central core of cells surrounded by a layer of palisade like elongated secretory cells are found. The insectivorous plants have digestive glands which are another type of glandular trichome.

In *Urtica dioica,* specialised glandular trichomes referred to as **stinging hairs** are found. These possess a peculiar type of cell structure for releasing the contents of the glands. These are unicellular structures with a broad calcified lower end and long, fine-capillary like silicified upper end. Their base (lower end) remains covered with collar-like growth of epidermal cells. The tip of the hair breaks off, when it comes in contact with the skin of an animal. The sharp edge penetrates into the skin and injects the irritating cell contents, which is rich in histamine and acetylcholine.

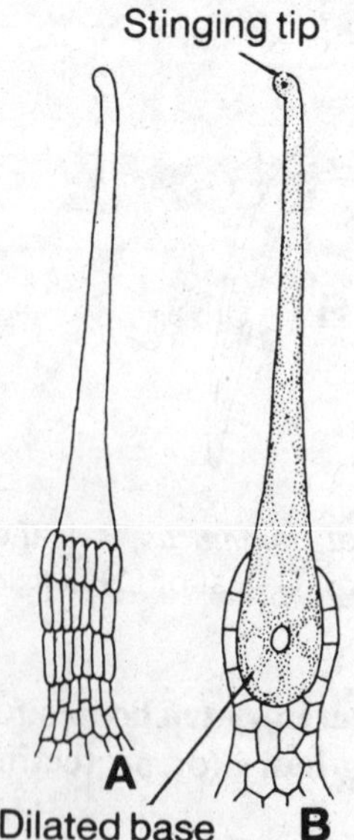

Fig. 6.8. Stinging hairs of *Urtica.*

In leaves and young stem of *Mesembryanthemum crystallinum* (so called ice plant), some of the epidermal cells become greatly distended and serve as water reservoir. These are often called **ice-beads.**

Certain halophytes such as *Distichlis spicata* have interesting glandular trichomes which are called **salt glands.** Here, the salt gland consists of two cells, a large basal cell and a cap cell that protrudes above the surface. Salt glands probably help in the removal of excess salt. In *Atriplex,* the salt gland has a narrow stalk cell and a bladder cell. Salt glands of *Avicennia* (a mangrove plant) consist of two-celled foot,

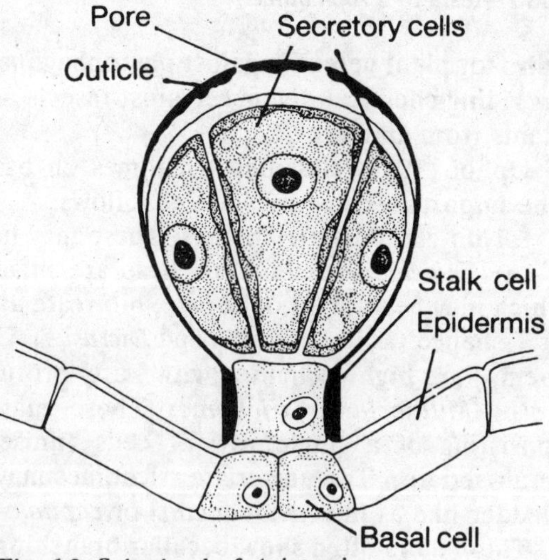

Fig. 6.9. Salt gland of *Avicennia.*

single celled stalk and a head of three to four salt secreting cells (Fig. 6.9).

(7) **Root hairs.** Root hairs are unicellular, elongated structures arising from epidermal cells of a root. These are usually absent in aquatic roots but most well developed in roots growing under humid conditions. Root hairs are concerned with absorption of water.

II. THE GROUND OR FUNDAMENTAL TISSUE SYSTEM

The tissue systems found in the region between epidermis and vascular bundles are collectively referred to as **ground meristem.** It is heterogenous in nature, composed of different types of cells concerned with different functions. This system forms the major mass of the primary body. In a stem, it basically includes **cortex, pericycle, pith** and **medullary rays** and in a leaf, **mesophyll tissues** are included in this system. It is partly derived from periblem and partly from plerome.

1. Cortex

Cortex is the extra-stelar zone, which forms that part of the axis which spreads from the layer next to epidermis to the layers enveloping the stele. In a dicotyledonous stem, cortex is generally differentiated into *hypodermis, general cortex, endodermis* etc.

(a) **Hypodermis.** It is situated just below the epidermis. In a dicot stem, hypodermis is characteristically composed of collenchymatous cells while in a monocot stem, it consists of sclerenchymatous cells. This region is generally uniformly developed except in certain angular stems e.g., *Cucurbita,* where it develops in the form of prominent patches, particularly below the ridges. In herbaceous green parts, collenchymatous cells contain chloroplasts and participate in photosynthesis. However, chloroplasts may not be found in all collenchymatous hypodermis. Similary, in case of green stems of monocots (cladode or phylloclade), hypodermis may be chlorenchymatous instead of sclerenchymatous or both may be present as in *Asparagus* and *Ruscus.*

The main function of hypodermis is to provide protection to the internal tissues but it also gives mechanical support to the peripheral region of the plant.

(b) **General cortex.** The region between hypodermis and endodermis is called general cortex which is usually composed of loosely arranged parenchyma cells with intercellular spaces. The cells are isodiametric in shape. Starch grains and different types of crystals are found in these cells. Chloroplasts may or may not be present. Sclereids, resin ducts, tannin cells, oil cavities and laticiferous tissues may also occur in this region. In stems, this region remains distinct from hypodermis while in roots, cortex is more or less homogenous, as it consists of only parenchyma cells. In case, where cortex and hypodermis are indistinct, term ground tissue is used to refer to this region. Hypodermis, thus, remains absent in roots.

The main function of general cortex is to store food materials and to participate in metabolic activities. It also gives mechanical support upto some extent.

(c) **Endodermis.** The innermost layer of cortex is referred to as endodermis. Some scientists consider that it is the limiting layer of stele. So its origin remains controversial. It is a single layer composed of compactly arranged modified parenchyma cells lacking intercellular spaces. Cells are vertically elongated which appear tubular or barrel-shaped in a cross section.

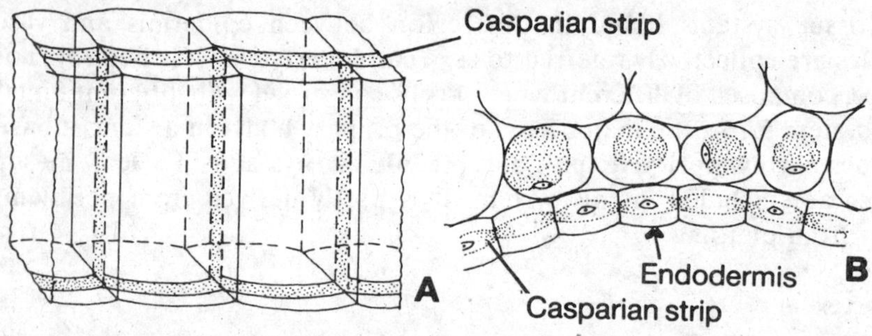

Fig. 6.10. Endodermal cells with wall thickening or Casparian strips in two different views.

The endodermal cells are characterised by the presence of waxy substance, like suberin, in the form of bands or strips on their radial and transverse walls. These are commonly called *Casparian strips* or bands named after its discoverer, Caspary (1865). Lignins may also be found present in these thickening strips. In some cases, thickening by suberin deposition is so pronounced that lumen of the cells may be obliterated. In such cases, thin walled *passage cells* or *transfusion cells,* to facilitate flow of water and solute, are found here and there. In roots, passage cells are found just opposite the protoxylem patches. The endodermis in pteridophytes and most dicots are thin walled while in roots of monocots, it consists of thick walled cells (Fig. 6.11). Some scientists use the terms, primary and secondary, to thin walled and thick-walled endodermis, respectively. In an endodermal cell, the cytoplasm is highly viscous and remains firmly attached to the Casparian strip, so much so, that it does not easily separate even during plasmolysis.

Endodermis is distinctly present in the stems of majority of herbaceous angiosperms, in aquatic plants, in creepers and rhizomes. It remains absent in woody stems of dicotyledons and gymnosperms, and leaves of angiosperms.

The exact and basic function of endodermis is still not clear. Some functions, on basis of its relation to water and vascular tissues have been attributed to it. (*i*) It is considered as a *water-tight jacket* between xylem and cortex. (*ii*) It acts like an *air dam* that prevents diffusion of air into xylem, which would otherwise get clogged. (*iii*) It is a *protective layer,* sometimes referred to as accessory epidermis. (*iv*) It regulates and maintains root pressure. (*v*) It acts as a *water dam* due to presence of compactly set

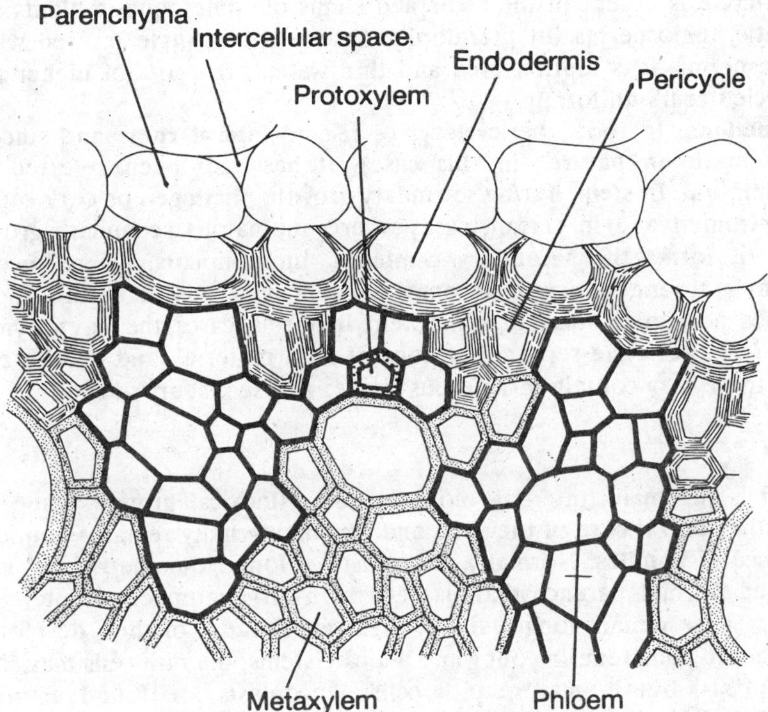

Fig. 6.11. T.S. of a portion of maize root showing position of endodermis and pericycle.

thickened cells. (*vi*) It provides regulated and systematised flow of water through its passage cells. (*vii*) The thin walled cells retain power of division, as such, the lateral branch and adventitious buds often originate through this layer.

2. Pericycle

Pericycle may consist of one or several layers of cells found between the endodermis and vascular bundles. In fact, pericycle region surrounds the vascular tissue and is regarded as the limiting layer of the stele. It forms the outermost part of intrastelar ground tissue, the other parts being the pith and medullary rays. Pericycle region may be homogenous and consists entirely of parenchymatous cells as found in all roots and pteridophytes. It may also be heterogenous in nature consisting of parenchyma and sclerenchyma or sclerenchyma may be found in the form of patches as in several dicotyledonous stems, e.g. *Helianthus annuus* or as a continuous circular band as in *Cucurbita* stem. In several cases, the sclerenchyma forms small patches intermixed with parenchyma cells. However, recent studies have revealed that so called pericycle fibres found in stem are actually phloem fibres. The commercial fibres obtained from stems of hemp (*Cannabis sativa*) and flax (*Linum usitatissimum*) are infact phloem fibres and not pericylic fibres.

Pericycle is absent in the roots and stems of some aquatic plants and in parasitic angiosperms. In pteridophyte stems, it is single layered while in gymnosperms, it is multilayered and thin walled. In roots of higher plants, pericycle occurs uniformly.

Functions. In roots, pericycle gives rise to **lateral roots** and since it is meristematic in nature (in this case), it has also been referred to as **pericambium.** In stem, during secondary growth, **phellogen** or cork cambium arises from it and in certain cases where anomalous secondary growth is found, it forms the **secondary cambium.** In monocots, where secondary growth is absent, pericycle becomes sclerenchymatous in nature and thus provides mechanical strength to the stem. In normal cases, the parenchymatous cells of the pericycle serve for storage of food material, and only in certain cases, these may contain laticiferous tissues and secondary ducts.

3. Pith

Pith is the main, internal, more or less cylindrical ground tissue which forms the central core of the stem and which externally remains bounded by the vascular bundles. Normally, it consists of long isodiametric and loosely arranged normal parenchymatous cells with large number of intercellular spaces. These remain longitudinally arranged because of their development from the rib meristem. In young and delicate stems, the pith cells may contain chloroplasts, but in mature pith cells, leucoplasts are found abundantly together with starch, fatty substances, tannins and crystals. Sometimes, secretory tissues, fibres, sclereids etc. may also be present. Thus, pith serves as an organ concerned with food storage.

In dicotyledonous stems, the pith is normally large and well developed except in certain cases such as *Cucurbita* stems, where it is hollow. The hollow pith is found in several herbaceous dicots, where it is destroyed during rapid elongation coupled with sudden radial expansion. In such cases, the broken walls lining the cavity may be seen. In monocotyledonous stems, pith is not distinct and the entire tissue is called ground tissue in which vascular bundles remain scattered. Pith is absent in many dicotyledonous roots, whereas it is found well developed, often sclerenchymatous, in monocot roots. In some plants, the outerpart of the pith consists of smaller cells with dense cytoplasm and thick walls and becomes morphologically distinct from rest of the pith. This region of pith is then referred to as **premedullary zone** or **medullary sheath.**

Functions. Pith, where it is present, basically serves as a region for storage of various substances such as starch, fatty substances, mucilage, tannin etc.

4. Medullary rays

The vascular bundles remain separated due to the presence of long strips of parenchymatous tissues passing through in between them. Such strips are called medullary rays or pith rays. These are actually considered as extension of pith cells occupying interfascicular regions. In roots, medullary rays are not distinct. Due to their origin from primary meristems, these are referred to as **primary medullary rays.** During secondary growth, some cells of these rays become meristematic and produce **secondary medullary rays.** The chief function of medullary rays is to help in conduction of food and water from cortical region to pith or central region.

Ground tissues of leaves

The ground tissue system of leaves is commonly called **mesophyll tissues,** which is remarkably different from that of stem and roots. It shows variation in different types of leaves. In **isobilateral leaves** (monocotyledonous leaves with parallel venation), mesophyll tissues consist of more or less isodiametric, parenchymatous cells with intercellular spaces. In **dorsiventral** (dicotyledonous leaves) and concentric leaves, having reticulate venation, mesophyll tissues are found differentiated into two types of cells, as follows:

(a) **Palisade cells**–These are usually found beneath the upper epidermis (adaxial surface) and are elongated or columnar shaped, arranged more or less at right angles to the epidermis. However, in certain cases such as *Nerium* (oleander) and *Eucalyptus,* palisade parenchyma are present below the upper and lower epidermis.

(b) **Spongy cells**–These occur towards lower epidermis (abaxial surface) and are characterised by isodiametric or irregular shaped cells with large number of intercellular spaces.

The mesophyll cells contain chloroplasts and perform photosynthetic function. In certain xerophytic or succulent plants such as *Bryophyllum, Agave, Aloe* etc., these cells store water and food substances in soluble forms. In some cases, sclereids and secretory cells are also found in this region. In aquatic and marshy plants, mesophyll tissue consists of aerenchyma which in addition to photosynthesis, provide buoyancy to the leaves.

III. THE VASCULAR TISSUE SYSTEM

Vascular tissue system consists of complex tissues, the *xylem* and the *phloem* which together make a unit strand called **vascular bundle.** Vascular bundles remain distributed in stelar region. A **stele** is considered as a central cylinder of root, stem or leaf which consists of vascular bundles, pith and pith rays enclosed in a pericycle. The pericycle remains surrounded by endodermis.

Origin. The elements of vascular bundle arise from procambium of apical meristem. During longitudinal divisions in the apical meristem, some cells are distinguished from other cells due to presence of dense cytoplasm in them. These cells are collectively called procambium which soon gets differentiated into two parts, the **phloic procambium** which forms phloem and the **xyloic procambium** which forms xylem.

The phloem elements mature earlier and these develop only in acropetal manner but xylem elements develop both in acropetal and basipetal directions. The phloem elements always differentiate centripetally, i.e. towards the centre of the axis. However, differentiation in xylem shows three different patterns:

(a) *Exarch*–In this pattern, the course of development is centripetal or towards the axis and thus, the initial xylem elements are located far from the axis. Exarch xylem is characteristic of roots.

(b) *Endarch*–In this pattern, the course of development is centrifugal and thus, the initial xylem elements are nearer the axis, i.e. towards the periphery. The youngest xylem elements are towards the centre. Endarch xylem elements are characteristic of the stem of phanerogams.

(c) *Mesarch*–In this pattern, the course of development proceeds in two directions, i.e. both centrifugally and centripetally, and thus, it leads to a condition where protoxylem is found in the centre and metaxylem towards its periphery. Mesarch xylem is characteristic of rhizome and rachis of certain ferns.

Elements of vascular bundles

A typical primary vascular bundle of the dicot stem consists of three types of elements, (i) primary xylem, (ii) primary phloem and (iii) cambium.

Primary xylem. Primary xylem is composed of **protoxylem** and **metaxylem** elements. Protoxylem are the first formed xylem and metaxylem are lately formed. Protoxylem differentiates from procambium when the plant organ is subjected to elongation and stretching. It consists of tracheids, vessels and parenchyma. The fibres are usually absent. The secondary wall layers are deposited in the form of rings (annular) and spirals which provide mechanical strength to the cell and thus prevent a collapse during elongation. So, annular and spiral, rarely scalariform, are characteristic thickenings of protoxylem. Sometimes, protoxylem due to stress and rapid elongation gets destroyed leaving a canal like body called *protoxylem lacuna* as found in some monocot stems.

The metaxylem elements are characterised by the presence of thickenings in the form of network (reticulate) or pits in their secondary walls. Rarely, scalariform or ladder like thickening may also occur. These are not subjected to much strains or stretching as these mature only when organs have reached maturity in length.

The type of primary xylem, whether exarch, endarch or mesarch is determined by the position of protoxylem in respect to metaxylem. If protoxylem is towards the outer side, i.e. circumference and metaxylem towards the centre, it shows **exarch** condition as in case of roots. In **endarch** condition, the situation is reverse, i.e. protoxylem towards centre and metaxylem towards periphery as in case of stems of phanerogams. The **mesarch** protoxylem remains flanked on two sides or found surrounded by metaxylem as in rhizome and rachis of ferns.

Primary phloem. Primary phloem consists of protophloem and metaphloem, both recognised on basis of differentiation. The first formed phloem elements are called **protophloem** while lately formed are called **metaphloem.** The elements of protophloem are slender and elongated, with their cell walls made up of cellulose, as these develop under stress due to rapid elongation of the organ. The metaphloem elements differentiate after the elongation phase of the organ is completed, so these are not subjected to elongation.

The protophloem consists of sieve tubes in angiosperms and sieve cells in gymnosperms and pteridophytes. Companion cells are rare or altogether absent. Metaphloem elements are complex in nature and consist of sieve tubes or cells, companion cells, parenchyma and fibres.

In monocots and in certain dicots such as *Cucurbita,* the primary phloem remains persistent throughout the life of plant organ, as secondary growth is rarely found in them and hence secondary phloem is not developed. In other cases, where secondary growth is found, it takes the function of primary phloem which often gets crushed.

The phloem is usually found on the outer side (external) of xylem of vascular bundles in stem and on the abaxial side in case of leaves and leaf-like organs, but in members of Solanaceae, Cucurbitaceae, Compositae, Apocynaceae, Asclepiadaceae and Convolvulaceae of dicotyledons, a part of phloem may be internal in addition to external. The **internal phloem** is present on the inner side of xylem and is normally called **intraxylary phloem.** Both types of phloem are structurally similar but internal phloem differentiates later than the external phloem. Yet another type of phloem, often referred to as **interxylary phloem** or **included phloem** is found in members of families like Combretaceae, Loganiaceae, Acanthaceae etc. but it is a secondary phloem formed due to abnormal activity of cambium (It is discussed in a separate chapter).

Cambium. Cambium occurs in the form of a thin strip of cells between xylem and phloem, in vascular bundles of stems of dicots and gymnosperms. It is a part of procambium or primary meristem and is classified as lateral meristem occurring parallel to the plant axis. It consists of two types of cells: (i) the elongated cells with tapering ends called **fusiform initials** and (ii) small, more or less isodiametric cells called **ray initials.** The cambium by its meristematic activities produces secondary tissues and brings about secondary growth. The fusiform initials form secondary xylem and secondary phloem, whereas ray initials form ray cells, found in xylem and phloem, which occur in transverse or horizontal series.

The cambium consists of one to few layers of rectangular or roughly brick-shaped cells. The cell wall is thin and cellulosic in nature which surrounds highly vacuolated cytoplasm with a large nucleus. The cell wall often contains primary pit fields.

Transfusion tissues. Transfusion tissues are found in leaves of certain gymnosperms, e.g. *Cycas* and *Pinus.* These are special types of conducting tissues found present around the vascular region. In *Pinus* needle (leaf), these are of two types, (i) the transfusion parenchyma cells and (ii) transfusion tracheids. **Transfusion parenchyma** are thin walled and non-lignified and contain dense cytoplasm and a prominent nucleus. These frequently occur near the phloem region. **Transfusion tracheids** are short, tracheid-like cells with lignified walls having bordered pits, scalariform or reticulate thickenings. These do not contain protoplasm, hence non-living and are found adjacent to xylem. The nature of origin of transfusion tissue is disputed. Some believe them to be derived from centripetal xylem, while others consider them as metamorphosed parenchyma located outside the vascular region. Their function is also uncertain, probably these translocate water or food substances from mesophyll to vascular bundle or vice versa. Some consider them as water storage tissue.

Types of vascular bundles

According to the mode of occurrence of vascular elements, vascular bundles are classified as follows:

1. **Radial.** In this type, the xylem and phloem are found as separate patches on alternate radii. These are most primitive in nature and are characteristic of roots (Fig. 6.12, F and 6).

2. **Conjoint.** In this type, two complex tissues (xylem and phloem) occur side by side on the same radius. According to the position of phloem, these are recognised into two types, (a) Collateral and (b) Bicollateral. The **collateral** V.B. are most common type found in stem and leaves of phanerogams. In this case, xylem and phloem lie together on the same radius, the phloem being external and xylem being internal (towards pith) in position. The collateral V.B. are referred to as **open type** if a strip of cambium between xylem and phloem is present as in case of dicot stems, and **closed type,** if cambium is not found, as in monocot stems. In **bicollateral** V.B., two phloem patches, one on outer side (external) and other on inner side (internal) of xylem, are found, as in stems of members of Cucurbitaceae. In these vascular bundles, the sequence would be outer phloem, outer cambium, xylem, inner cambium and inner phloem. These are always open type (Fig. 6.12).

3. **Concentric.** In this type, one complex tissue remains completely surrounded by the other. These are of two types, (i) **amphivasal** or **leptocentric** bundles, where phloem is surrounded by xylem, such as secondary vascular bundles of *Dracaena, Yucca* and medullary bundles in dicotyledons and (ii) **amphicribal** or **hadrocentric,** where xylem is surrounded by phloem, as in rhizomes of ferns, stems of *Lycopodium* and *Selaginella* and small bundles of flowers, fruits and some leaves of dicotyledons. The concentric bundles are always closed.

These are common types of vascular bundles but there can be several variations. The presence of Y-shaped arrangement of xylem in monocotyledonous stems, cortical vascular bundles of Oleaceae and Cactaceae and medullary bundles of Piperaceae and Amarantaceae are some examples of variations.

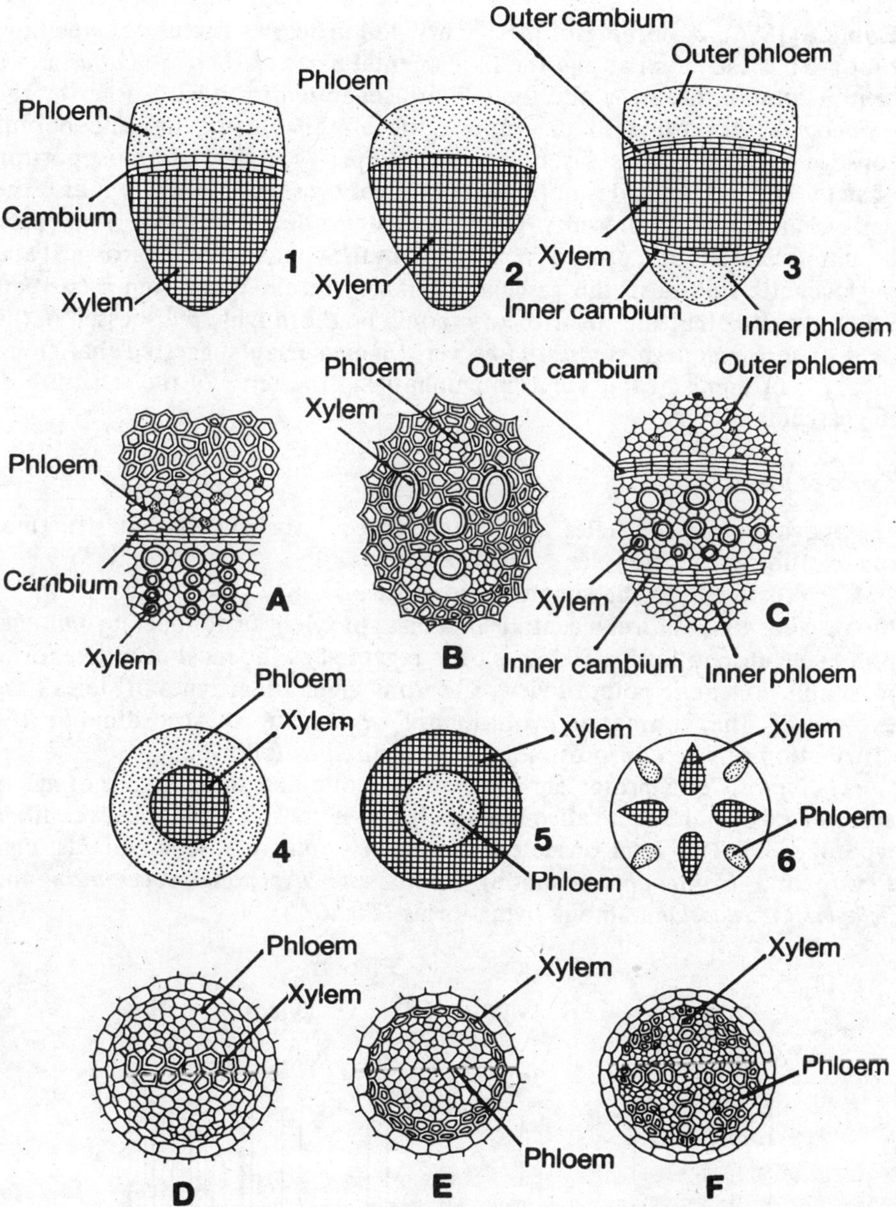

Fig. 6.12. Various types of vascular bundles. 1 and A–conjoint, collateral and open vascular bundle, 2 and B–conjoint collateral and closed V.B., 3 and C–conjoint, bicollateral and open V.B., 4 and D–concentric and amphicribal V.B., 5 and E–concentric and amphivasal V.B. and 6 and F–radial V.B.

Concept of stele

At one time, the vascular bundle was considered as fundamental unit in the vascular system. Later, a *stelar theory* was proposed by Van Tiegham and

Douliot (1886). According to this theory, the primary structure of stem and root are basically similar and the fundamental parts of each are *cortex* and a central column, the *stele*, and the two are separated from each other by the *endodermis*. According to the proponents of this theory, the endodermis constitutes the innermost layer of cortex, and pericycle the outermost portion of stele. The stele includes not only the primary vascular tissues, but also the pericycle and pith (if present). The idea of stelar theory was widely accepted by many scientists and proved to be useful in the comparative anatomical and phylogenetic studies of the vascular plants. In recent years, some scientists have suggested that the stelar theory should be thoroughly rediscussed in the light of some modern studies. However, Madam Esau suggested that stelar theory is of unmistakable value in emphasising the unity of the structure of the vascular system.

Types of steles

Following types of steles are recognised on basis of different structural organisation of the complex vascular elements.

1. **Protostele.** It is the simplest and most primitive type of stele in which the vascular tissues form a central solid mass of xylem surrounded by phloem, pith being absent. Such a stele has been regarded as the most primitive form, from a phylogenetic point of view, and from which other types of steles have evolved in the course of evolutionary organisation. According to the distribution of xylem, protosteles are recognised as following.

(a) *Haplostele*–A protostele with central, solid and smooth core of xylem surrounded by phloem is called **haplostele** and is considered as most primitive type of stele. This kind of stele is found in primitive vascular plants, e.g., *Rhynia* and *Horneophyton* among fossils and *Lycopodium cernuum* and *Selaginella kraussiana* among living forms (Fig. 6.13).

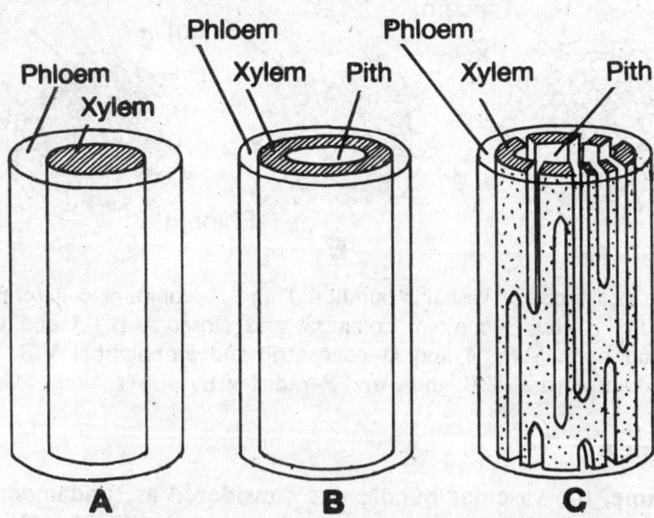

Fig. 6.13. Various types of steles. A–protostele, B–siphonostele and C–dictyostele.

(b) *Actinostele*–In this type, the central xylem core is not smooth, but consists of radiating ribs with phloem alternating with its rays (Fig. 6.14). It is found in *Psilotum triquetrum* and *Lycopodium serratum.*

(c) *Plectostele*–It is the most advanced type of protostele in which xylem and phloem intermingle and xylem occurs as separate plates usually lying parallel to one another (Fig. 6.14). It is found in *Lycopodium clavatum* and *L. volubile.*

(d) *Mixed protostele*–In certain cases, xylem appears in the form of irregular patches embedded in the ground mass of phloem, as in *Lycopodium cernuum.*

Protosteles are common in lower vascular plants and some aquatic angiosperms.

2. **Siphonostele.** A kind of stele in which pith is present in the central region surrounded by xylem is called *siphonostele* or *medullated protostele.* This type of stele is said to have evolved from the protostele and represents a stage in evolutionary advancement. In a siphonostele, the vascular tissues are arranged in the form of a hollow cylinder, with a distinct pith in the centre. This type of stele is commonly found in ferns, gymnosperms and angiosperms.

There is a general acceptance that the siphonostele has evolved from a protostele. Two theories have been proposed accounting for the phylogenetic origin of pith, the intrastelar theory and the extrastelar theory. **Intrastelar theory,** which is supported by Boodle (1901), Gwynne-Vaughan (1908), Bower (1911) and others, holds that pith originates by metamorphosis of inner vascular elements (such as xylem) into parenchyma. Support for this theory is furnished by a number of plants like *Botrychium virginianum, B. lunaria* and *Osmunda regalis* in which central region of protostele consists of both tracheids and parenchyma cells. According to **extrastelar theory,** which was put forth by Jeffrey (1897), the pith is extra stelar in origin. The theory holds that the pith originated as a result of the invasion of cortical tissues into the stele through the leaf gaps and branch gaps in the course of phylogenetic development of the vascular plants. The pith is therefore cortical in origin. This theory may also be called **invasion theory.**

According to the distribution of vascular tissues, siphonosteles are recognised into two types–(a) Ectophloic siphonostele and (b) Amphiphloic siphonostele.

In **ectophloic siphonostele,** the xylem is in the form of a hollow cylinder surrounding the pith, the phloem occurring only outside the xylem as in *Equisetum, Osmunda* and *Schizaea.* In **amphiphloic siphonostele,** the xylem forms a hollow cylinder enclosing the pith with the phloem occurring on both the inner and outer side of the xylem (Fig. 6.14) as found in *Adiantum* and *Marsilea.*

In the lower vascular plants, the leaf gaps are absent as in case of *Lycopodium* and *Selaginella* where stele shows a continuous ring of tissues. A siphonostele which has no leaf gap is termed as **cladosiphonic siphonostele.** In ferns and higher plants, leaf gaps are, however, found and then such a siphonostele is called **phyllosiphonic siphonostele.** According to position of

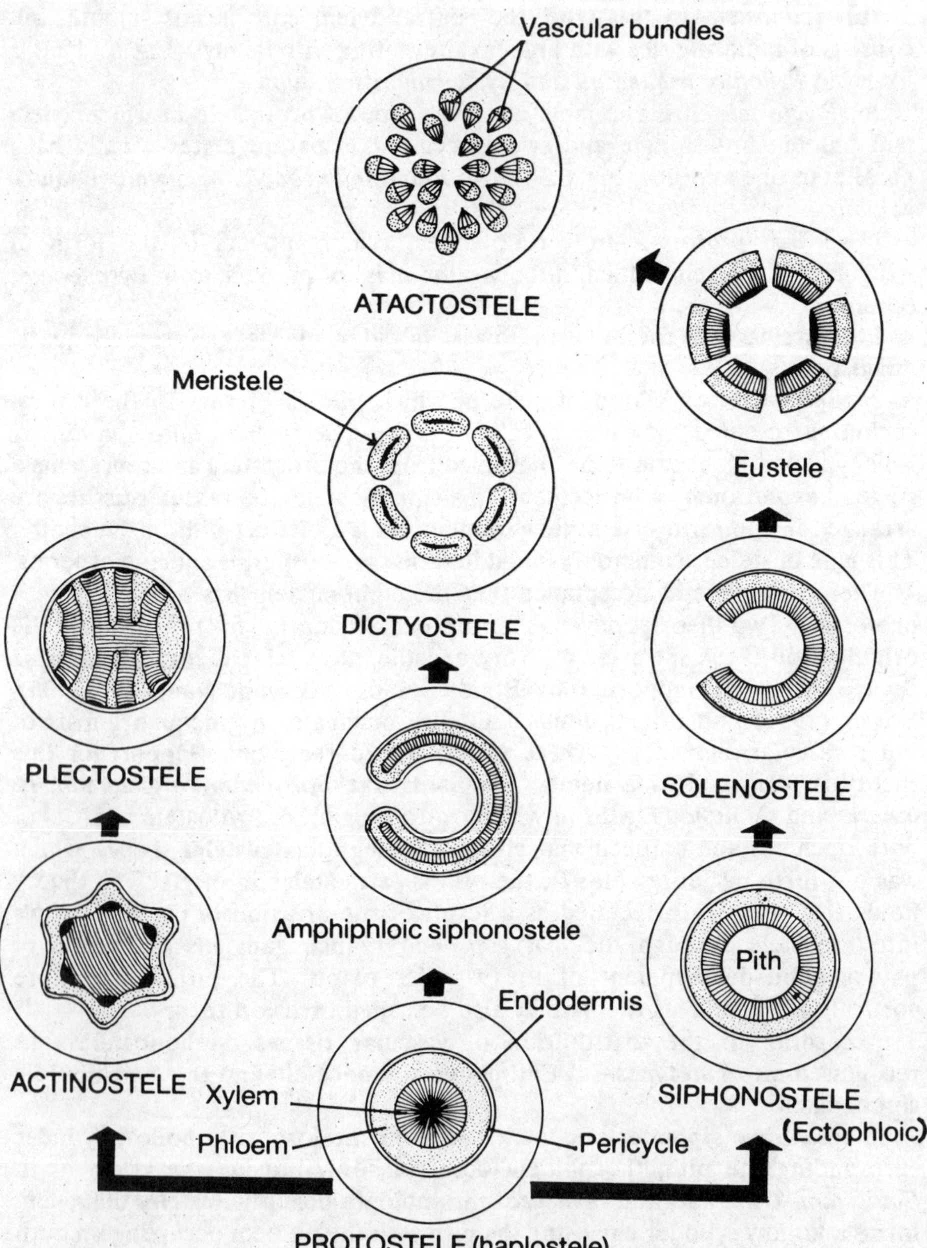

Fig. 6.14. Various types of steles showing evolutionary sequence.

leaf gaps, two types of phyllosiphonic siphonosteles are recognised, the **solenostele** and the **dictyostele**. In **solenostele,** overlapping leaf gaps are not found as in *Marsilea* and some ferns and in a cross section, it shows a continuous ring of vascular tissues. **Dictyosteles** are characterised by the presence of large number of overlapping leaf gaps, so that the whole stele is dissected into a net-like structure. In cross section, dictyostele shows separate vascular strands, each called **meristele.** Each meristele is a miniature protostele which is concentric in structure, consisting of a central strip of xylem surrounded by phloem. Such a stele is characteristic of fern rhizome (*Pteris* and *Pteridium*) and rhizome and stems of dicotyledons and gymnosperms. In some pteridophytes, the gaps occur in the stele which are not associated with leaf traces. Such gaps are called perforations and the perforated stele is called a dictyostele. Some modern workers prefer to use the term **eustele,** meaning true stele, for dissected siphonostele (dictyostele). The stele found in stems of dicotyledons and gymnosperms is eustele which may also be referred to as **polyfascicular siphonostele.** In most stems of monocotyledons, the vascular bundles remain scattered in the ground tissues, so that collective appearance of stele is lost. This complex type of stele is called **atactostele** (*atactos* means without any order).

In certain pteridophytes, complex type of stelar structure is seen in which two or more concentric rings of vascular tissues are present. This type of stele is known as **polycyclic stele.** In *Pteridium aquilinum,* there are two concentric rings of vascular tissues, in which the inner ring forms a siphonostele and the outer ring forms a dictyostele consisting of numerous meristeles. In *Matonia pectinata,* there are three rings of vascular tissues.

GENERAL QUESTIONS

1. Write short notes on: (i) Epidermis, (ii) Ground tissue and (iii) Vascular organisation.
2. What is stele? Describe different types of steles found in plants.
3. Write a brief account on special structures found in epidermis.
4. Write short notes on (i) Stomata, (ii) Trichomes, (iii) Siphonostele, (iv) Dictyostele, (v) Primary phloem, and (vi) Primary xylem.

CHAPTER 7

The Stem–Primary Structure

There are two important organs of a vascular plant, the shoot and the root. The shoot consists of axis (stem) and its lateral appendages called leaves. The anatomical features of all these, i.e. stem, root and leaves are complicated and are discussed here in separate chapters. In this chapter, anatomical features of stem, as seen in its primary structure is accounted.

Both stem and roots have several characteristics and distinguishing features that help in identifying them. (Table 7.1)

Table 7.1. Characteristics of Root and Stem.

Root	Stem
A. Morphological Characters	
1. Roots are misty in colour and of irregular shape.	1. Stems are usually green and straight.
2. Nodes and internodes are not distinct.	2. Nodes and internodes are well marked.
3. Spines or thorns are not found.	3. Spines, thorns and marks of leaf bases are present.
B. Anatomical Characters	
4. The epidermis possesses unicellular hairs.	4. Hairs, if present are multi-cellular.
5. Cuticle layer and stomata are absent.	5. Cuticle layer is present and stomata may also be present.
6. Epidermis takes the function of water absorption from soil.	6. Epidermis is only protective in nature.
7. Hypodermis is not differentiated.	7. Hypodermis is well marked.
8. Cortex is broad and well developed.	8. Cortex is relatively less developed.
9. Endodermis is quite distinct. Endodermal cells have radially thickened walls intermingled with few, thin walled passage cells.	9. Endodermis is indistinct and passage cells are not found.
10. The type of stele is actinostele.	10. Stele is polyfascicular siphonostele in dicots and atactostele in monocot.

Root	*Stem*
11. Pericycle is single layered and composed of parenchymatous cells.	11. Pericycle is several layered thick and usually consists of sclerenchymatous cells.
12. The vascular bundles are of radial type.	12. The vascular bundles are of collateral type.
13. Xylem is exarch in nature.	13. Xylem is endarch in nature.
14. Lateral branches arise endogenously, i.e., from pericycle.	14. Lateral branches arise exogenously.

As regards, anatomical features, stems of dicotyledons and monocotyledons show remarkable variations. Some comparative characteristics of both these types are accounted here (table 7.2).

Table 7.2. Characteristics of dicot and monocot stems.

Dicot Stem	*Monocot Stem*
1. Several multicellular hairs usually arise from epidermis in dicot stem.	1. Epidermal hairs are usually not found.
2. Hypodermis consists of collenchymatous cells.	2. Hypodermis is usually sclerenchymatous.
3. Cortex is distinct.	3. Hypodermis and cortex are not well distinguishable and ground tissue is present next to hypodermis.
4. Endodermis and pericycle are distinctly present.	4. Endodermis and pericycle are not present.
5. Vascular bundles are arranged in a ring.	5. Vascular bundles are found scattered in the ground tissue.
6. Vascular bundles are conjoint, collateral and open type.	6. Vascular bundles are conjoint, collateral and closed type.
7. Sclerenchymatous bundle sheath is not found.	7. Each vascular bundle remains surrounded by sclerenchymatous sheath.
8. In vascular bundles, several vessels are present and are linearly arranged.	8. Vessels in each vascular bundle are arranged in V-Shape and are fewer in number.
9. Phloem parenchyma is found.	9. Phloem parenchyma is absent.
10. Lysigenous cavity in vascular bundle is absent.	10. Lysigenous cavity in each vascular bundle is present.
11. Both pith (medulla) and medullary rays are present.	11. Medulla and medullary rays are absent.
12. Secondary growth is found.	12. Secondary growth is not found.

ANATOMY OF DICOTYLEDONOUS STEMS

Anatomical features of dicotyledonous stems show considerable variations and therefore stem of *Helianthus annuus* is considered as a type example.

Stems with abnormal features and secondary growth are discussed separately. For general study, the young stems of sunflower (*Helianthus annuus*) and *Cucurbita* are taken. Transverse and longitudinal sections are cut and suitably stained for the study of internal structure.

I. Young Stem of Sunflower

A section of young stem of sunflower (*Helianthus annuus*) reveals following important structures:

1. **Epidermis.** It is the outermost single layer composed of compactly arranged tubular parenchymatous cells. The cells are attached end to end along their radial walls without intercellular spaces and appear rectangular in cross section. Cells are living with vacuolated protoplasm which is usually without chloroplasts. The outer wall is cuticularised to check the loss of water. A large number of multicellular epidermal hairs and some times a few stomata are found on the epidermis. Stomata are ranunculaceous type.

2. **Ground tissue.** Ground tissue is differentiated into two distinct zones, the hypodermis and the cortex.

(i) **Hypodermis.** Hypodermis is found immediately below the epidermis which consists of a few layers of collenchymatous cells showing angular thickness. Hypodermis forms a continuous band providing mechanical support to the growing stem. Collenchymatous cells may have a few chloroplasts.

(ii) **Cortex.** It is found next to hypodermis and consists of loose parenchymatous cells with fairly large number of intercellular spaces. The cells are thin walled, oval or rounded and living with distinct nucleus and cytoplasm. Some isolated mucilaginous canals lined with epithelial cells are found in this region. The cells of cortex may contain some chloroplasts and thus may function as assimilatory cells. Cortical cells mainly serve for food storage.

(iii) **Endodermis.** The innermost layer of cortex which separates cortex from stele is the endodermis. It is a kind of wavy band made of compactly-set barrel-shaped parenchymatous cells. Since these cells are rich in starch grains, the endodermis is some times referred to as **starch sheath.** Their radial and inner walls are thickened due to deposition of lignin and suberin forming casparian strip, a characteristic feature of endodermis.

3. **Stele.** It includes pericycle, vascular bundles, pith and medullary rays. The stele is polyfascicular siphonostele or dissected siphonostele.

(a) **Pericycle.** Pericycle is found just below the endodermis and is composed partly of parenchymatous and partly of sclerenchymatous cells. Sclerenchymatous tissues are found in the form of crescent shaped patches outside the vascular bundles. This has also been called *bundle cap.* Recent ontogenetic studies have revealed that these sclerenchymatous patches or fibres belong to phloem and not to pericycle, hence the term **hard bast** is also used to refer these patches. However, these provide mechanical support to the plant parts.

(b) **Vascular bundles.** The vascular bundles are arranged in a ring, each

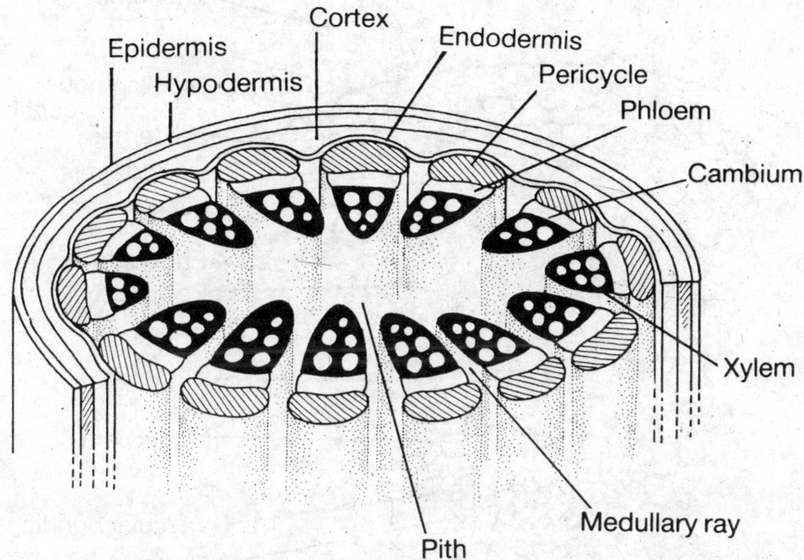

Fig. 7.1. Diagrammatic representation of T.S. of stem of sunflower (*Helianthus annuus*).

vascular bundle being wedge-shaped and typically conjoint, collateral and open type. Each bundle has a patch of xylem towards centre, a patch of phloem towards the periphery and a strip of cambium in between them.

(i) *Xylem*–It lies inner to the cambium and consists of tracheids, vessels, fibres and parenchyma. The tracheids are found around the vessels and lying in between. Vessels are arranged in radial rows, with protoxylem lying towards centre and metaxylem towards cambium, i.e., endarch arrangement of xylem. The protoxylem consists of annular and spiral or some times, reticulate vessels and metaxylem has reticulate and pitted vessels. Xylem fibres are irregularly intermixed while parenchyma are found around this protoxylem.

(ii) *Phloem*–Phloem is found just below the sclerenchymatous patches and consists of sieve elements, companion cells and phloem parenchyma. The sieve elements appear as large cavities and dispersed in isolated manner. Companion cells are relatively smaller in size and are found associated with sieve tubes. The remaining cells of phloem are parenchyma; the bast fibres being absent.

(iii) *Cambium*–It usually consists of number of layers of meristematic cells and lies inner to the phloem. The cells are thin walled and appear rectangular or brick-shaped in cross-section. This cambium is called intra-fascicular or fascicular type (*fascicle*-bundle) as it is embedded in the vascular bundle. Cambium is a type of lateral meristem.

(c) **Pith.** Pith occupies the centre part of stem which consists of parenchymatous cells. The cells are rounded or polygonal, thin walled with several inter-cellular spaces. Cells of pith store food material.

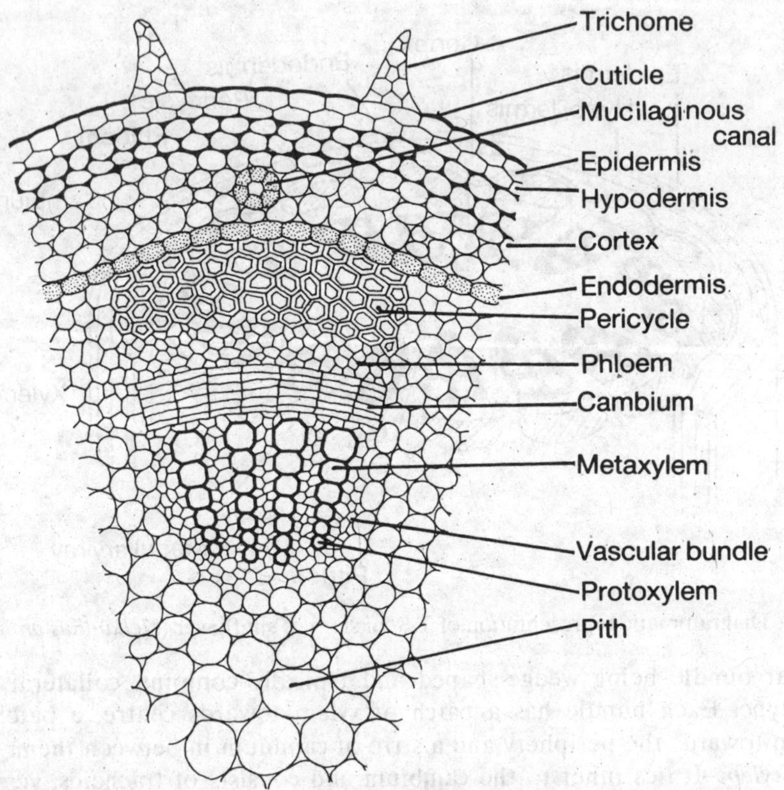

Fig. 7.2. A cellular enlarged portion of T.S. stem of sunflower.

(d) **Medullary rays.** These are also called pith rays. The rays form elongated structure which extends in between the two vascular bundles. These are composed of thin walled, radially elongated parenchymatous cells which participate in lateral translocation of food, water and other substances and also some times in storage of food material.

II. Young Stem of Cucurbita

Cucurbita maxima is a weak stemmed plant which climbs with the help of tendril. Following are important internal structure of a young stem.

1. **Epidermis.** It is the outermost single layer composed of thin walled compactly arranged tubular parenchymatous cells without intercellular spaces. The outer surface of the cells remains cutinised. Numerous unbranched multicellular hairs arise from epidermis. Stomata may be present. Cells are living and contain vacuolated protoplast. The epidermis is not circular but is wavy, showing prominent ridges and furrows.

2. **Hypodermis.** It lies next to epidermis and consists of six to eight layered collenchymatous patches below the ridges and one to four layered parenchymatous patches below furrow regions. The parenchyma often contain

chloroplast, hence are photosynthetic in nature but collenchyma remain more concerned with providing protection and mechanical support.

3. **Cortex.** It lies just below hypodermis and consists of two to three layered thick parenchymatous narrow zone around the central stele. Chloroplasts may be found in these cells which carry out the process of food manufacture. Cortical cells usually store the food material.

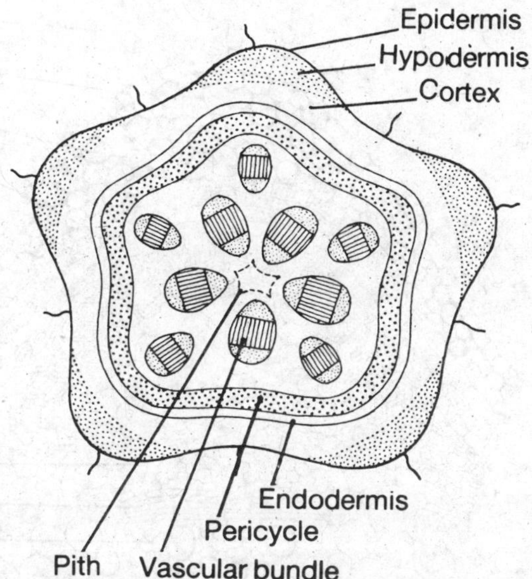

Fig. 7.3. T.S. of *Cucurbita* stem (diag.).

4. **Endodermis.** The innermost layer of cortex is popularly called endodermis which is a uniseriate layer composed of barrel-shaped cells containing starch grains. That is why it is some times also referred to as starch sheath.

5. **Stele.** Stele mainly includes pericycle, vascular bundles, pith and intrastelar ground tissue. It is polyfascicular siphonostele type.

(a) **Pericycle**–Pericycle consists of few layers of sclerenchymatous cells found on the innerside of the starch sheath and is in the form of continuous band. Some parenchymatous cells are also found between vascular bundles and sclerenchymatous band and these are also considered as a part of pericycle.

(b) **Vascular bundles**–In all ten vascular bundles are found in the stem which are organised into two rows. Vascular bundles of outer row are smaller and occur below the ridge regions while those of inner rows are larger and are placed in furrow regions. Vascular bundles are conjoint, bicollateral and open type having endarch xylem. Sequence of tissue in vascular bundle is outer pholem, outer cambium, xylem, inner cambium and inner phloem. The outer phloem is relatively better developed and consists of sieve tubes, companion cells and phloem parenchyma. Sieve tubes are well developed and show perforated sieve plates here and there. The outer phloem is somewhat convex and inner phloem has semi-lunar shape. The phloem parenchyma are small and thin walled.

In xylem, metaxylem vessels are of quite large size which shows characteristics of a climber. They have pitted thickenings. Protoxylem vessels are narrower and are found towards centre. They have usually spiral and annular thickenings. The tracheids and fibres are either few or lacking but xylem parenchyma are abundant. Vessels are not found radially arranged.

Cambium occurs on either side of the xylem and are accordingly called

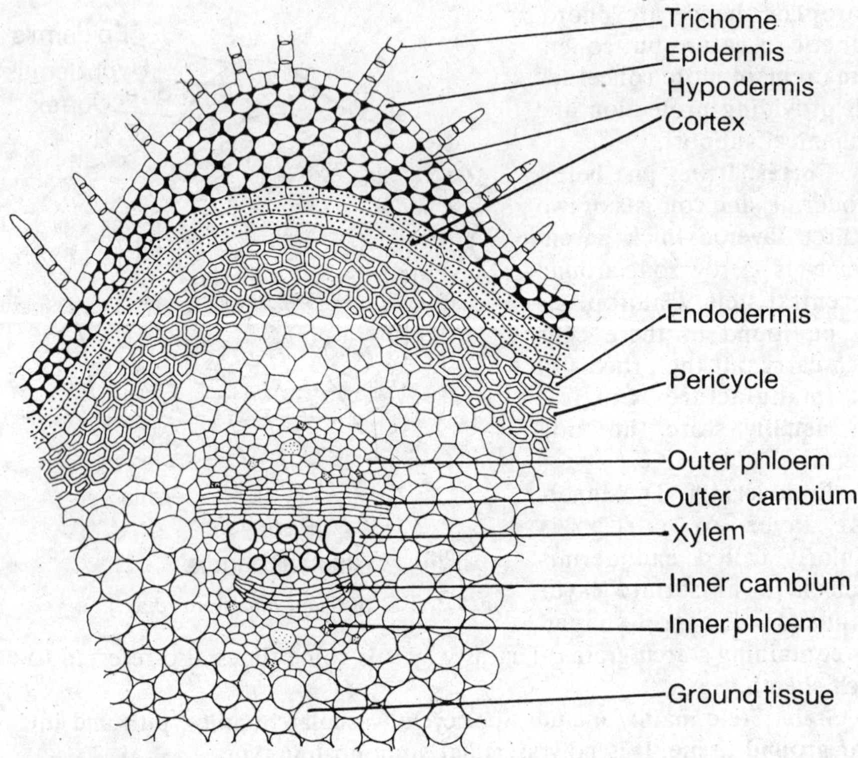

Fig. 7.4. T.S. stem of *Cucurbita* (an enlarged cellular position).

outer and inner cambium. The outer cambium is many layered while inner consists of only few rows of cells.

In young stem, pith lies in the centre and is composed of thin walled parenchymatous cells. In mature stem, the cells of pith are broken and disorganised leaving a hollow pith cavity in the centre. Medullary rays are parenchymatous in nature.

The vascular bundles of outer row represent the leaf traces and those of inner row correspond to true stem bundles, i.e. cauline bundles. These vascular bundles develop from procambial tissues of apical meristem. Secondary growth in climbing *Cucurbita* is restricted to individual vascular bundles and secondary vascular tissues are hardly distinguishable. In such cases, outer phloem towards pericycle consists of primary phloem and towards cambium the secondary phloem. The inner phloem remains entirely made up of primary elements. The secondary xylem lies just below the outer cambium and have much large vessels as compared to primary xylem. A continuous band of secondary wood and secondary phloem is not found in *Cucurbita*.

General Account of Dicot Stem

The stem consists of three distinguishable tissue systems, the *dermal,* the *ground tissue* and the *vascular system.*

Dermal tissues

The stem remains externally bounded by single layer of closely arranged rectangular cells, called **epidermis.** It develops from **dermatogen.** Epidermal cells in aquatic and shade loving plants are thin walled but those of plants growing under adverse conditions are narrow and thick walled. These are living cells, however, chloroplast is not found in them except in aquatic and allied plants. The outer radial and tangential walls are commonly impregnated with fatty or waxy substances, called cutin and the deposited layer is referred to as *cuticle.* At places, epidermis remains interrupted due to presence of stomata in green stems. In certain cases, epidermal cells also give rise to various types of multicellular hairs, also called *trichomes.*

Epidermis is protective in nature which protects the inner tissues from dessication, excessive heat and from microbial attacks. Due to deposition of cutin, it highly minimises the loss of water due to transpiration. It may persist through out the life of stem as in case of ephimeral and herbaceous plants or may be destroyed or sloughed off, as a result of secondary growth and formation of periderm as in woody plants.

Ground tissue system

In dicotyledonous stems, the ground tissue remains differentiated into hypodermis, cortex and pith. Such a differentiation is not found among monocotyledonous stem. **Hypodermis** in herbaceous stems is usually collenchymatous type which may be in the form of a continuous ring as in *Aristolochia* or in the form of patches in angular stems such as *Cucurbita* and *Peristrophe.* Chloroplasts may or may not be present in these cells. Their primary function is to provide mechanical strength and photosynthesis if found is a function of secondary importance.

The **cortex** is often parenchymatous and consists of thin-walled, round or oval cells with large number of intercellular spaces. Outer few layers of cortical cells may be assimilatory in nature due to presence of chloroplasts in them. The chlorenchymatous cortex is well marked in phylloclade (stem) of *Euphorbia tirucalli,* as in this case due to absence of leaves, stem takes the function of leaves. The cortex in most of the aquatic plants mainly consists of aerenchyma with a system of large intercellular spaces. In certain plants, laticiferous cells or ducts, mucilaginous canals, resin canals, tannin cells etc. may also be found.

The innermost layer of cortex is called **endodermis** which consists of cells with Casparian strip thickenings. However, in certain cases, the cells of innermost layer are rich in starch grains and then this layer is called **starch sheath.**

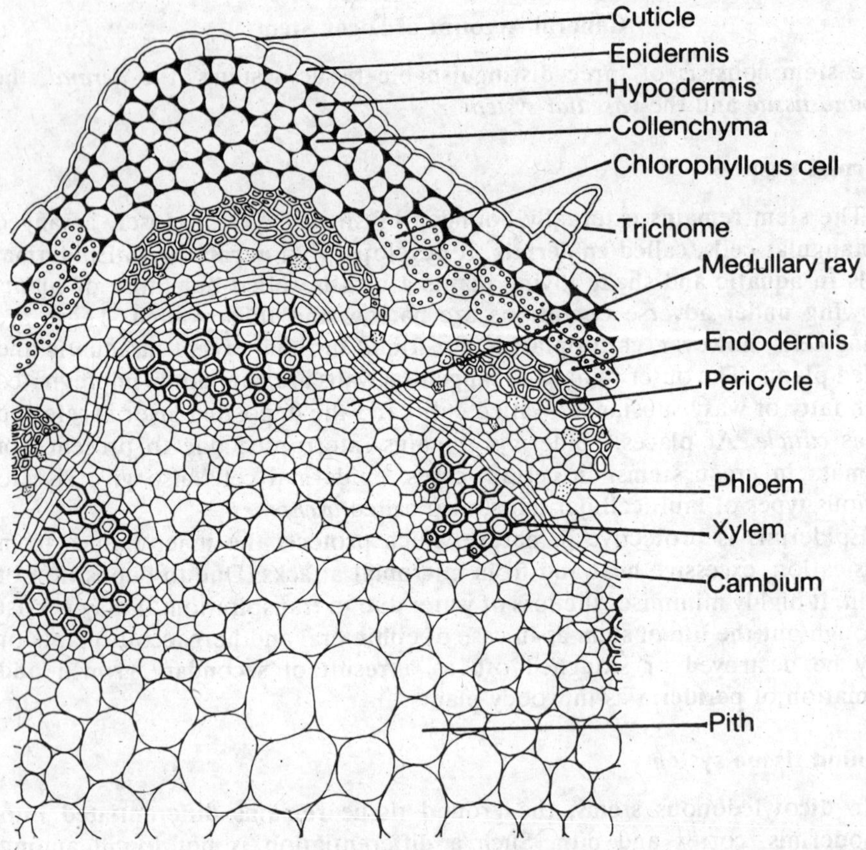

Fig. 7.5. An enlarged cellular portion of T.S. of young stem of *Medicago*.

Vascular systems

The most important constituent of vascular system is the stele, which consists of vascular bundles, but in addition, pericycle, pith, medullary rays may also be found in this system.

(a) **Pericycle**–The tissues lying inner to endodermis and outer to phloem are collectively referred to as pericycle and these are regarded as non-phloic in origin. Pericycle is distinct and well marked among dicotyledonous stem. Often it consists of sclerenchymatous fibres which may be found in the form of a solid spherical band as in *Cucurbita* or in the form of patches as in *Helianthus* and *Medicago*. Some scientists call them as perivascular fibres. Quite many scientists believe these fibres are phloic in origin. Thus, pericyclic fibres remain a matter of controversy.

(b) **Vascular bundles**–The stele in dicotyledonous stem is polyfascicular siphonostele or dissected siphonostele, consisting of vascular bundles arranged in a form of a ring and intrastelar ground tissues. The vascular bundles consists of primary phloem, cambium and primary xylem. The xylem is endarch, protoxylem lies towards centre and metaxylem towards periphery.

In young stem, the vascular tissues differentiate from procambium, also called pro-vascular meristem, in the stem apex. The primary xylem is formed centrifugally, i.e. protoxylem is formed first and metaxylem later on. The cambium found in between xylem and phloem is the persisted residual meristem which is called intrafascicular cambium.

The vascular bundles in general are **conjoint, collateral** and **open** type. However, in cucurbits, these are conjoint, bicollateral and open type. The open type of vascular bundles are capable of undergoing secondary growth. The stems are usually monostelic type but polystely is found in rhizomes of several members of Nymphaeaceae, where each stele consists of two to twenty or even more vascular bundles.

(c) **Pith**–Pith forms the central part of stem and also of the stele which remains enclosed by the vascular tissues. It usually consists of loosely arranged parenchymatous cells with abundant intercellular spaces. In certain cases, such as in *Cucurbita,* pith cells are destroyed and a hollow pith cavity is formed. The cells may store starch and sometimes waste products. Laticifers, secretory cells, sclereids, idioblasts, crystals of various kinds and ergastic substances may be found in pith. The outer few layers of pith may be composed of small thick walled cells and such a distinct outer pith is called perimedullary zone or medullary sheath.

The origin of pith is controversial and both opinions are taken as equally important. Some scientists believe, it is extrastelar while others believe that it originated from the stele, i.e. intrastelar.

(d) **Medullary rays**–These are narrow, vertically elongated bands of thin walled parenchymatous cells, traversing from pith towards periphery. In a cross section, these look like radiating bands of parenchyma separating the vascular bundles. Their main function is the lateral or transverse conduction of food material and water. Sometimes, these perform the function of food storage. Medullary rays are very prominent in *Aristolochia.*

ANATOMY OF MONOCOTYLEDONOUS STEMS

For studying the anatomy of monocotyledonous stem, *Zea mays* (maize) and *Triticum aestivum* (Wheat) are taken as type examples.

I. Stem of Maize *(Zea mays)*

Following important features are found. It also shows three distinct regions, the epidermis, the cortex and the stele.

1. **Epidermis.** It is the single outermost layer composed of small, compactly arranged, thin walled somewhat barrel-shaped parenchymatous cells. The external surface of cells contain the deposition of cutin, the layer is called cuticle. Sometimes stomata are also found but epidermal hairs are absent.

2. **Cortex.** Ground tissues do not show very distinct demarcation of hypodermis and cortex.

(a) **Hypodermis**–A few layers on the inner side of epidermis are typically sclerenchymatous and this region is called hypodermis. The sclerenchymatous cells are compactly arranged without intercellular spaces.

(b) **Ground tissue**–The region which lies below the hypodermis consists of thin walled parenchymatous tissues with a large number of intercellular spaces. It is commonly called ground tissue. There is no differentiation of general cortex, endodermis, pericycle, pith and pith rays.

3. **Stele.** Stele is **atactostelic type** in which number of vascular bundles are found irregularly scattered. The vascular bundles are conjoint, collateral and closed type. The peripheral bundles are relatively smaller and compactly set but those of central region are larger in size and widely placed. But all these have similar structure. Few bundles remain partially covered with sclerenchymatous hypodermis.

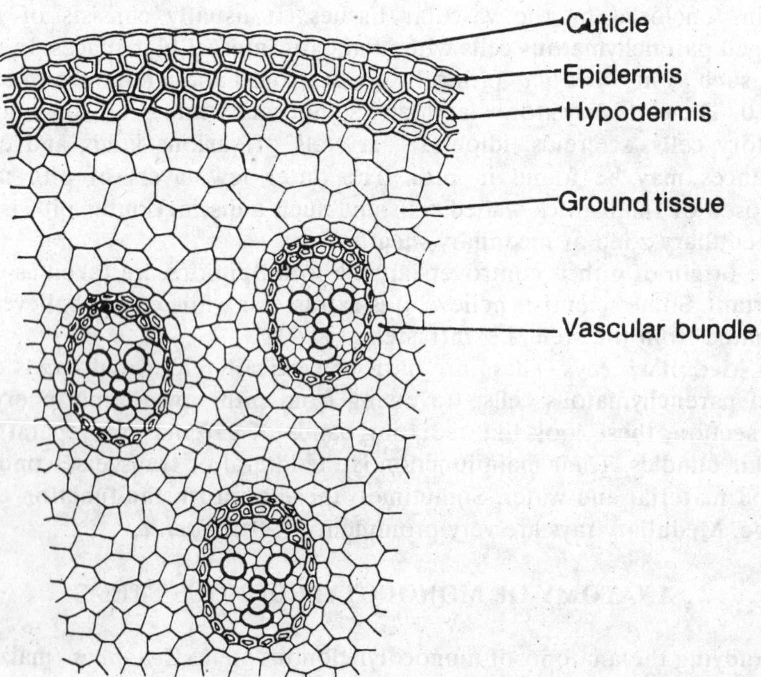

Fig. 7.6. An elarged cellular portion of T.S. stem of maize (*Zea mays*).

Vascular bundles are nearly oval in shape and are found surrounded by sclerenchymatous bundle sheath. The xylem is Y-shaped. The two metaxylem vessels with wider cavities and pitted thickening lie at the two lateral arms and protoxylem usually one or two, with narrow cavities and spiral or annular thickening at the base. Below the protoxylem vessels, there lies a large water containing cavity formed lysigenously by disintegration of lowest protoxylem and it is called protoxylem cavity or lysigenous cavity. Such a cavity is conspicuous in mature bundles. A few tracheids and xylem parenchyma lie around protoxylem vessels.

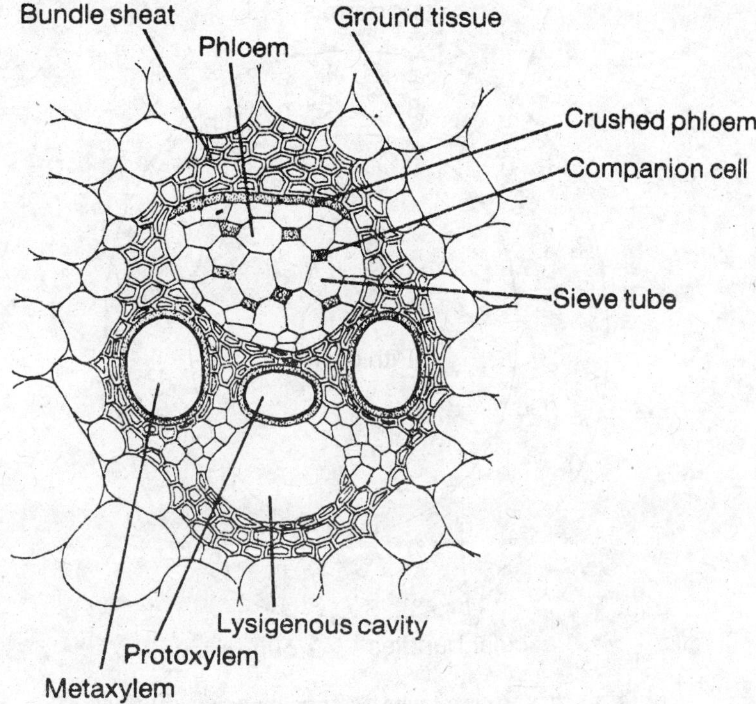

Fig. 7.7. A single enlarged vascular bundle of stem of maize (*Zea mays*).

Phloem lies outside the xylem and is rather small. It is composed of only sieve tubes and companion cells, phloem parenchyma being absent. In a mature bundle, the protophloem cells get crushed due to internal pressure so the inner phloem is the metaphloem.

II. Stem of Wheat (*Triticum aestivum*)

The cross-section of wheat stem reveals that it is hollow at the internodes and vascular bundles appears to be arranged in two rows. The following imporant features are noted.

1. **Epidermis.** It is the single outermost layer of compactly arranged barrel-shaped cells. A distinct layer of cuticle is found on its outer surface. At places, stomata are found.

2. **Ground tissue.** The sub-epidermal or hypodermal region consists of patches of sclerenchyma and chlorenchyma, so it is partly photosynthetic in nature. The remaining part of ground tissue consists of parenchymatous tissues which are somewhat loosely arranged with abundant intercellular spaces. The central portion is hollow.

3. **Stele.** It is **atactoslelic** in nature. The vascular bundles occur in two rows. Vascular bundles of outer ring remain associated or embedded in sclerenchymatous patches. The vascular bundles of second ring are relatively larger.

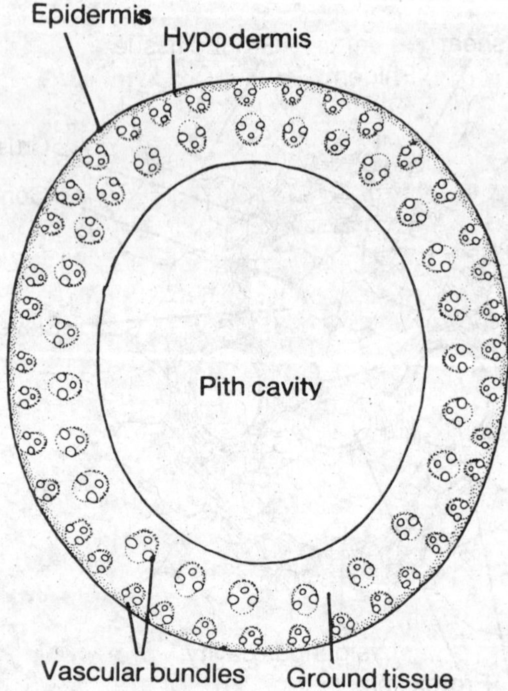

Fig. 7.8. T.S. stem of wheat (*Triticum aestivum*) (diag.).

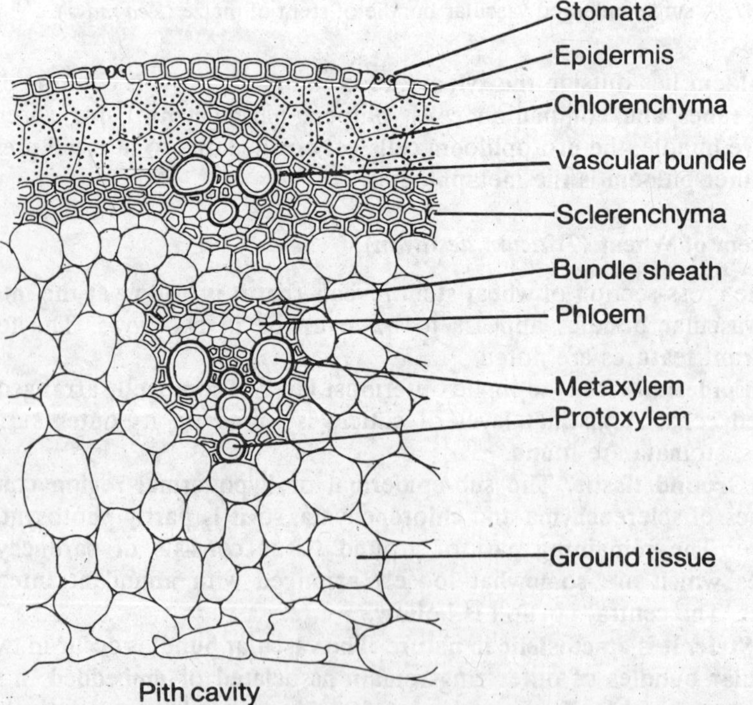

Fig. 7.9. An enlarged portion of T.S. of stem of wheat (*Triticum aestivum*).

Vascular bundles are conjoint, collateral and closed type and each remains surrounded by sclerenchymatous sheath.

General Account of Monocot Stem

The general plan of structural organisation of monocotyledonous stems is fundamentally similar to that of dicotyledonous stem. The difference between two types have been outlined in Table 7.2.

The epidermis is almost similar in construction. The ground tissue however, does not show differentiation into hypodermis, cortex and pith. Endodermis, pericycle, pith and medullary rays are absent.

Stele is atactostelic type characterised by irregularly scattered vascular bundles. In some grasses, such as *Triticum,* the central region of stem is hollow and this central cavity is called pith cavity.

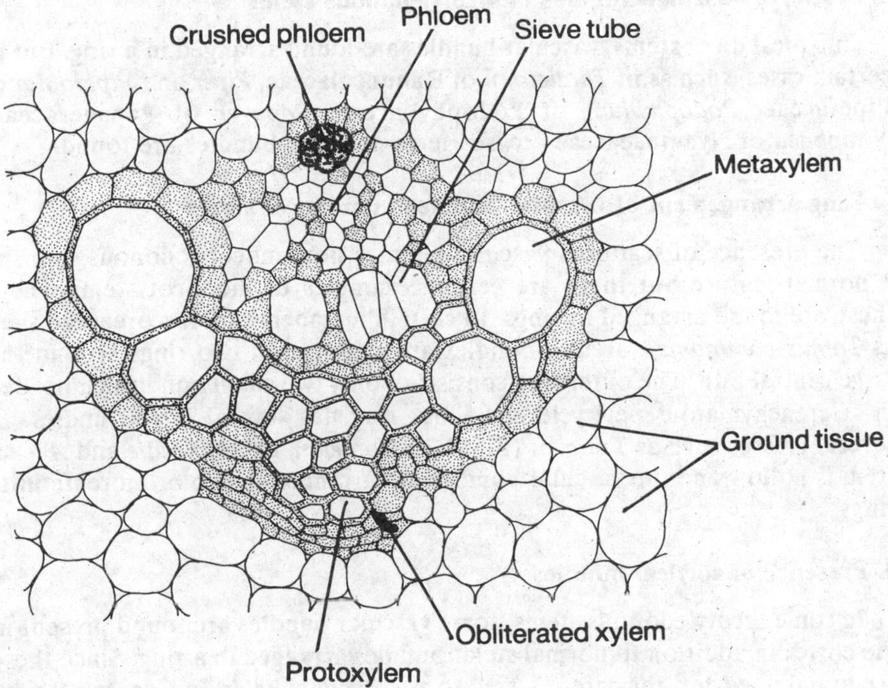

Fig. 7.10. A cellular enlarged figure of vascular bundle of a monocot stem.

Vascular bundles are conjoint, collateral and closed (V.B. without cambium). These remain surrounded by sclerenchymatous bundle sheath. All vascular bundles may be of uniform size as in case of *Sorghum* and *Saccharum* or outer bundles are smaller and becomes gradually larger towards centre as in *Triticum, Hordeum, Avena* and *Oryza.* The xylem elements show Y-shaped arrangement and it is composed of two metaxylem vessels situated at the tip of arms of 'Y' and the two protoxylem vessels at the base of 'Y'. The tracheids and xylem parenchyma are usually present around protoxylem. In certain cases, such as maize, the protoxylem vessels disintegrate due to rapid elongation of stem to form lysigenous cavity. Phloem is situated in between the arms of 'Y' or little external to xylem. It consists of sieve tubes and small companion cells, phloem parenchyma are characterically absent. However, a conspicuous sclerenchymatous bundle cap is usually present.

PRIMARY ANOMALOUS STRUCTURES

So far, the internal organisation of typical stems was taken into account but in several dicotyledonous as well as monocotyledonous stems, primary abnormal structures are present. Some of the common types of anomalies found in them are as follows:

1. Scattered vascular bundles in dicotyledonous stems

In typical dicot stems, vascular bundles are found arranged in a ring. But in certain cases, such as in *Thalictrum* of Ranunculaceae, *Piper* and *Piperomia* of Piperaceae, *Podophyllum* of Podophyllaceae, *Papaver* of Papaveraceae, *Nymphaea* of Nymphaeaceae etc. scattered vascular bundles are found.

2. Ring arrangement of vascular bundles in monocot stems

The presence of scattered vascular bundles in monocotyledonous stems is a normal feature but there are certain examples of monocot stems where these are found arranged in rings. In certain members of Dioscoreaceae such as *Tamus communis,* vascular bundles are arranged in two rings around the large central pith. The outer ring consists of only two small bundles embedded in sclerenchymatous pericycle and inner ring has several large bundles. In several grasses such as *Triticum* (Fig. 7.7), *Hordeum, Oryza, Secale* and *Avena,* stem is hollow and the vascular bundles are arranged in two or more definite rings.

3. Presence of cortical bundles

In some dicotyledonous stems, some vascular bundles are found present in the cortex in addition to normal stelar bundles arranged in a ring. Since these are found in cortex, they are referred to as *cortical bundles.* In *Casuarina,* such cortical bundles are found below the ridges, though these are normally oriented but are additional in number and position. In *Nyctanthes,* four inversely oriented cortical bundles are found at the four prominent ridges of the stem. Inversely oriented bundles are also found in *Calycanthus.* In *Limonium,* several irregularly scattered cortical bundles are found. In these forms, the bundles are conjoint and collateral. However, cortical bundles of concentric nature are also found. In certain members of Crassulaceae such as *Kalanchoe,* and *Aeonium,* these are amphicribal type whereas in *Lobelia,* these are amphivasal type.

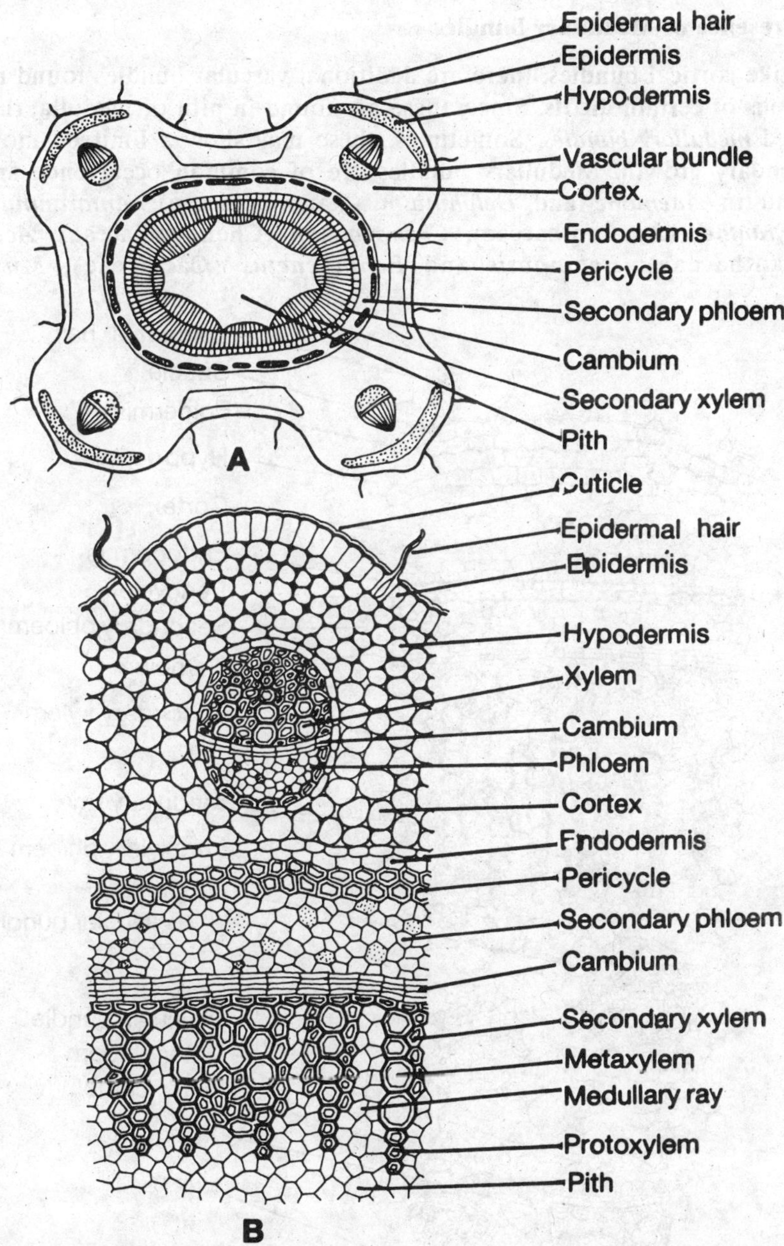

Fig. 7.11. T.S. stem of *Nyctanthes* showing the presence of cortical vascular bundles. A–diagrammatic and B–an enlarged cellular portion.

Cortical bundles are leaf trace bundles which traverse through cortex before entering into the petiole.

4. Presence of Medullary bundles

Like cortical bundles, there are additional vascular bundles found in pith regions of certain dicots. Since, these are found in pith or medulla, they are called *medullary bundles.* Sometimes, these may show a limited amount of secondary growth. Medullary bundles are of common occurrence and are found in *Anemone* and *Delphinium* (Ranunculaceae); *Amaranthus* and *Achyranthes* (Amaranthaceae); *Chenopodium* (Chenopodiaceae); *Acanthus* (Acanthaceae); *Echinopsis* and *Echinocactus* (Cactaceae); *Mirabilis,*

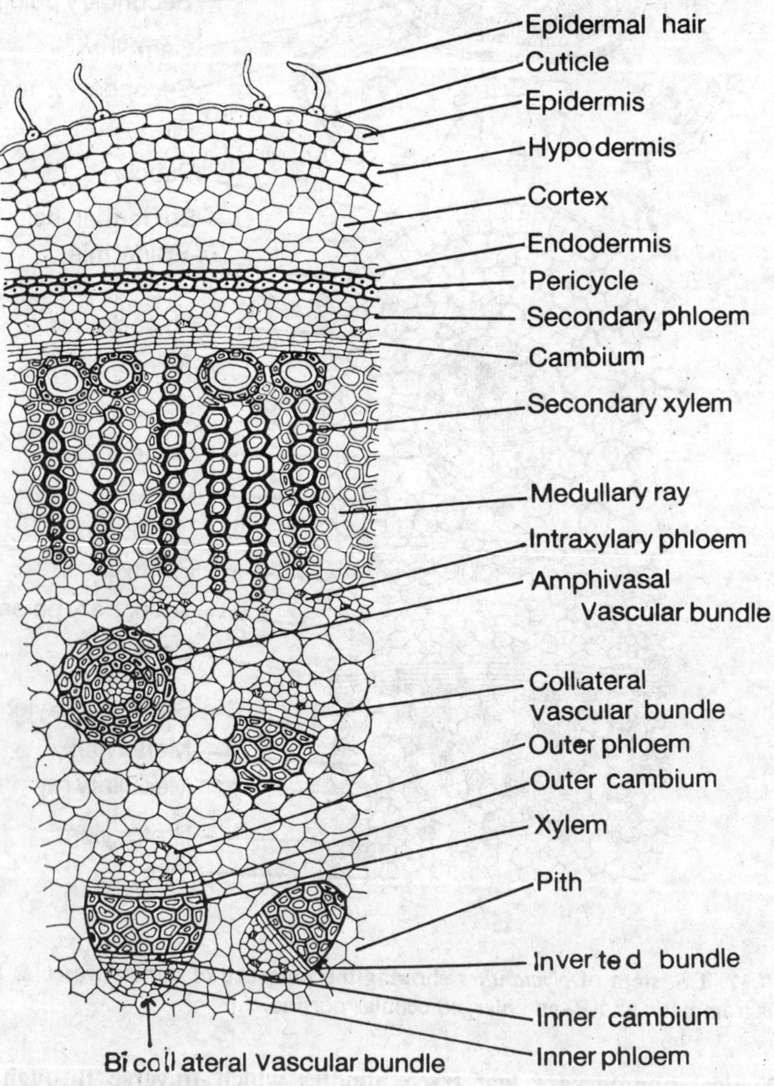

Fig. 7.12. An enlarged cellular portion of T.S. stem of *Argyreia* showing different types of medullary vascular bundles.

Bougainvillea and *Boerhaavia* (Nyctaginaceae) and *Argyreia* (Convolvulaceae) (see figures in chapter in *Anomalous Secondary Growth).*

Medullary bundles show variation in arrangement and may be arranged in one or more rings or even irregularly scattered. Similarly, number may also vary. *Achyranthes aspera* is a typical example where number of bundles are not constant throughout the length. There are two distinct, decussately arranged medullary bundles below and above nodes but these unite to form a single amphixylic bundle in the nodal region.

In *Boerhaavia diffusa,* medullary bundles are arranged in two rows and inner ring consists of two large bundles whereas outer ring has six to eight relatively smaller bundles. These bundles have fascicular cambium which produces a little amount of secondary tissue.

In *Amaranthus, Piper, Mirabilis* and *Bougainvillea,* several medullary bundles are found scattered in the pith region.

A considerable variation in the arrangement of medulary bundles is found among different members of Araliaceae. In *Aralia* and *Eremopanex*, medullary bundles are arranged in a ring but are inversely oriented. In *Cussonia,* a ring of normally oriented medullary bundles is found. In *Didymopanex* and *Kissodendron,* the peripheral medullary bundles are arranged in a ring while those found in the central region are scattered and oriented in different directions. In *Trevesia* and *Polyscias,* medullary bundles are scattered in pith.

In *Argyreia,* following types of medullary bundles are found scattered in the pith (Fig. 7.12).

(i) inversely oriented, i.e. conjoint, collateral, open and exarch;
(ii) normally oriented, i.e. conjoint, collateral, open and endarch;
(iii) Bicollateral vascular bundles; and
(iv) concetric and amphivasal bundles.

Morphological nature of medullary bundles is not clear. Maheswari (1930) stated that these may represent a transitional stage between the ring arrangement of dicots and scattered arrangement of monocots and this situation could have been brought about by weakening of cambial activity. These bundles run throughout the length of stem without any branching, so these cannot be concerned with lateral supply to leaf and branch. Instead, these may be related with vertical translocation coupled with mechanical strength.

5. Presence of intra-xylary phloem

Intraxylary or internal phloem also called medullary phloem is found present on the inner side of vascular bundles. It may occur in the form of isolated patches as in *Solanum, Capsicum, Calotropis, Argyreia* and *Leptadenia* or in the form of a ring as in *Asclepias, Convolvulus, Eucalyptus* and *Punica.* The internal phloem is primary in nature which develops from provascular tissues. Structurally, there is no much difference between outer and internal phloem except latter one is small, with inconspicuous fibres or even absent and fewer sieve elements surrounded by parenchyma (see figures in chapter on *Anomalous Secondary Growth).*

6. Presence of separate xylem and phloem bundles

In certain dicotyledons, some vascular bundles remain incomplete consisting of either exclusively xylem or phloem. For example, only phloem bundles, in addition to normal vascular bundles are found in *Cuscuta, Entada, Rumex, (R. crispus), Boerhaavia (B. diffusa), Ricinus (R. communis)* and *Antigonon (A. leptopus)*. Similarly, in *Paeonia,* only xylem bundles in addition to normal ones are found. The stem of *Circaeaster*, furnishes an interesting example where the stem gives the appearance of a diarch root due to the presence of separate xylem and phloem strands.

7. Presence of polystelic condition

Monostelic condition with vascular bundles arranged in a ring is a characteristic feature of dicotyledonous stems. There are certain plants which show anomaly in this respect. Presence of polystelic condition in angiosperm (dicots) has been reported in *Bryocarpum, Dodencantheon, Primula, Dianthera, Parnassia* and *Nymphaea*. In *Dianthera americana,* in addition to central stele, six peripheral steles, each resembling a meristele are found. In *Primula aerecula* numerous additional concentric strands, each representing an individual stele, having individual pericycle and endodermis, are found. However, in this case, these are formed by division of normal ring or vascular bundle of the young stem.

GENERAL QUESTIONS

1. Describe the primary anatomical features of a dicotyledonous stem with the help of suitable examples.
2. What are characteristic features of monocotyledonous stem? State its differences from typical dicotyledonous stems.
3. What are primary abnormal structures found in stem with special reference to dicots.
4. Write short notes on:
   - i. Polyfascicular siphonostele
   - ii. Differences between stem and root
   - iii. Cortical bundles
   - iv. Medullary bundles
   - v. Polystely in angiosperms.

CHAPTER 8

The Root–Primary Structure

The root develops from the radicle of the embryo and grows down into the soil to form the primary root. In dicotyledons, the primary root grows further deep into the soil, developing a number of secondary and tertiary branches to form a well developed tap root system. In monocotyledons, the primary root is soon replaced by a bunch or cluster of roots, forming the adventitious root system. In some cases, the roots become specialized for specific function, e.g. in epiphytes and mangrove plants, aerial and respiratory roots are formed and in case of turnips, carrots and radish fleshy roots are formed for storage. In roots, the position of apical meristem is subterminal due to terminal position of root cap. The summarised differences between dicot and monocot roots are outlined in table 8.1.

Table 8.1. Comparison between dicot and monocot roots.

Dicot roots	*Monocot roots*
1. Epidermis is single layered and cortex is homogenous and parenchymatous in nature.	1. Epidermis and cortex are similar to dicots.
2. Endodermis is single layered, distinct and well developed.	2. Same as in dicot roots.
3. Pericycle is usually single layered.	3. Pericycle is usually two to three layered.
4. The number of phloem and xylem bundles are two to six, rarely eight.	4. The number of phloem and xylem bundles are usually more than eight.
5. Xylem vessels appear circular in outline.	5. These are usually polygonal in outline.
6. Metaxylem usually meets in centre due to which pith is either absent or highly reduced.	6. Pith remains distinct, well developed and large.
7. Secondary growth is found.	7. Secondary growth is not found.

ANATOMY OF DICOT ROOT

For studying anatomical features of a typical dicot root, young root of sunflower *(Helianthus annuus)* is selected. Section of the root is cut and double stained and studied.

I. Root of sunflower (*before secondary growth*)

The following important anatomical features are noted in the root of sunflower.

1. **Epidermis.** It is also called **epiblema** which is the outermost uniseriate layer composed of thin walled, compactly packed, parenchymatous cells without intercellular spaces. Here and there, the epidermal cells give rise to unicellular root hairs. Epiblema of roots is characterised by absence of cuticle and stomata. Root hairs absorb water and minerals from soil.

2. **Cortex.** It lies below the epiblema and is considerably homogenous and well developed. Cortex is composed of thin walled, loosely arranged, parenchymatous cells with large number of intercellular spaces. However, outer few layers may be compactly arranged. These cells contain amyloplasts (leucoplasts) and store the starch in the form of starch grains.

3. **Endodermis.** It is the innermost layer of cortex which surrounds the stele. Endodermis is composed of barrel-shaped, parenchymatous cells without intercellular spaces. Most of the cells, except those lying opposite the protoxylem point, possess strip-like thickenings of suberin and lignin on their radial and tangential walls which are known as Casparian strips. Cells lying opposite the protoxylem elements are thin walled and are called passage cells because these allow the passage of water from cortex to xylem. The endodermis behaves like water-tight jacket around the stele.

4. **Stele.** It is tetrarch actinostele (protostele) composed of pericycle and vascular bundles. In general following parts are observed.

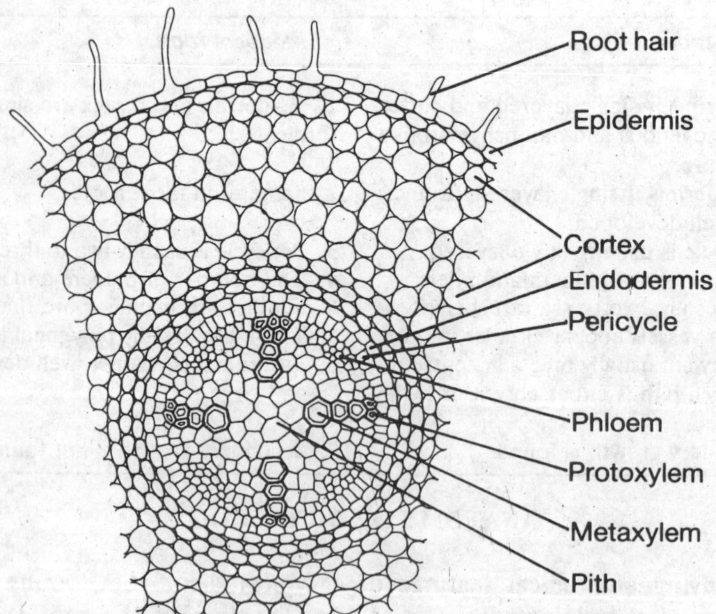

Fig. 8.1. A cellular enlarged portion of T.S. of root of sunflower (*Helianthus annuus*).

(i) *Pericycle*–The layer present on inner side of endodermis is the pericycle. It consists of a single uniseriate layer of thin walled parenchymatous cells containing dense protoplasm. It is usually two to three layered opposite the protoxylem elements which shows the point of origin of lateral roots.

(ii) *Vascular bundles*–These are four in number which are arranged in the form of a ring. Vascular bundles are radial type in which xylem and phloem are placed on different radii and alternate in their position.

Xylem is conical is shape with metaxylem vessels towards centre and protoxylem towards periphery. So the xylem is exarch in nature. In mature and well developed roots, the metaxylem elements meet in the centre and as a result, the pith gets obliterated. Fibres and xylem parenchyma are absent. Xylem tracheids are found present around the vessels.

Phloem lies alternate to xylem patches on separate radii. It consists of sieve tubes, companion cells and pholem parenchyma. The phloem fibres are absent. Phloem is present in the form of small patches, the outer part (towards pericycle) being *protophloem* and inner part (towards metaxylem) *metaphloem.*

(iii) *Conjunctive tissue*–These are parenchymatous in nature which lie in between patches of xylem and phloem.

(iv) *Pith*–In young root, the central part of stele is the pith which consists of few parenchymatous cells compactly arranged without intercellular spaces. In older root, pith may be altogether absent due to development of metaxylem.

ANATOMY OF MONOCOT ROOT

The root of *Zea mays* or *Commelina* may be selected to study the anatomical features of typical monocotyledonous roots. Sections are cut and double stained for study.

I. Root of maize *(Zea mays)* or *Commelina* (Fig. 8.2)

The internal structure of maize root shows following parts--

1. **Epiblema, Cortex** and **Endodermis** of monocot roots are structurally similar to those found in dicot roots. In maize, endodermal cells are thick walled and casparian strips are indistinct.

2. **Stele.** It is actinostele (protostele) which shows following features.

(i) *Pericycle*–It is single layered lying on the inner side of endodermis. It is partly composed of parenchymatous cells and partly of sclerenchymatous cells.

(ii) *Vascular bundles*–These are found in large number (more than eight) and each one is radial type (as described in dicot root). These are arranged in a ring around the central pith.

Xylem is composed of single rounded metaxylem vessel towards the pith and one or sometimes two small, polygonal or rounded thick-walled protoxylem vessels towards the periphery. So the xylem shows exarch arrangement.

Phloem lies in form of small patches alternate to xylem and is composed of

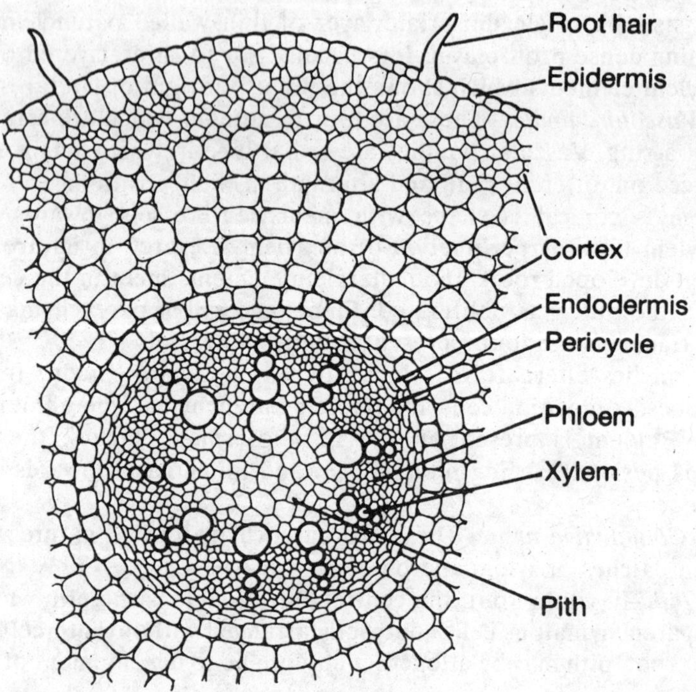

Fig. 8.2. An enlarged cellular portion of T.S. of root of *Commelina* showing cells of inner cortex arranged in concentric rings.

sieve tubes, companion cells and phloem parenchyma.

Parenchyma cells associated with xylem undergo sclerosis and thus become thick walled.

(iii) *Pith*–Pith occupies the central portion of stele. It is composed of thin walled parenchymatous cells containing abundant starch grains. Cells are loosely arranged.

GENERAL ACCOUNT OF ROOT ANATOMY

Anatomical features of roots shows relatively simpler organisation to that of stems and as such three distinct zones recognised in the primary root of growth are the *epiblema,* the *cortex* and the *stele.*

1. **Epiblema.** The epidermis is single layered and consists of uncutinised thin walled cells. These are compactly set and in younger region, epidermal cells here and there form root hairs. The root hairs are unicellular and thin walled and absorb water and minerals from the soil. These have a large central vacuole behaving as water balancing organ.

Multilayered epidermis is found in certain epiphytic plants such as orchids (*Vanda* species) and there it is called **velamen.** The velamen consists of compactly arranged spongy non-living cells, often with spiral or reticulate secondary thickening. Cells are specially adapted to serve as moisture absorbing tissue. The walls are usually porous, so that cells work like a sponge and secondary thickenings act like supporting ribs. In dry weather, the velamen cells remain filled with air while in rainy season, these become filled with water. Velamen also contains some specialised structures called **pneumatodes** which consist of group of cells having dense spiral thickenings and are helpful in gaseous exchange when roots are saturated with water. Some anatomists believe that the primary role of velamen is protective and it reduces the loss of water from cortex.

2. **Cortex.** Cortex, which lies internal to epidermis is rather simple and homogenous type and usually consists of thin walled, isodiametric, parenchymatous cells accommodating several intercellular spaces. In aquatic plants like *Hydrilla* and others, it is largely aerenchymatous enclosing air cavities. In certain cases, air cavities develop in the root by breaking down of some cortical cells.

Cortical cells usually store starch. Chloroplast is not found but it may be found in cortical cells of some aquatic plants such as in *Trapa*. Some of the aerial climbers such as *Tinospora* have got photosynthetic or assimilatory roots, and the chloroplasts are found in their cortical cells. Sometimes, fibres, sclereids, idioblasts and latex cells may also be found in root cortex. The arrangement of cells may also vary, for example, in *Avena sativa,* cortical cells are arranged in distinct radial rows. In *Commelina,* the cells of inner cortex are arranged in the form of concentric rings. The cortical cells in roots of several gymnosperms and some members belonging to families such as Cruciferae, Rosaceae and Caprifoliaceae have lignified reticulate thickenings. In most monocotyledons, some dicotyledons and gymnosperms, the sub-epidermal or outer few cortical layers differentiate into protective tissue, called **exodermis** which consists of suberised wall. Some of the exodermal cells show the presence of Casparian strip. Exodermis is distinctly developed in different species of *Iris*.

3. **Endodermis.** It is considered as innermost layer of cortex and is quite distinctive in roots. Endodermis is of universal occurrence in roots and is composed of closely packed living cells. Two types of cells are found, one which have characteristic casparian thickening in their radial and transverse walls and other which are thin walled and found opposite to the protoxylem points. These thin walled cells are also called passage cells. The Casparian thickening is formed due to deposition of suberin and lignin. (The various functions of endodermis have been discussed in chapter on tissue system.)

Endodermis together with cortex is cast off with secondary growth in dicot roots.

4. **Stele.** It includes pericycle, vascular bundles and pith in roots.

(i) Pericycle–It occurs on the inner side of endodermis and forms the outer boundary of primary vascular bundles. It is usually uniseriate but multiseriate pericycle is found in *Salix* and *Smilax*. However, it may be even absent in certain cases such as in some parasites and few hydrophytes. In most cases it is parenchymatous in nature but in *Smilax,* it consists of thick walled cells.

Bruch (1955) reported the presence of schizogenous secretory canals in pericycle of certain members of Umbelliferae.

The pericycle originates quite early from procambial strands which retains meristematic properties. It functions as site of origin of lateral roots, especially before secondary growth or in plants without secondary growth. During secondary growth, it forms a part of fascicular cambium and plays important role in completing the cambial ring, particularly opposite the protoxylem points. In addition to that, the entire initial cork cambium (phellogen) originates from it.

(ii) Vascular system–The vascular bundles found in roots show radial arrangement with alternating strands of xylem and phloem. Xylem elements show exarch condition, with protoxylem towards periphery and metaxylem towards centre. The development of xylem is centripetal. The number of vascular bundles ranges from one to eight (commonly less than eight) in dicotyledonous roots and eight to twenty in monocotyledonous roots. Depending upon number of xylem bundles or patches roots may be called *monarch* (with one xylem bundle), *diarch* (two bundles), *triarch* (three bundles), *tetrarch* (four bundles) or *polyarch* (many bundles). The number of protoxylem elements in each bundle is very few, usually, there is a single protoxylem vessel with annular or spiral thickening. The metaxylem is composed of polygonal vessels, in cross section, with pitted thickening. There may be some variation in number of protoxylem and metaxylem vessels, e.g. *Sisyrinchium* has a single large central metaxylem vessel surrounded by five to seven strands of protoxylem vessels. In *Gladiolus* and *Freesia,* there is present a single metaxylem vessel at the base of each protoxylem strand.

Phloem consists of sieve tubes, companion cells and phloem parenchyma. Their fundamental structure is similar to those of stem.

The vascular bundles are separated from each other by conjunctive tissues composed of parenchymatous cells.

(iii) Pith–Pith occupies a small central area of stele and usually consists of thin walled parenchymatous cells with intercellular spaces. In older roots of dicots, the pith is altogether obliterated (absent). It is well developed in monocotyledonous roots which may also contain a few tannin cells. Sclerenchymatous pith is found in certain members of Iridaceae.

DEVELOPMENT OF LATERAL ROOTS

The lateral roots of both gymnosperms and angiosperms arise endogenously from the pericycle and grow through the cortex. The position of lateral roots may show variation. It arises between the phloem and xylem strands in diarch roots, opposite the protoxylem in triarch to pentarch roots and opposite the protophloem in polyarch roots.

The method of formation of lateral roots is almost similar in higher plants. A few mature cells of pericycle become meristematic and show periclinal divisions. These initial cells divide both anticlinally and periclinally and a root primordium is formed. The primordium is some thing like a protrusion

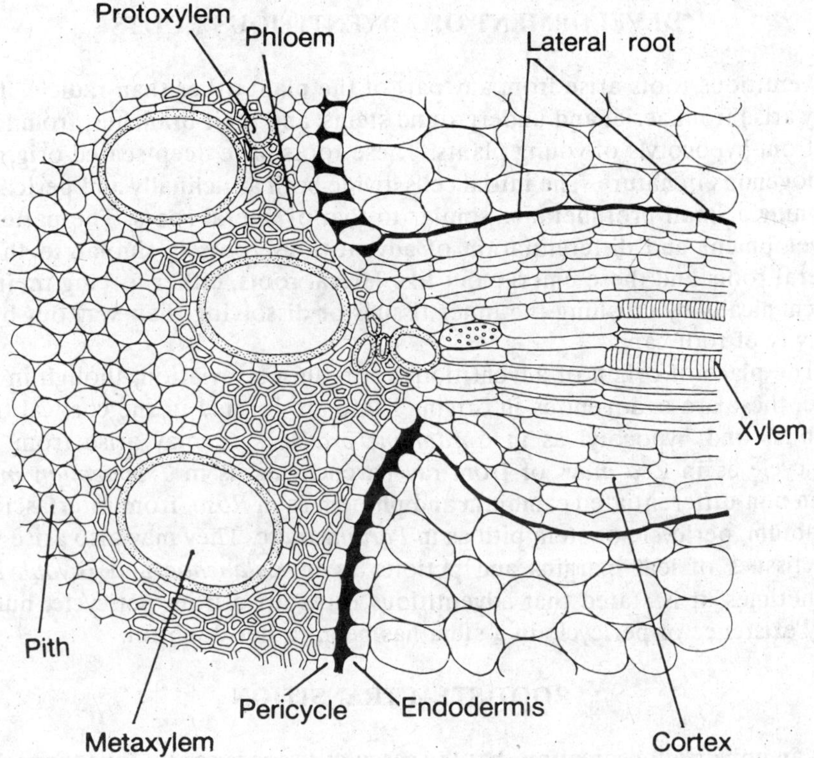

Fig. 8.3. A cellular enlarged portion of transection of maize root showing a lateral root.

which makes its way by piercing the cortex and epiblema. The passage of lateral roots through the cortex is by mechanical penetration but enzyme action may also facilitate the growth. During the process, cells of cortex are either crushed or pushed aside. That is how, it eventually comes out piercing all the tissues.

The primordium soon takes the shape of a growing point with its initial cells, the cap and other histogens. The root cap and the promeristem of the growing lateral root develop when it is still within cortex. The vascular elements differentiating in the lateral roots become connected with that of parent root through the cells of pericycle.

There are reports in some cases that endodermis also participates in the formation of root primordium. It divides preferably by anticlinal divisions followed sometimes by a few periclinal divisions and forms a layer surrounding the primordium. However, cells of this layer die and are shed when lateral roots come out.

The lateral roots usually develop in acropetal order, i.e. older at the base and younger towards the apex. But, there is no regularity in this order with reference to each other.

DEVELOPMENT OF ADVENTITIOUS ROOTS

Adventitious roots arise from any part of the plant other than radicle. These may arise from aerial and underground stems, and their branches, from leaves or from hypocotyls of young plants. These roots have deep seated origin, i.e. endogenous in nature. The initial cells divide both anticlinally and periclinally forming a small protuberance similar to that of lateral roots. The pattern of development and differentiation of adventitious roots is similar to that of lateral roots and these emerge out like lateral roots, either forcing their way mechanically by crushing the inner tissues or dissolving their way out by the activity of root cap.

The place of origin of adventitious roots shows variation, though in each case, these are endogenous in origin. These are formed in the cortical tissue of buds and hypocotyl as in *Cardamine pratensis* or may arise from stem pericycle as in *Zea mays* or from ray parenchyma as in *Tropaeolum majus,* from non-differentiated cambium and phloem as in *Rosa,* from interfascicular cambium, pericycle or stem pith as in *Portulaca* etc. They may also arise from the tissues of leaf margins and petioles as in *Kalanchoe* and *Bryophyllum.* Sometimes, it is stated that adventitious roots arise from pericycle, but the very existence of pericycle in a stem has been rather doubtful.

ROOT-STEM TRANSITION

It is an universal observation that the vascular tissue is continuous throughout the root and stem, however, the vascular bundles are differently oriented in the two parts. The region where stem and root meet is called *root-stem transition zone.* The epidermis, cortex, endodermis, pericycle and the secondary vascular tissue are in direct continuation, hence no significant change in them takes place. But two important changes in primary vascular organisation are noted, i.e. vascular bundles from radial in roots to conjoint collateral type in stem and the xylem from exarch in roots to endarch in stems. The stele also increases in diameter. Eames and Mac Daniels (1947) proposed that these changes in transition zone involve multiplication, forking, rotation and fusion of strands accompanied with increase in diameter.

Such transition zone having these reorienting changes show variation in position and may be present either at the tip of radicle or at the base, middle or top of the hypocotyl. In this transition, the metaxylem strands of root rotate through 180° around the longitudinal axis. Protoxylem, however, takes no part in this rotation. The length of this region and the time involved in these changes also vary. The transition may be gradual or abrupt type.

Eames and Mac Daniels (1947) proposed four different types of transitions in various plants.

1. **Fumaria type.** It is characterised by a radial splitting of primary xylem strands as found in *Fumaria, Mirabilis* and *Dipsacus.* Each xylem strand divides radially into two branches, one turns left and other towards right. These turn through 180 degrees and meet the phloem strands on their inner

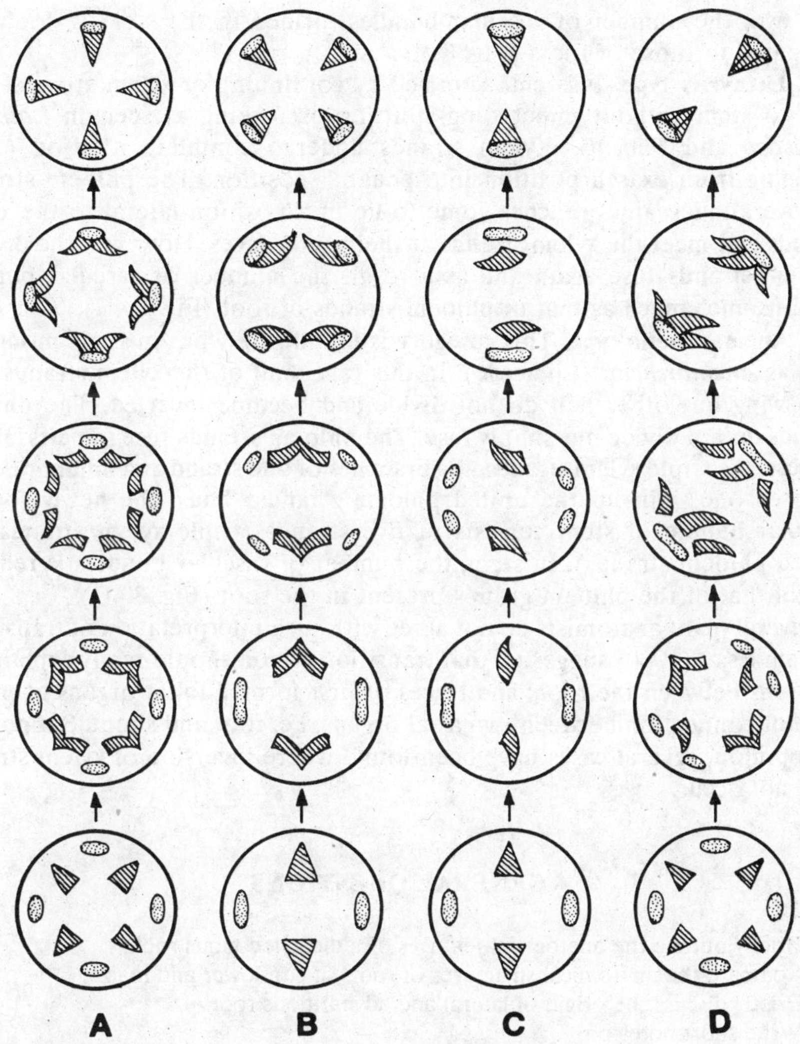

Fig. 8.4. Schematic illustration of different types of root-stem transitions. A–*Fumaria* type, B–*Cucurbita* type, C–*Lathyrus* type and D–*Anemarrhena* type (Redrawn after Eames and Mac Daniels, 1947)

face. The phloem, however, does not change the position and passes as such from root to stem. In this type, the number of vascular bundles remains same as number of phloem strands remains unchanged (Fig. 8.4A).

2. **Cucurbita type.** It is characterised by forking of both xylem and phloem strands as observed in *Cucurbita, Phaseolus, Acer* and *Tropaeolum.* During the course from root to stem, the branches of each xylem strand turn laterally and become inverted (from exarch to endarch) and join the inner face of phloem strands. Phloem strands do not show re-orientation but because of forking

into two, the number of vascular bundles formed in the stem is double as compared to those of roots (Fig. 8.4B).

3. **Lathyrus type.** It is characterised by continuing of xylem strands from root to stem without undergoing splitting or forking as seen in *Lathyrus, Medicago* and *Phoenix*. Xylem strands undergo complete rotation (180°) changing from exarch position into endarch position. The phloem strands, however, divide and branches come to lie in a position lateral to the xylem strands and meet the xylem strands at their outer faces. However, the divided phloem strands fuse again and as a result the number of vascular bundles remains unchanged as that of phloem strands of root (Fig. 8.4C).

4. **Anemarrhena type.** This category is found rarely but in some monocots such as *Anemarrhena* (Liliaceae). In this case, half of the xylem strands fork and swing but other half do not divide and become inverted. The phloem strands do not divide but simply fuse. The phloem strands fuse in pairs. In the process, the triple xylem strands, the branches of one strand and an unbranched inverted one, join up the united phloem strands. Thus, the newly formed vascular bundle of stem consists of five strands (triple xylem strands and united phloem strands). In stem, the number of vascular bundles is reduced to half that of the phloem groups present in the root (Fig. 8.4D).

Several plant anatomists do not agree with such interpretation of transition. Madam Esau (1960) suggested that transition region should really explain the relations between the roots and traces of first formed foliar organs of plants and not connection between two axial organs, i.e. root and shoot. Supporting her opinion, several cases have been found where inversion of xylem strands does not occur.

GENERAL QUESTIONS

1. Briefly outline the anatomical features of a dicotyledonous root.
2. Compare the anatomical structures of roots of sunflower and maize.
3. Briefly discuss the origin of lateral and adventitious root.
4. Write short notes on
   (i) Root-Stem transition
   (ii) Vascular tissues in roots
   (iii) Difference between root and stem
   (iv) Velamen.

CHAPTER 9

The Leaf

Leaves are lateral expansions of the axis which are green in colour and characterised by broad or thin lamina. These are the organ of limited growth which develop as protrusion from shoot apex. At maturity, the foliage leaf of angiosperm consists of a more or less expanded portion called **blade** or **lamina** and a stalk through which lamina remains attached with the stem, called **petiole.** In certain cases, the petiole may be absent and then leaves are termed as sessile.

As such leaves are classified into three groups on the basis of their position on the stem, **radical** (arising as if from root as in *Raphanus*), **cauline** (arising from unbranched axis directly as in maize and sugarcane) and **ramal** (arising from the branches as in most dicot trees). Leaves may also be classified as (i) **dorsiventral** or *bifacial type* with dissimilar upper and lower leaf surfaces, (ii) **isobilateral** or *unifacial* with similar type of upper and lower surfaces and (iii) **centric** which are circular in section and are with tissue distributed evenly as in *Allium cepa.* Dorsiventral leaves are characteristic of dicots while isobilateral of monocots.

ONTOGENY OF LEAVES

The leaves arise as protrusion from the promeristem of shoot apex. This protrusion is called *leaf primordium* which originates by anticlinal and periclinal divisions of outer layer of apical meristem. The meristematic cells found at the tip of a leaf form the apical meristem of leaf. The apical growth in leaves continues only at the young stage but stops later. Further expansion of leaf is achieved by intercalary growth and rapid enlargement of cells. The basal part of leaf remains unexpanded and forms the narrow base which remains meristematic and gives rise to petiole. During expansion process, a row of meristematic cells differentiates along each surface of leaf axis and is called marginal meristem. Marginal meristem is thus found on both the margins of leaves. Meristematic cells below them are called submarginal initials which divide in all planes and produce internal tissues of the leaf.

The following scheme briefly elucidates the plan of origin and differentiation of different parts of a typical dicotyledonous leaf.

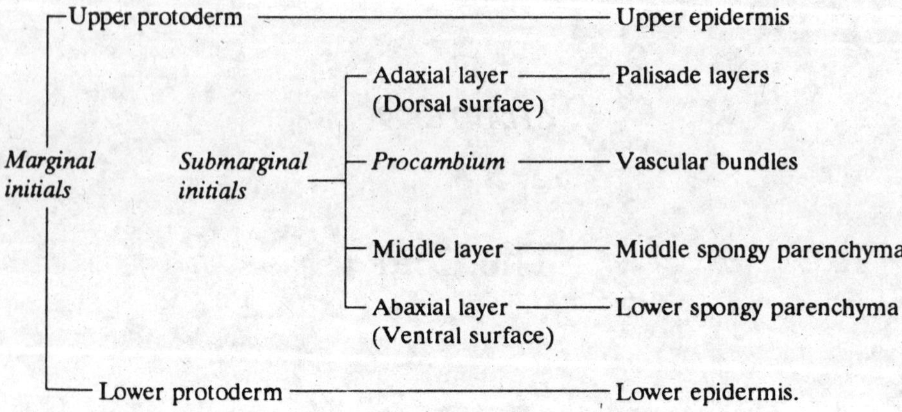

Fig. 9.1. Differentiation in a normal dicot leaf.

The marginal initials are cells of the outermost layer. These divide anticlinally and add new cells to the adaxial and abaxial protoderm of the leaf and differentiate into upper and lower epidermis, respectively. In certain monocots, the periclinal divisions in marginal initials are also found which contribute to part of internal tissues. The behaviour of submarginal initials is not always uniform. Foster (1936) proposed two common modes of their future differentiation as outlined in figure. 9.2.

Type A (as found in *Nicotiana tabacum*)

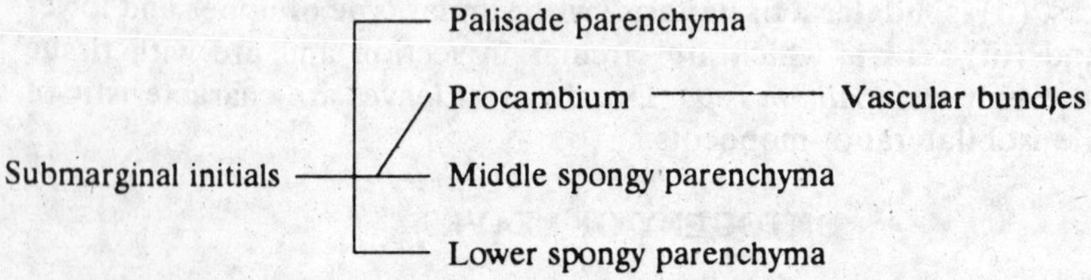

Type B (as found in *Carya buckleyi*)

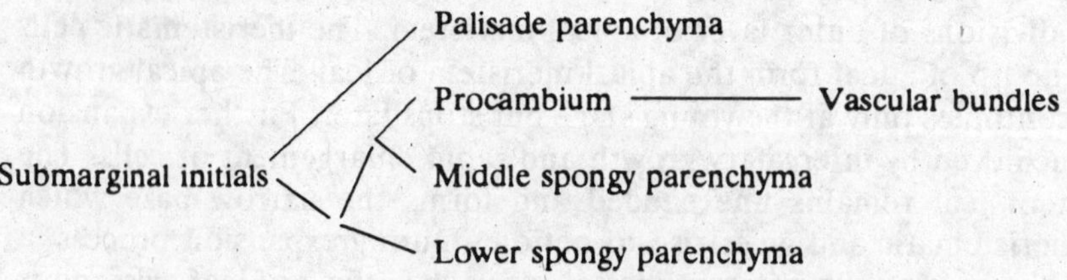

Fig. 9.2. Two common modes of activities of submarginal initials (after Foster, 1936).

The development of vascular system begins with the differentiation of procambium of mid vein during initial stage of leaf development. The lateral procambial strands giving rise to lateral veins develop from midrib acropetally.

The differentiation of reticulate venation occurs mainly during the intercalary growth of the leaf. The differentiation of both xylem and phloem is acropetal and the differentiation of phloem precedes the xylem. Some variations may be found in vascular differentiation in certain monocot leaves, for example, in *Triticum,* O'Brein and Carr (1970) found that procambial strands of the longer lateral veins develop acropetally, but those of smaller veins, which alternate with the longer veins, develop basipetally. The cross veins or interconnections also develop basipetally.

DORSIVENTRAL LEAVES

For studying the internal structure of a dorsiventral leaf (also called dicotyledonous leaf or bifacial leaf), leaves of mango, banyan or oleander may be taken. A thin section of lamina is cut and studied.

1. Internal structure of Mango leaf (*Mangifera indica*)

It shows following important structure:

(i) **Epidermis.** There are two epidermal layers, one on adaxial (dorsal) and other on abaxial (ventral) surfaces and are accordingly called upper epidermis and lower epidermis, respectively. Epidermis is uniseriate, composed of a row of compactly arranged, thin walled, parenchymatous cells. A thick and distinct cuticle layer is found on cells of the outer surface of epidermis which is relatively thinner on lower epidermis. Stomata are present on lower epidermis and thus, the leaf is **hypostomatic.** Guard cells are kidney-shaped. A distinct substomatal cavity is found below the stomata.

(ii) **Mesophyll tissues.** The ground tissue forming the mesophyll is differentiated into palisade and spongy cells. Palisade cells are found below

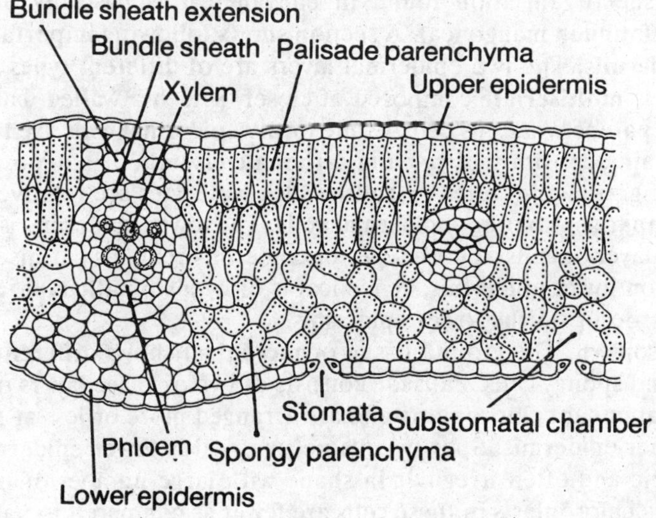

Fig. 9.3. A portion of V.S. of leaf of mango (*Mangifera indica*).

the upper epidermis and are arranged at right angles to its cells. They are columnar in shape with scanty intercellular spaces. Each cell contains a large number of discoid chloroplasts along the radial walls of the cells. Chloroplasts may change their position in response to intensity of light. In mango leaf, the palisade cells are arranged in two layers.

The **spongy cells** lie below the lower epidermis and are polygonal in shape. These are quite loosely arranged with large number of intercellular spaces. Number of chloroplasts in each cell is lesser than that of palisade and thats why lower surface of leaf appear pale green. The main function of mesophyll cells is to manufacture food materials by photosynthesis.

(iii) **Vascular bundles.** These are uniformly present in the same plane and embedded in mesophyll tissues. Bundles are conjoint, collateral and closed and are of different sizes according to their position. Mid-vein bundle is large while others are smaller. The bundles remain surrounded by a row of colourless parenchymatous cells. The row may be called **bundle sheath** or border parenchyma due to which the bundle is not in direct contact with the mesophyll cells. Parenchyma as well as few collenchymatous cells occur on outer and inner sides of bundle and spread upto upper and lower epidermis. These cells may collectively be called **bundle sheath extension.**

Xylem lies towards upper epidermis and is composed of annular and spirally thickened vessels, tracheids, fibres and xylem parenchyma. In larger vascular bundles, protoxylem lies towards upper surface and metaxylem towards lower side. In smaller vascular bundles, the two types of xylem elements remain indistinct.

Phloem which lies towards lower epidermis consists of sieve tubes, companion cells and phloem parenchyma.

2. Internal structure of banyan (*Ficus bengalensis*) leaf

The tissue organisation found in banyan leaf is basically of the same pattern as found in mango leaf. A section shows following important features:

(i) **Epidermis.** The two epidermal layers are of different types. The upper epidermis is **multiseriate** composed of closely set, thin walled, barrel-shaped or tabular parenchymatous cells. Some of the epidermal cells are lithocytes as these contain deposition of calcium carbonate crystals. The deposition is like a bunch of grapes and called **cystolith.** Such lithocytes are relatively larger in size in comparison to other epidermal cells. The lower epidermis is uniseriate. The outer layer of cells of both types of epidermis remains cutinised. Stomata are found on lower epidermis, hence leaf is of **hypostomatic type.** Sometimes, wax glands occur on the lower epidermis.

(ii) **Mesophyll.** The ground tissue or mesophyll shows differentiation into palisade and spongy cells. Palisade consists of two or three layers of columnar cells with abundant chloroplasts and are arranged more or less at right angles to the upper epidermis. Spongy cells occur towards lower epidermis and are isodiametric and often irregular in shape with large number of intercellular spaces. The chloroplasts in these cells are fewer as compared to palisade cells.

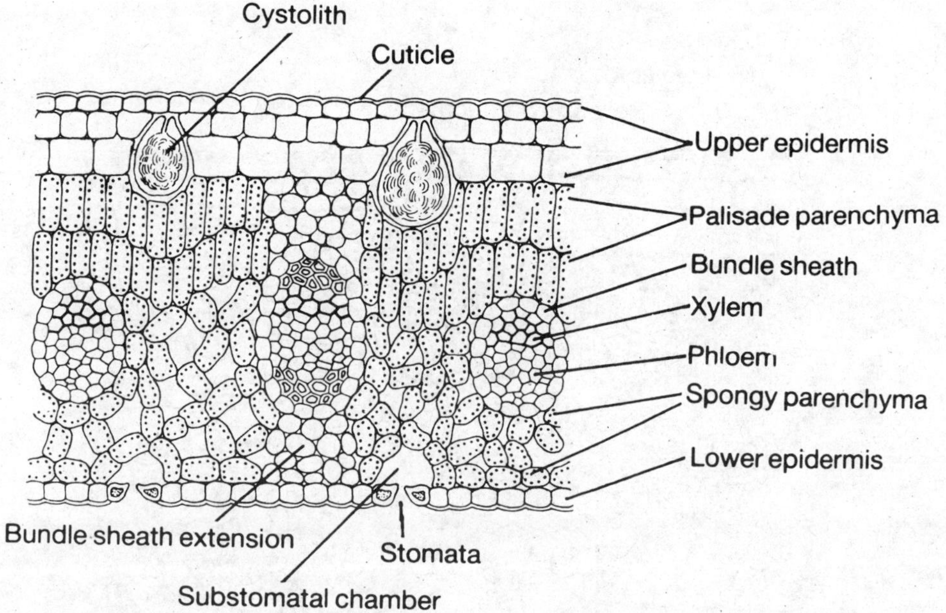

Fig. 9.4. A portion of V.S. of leaf of banyan (*Ficus bengalensis*).

The main function of mesophyll cells is to manufacture food by the process of photosynthesis.

(iii) **Vascular bundles.** The vascular bundles are conjoint, collateral and closed with xylem towards upper and phloem towards lower side. Each vascular bundle remains surrounded by a bundle sheath composed of colourless parenchymatous cells. The basic structure of xylem, phloem and bundle sheath extension remains similar to those of mango leaf. The mid vein is swollen towards the lower epidermis and in mature leaf, sometimes in the bundle sheath extension sclerenchymatous fibres are found which provide mechanical strength to the bundle.

3. Internal structure of leaf of oleander (*Nerium odorum*)

The leaf of *Nerium* is dorsiventral but with parallel venation so, it is slightly different from normal dorsiventral leaf and more adapted to face adverse dry (xeric) conditions. The internal organisation of tissues shows following important features:

(i) **Epidermis.** Both upper (adaxial) and lower (abaxial) epidermis are multiseriate (multiple) type and composed of few layers of compactly arranged, isodiametric thin walled parenchymatous cells. On the outer surface of the cells of outermost layer of epidermis, thick cuticle is found deposited, which is comparatively, more prominent on upper epidermis than the lower epidermis. Epidermal outgrowths are present.

Presence of sunken stomata that too in cup-shaped stomatal pits

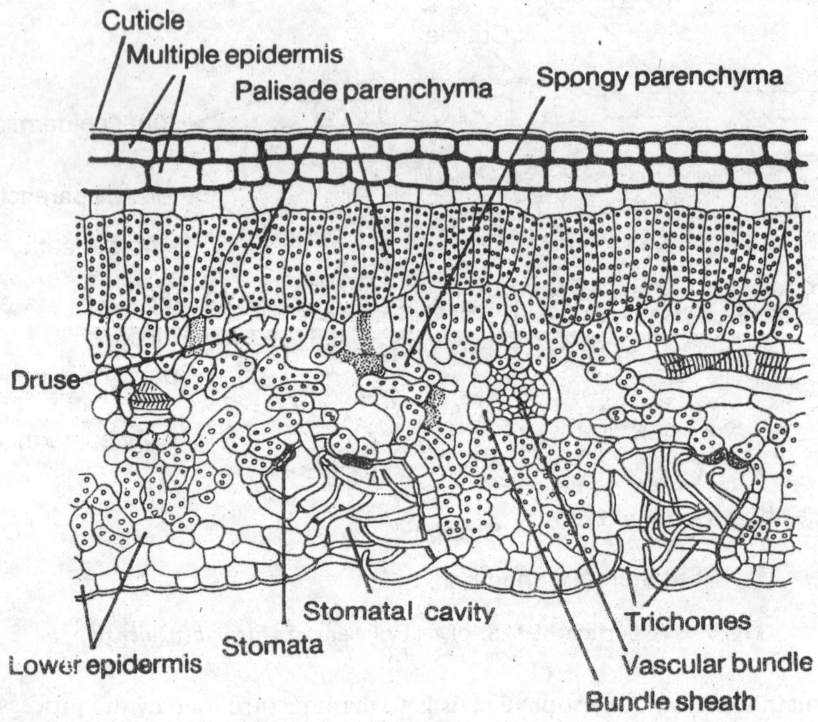

Fig. 9.5. A portion of V.S. of leaf of oleander (*Neriüm odorum*).

(depression) known as stomatal crypts on lower epidermis is an important ecological feature of the leaf. A fairly large number of glandular and non-glandular trichomes develop from the cells bordering the depression. The position and nature of surrounding tissue of stomata are such which reduce the rate of transpiration. Cells containing crystals of calcium oxalate (idioblasts) and silica are of common occurrence.

(ii) **Mesophyll.** It remains differentiated into palisade and spongy cells. **Palisade cells** are found below both upper and lower epidermis but these are compactly arranged in two or more layers below upper epidermis but in a single layer below lower epidermis. Lower palisade layer consists of comparatively smaller cells. Chloroplasts are abundantly present along the radial walls of cells.

Spongy cells occur between the two palisade layers and are composed of loosely arranged, smaller, isodiametric or polygonal chlorophyllous cells with profuse intercellular spaces. Calcium oxalate crystals (sphaeroraphides) are present here and there in spongy tissues. The main function of palisade is to manufacture food by photosynthesis.

(iii) **Vascular bundles.** Vascular bundles are conjoint, collateral and closed with xylem on the upper and phloem towards lower side. Each bundle remains surrounded by parenchymatous bundle sheath. The structure, elements and functions of xylem and phloem tissues are same as that of mango.

ISOBILATERAL LEAVES

For studying internal organisation of isobilateral leaves, leaves of maize or bamboo may be studied. Leaves of certain monocots such as *Musa* and *Canna* are dorsiventral type. Here, leaves of maize and bamboo are taken into account.

1. Internal structure of leaf of maize (*Zea mays*)

(i) **Epidermis.** The upper and lower epidermis are uniseriate and composed of compactly arranged, oval or barrel-shaped thin walled parenchymatous cells. Their outer walls are thickly cutinised. Few cells of upper epidermis are larger and empty, hence distinct and called **bulliform cells** or motor cells. They are considered to help in rolling of leaves in dry weather. Stomata are found on both the epidermal layers, hence leaf is of **amphistomatic type.**

(ii) **Mesophyll.** Mesophyll or ground tissue are not differentiated into palisade and spongy cells, rather composed of nearly compactly arranged isodiametric cells. At places, small intercellular spaces are found.

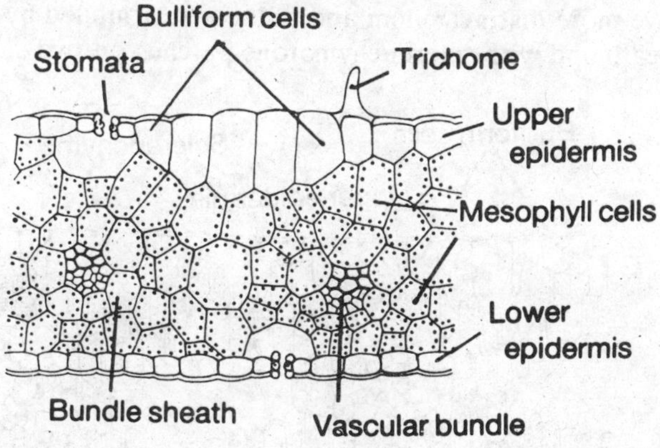

Bulliform cells
Stomata
Trichome
Upper epidermis
Mesophyll cells
Lower epidermis
Bundle sheath
Vascular bundle

Fig. 9.6. A portion of V.S. of leaf of maize (*Zea mays*).

In the mid vein region, mesophyll tissues are not found. The space is occupied by sclerenchymatous patches which extend from vascular bundle to upper and lower epidermis.

(iii) **Vascular bundles.** Vascular bundles are conjoint, collateral and closed type which are arranged in parallel series. Large and small bundles are arranged at regular intervals.

The **smaller bundles** have xylem on upper and phloem on lower side. Each bundle remains surrounded by a bundle sheath composed of thin walled, large, compactly arranged parenchymatous cells containing plastids and starch grains. The **larger bundles** are similar to that of stem. Sclerenchymatous patches occur on both upper and lower side of bundles, obviously to provide mechanical strength.

The cells of bundle sheath serve as temporary storage cells, apart from participation in conduction of photosynthetic products from mesophyll to phloem.

Xylem consists of usual tracheary elements, i.e. vessels, tracheids and xylem parenchyma and **phloem** of sieve tubes, companion cells and phloem parenchyma.

2. Internal structure of bamboo leaf (*Bambusa* species)

It reveals following organisation of internal tissues.

(i) **Epidermis.** The upper and lower epidermis are distinct. A number of bulliform cells are found in upper epidermis. Lower epidermis has well developed cuticle and irregularly distributed stomata. At places stiff and sharply pointed hairs are present. There is considerable deposition of silicon on epidermal surface.

(ii) **Mesophyll.** It is not differentiated into palisade and spongy cells but composed of compactly arranged, small, palisade like cells. At places, small but distinctive air cavities are found in this region.

(iii) **Vascular bundles.** Vascular bundles are conjoint, collateral and closed and are arranged in parallel series. Xylem is found towards upper side (adaxial side) while phloem towards lower side (abaxial side). Larger bundles have more distinct xylem and phloem surrounded by parenchymatous bundle sheath and with sclerenchymatous patches on two sides.

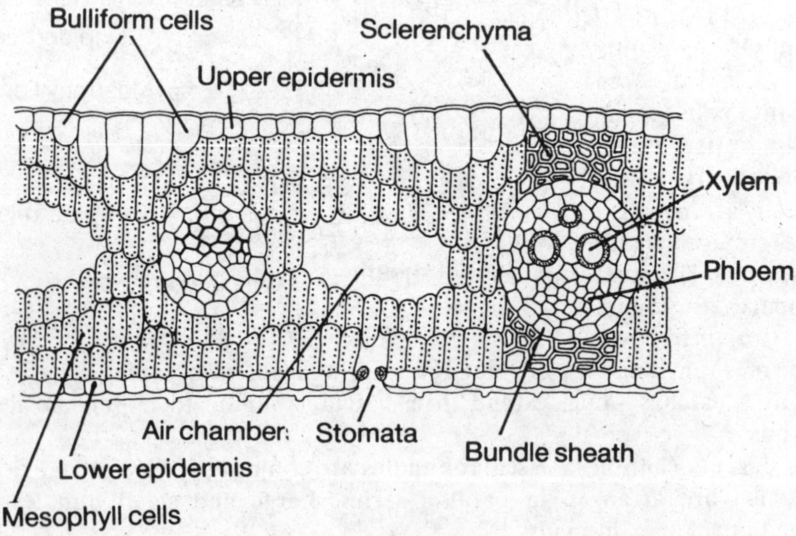

Fig. 9.7. A portion of V.S. of leaf of bamboo (*Bambusa*).

CENTRIC LEAVES

For the study of internal structure of a centric leaf, the leaf of onion (*Allium cepa*) is selected.

1. Internal structure of leaf of onion (*Allium cepa*)

These leaves are acicular in shape and more or less rounded in cross section. It reveals following important structures:

(i) **Epidermis.** It forms one cell thick (uniseriate) layer composed of small and closely packed parenchymatous cells. Outer walls are strongly cuticularised. Stomata are present which are slightly sunken in position due to projected susbsidiary cells.

(ii) **Mesophyll.** Mesophyll or ground tissue consists of two types of cells, the palisade cells and spongy cells. The palisade tissue consist of a continuous layer of columnar cells arranged in two rows. Cells have abundant chloroplasts. The spongy cells are elliptical, isodiametric or slightly oval in shape arranged slightly loosely with few small intercellular spaces. Cells have few chloroplasts. The central part of leaf is hollow. The cells bordering the cavity are colourless and without chloroplasts.

(iii) **Vascular bundles.** The vascular bundles are found scattered in the lower part of the mesophyll. Each bundle is conjoint, collateral and closed.

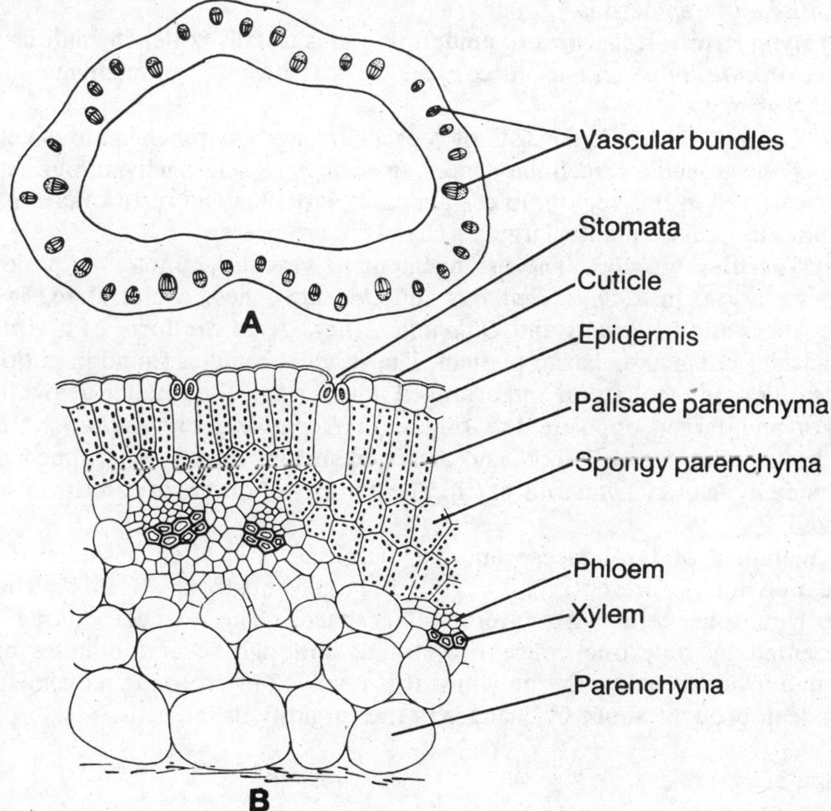

Fig. 9.8. V.S. of leaf of onion (*Allium cepa*). A--diagrammatic and B--an enlarged cellular portion.

Xylem and phloem elements have usual structure and remains surrounded by parenchymatous bundle sheath. The bundle sheath is single layered.

GENERAL ACCOUNT OF LEAF ANATOMY

The two parts of leaf, the petiole and the lamina, are considered for studying the internal structure of complete leaf.

Petiole

In a cross section, petiole may be of different shapes. In *Nymphaea,* it may be ciruclar while it is triangular in *Mangifera,* with an adaxial groove in *Prunus* and *Citrus* and with distinct ridges and furrows as found in *Cucurbita.* The internal organisation shows following important features:

(i) **Epidermis.** It is single layered composed of compactly arranged barrel-shaped cells. However, cells may be radially elongated in certain aquatic plants such as *Eichhornia* and *Nymphaea.* Epidermis is as usual covered with cuticle in terrestrial plants. In *Banksia* and *Cucurbita,* several multicellular hairs arise from epidermis.

(ii) **Hypodermis.** It lies next to epidermis and is usually collenchymatous in nature. In *Mirabilis jalapa,* it consists of patches of collenchyma and sclerenchyma.

(iii) **Ground tissue.** It consists of loosely arranged parenchymatous cells with well developed intercellular spaces. In *Mangifera,* sclerenchymatous cells occur scattered in this region. In certain cases, laticifers and resin canals may be found. In aquatic plants, large air chambers are present.

(iv) **Vascular bundles.** The arrangement of vascular bundles in petiole shows variation. In *Ricinus, Platanus* and *Quercus,* these are arranged in a single ring while in *Prunus* and *Galphimia,* they are in the form of a girdle with adaxial xylem and abaxial pholem. The vascular bundles found in petiole of *Cucurbita* are bicollateral and arranged in two rings, the smaller below the grooves and larger opposite the ridges. In *Heptapleurum venulosum,* the vascular bundles are collateral and arranged in two rings, each capped on outer side by sclerenchymatous patch. The bundles of outer ring are inversely oriented.

In members of Oxalidaceae and Leguminosae, a swollen basal portion called pulvinus is present. This region is largely made up of thin walled parenchymatous cells with intercellular spaces. The vascular tissue is represented by only one concentric strand, although several bundles are present above and below the pulvinus. It is involved in drooping movements of the leaf, brought about by changes in the turgidity of the cells.

Lamina

The internal organisation of leaf lamina shows differentiation into epidermis, mesophyll and vascular tissues.

Epidermis. Epidermis forms the dermal tissue system of leaf. Both upper and lower epidermis are usually uniseriate composed of compactly arranged cells. In *Ficus*, the upper epidermis is multiseriate whereas in *Nerium*, both upper and lower ones are multiseriate. In *Begonia* (a succulent plant), the upper epidermis is three layered while lower one is two layered. Usually, all cells of epidermis are of similar type but in certain grasses, groups of large cells but with thin flexible walls are found in longitudinal furrows which are called bulliform cells or motor cells. These help in rolling of leaves during dry weather, thus preventing the loss of water via transpiration.

The outer walls of epidermal cells have deposition of cutin, which is very thick and distinct among xerophytic leaves and relatively less distinct in mesophytic leaves. Cuticle is absent in case of submerged, aquatic leaves.

At places, stomata are found. It is an important feature of the epidermis. If these are present only on upper surface, the leaf is called **epistomatic** as in case of *Nymphaea*, if on only lower surface it is **hypostomatic** as in mango, oleander and banyan and if on both the surfaces, it is **amphistomatic** as in isobilateral leaves. Stomata remain surrounded by two highly specialised cells called guard cells which may be reniform or kidney-shaped as in most dicots or dumble-shaped (double headed) as in grasses or monocot leaves. The stomata may be present at the same level to epidermis as in mesophytic leaves, or may be sunken as in *Cycas, Pinus* or *Allium* or in stomatal cavities as in *Nerium*. In certain cases, glandular and non-glandular hairs arise from epidermis. Some epidermal cells are idioblasts which contain **cystolith** as in *Ficus* and *Momordica*. Some silica cells may be present in epidermis as in certain grasses.

Mesophyll. It is the most important part of leaf and forms the bulk of internal tissue of leaf extending from upper epidermis to lower epidermis. In most dicotyledonous leaves, it shows remarkable differentiation into palisade and spongy cells. **Palisade cells** are columnar and more or less compactly arranged with their long axis at right angles to the leaf epidermis. The amount of palisade varies in different plants which also correlates with external environment. The xerophytic plants have well developed palisade tissue. In *Rosa* and *Lawsonia*, the palisade cells are arranged in a single layer but these are arranged in two to three layers in *Musa, Mangifera, Ficus* and *Eucalyptus* and many layered in *Malus* and *Nerium*. In centric leaves as in *Allium* and *Juncus*, the palisade occurs all around the periphery of leaf. In number of dicotyledons such as *Caltha, Anemone* and *Aconitum*, and monocotyledons such as *Bambusa* and *Arundinaria*, the palisade cells have peculiar folds projecting into the cell cavity and these are called **arm-palisade.** In certain members of Piperaceae and Gesneraceae, the palisade cells may be funnel-shaped and hence called funnel cells. In some members of Liliaceae, the palisade cells appear branched due to presence of large lobes. The arrangement of palisade cells at the right angles to the leaf surface protects chloroplasts from excessive heat of sun rays.

Spongy cells are of variable shapes and are loosely arranged with abundant intercellular spaces. The intercellular spaces which maintain continuity with stomatal chamber facilitate the gaseous exchange.

In some monocotyledonous leaves, such as grasses, mesophyll tissues are not differentiated into palisade and spongy cells and there occur only one type of cells. These cells contain abundant chloroplasts and intercellular spaces. In aquatic plants, mesophyll region accommodates large air chambers. The cells surrounding them have fewer chloroplasts. Sometimes, various types of idioblasts are found in mesophyll region. Presence of lithocysts in *Ficus,* oil glands in *Citrus* and laticifers in *Calotropis* are few such examples. In certain plants, stellate cells and trichosclereids are present to provide mechanical strength.

Vascular tissues. The vascular tissues in leaf show no much variation. The mid rib in most dicot leaves have large bundles with adaxial xylem and abaxial phloem. This bundle is conjoint, collateral and open. A thin strip of cambium is present between xylem and phloem. In *Liriodendron* and *Vitis,* the mid rib region has several bundles arranged in a ring while in *Helianthus* several, but irregularly distributed bundles are found. Vessels in xylem and sieve tubes in phloem are found in larger bundles but in smaller bundles, xylem is represented by tracheids and phloem by some sieve elements. It is interesting to note that larger veins in case of some dicotyledons such as *Ficus religiosa* show a limited amount of secondary growth as a result of activity of cambium.

In monocotyledonous leaves also, the bundle is conjoint, collateral and closed.

The vascular bundles in leaf have characteristic bundle sheath. The bundle sheath is usually composed of parenchymatous cells which are colourless and morphologically distinct. In dicotyledons, the parenchymatous cells extend in the direction parallel to veins. Mostly the bundles have uniseriate sheath of parenchyma cells but sometimes a sheath of sclerenchyma is present around the vein as in members of Melastomataceae and Winteraceae. The sheath in dicots usually extends upto epidermis on one or both sides of leaf.

In monocotyledons, the bundles may remain partly or completely surrounded by one or two bundle sheaths, each consisting of a single layer of cells. The cells of inner sheath are smaller and achlorophyllous and have thick walls while cells of outer layer have abundant chloroplasts and are thus usually darker than other mesophyll cells. The chloroplast containing bundle sheath cells have been studied in detail in certain monocotyledons such as maize and sugarcane and it has been seen that these contain chloroplasts in which grana are absent. These participate in C_4 Cycle, an alternative pathway of C_3 cycle of photosynthesis. This cycle enables these plants to carry on photosynthesis at low carbon dioxide concentration and at relatively higher temperature. As such C_4 plants are considered to be more efficient in manufacturing the food via photosynthesis. From evolution point of view, it is an advanced physiological character of C_4 plants (monocots) over C_3 plants (dicots).

In general, the leaves of dicotyledonous and monocotyledonous plants show following important differences, as summarised in Table 9.1.

Table 9.1. Differences between Dicot and Monocot leaves.

	Dicot leaf		*Monocot leaf*
1.	The stomata on epidermis are generally present only on lower surface, thus leaves are hypostomatic types.	1.	Stomata are found on both the surfaces of leaves, thus leaves are amphistomatic types.
2.	Guard cells are kidney-shaped.	2.	Guard cells are dumble-shaped.
3.	In epidermis, bulliform cells are not found.	3.	Bulliform cells are generallly present.
4.	Mesophyll tissues are generally well differentiated into palisade and spongy cells.	4.	Mesophyll tissues are not differentiated into palisade and spongy cells.
5.	Intercellular spaces are large and numerous.	5.	Intercellular spaces are fewer and smaller.

GENERAL QUESTIONS

1. Discuss briefly the internal organisation of dicotyledonous leaves.
2. Write short notes on internal organisation of leaf of:
   (i) *Ficus* (ii) *Nerium* (iii) *Zea mays*.
3. Write short notes on:
   (i) Differences between dicot and monocot leaves
   (ii) Bulliform cells
   (iii) Mesophyll tissues.

CHAPTER 10

Secondary Growth

The growth in main axis in length which takes place by the activity of primary meristem (apical meristem) is called primary growth. Most of the monocots, many annuals and herbaceous dicots have only primary growth. In majority of gymnosperms and dicotyledons, stem increases in girth due to secondary growth which is brought about by the activity of lateral meristems. It occurs in the main stem and its branches. Secondary growth is not found in monocot stem and roots.

SECONDARY GROWTH IN DICOTYLEDONOUS STEM

The primary organisation of a typical dicot stem shows the presence of epidermis, hypodermis, cortex, stele (vascular bundles) and pith. Each vascular bundle is conjoint, collateral and open having intra-fascicular or fascicular cambium. The fascicular cambium is a type of lateral meristem which is found in stelar region and produces secondary xylem and phloem. Yet another lateral meristem is phellogen that forms the periderm. It is the activity of fascicular cambium which is mainly taken up in this account.

Activity of Cambium

In the process of secondary growth, the first step is the formation of cambium ring.

Cambium ring. The intra-fascicular cambium or cambium of vascular bundles becomes active and meristematic. At the same time, some cells between the two vascular bundles (may be called cells of medullary rays) lying in the line with fascicular cambium also become meristematic and form interfascicular cambium (the cambium between the two vascular bundles). The interfascicular cambium joins the strips of intra-fascicular cambium from the sides to form a complete circular ring of cambium (Fig. 10.1B). This is known as **cambium ring** which gives rise to secondary tissues in the stelar region. In certain cases, such as *Linum* and *Tilia*, the vascular bundles are found very close to each other, and as such the intra-fascicular cambium forms the complete ring without the formation of interfascicular cambium.

The vascular cambium consists of two types of cells, the *fusiform initials* and the *ray initials.* Cells of fusiform initials are vertically oriented and produce different elements of xylem and phloem while the cells of ray initials are isodiametric which produce rays of parenchyma in secondary xylem and phloem.

Secondary vascular tissues. Secondary vascular tissues are formed by the activity of cambium. The fusiform initials of cambium divide vertically producing two daughter cells. One of these remains meristematic while the other forms xylem or phloem mother cell which directly differentiates into xylem or phloem, as the case may be (Fig. 10.2). The new cells formed towards the outer side facing the primary phloem get gradually modified into the elements of secondary phloem while those towards the inner side facing the primary xylem are modified into elements of secondary xylem. However, in gymnosperms, xylem mother cell undergoes several transverse divisions to form tracheids, fibres and xylem parenchyma.

The **ray initials** of cambium cut off parenchymatous cells both outwards and inwards and thus form secondary medullary rays. Such rays extend from secondary phloem to xylem and pith.

In the beginning, secondary tissues are formed more in the region of vascular bundles, as intra-fascicular cambium first becomes active, but soon,

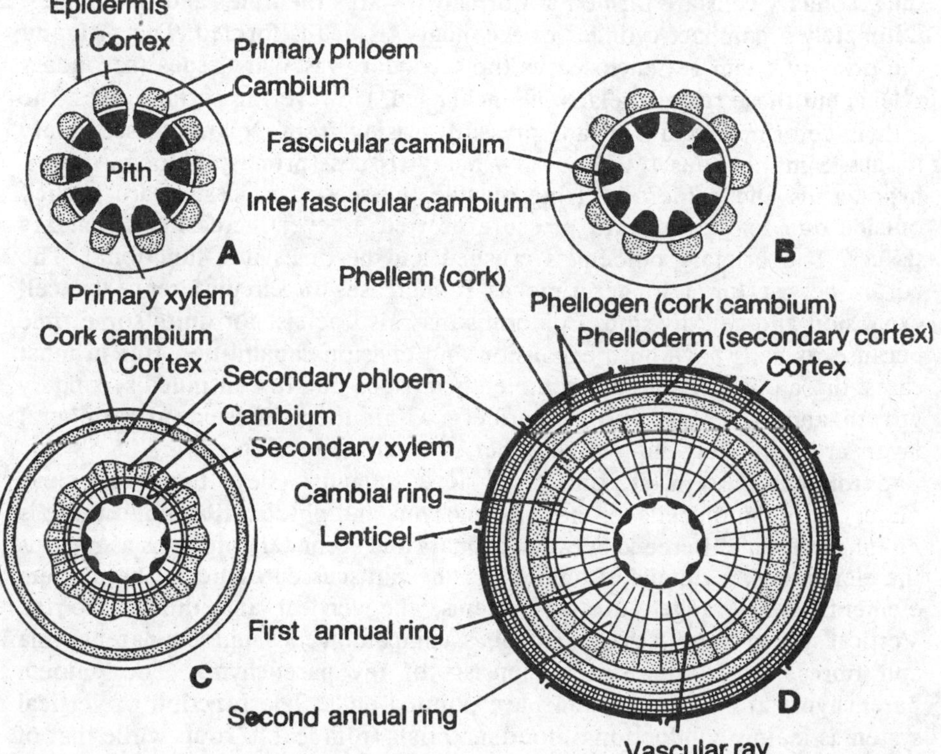

Fig. 10.1. Schematic representation of secondary growth in thickness in a dicot stem. A to D–different stages in the secondary growth.

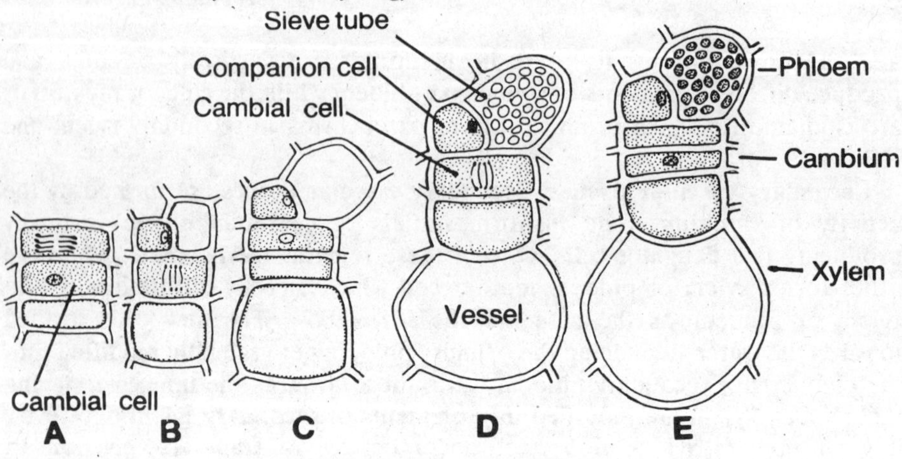

Fig. 10.2. Schematic representation of different stages of differentiation of conducting tissues (xylem and phloem) from a combial cell (Redrawn after Holman and Robbins, 1934).

the cambial activity becomes uniform throughout the ring and subsequently the secondary cells are formed uniformaly towards the outer and inner sides. Ultimately a compact cylinder of secondary tissues is formed. The pith and the primary xylem get enclosed by the secondary vascular tissues (secondary xylem) but these remain relatively unchanged. However, in certain cases, the pith is deformed by an inward pressure arising from expanding secondary tissues. Simultaneously, the outer primary tissues (primary phloem, cortex, hypodermis and epidermis) lying outside the secondary tissues are pushed outside due to the outward pressure exerted by ever increasing secondary tissues. The primary phloem is crushed and becomes non-functional. The cortex persist for a longer time as it increases in circumference by cell expansion and cell division. Epidermis may also persist for quite some time because its cells have both expansion and division capabilities. But in most cases, these outer tissues are completely crushed due to continued secondary growth–and by the formation of periderm within them. These dead peripheral layers are ultimately eliminated either by decay or by abscission.

Secondary phloem. It consists of sieve elements (sieve tubes and sieve plates), companion cells, phloem parenchyma and phloem fibres and there is no fundamental difference between primary and secondary phloems as regards the elements (the details of phloem has been discussed earlier). The phloem elements are arranged in two systems, the vertical and the horizontal. Vertical system comprises of sieve tubes, companion cells, phloem parenchyma and fibres. Horizontal system consists of ray parenchyma. The phloem parenchyma are elongated and have pointed ends. The function of vertical system is mainly conduction of food materials from leaf to roots while that of horizontal system is to conduct food material laterally to cortex or pith. Phloem fibres provide mechanical strength to the plant and are highly valued.

Many of the textile fibres of commercial uses, e.g. jute, hemp, flax etc. are phloem fibres or bast fibres.

Secondary xylem. The xylem elements formed by the activity of cambium are collectively called secondary xylem. It consists of tracheids, vessels, fibres and wood parenchyma. Tracheids and vessels have scalariform, reticulate and pitted thickenings. In general, the structure of both primary and secondary xylem is too similar to be distinguished.

The main difference between the two xylems is that the primary xylem has usually long tracheary elements in comparison to those of secondary xylem. Secondly, the secondary xylem has two systems of tissues, i.e. axial system or

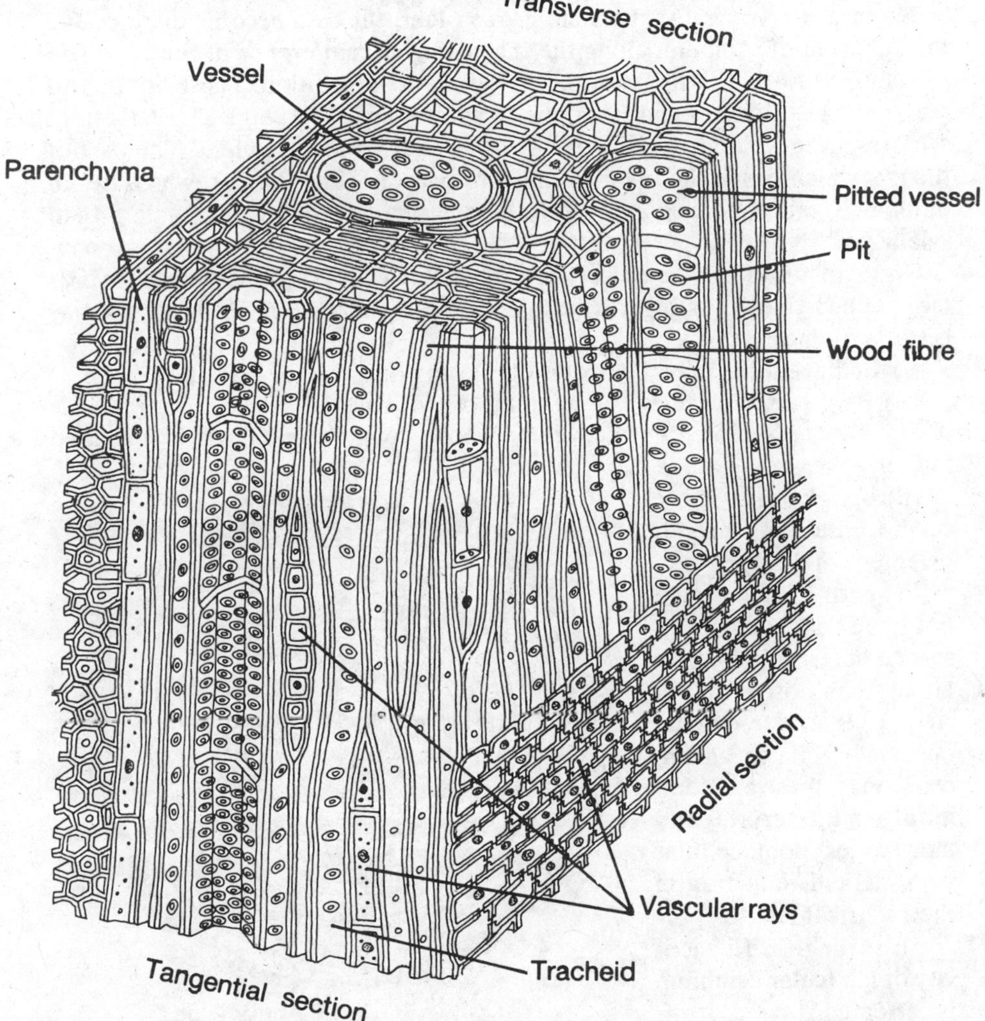

Fig. 10.3. Three dimensional views (transverse, radial and tangential) of a piece of oak wood (Redrawn after Fuller and Tippo, 1954).

vertical or longitudinal system and radial system which remains absent in primary xylem. The axial system is made up of vertical rows of cells with their long axis parallel to the long axis of the organ. It consists of tracheids, vessels (trachea), fibres and wood parenchyma. The radial system consists of rows of parenchymatous cells oriented at right angles to the longitudinal axis of the plant and forms the xylem rays.

Structure of Wood

Secondary medullary rays. Ray initials of cambium form some narrow bands of parenchymatous cells which in transverse section appear as radial rows of cells extending radially from pith to the phloem. These bands are called medullary rays. As the stem grows older, the rays become distinct due to expansion of cambium along the increasing circumference of the axis. The rays show variation in their width and cellular composition. These are one to few layers in thickness and from one to several layers in height. Usually both are present in a wood, though one or other may be less frequent. The part of the ray, which lies in the xylem is called as *xylem ray* and the part which lies in phloem is called *phloem ray.* The phloem rays are relatively smaller than xylem rays.

The xylem rays may be uniseriate (one celled wide) or multiseriate (two or more celled thick). As regards cellular composition, the medullary rays may be of two types:

(i) **Homocellular rays,** i.e. made up of only one type of cells, and

(ii) **Heterocellular rays,** i.e. made up of two types of cells–

(a) *Procumbent cells*–Those cells whose longest diameter is oriented radially.

(b) *Upright cells*–Those cells whose long axis is oriented vertically.

In **primitive woods,** both uniseriate and multiseriate heterocellular rays largely composed of upright cells are frequently observed while in advanced woods, these are commonly multiseriate and more or less homocellular with marked reduction in their vertical length.

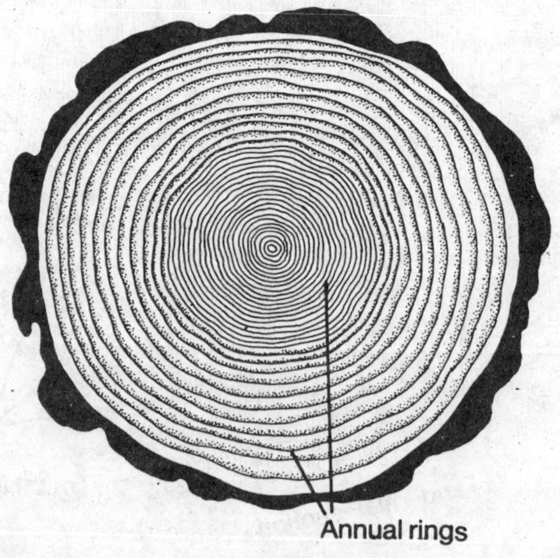

Fig. 10.4. T.S. of old dicot stem showing annual rings (diag.).

Annual rings. The activity of fascicular cambium is affected by environmental fluctuations. Such fluctuations are pronounced in temperate regions

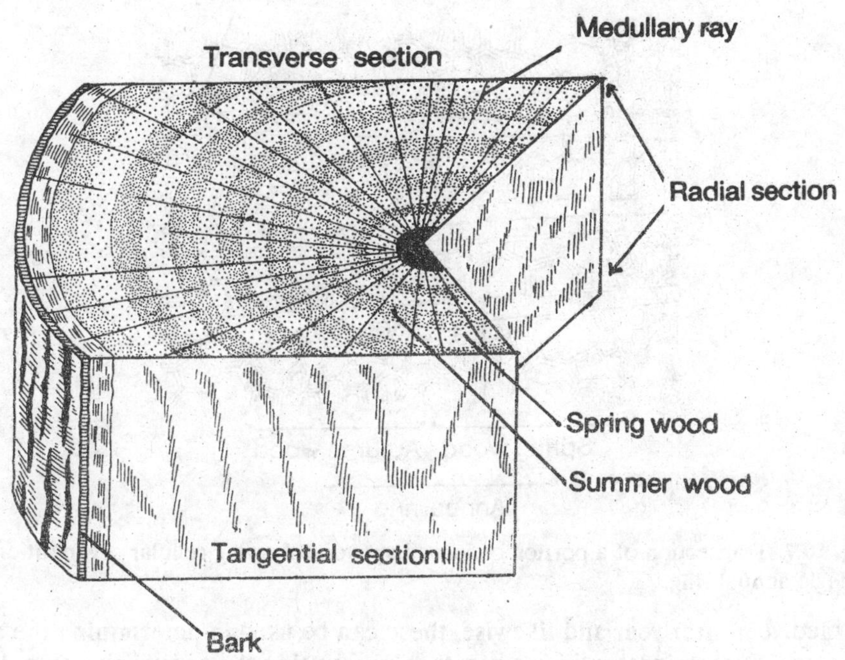

Fig. 10.5. Three dimensional structure of dicot stem.

in a span of a year where spring and autumn are well marked and extreme seasons. During these seasons, the cambium shows marked variation in the activity, and as such there is annual alternation of growing and quiscent (very slow growth) periods. The cambium remains inactive in winter but most active in the spring season. (Fig. 10.4).

The secondary wood formed during spring season is profuse and consists of large vessels with wider lumen. This part of the wood is called **spring wood** or **early wood.** In summer or autumn, the activity of cambium is highly reduced, therefore, secondary wood formed is not profuse and consists mainly of fibres and narrow vessels. This part of the wood is called **summer wood** or autumn wood or **late wood.** In a transverse section of stem, early and late woods appear in the form of distinct concentric rings which are commonly called **annual rings** (Fig. 10.4 and 10.7). Hence, an annual ring consists of two parts, the spring ring and the autumn ring which are formed in a span of one year. The successive annual rings are

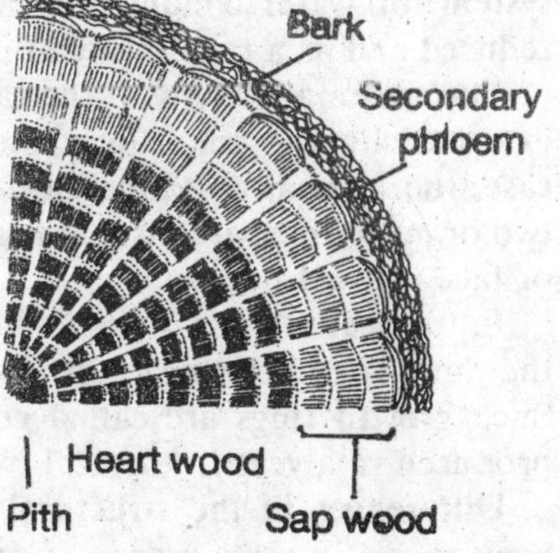

Fig. 10.6. Transection of a portion of the old dicot stem showing different structures (diag.).

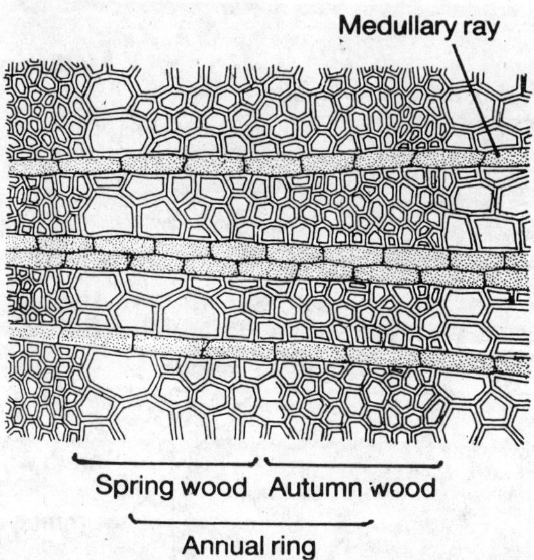

Fig. 10.7. Transection of a portion of secondary wood showing cellular organisation of a single annual ring.

formed year after year and likewise, these can be used in determining the age of a tree. Determination of age of a tree by counting the annual rings is called **dendrochronology.**

The variation in structure of spring and autumn wood is according to the requirments of the plant during these periods. In spring season, the vegetative activity is at its peak and there is an increase in transpiring surface. So there is a great demand for rapid conduction of water and mineral salts and to meet it the cambium actively divides to form additional water channels with broader vessels. On the contrary, during autumn season (summer), there is a leaf fall and less vegetative growth. There is no need of having well developed system for water conduction. Therefore, the activity of cambium is highly reduced and as a result, the autumn wood has relatively narrow vessels.

The seasonal activity of the cambium may be disturbed by the influence of external climatic conditions like drought, defoliation and diseases. In that case, the normal development of secondary vascular tissues is checked and two or more rings are formed in a year, these are called **multiple annual rings** or false annual rings.

In tropical regions, the climate is normally uniform and is generally wet so the number of growing rings formed does not correlate with age of the tree. Such growth rings are called **growth marks.** Sometimes, two rings may be produced in a year in tropical trees.

Differences in the structural components of the annual rings have also been found in certain cases. For instance, in 130 years old stem of *Quercus pedunculata,* the fibres of first ring were 0.42 mm in length and those of outer few rings, 1.22 mm. in length. Similar differences in the length of tracheids have also been found. However, vessels and parenchyma do not show any change in their length. Similarly, variation in the thickness of annual ring has also been found which depends upon fluctuations of general condition. Such variations may be used in determining the meteorological conditions which existed in the past.

Heart wood and Sap wood. In a transverse section of a woody trunk, two regions, the outer light coloured and central dark coloured regions, are distinct. The light coloured region is called sap wood and dark coloured region is called heart wood (Fig. 10.6). The **sap wood** (also called *albernum*) consists of living and functional cells involved in conduction of water. In fact, the outer few layers carry on conduction process while cells of inner rings function as storage organ.

The **heart wood** (also called *duramen*) is composed of dead cells with their walls heavily impregnated with various compounds such as resins, gums, tannins, pigments or phenolic compounds and hence becomes unsuitable for conduction. The pigments found in heartwood in certain cases are of commercial importance, e.g. haematoxylin (*Haematoxylon campechianum*), Brasilin (*Caesalpinia sappan*), santalin (*Pterocarpus santalinus*) etc.

As the growth process continues, the rings of sap wood bordering the heartwood keep on converting into heart wood. In the process, the living cells of sap wood lose their protoplast, water contents are reduced and vessels are blocked by tylosis (Fig. 10.8). Simultaneously, deposition of various substances like gums, resins or tannins also occurs. So with the increase in the growth amount of heart wood keeps on gradually increasing.

In certain cases, distinction between sap wood and heart-wood is not sharp as in *Holoptelea integrifolia* and *Abies pindrow*.

Porous and Nonporous wood. These two types of wood are recognised on basis of presence and absence of pores of commerce, i.e. vessels. The wood in which vessels are not found is called **non-porous wood** such as that of gymnosperms. Whereas the dicotyledonous or angiospermic wood which has vessels in xylem is known as *porous wood.* The non-porous wood is technically called **soft wood** and porous wood is called **hard wood.** Soft wood or hard wood terms are not at all related with softness or hardness of timber, e.g. hard wood of *Bombax cciba* (angiosperm) is very soft and soft wood of *Pinus* and *Cedrus* (both gymnosperms) is very hard. The wood of balsa (*Ochroma lagopus*), an angiosperm, is the lightest yet it is classified as hard wood.

The structure and composition of porous wood is more complex than non-porous wood. In a transverse section, the pores (or vessels) may appear circular or oval in outline. Several woods of commercial value are identified on basis of arrangement of pores. On basis of distribution of pores the woods can be classified into *ring porous* and *diffuse porous.* In the **ring porous wood,** pores of early wood are larger than those of late wood and therefore, these appear distinctly arranged in the form of a conspicuous ring, such as in most temperate angiospermic trees. Ring porous wood is characteristic of *Lagerstroemia, Morus, Tectona, Tilia, Tamarix, Ulmus, Toona* etc.

In **diffuse porous wood,** the pores are more or less uniformly distributed throughout the ring because there is no appreciable difference between early and late woods such as in most tropical angiospermic trees. Most of the Indian hard woods come under this category. Diffuse porous wood is characteristic of *Artocarpus, Acacia, Pterocarpus, Michaelia, Adina, Hopea, Albizzia, Terminalia, Mangifera, Dalbergia, Shorea, Bombax* etc.

Reaction wood. Reaction wood develops under stress conditions particularly to counteract the forces which induced the inclined position of branches. Under such circumstances, the growth of wood is unequal and cross sections of stems show eccentric position of secondary xylem and phloem indicating the difference in the cambial activity in various regions. In the formation of reaction wood, stimulus of the gravity and the distribution of auxins play an important role.

In gymnosperms, reaction wood develops on the lower side of branches and is called the **compression wood** while in dicotyledons (angiosperms), it develops on the upper side of branches, and is called the **tension wood.**

The reaction wood shows some structural differences from the normal wood. In gymnosperms, the tracheids of compression wood (reaction wood) are relatively shorter and rounded with their cell walls heavily lignified as compared to those of normal wood. In angiosperms, the vessels of tension wood are reduced in number and width. The wood has also got characteristic presence of gelatinous fibres which have a thick highly refractive gelatinous layer in the cell wall.

Wood parenchyma. The parenchyma found in wood arise either from fusiform initials or from ray initials of the cambium. The former is called **axial parenchyma** and latter is called **ray parenchyma.** The ray parenchyma are usually shorter than axial parenchyma and may some times develop secondary thickening in their walls. The distribution of axial parenchyma in the wood shows considerable variation in different plants and that is why these are quite important for taxonomic studies (the details have been dealt in a separate chapter). In most of the woods, axial parenchyma forms a larger portion of wood but in certain cases, it may be entirely absent as in *Homalium* and *Sonneratia* or if present, these are too less to be detected as in *Boswellia serrata* and *Garuga binnata* (for details on wood parenchyma, see chapter on *Systematic Plant Anatomy*).

Tyloses. In many plants, wood parenchyma cells develop balloon like protrusions into the tracheary elements. Such ingrowths found in the lumen of either tracheids or vessels are called tyloses (Fig. 10.8). These are usually found in secondary wood but at times, tyloses may also develop in primary xylem. The development of tyloses is simple. The parenchyma adjoining the pits of vessels, penetrate into the vessel in the form of short protuberances, each gradually enlarges to form a balloon-like structure (Fig. 10.8A-F), and become so much enlarged that the lumen becomes almost blocked. Soon the wall of parenchyma also gets thickened due to gradual lignin deposition. In a tylose, a nucleus and some starch, resin or other materials remain present. Usually tyloses occur singly but in certain cases, they may divide to form a multiple structure.

Tyloses are frequently found in several angiospermic heart wood. These

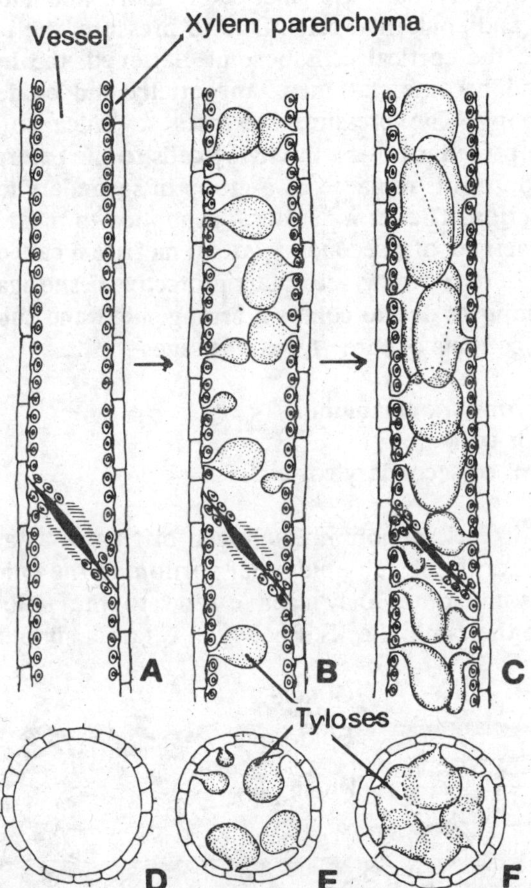

Fig. 10.8. Different stages in the development of tyloses in xylem vessel. A to C–longitudinal sections and their D to F–transectional views.

block the passage of tracheids and vessels and prevents the flow of fluids and infectious agents from the soil such as fungal spores or hyphae, that destroy the wood. In this way, these significantly induce maturation and durability of wood. The presence of tyloses in certain herbaceous plants such as *Cucurbita, Coleus* and *Canna* is also known.

In gymnosperms, the tylose-like intrusions are found and these are called **tylosoids.** These are common in the wood with resin ducts, in vertical and horizontal systems. Here, the epithelial cells of the wall of resin ducts enlarge, block the duct and look like a tylose intrusion. These are not true tyloses because they never grow through pits.

PERIDERM

Because of continued secondary growth, the stem gradually keeps on increasing in width and secondary tissues formed exert more and more pressure on phloem, cortical and epidermal cells. Due to pressure, the phloem cells are pushed outwards, the cortical cells become flattened, the hypodermal cells are disturbed and become flattened tangentially and epidermis is put to maximum stretching. The pressure increases so much so that epidermis ultimately gets ruptured exposing the living cells to the external atmosphere. At this stage epidermis is replaced by a group of secondary tissues which are protective in function. The term periderm is applied to these cells. **Periderm** is formed by the activity of a secondary lateral meristem called **cork cambium** or **phellogen.** Periderm serves as secondary protective tissue against dessication and mechanical injuries due to compact arrangement and suberised wall.

The periderm consists of three types of tissues–

(i) *phellogen* or cork cambium,
(ii) *phellem* or cork, and
(iii) *phelloderm* or secondary cortex.

(i) **Phellogen** or **cork cambium**–Phellogen originates as a single layer of initiating cells either in the sub-epidermal portion, in the epidermis itself, in the cortex and some times may even extend to the phloem. These are arranged just like the cambial cells. Phellogen is rather simple in comparison

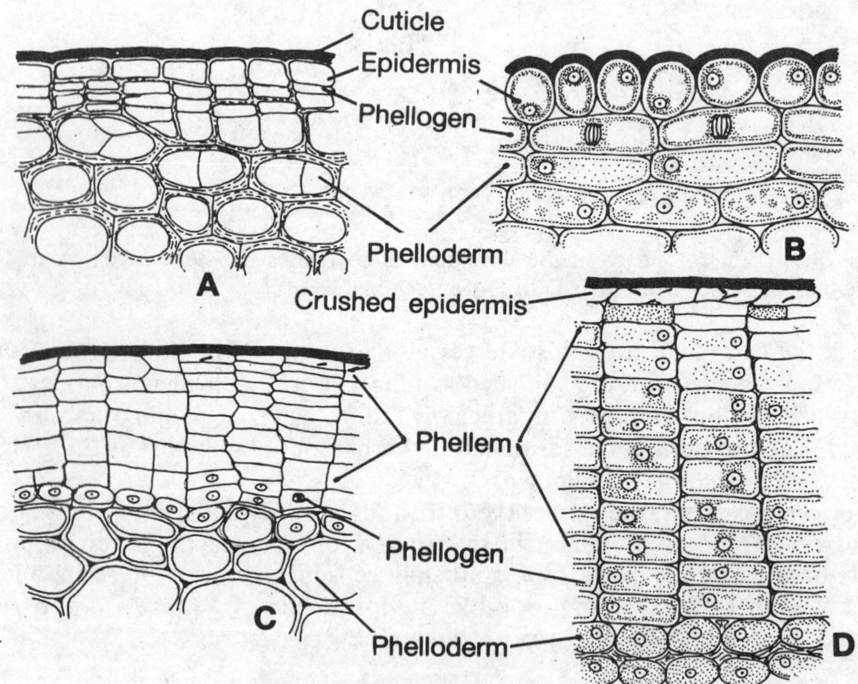

Fig. 10.9. Origin of periderm. A and B–in *Pyrus* and C and D–in *Prunus* (Redrawn after Esau, 1965).

to stelar cambium as it consists of only one type of initials. The cells are narrow, thin walled and radially flattened or rectangular in shape and are compactly arranged. These have vacuolated protoplasts.

Cells of phellogen behave like cambial cells which divide tangentially producing new tissues both towards the inner as well as towards the outer side. These secondary cells usually remain arranged in radial rows. The cells towards inner side form the secondary cortex and those towards outer side form the phellem or cork (Fig. 10.9).

(ii) **Phelloderm** or **Secondary cortex**–Secondary cortex is formed towards the inner side of the cork cambium. The cells of this region are thin-walled and living and have pits on their cellulose wall. So these cells resemble those of cortex in their nature, wall structure and contents. These are arranged in radial rows and isodiametric in shape. Some intercellular spaces are found between them. The cells serve in storage of food. Sometimes, these may contain chloroplast and may carry on photosynthesis.

(iii) **Phellem** or **cork**–Phellem arises towards the outerside of phellogen and are produced in more quantity than phelloderm. The cork cells are subersied and thick walled, dead, compactly packed and arranged in radial rows without intercellular spaces. These characteristically appear rectangular in transverse section. They lose protoplasts after differentiation and become dead and remain filled with air and coloured organic matters.

The wall characters of cork cells are quite important and useful. Due to deposition of suberin these become impervious to water and gases and thus they have more commercial value. Cork is commonly used as bottle stoppers. Sometimes, structure of cork also shows variation. In *Eucalyptus,* cork cells remain filled up with resinous and other materials while in *Betula*, the cork cells are thin walled and two types of phellem are formed in alternating layer due to which cork is peeled off like thin sheets of paper. These thin sheets were used in writing in ancient times and were popularly called **bhoj patra.**

BARK

Due to continuous activity of phellogen, the phellem becomes multilayered. Cells of outer layers fail to get sufficient supply of water and food and eventually all of them become dead. These dead cells outside the active cork-cambium constitute the bark of trees. The bark becomes conspicuous when cork-cambium is deep seated and then it may be composed of dead cells derived from epidermis, cortex and phloem. Bark may be found in stem as well as roots.

The term bark has been loosely used and thus causes considerable confusion. In non-technical sense, the term bark means all the tissues outside the cambium. Some anatomists prefer to use the term **rhytidome** for outer bark and the term bark for all the tissues external to the vascular cambium. Some use the term rhytidome for the tissue cut off outside the periderm.

Two types of barks are found, the *ring bark* or the *scale bark.* If the bark is found in the form of concentric rings surrounding the entire stem, it is called ring bark as in *Vitis, Betula* and *Clematis* and if the bark is found in the form of separate strips or patches, it is called scale bark as in *Ficus.*

LENTICELS

Due to continued secondary growth, the protective tissues or periderm are formed, of which, the cork cells are impervious to water and gases. Thus, the gaseous exchange between the internal living cells and, outer atmosphere becomes difficult and is brought about by lenticels. Lenticels are pore-like openings which look-like lens-shaped raised spots on the surface of stem. These take up the function of gaseous exchange. Lenticels are found in most woody plants where periderm is developed, except a few climbers.

Lenticels are usually formed below the stomata and develop simultaneously, rarely before or after, with the formation of periderm. During their origin, the substomatal parenchymatous cells become active and divide in different planes to form a mass of rounded cells (Fig. 10.10). These cells become enlarged, loose and light coloured. Simultaneously, the cork cambium in this region instead of giving rise to normal cork cells contributes some loosely arranged cells on outer side. Both these loosely arranged cells derived from division of parenchyma and cork cambium are collectively called **complementary cells.** The complementary cells are thin walled, somewhat oval in shape and loosely arranged with distinct and large intercellular spaces. Their continued growth causes the epidermis to rupture and as a result the underlying living cells are exposed. Rarely, lenticels develop independent of stomata.

Often, alternating with loosely arranged cells, masses of more dense and compact cells are formed. These are called closing cells and layer formed by

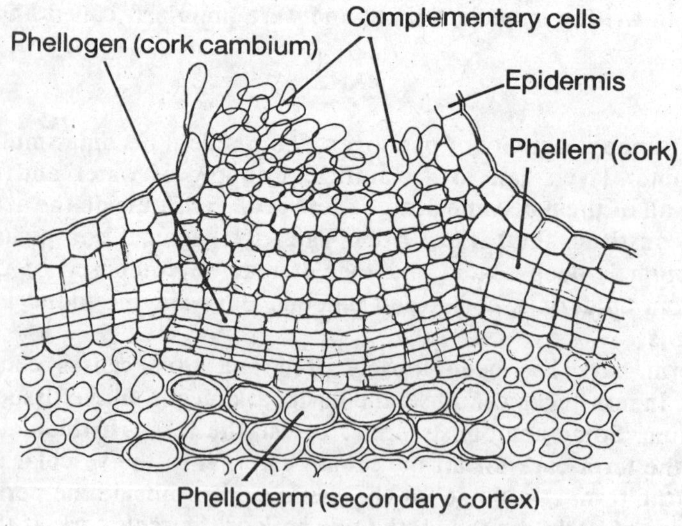

Fig. 10.10. Structure of a lenticel.

them is called *closing layer.* A closing layer is helpful in keeping the loosely arranged complementary cells in position.

Arrangement and deposition in the cell wall of complementary cells show variation in some cases, e.g. in *Magnolia,* these are somewhat compactly arranged with less distinct intercellular spaces, in *Quercus* and *Sambucus,* the cells are thin walled, unsuberised and loosely arranged and in *Prunus,* both suberised and unsuberised complementary cells are found. The number of lenticels formed in a stem is variable. They may remain scattered or arranged in longitudinal or vertical rows. Rows of lenticels may occur opposite the medullary rays, thus facilitating free interchange of gases between living cells and atmosphere.

SECONDARY GROWTH IN DICOTYLEDONOUS ROOT

Secondary growth takes place in the roots of gymnosperms and most of the dicots while it is absent in most of the monocot roots. It occurs to meet the growing needs of absorption and conduction of food materials.

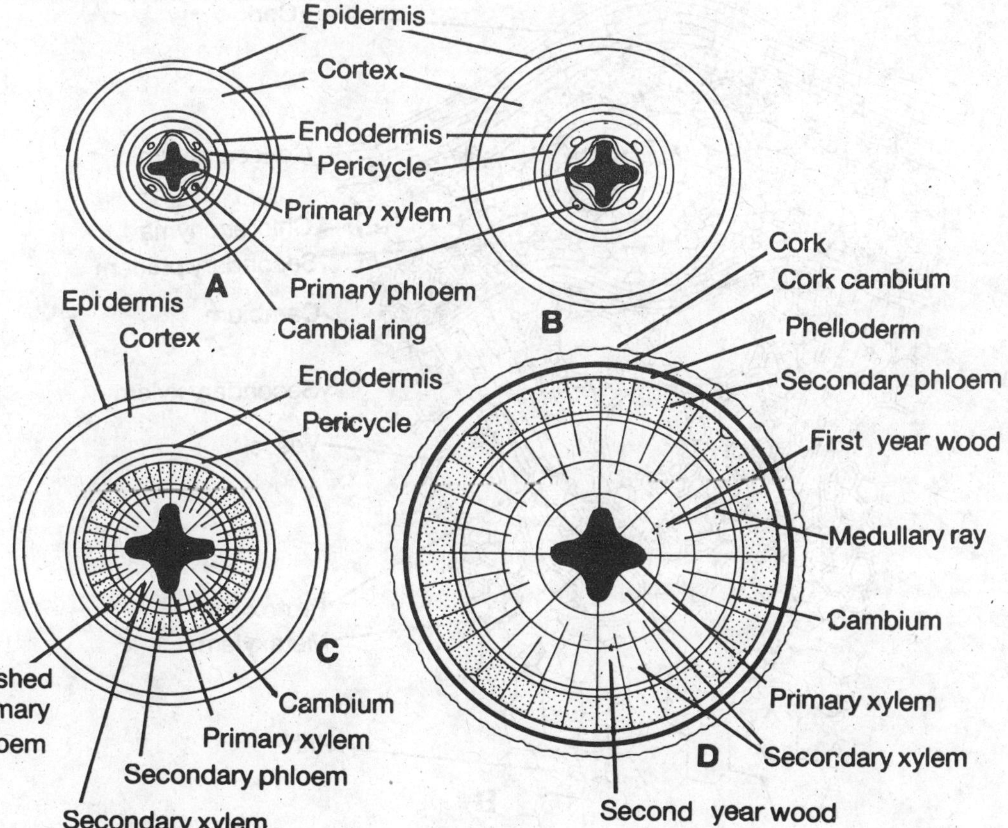

Fig. 10.11. Diagrammatic representation of different stages of secondary growth in a typical dicot root (Redrawn after Fuller and Tippo, 1954).

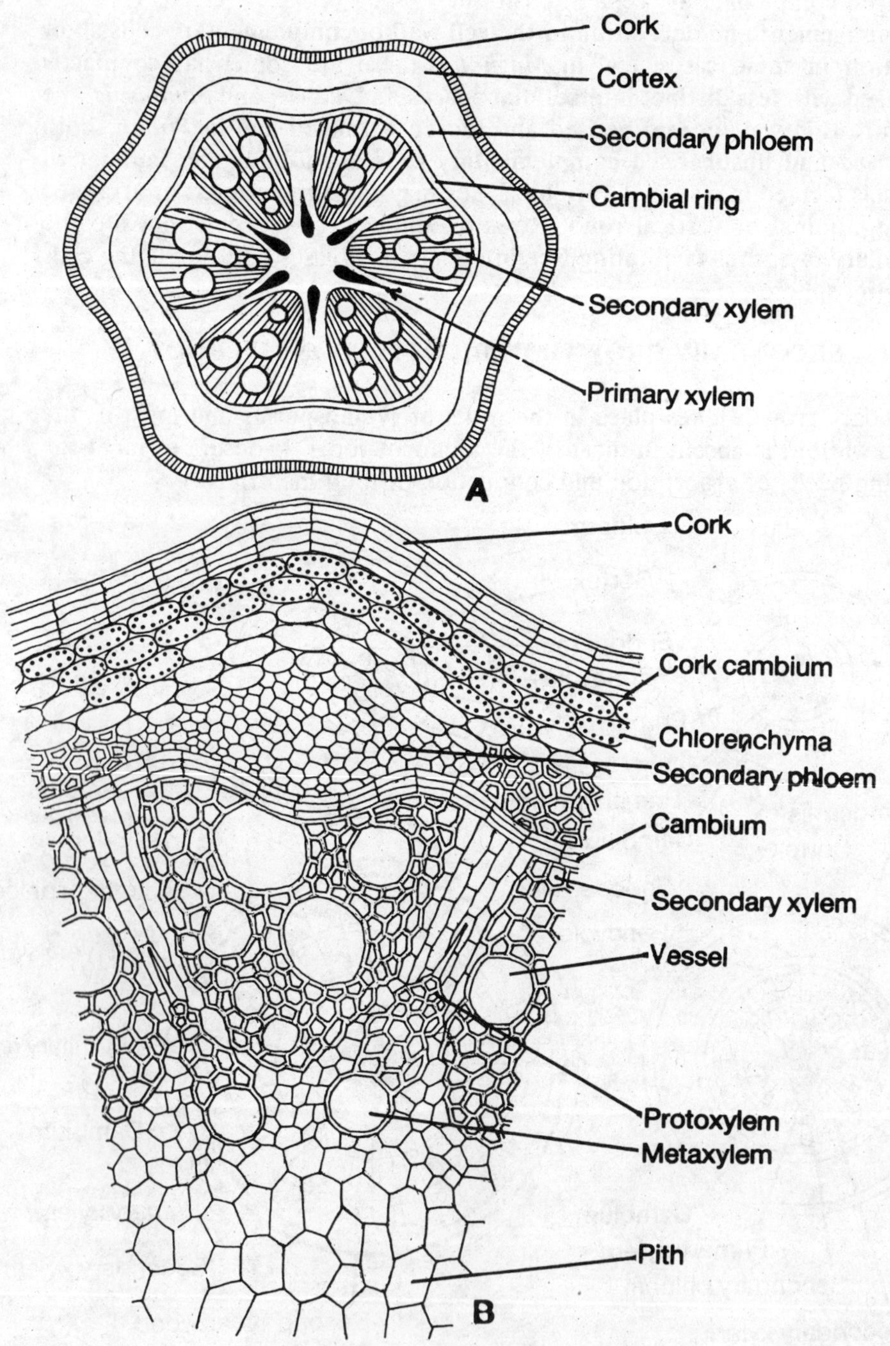

Fig. 10.12. Transection of *Tinospora* root showing secondary growth. A–diagrammatic and B–an enlarged cellular portion.

During secondary growth in roots two secondary meristems develop, one in the *stelar region*, called **fascicular** or **vascular cambium** which gives rise to secondary xylem and phloem and other in the *extra-stellar* region, called **phellogen** or **cork cambium** which forms periderm. Though the process of secondary growth in roots is initiated in somewhat different manner, the fundamental structure of secondary tissues formed remains similar to that of stem. It may be studied as follows:

Formation of cambium. In roots, a ring of cambium is formed mainly by parenchymatous cells situated along the inner edge of phloem groups which becomes meristematic and form crescent shaped strips of cambium and partly by pericycle cells lying against the protoxylem elements which becomes meristematic and forms small strips of cambium outside the xylem. These strips meet laterally and form a complete wavy band of cambium (Fig. 10.11A). The parenchymatous cells below the phloem may also be called *conjunctive tissues.* In many roots, the cambium which arises in the pericycle functions as ray initials and produce wide medullary rays.

Formation of secondary tissues. The first formed cambial strips present along the inner edge of phloem starts functioning earlier than the later formed cambium from the pericycle. It begins to cut off secondary xylem centripetally (towards inner side) and secondary phloem centrifugally (towards outer side), with the result the primary phloem is pushed outwards and soon the wavy band of cambium becomes a circular ring. After the formation of a circular ring of cambium, all cells of its become equally active and produce secondary xylem towards inner side and secondary phloem towards outer side.

Secondary phloem. The fundamental composition of secondary phloem formed in the root is similar to those found in a dicot stem. It comprises of sieve elements, companion cells, phloem parenchyma and phloem fibres. The primary phloem is usually crushed. However, the phloem in roots has less sclerenchyma and more storage parenchyma as compared to that of stem.

Secondary xylem. The secondary xylem elements found in roots are also similar to those found in a dicot stem. It consists of vessels, tracheids, xylem parenchyma and a few fibres. The secondary xylem can be distinguished from the primary xylem by the comparatively larger size of the vessel and thinner walls. The secondary xylem of roots have more parenchyma, fewer fibres, larger or more abundant rays and vessels with thinner walls as compared to those found in stem.

Annual rings. Because of the fact that soil temperature usually remains uniform throughout the year, annual rings are not well marked in roots. However, in very few plants like *Salix* (willow), the spring wood and autumn wood in the root remain clearly-distinct.

Medullary rays. In roots, prominent medullary rays are formed by that part of cambium which arises from pericycle. These remain radiating outwards through secondary xylem and phloem from the tips of protoxylem, e.g. *Ficus* root (Fig. 10.13). Such medullary rays may be called primary medullary rays. Some medullary rays are also formed by the cambium present along the inner edge of phloem and then these may be called secondary medullary rays.

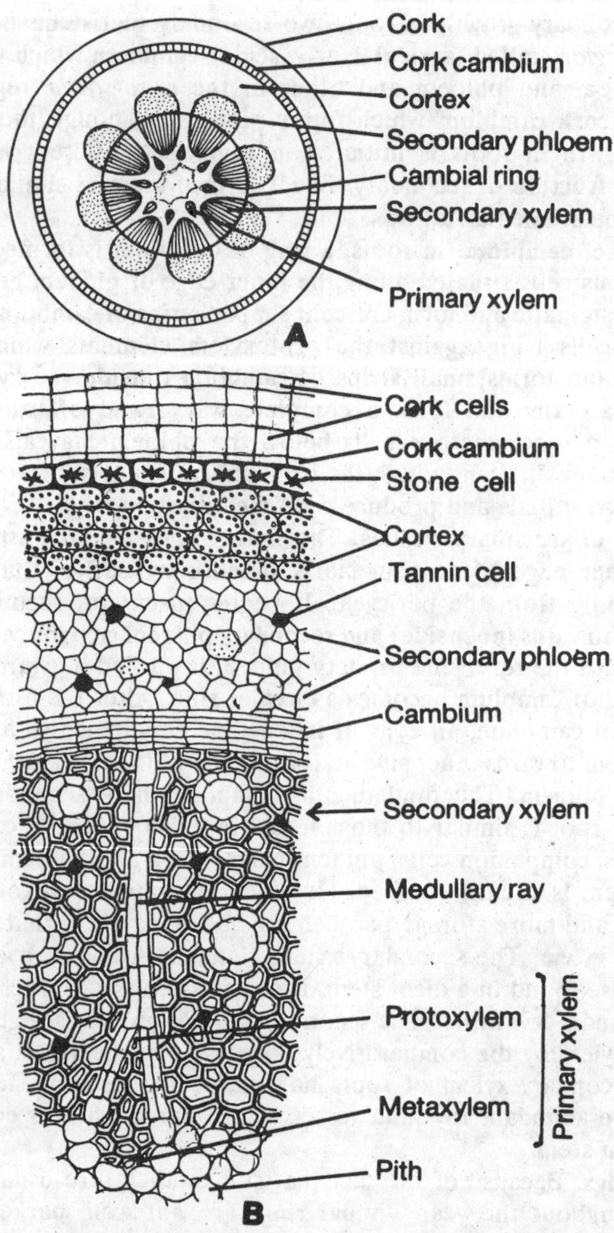

Fig. 10.13. Transection of *Ficus* root, A–diagrammatic and B–an enlarged cellular portion.

Periderm. The formation of periderm is quite similar to that of stem (Fig. 10.13).

Lenticels. Lenticels may be developed for the gaseous exchange almost in the same pattern as studied in stem.

GENERAL QUESTIONS

1. **Explain how does secondary growth take place in stem?**
2. **Write short notes on (i) annual rings, (ii) periderm, (iii) Tyloses, (iv) early wood and late wood and (v) lenticels.**
3. **What is the difference between:**
   (i) Cambium and cork cambium, (ii) Spring wood and autumn wood,
   (iii) Heart wood and sap wood, (iv) Annual rings and growth marks.
4. **Explain briefly how does secondary growth take place in a dicot root with the help of suitable diagrams.**

CHAPTER 11

Abscission and Healing of Wounds

ABSCISSION

The term abscission of leaf refers to separation of leaf from stem without causing any injury to stem. It is considered as a natural and special phenomenon which is commonly found in the leaves of woody dicotyledons and gymnosperms and less commonly among herbaceous angiosperms. Abscission is considered as a vital phenomenon involving several special changes, both structural and biochemical in nature occurring in the cells present at the base of petiole. It is supported by the fact that abscission occurs only in living branches while in dead branches leaves are not shed rather they get dried up and decayed in course of time and then fall off.

In deciduous species, the leaf fall is a seasonal phenomenon while in evergreen species, the individual leaf abscises after maturation and they may shed their leaves in any season. In deciduous plants such periodic defoliation occurs at the approach of winter, i.e. hardy or very unfavourable season. Before leaf fall, leaves undergo several biochemical changes and there is yellowing of tissues due to disintegration of chlorophyll and preponderance of carotenoid pigments. The waste products are also transferred to the leaves, so that plants may get rid of them with the leaf fall.

The leaf fall occurs because some structural modifications take place at the base of leaf and finally they are abscissed. Structural modifications include the formation of abscission zone which is usually initiated before the leaf fall. It is usually formed at the base of petiole in case of simple leaves and at the base of rachis and also at the base of individual leaflets in case of compound leaves. The **abscission zone** which some times can be distinguished externally by its paler colour and slightly constricted walls, consists of two distinct regions: the **separation** or **abscission** layer where local break down of cells occur and the **protective layer** which develops around the scar. The abscission or separation layer is considered to be the immediate structural reason for the fall of leaf. The protective layer, as the name indicates, protects the exposed surface after leaf fall from loss of water and attacks of micro-organisms.

Formation of abscission zone

The abscission layer gets differentiated few days or even few weeks before actual leaf fall takes place. It consists of a few to several layers of smaller parenchymatous cells with dense cytoplasm and abundant starch grains. Vessels usually become clogged by gums and tyloses below the zone. The pericycle fibres found in this region lack lignification and vascular elements have increased amount of parenchyma. The conduction is, however, maintained by xylem elements to keep the leaf turgid until it falls.

Several changes take place in the parenchymatous cells of abscission layer. The immediate change is the swelling of pectic material of the cell wall which is first enzymatically converted into pectic acid and then into water soluble pectin. It is followed by the break or dissolution of other cell wall components including non-cellulosic and cellulosic polysaccharides. The dissolution may be of three types:

(i) the middle lamella dissolves between two cells and the primary wall remains intact,

(ii) both, the middle lamella as well as primary walls of both the adjacent cells may dissolve, and

(iii) all the cells found in abscission layer may dissolve completely.

The break down of cell wall results in collapse of cells of the abscission layer. After the formation of the separation layer, which takes place from

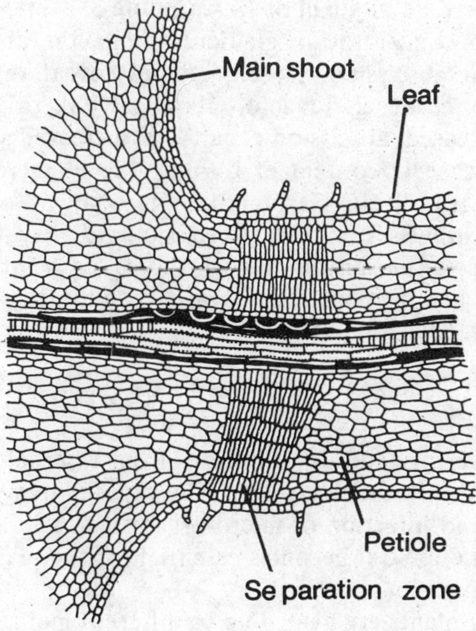

Fig. 11.1. Longitudinal section of leaf stalk of *Coleus* showing the separation zone (Redrawn after Torrey, 1968)

periphery to centre, the leaf remains attached to the stem only by the vascular elements. It breaks in due course, either by slight swaying due to wind or due to the mechanical weight of the leaf. The reduced vascular elements break by mechanical weight after the separation layer has been formed and ultimately leaf fall results.

After the abscission of leaf, the tissues of the scar are exposed to the air, dessication, outside injury and infection. To protect them, a protective layer is developed just beneath the separation layer which forms a cover around the scar caused by leaf fall. It consists of one to several cells deep tissue. The protective layer is either developed by suberisation and possibly partial lignification of already existing parenchyma cells (*the primary protective layer*) or a typical periderm is formed (*the secondary protective layer*) and thus protect the exposed surface. Some workers believe that this lignin is really a wound gum giving reactions similar to lignin. In most cases, phellogen is formed on the inner side of protective layer and a periderm is ultimately formed which during the course of time becomes continuous with the periderm of stem.

Cause of abscission

Since abscission is an important phenomenon found in certain plants efforts have been made to find out the physiological reasons and nature of stimulus that brings about initiation or formation of separation layer and ultimately leaf fall. Both auxin and abscissic acid (ABA) play a decisive role in this respect. A decrease in auxin or IAA (indole 3, acetic acid) causes leaf fall. A critical auxin concentration gradient across, the abscission zone, is necessary to prevent abscission. Recent studies reveal that an abscission accelerating hormone, abscissic acid, plays a vital role and when its concentration is increased, abscission occur. The increased synthesis of ABA may be due to decrease in content of IAA or due to increased amount of carotenoids in the leaf itself. Carotenoids are known to play a role of precursor in the synthesis of ABA. Abscissic acid may be formed from xanthophyll, such as violaxanthin which serves as a precursor for the biosynthesis of ABA and by oxidative cleavage, each molecule of violaxanthin yields two fragments, each similar to ABA.

HEALING OF WOUNDS

Wounds in plants are formed due to injuries from external sources (natural or induced) which exposes the living cells. The exposed living cells may suffer from loss of water and infection of micro-organisms such as bacteria, fungi, viruses, mycoplasma etc. So it becomes essential for the plant to heal up the wound as early as possible.

Wounds in higher plants are healed up by different methods depending on the nature of wound. In case, if the wound is **superficial,** the exposed underlying cells die and dry up. The cells present beneath the wound get impregnated with protective substance such as suberin. The process of healing of superficial wounds in plants having laticifers becomes more simple. They heel up simple wounds by secretion of latex which coagulates and serves as a protective cover.

In case of **deep wounds**, the exposed uninjured living cells beneath the wound become meristematic and begin to produce a mass of undifferentiated parenchyma cells. This amorphous growth of parenchyma cells is called callus and by which a wound is filled up. Soon after sufficient callus formation has taken place a cork cambium arises in its peripheral layers. The cork cambium produces a layer of cork on its outer surface, which may also be called *wound cork.* By the formation of wound cork the wound is healed up. The callus may some times over grow the wound and form characteristic swelling or knots on the stem. Such knots often get embedded within the secondary tissues during secondary increase in thickening in a course of time. Knots in the secondary woods are also formed when a branch is either cut off or broken down and to cover this injury the same procedure is followed. Callus develops and wound cork is formed and ultimately during long course of time, these get embedded in the secondary wood. They appear as knots in the timbers. Knots which frequently appear in boards (timbers) are the basal portions of branches which have become burried in the wood of the main trunk by the growth over a period of years of new annual rings over the branch base. The presence of knots weakens the board.

Sometimes, the wound is deep enough to injure the cambium of vascular tissues. In such cases also, callus is similarly formed and wound is filled up. After some time, a few parenchymatous cells of the callus becomes meristematic to form a new cambium . The new cambium joins up with the portions of uninjured cambium and thus, a continuity is achieved and the formation of secondary tissue is not disturbed. Such a vast capability of plants has been regularly exploited by horticulturists and they have encashed the advantage of this process in budding and grafting practices.

GENERAL QUESTIONS

1. Give a brief account of abscission of leaf and healing of the wounds.
2. Write short notes on:
   (i) Leaf fall (ii) Wound healing
   (iii) Knots (iv) Wound cork.

CHAPTER 12

Anomalous Secondary Growth

ANOMALOUS SECONDARY GROWTH IN DICOT STEM

In many dicotyledonous plants, the pattern of secondary growth shows a deviation from the normal type. The term anomalous secondary growth is given to this deviated pattern of secondary growth. In fact, the term anomalous simply serves to assemble growth patterns that appear to be less common (Esau, 1965). Anomalous type of growth may be more common than is known at present because the tropical flora in which it is frequently found has not been adequately studied anatomically.

Anomalous or abnormal secondary growth may be caused due to following reasons:

(i) The normal cambium behaves peculiarly or irregularly, resulting in abnormal arrangements of vascular tissues.
(ii) The normal cambium is situated in an abnormal position hence the tissues cut off are placed abnormally.
(iii) The normal cambium does not develop or if present, is replaced by additional or accessory cambial rings.

Abnormal secondary growth in plants may be studied under two headings–

(i) Plants in which cambium is of normal type but its activity is abnormal, resulting in abnormal arrangement of vascular tissues.
(ii) Plants in which the normal cambium is not formed or if formed, is soon replaced by accessory cambium. This accessory cambium may develop either from the cortex or pericycle region and exhibits abnormal activity.

However, the abnormal secondary growth may be studied as follows:

I. Cambium normal but showing abnormal activity

The cambium ring is formed normally but behaves in an abnormal manner. The abnormal activity of the cambium mav be categorised into the following types:

1. The cambium forms vascular tissues only in the region of vascular bundle. This type of anomaly is found typically in many woody climbers or lianas. The normal cambium ring is formed by the union of fascicular and interfascicular cambium, but the activity is abnormal in the sense that either the fascicular cambium or the interfascicular cambium is active. In case of the stems of *Vitis* and *Clematis,* the interfascicular cambium is active cutting off parenchyma only, whereas in case of *Cucurbita,* the fascicular cambium is active cutting off vascular tissue. However, in *Aristolochia,* the fascicular cambium cuts off secondary vascular tissue and the interfascicular cambium cuts off parenchyma cells.

(a) **Cucurbita stem** (Fam.–Cucurbitaceae)–The stem of *Cucurbita* has ten vascular bundles arranged in two rings of five each. Each vascular bundle is conjoint, open and bicollateral having an outer and inner cambium. The outer cambium of both inner and outer bundles becomes active along with the parenchymatous cells of the ground tissue and the two combine to form a more or less wavy ring of cambium. This normal cambium behaves abnormally as it cuts off secondary xylem and phloem in vascular bundle region only, resulting in the increased size of the bundle.

(b) **Aristolochia stem** (Fam.–Aristolochiaceae)–The stem in young conditions exhibits a typical dicotyledonous structure. The epidermis encloses the multilayered collenchymatous hypodermis, followed by the cortex made up of collenchyma and parenchyma. The endodermis known as starch sheath is followed by multilayered, sclerenchymatous pericycle. The vascular system consists of a ring of open, collateral vascular bundles separated by interfascicular areas. The median and two lateral leaf traces arise from in between the vascular bundles.

At the time of secondary growth, the inter and intra-fascicular cambium become active and form a complete normal cambium ring. However, this normal cambium ring behaves abnormally, as it cuts off vascular tissue in the intra-fascicular region (within V.B. region) and parenchymatous cells in the interfascicular region. As such the secondary vascular tissue is not prominent and the bundles remain discrete. In older stems, the pith and the rays become somewhat crushed due to increasing pressure of secondary tissues and partly due to the resistance offered by the pericycle to the expanding vascular system. After sometime, the cumulative pressure breaks up the continuity of the pericycle and the adjacent parenchyma cells invade the breaks by intrusive growth (Esau, 1965), and may differentiate into sclereids. The intra-fascicular cambium may at times behave abnormally and cuts off parenchyma cells, resulting in the xylem becoming fissured and the bifurcation of the bundles.

In case of *Aristolochia triangularis,* the pith enlarges and the original vascular bundles assume a fan-like shape. The parenchyma present in between the bundles act like a shock absorber and also enables the stem to resist pulling and compression, resulting from pressure of high winds. This specific trait arising as a result of anomalous activity of the cambium is thus an adaptation to the climbing habit of the plant.

2. The cambium forms unusually large proportion of vascular tissues in the vascular region only. The cambium ring is formed normally by the union of the fascicular and interfascicular cambium. However, the cambial ring behaves abnormally, in the sense that it is more active in certain regions and less active or inactive in other regions. This peculiar growth pattern is well exemplified by *Bauhinia, Bignonia, Doxantha, Menispermum, Tinospora* etc.

(a) **Bauhinia stem** (Fam.–Caesalpinioideae)–In this plant, the cambial activity varies in different species. In *B. rubiginosa,* the cambium is more active at certain points but shows limited activity or no activity at other regions. In the regions of more activity, excessive secondary vascular tissue is produced, whereas in the regions of less activity, little or no vascular tissue is formed. Such an abnormal growth pattern results in the formation of ridged secondary vascular cylinder and the mature stem exhibits a ridged appearance, externally. In other species of *Bauhinia,* the activity of the cambium is limited to the two opposite poles only, resulting in an elongated flattened strap-like structure.

(b) **Bignonia stem** (Fam.–Bignoniaceae)–The young stem of *B. venusta,* a common garden climber, shows a wavy outline with prominent ridges and furrows. The thickly cuticularised epidermis is followed by the hypodermis which is collenchymatous or sometimes sclerenchymatous in the ridge region and sparsely chlorenchymatous beneath the furrows. The few layered parenchymatous cortex is followed by an ill-defined endodermis, supported by heterogenous pericycle. The vascular bundles are conjoint, collateral and arranged in a ring around a prominent pith.

At the start of secondary growth, the fascicular and interfascicular cambium become active and join to form a cambium ring. This cambium ring behaves normally in the beginning, cutting more secondary xylem towards inner side and less secondary phloem on outer side. However, after sometime the cambium ring develops unidirectional areas of cambium at four diagonal points. Unidirectional area of cambium is that portion of the cambial cylinder that produces little or no secondary xylem and extensive amounts of secondary phloem. Hence at these four points lesser amounts of xylem are cut off internally, whereas massive amounts of secondary phloem are cut off externally. These phloem masses intrude inwards forming four deep wedges of irregular width and supported by transverse bands of sclerotic cells. After a short period, the central portion of the cambial strip, which was performing normal activity, develops unidirectional area of cambium and starts cutting off more phloem on the outer side as compared to the inner side. This results in the formation of four or more wedges of phloem intruding into the xylem cylinders. The mature stem of *Bignonia* thus has four big-sized and four small-sized wedges of phloem projecting into the xylem.

The cork cells are formed from the cork cambium which arises from the cortical cells. In later stages, the pericycle fibres are cast off after its activity and the stem assumes circular shape. These wedges of phloem are an adaptation to the plants mechanical requirement functioning as shock absorbers.

(c) **Doxantha stem** (Fam.–Bignoniaceae)–The anomalous pattern of

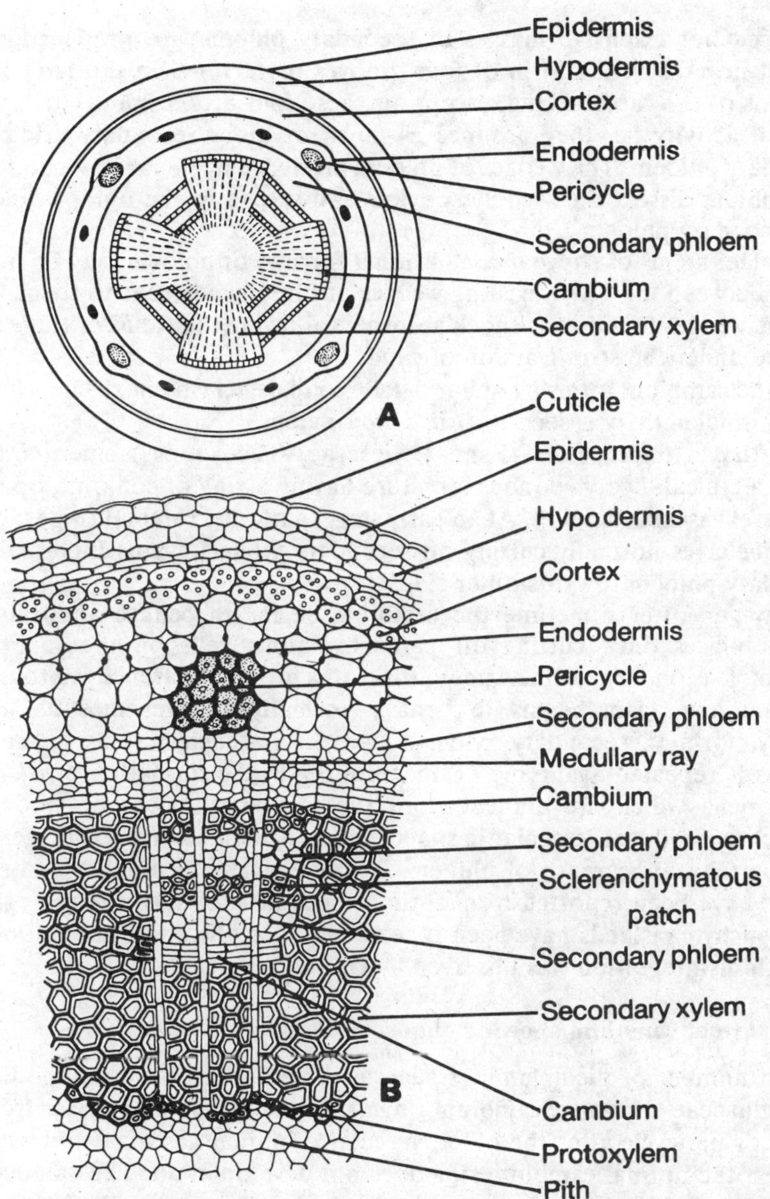

Fig. 12.1. T.S. stem of *Bignonia*. A--diagrammatic and B--an enlarged cellular portion.

secondary growth in *D. unguis-cati* was described by Dobbins (1971). A normal cambial ring develops in the earlier stages, which later develops unidirectional and bidirectional areas of cambium. Initially, the cambium behaves normally cutting off secondary xylem and phloem, that have a cylindrical configuration, but later four arcs of cambium are differentiated, one near each of the four major primary vascular bundles. These four arcs of cambium show unidirectional activity, i.e. little or no secondary xylem is produced but extensive masses of secondary phloem are produced, which project inwards in the form of four grooves or furrows. In the later stages, portions of the cambial ring, except for these four arcs, exhibit bidirectional cambial activity i.e. they produce as much or more secondary xylem than secondary phloem. The wedges of phloem produced by the unidirectional arcs of cambium dissect the secondary xylem and extend to the outer cylinder of the secondary phloem.

In older stems of *Doxantha*, additional furrows of phloem may be formed, either between the older existing wedges or directly adjacent to them. These phloem wedges function as shock absorbers allowing the stems of these lianas to bend under the strong action of winds.

3. The cambium forms irregular patches of parenchyma in xylem. This type of abnormal activity is seen in *Urtica dioica* stem.

(a) **Urtica dioica stem** (Fam.–Urticaceae)–The young stem of *U.dioica* shows a typical dicotyledonous structure having a ring of conjoint, open and collateral vascular bundles. At an early stage, a normal cambial ring is formed which behaves normally cutting off secondary xylem towards inner side and secondary phloem towards outer side.

However, after sometime the cambium begins to behave abnormally at certain places only, cutting off parenchymatous cells on the inner side, instead of secondary xylem. After cutting off a group of parenchymatous cells, the cambium again begins to behave normally and resumes its normal activity, forming secondary xylem above the parenchymatous group. The process is repeated again and again, resulting in the formation of *islands* of parenchyma which are embedded in the secondary xylem or wood. The parenchyma cells are irregular in shape, unlignified and elongated tangentially, giving a false appearance of phloem. Usually, increased number and size of islands have been reported from spring wood. In later stages distintegration of parenchyma islands have been reported by Metcalfe and Chalk (1960) but no such disintegration was observed by Gupta (1960).

II. Abnormal cambium showing abnormal activity

In a number of dicot families such as Amaranthaceae, Chenopodiaceae, Nyctaginaceae etc. a new cambium ring or accessory cambium arises from the cortex or pericycle after the older, vascular cambium has ceased functioning or when the normal cambium ring does not develop at all. The cambial ring being placed abnormally functions in an abnormal manner, which may be of the following types:

(i) **Formation of rings of vascular bundles embedded in parenchymatous tissue.** This type of secondary growth has been observed in *Amaranthus, Mirabilis, Bougainvillea* etc. and occurs in the following manner.

(1) The first ring of cambium arises *de novo* in the pericycle region. Such a type of cambium, arising outside the vascular bundle is known as extra-stelar cambium.

(2) The extra-stelar cambium behaves abnormally, cutting off xylem patches alternating with parenchymatous patches on the inner side, whereas on the

outer side, it cuts off continuous layers of parenchyma initially and later phloem patches over the xylem and parenchyma at the remaining places.

(3) After a complete ring of vascular bundle has been formed, the cambial ring becomes passive and ceases to function.

(4) A new ring of cambium (accessory cambium) is now formed from the parenchymatous cells cut off externally by the first cambium (extra-stelar).

(5) This cambium also behaves abnormally, similar to the previous one and cuts off a ring of vascular bundles separated by parenchyma. It then ceases to function and becomes passive.

(6) Similarly, more accessory cambia are formed consecutively which give rise to consecutive rings of vascular bundles.

(a) **Amaranthus stem** (Fam.–Amaranthaceae)–The stem of *Amaranthus* exhibits abnormal activity by an abnormally placed cambial ring. The young stem has an outermost single layered epidermis followed by a few layered collenchymatous hypodermis, parenchymatous cortex, endodermis and pericycle in small irregular sclerenchymatous patches. The stele is represented by two medullary bundles only. The normal ring of vascular bundles, typical of dicots, being absent, the cambial ring is not formed. However, a layer of cells from the pericycle becomes active and forms a cambium ring (extra-stelar). This extra-stelar cambium shows abnormal activity, as it initially cuts small amounts of parenchyma on the outer side and then starts cutting alternating patches of secondary xylem and parenchyma on the inner side. On the outer side, it cuts secondary phloem above the secondary xylem and parenchyma above the secondary parenchyma. This results in the formation of conjoint, collateral vascular bundles arranged in a ring and embedded in parenchyma.

After the formation of these vascular bundles the cambium becomes inactive and stops functioning. However, a second accessory cambium develops from the secondary parenchymatous cells cut off by the first extra-stelar cambium. It behaves in a manner similar to the first cambium ring, i.e. cutting off vascular bundles embedded in parenchyma. The second ring of V.B. alternate in position to the first V.B.

Similarly, more accessory cambia develop one after the other forming concentric rings of vascular bundles which appear as if scattered in ground tissue. The last cambial ring, however, behaves in a slightly different manner, as it cuts off secondary xylem alternating with sclerenchyma on the inner side. Hence, the last ring of vascular bundles appear to be embedded in sclerenchyma.

In mature stems, the two medullary bundles along with small amounts of parenchyma disintegrate forming a cavity. The cork cambium functions normally forming cork and secondary cortex.

(b) **Mirabilis stem** (Fam.–Nyctaginaceae)–The young stem of *Mirabilis* has two grooves opposite to each other, bearing numerous multicellular shoot hairs. Internally, the stem has a typical dicotyledonous structure with a single layered, cuticularised epidermis, a few layered collenchymatous hypodermis followed by multilayered parenchymatous cortex. The extra-stelar region is demarcated from the stelar region by a continuous layer of epidermis and a pericycle with patches of sclerenchyma. Numerous vascular bundles are scattered in the stelar region.

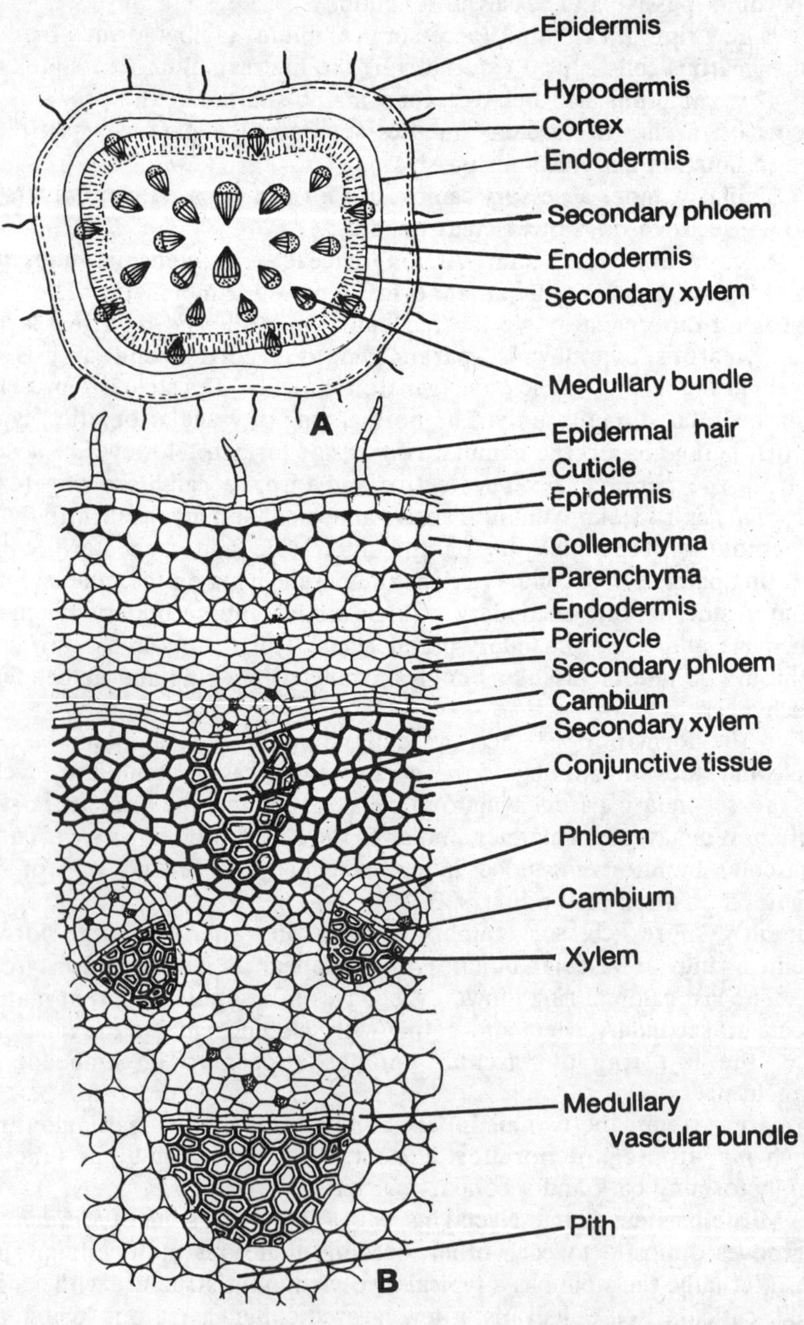

Fig. 12.2. T.S. stem of *Mirabilis jalapa.* A--diag. and B--an enlarged cellular portion.

De Barry (1884) stated that a normal cambium ring is not formed but a completely new ring develops from the pericycle cells lying outside the vascular region. This view is supported by Mikesell and Popham (1977) who observed that the first cambium (extra-stelar) originates from the cells of the inner cortex or the pericycle. However, Maheshwari (1930) differs in his view and stated that the first cambium ring develops in the region of outermost vascular bundle.

This extra-stelar cambium forms secondary xylem alternating with parenchyma, on the inner side, and secondary phloem over the secondary xylem and outer parenchyma above the inner parenchyma, on the outer side. A ring of conjont, collateral vascular bundles is thus formed, separated by parenchymatous cells. The first cambium ceases its function after forming the ring of vascular bundles.

Successive rings of accessory cambia are formed, from the secondary parenchyma, cut off by the previous cambial ring and they behave in a manner similar to the first one. The last cambial ring cuts off sclerenchyma alternating with secondary xylem on the inner side and secondary phloem and parenchyma on the outer side (Fig.12.2). Hence the last ring of vascular bundles are embedded in lignified tissue and provide mechanical strength to the stem.

(ii) **Formation of rings of vascular bundles embedded in conjunctive tissue.** The cambial ring originates in the extra-stelar region and cuts off secondary xylem and thick-walled conjunctive tissue on the inner side and secondary phloem and conjunctive tissue on the outer side. The vascular bundles appear to be embedded in a mass of conjunctive tissue, e.g. *Bougainvillea, Boerhaavia* etc.

(a) **Bougainvillea stem** (Fam.–Nyctaginaceae)–The young stem of *Bougainvillea* has a circular outline formed of a single layer of thickly cuticularised epidermis, followed by a few layered collenchymatous hypodermis and a well developed parenchymatous cortex.The endodermis is indistinct and the pericycle is an heterogenous assemblage of parenchyma alternating with sclerenchymatous patches.

The primary vascular bundles, developing from the procambium strand, appear to be irregularly distributed in the central ground tissue. These bundles are actually the leaf trace bundles associated with the leaves, and may exhibit limited secondary growth accompanied by slight increase in size. The first cambial ring is extra-stelar in origin and arises from the pericycle (Esau and Cheadle, 1969 and Stevenson and Popham, 1973), followed by successive rings of cambia, however, Belfour (1965) stated that the entire secondary tissue is formed from a single cambium.

The cambial ring cuts off secondary xylem alternating with secondary parenchyma on the inner side and secondary phloem above secondary xylem and parenchyma above parenchyma on the outer side. The cells of the secondary parenchyma are fusiform, more or less radially arranged and usually develop lignified, thickened walls. These are referred to as conjunctive tissue or ground tissue (Metcalfe and Chalk, 1950).

Successive rings of cambia are formed which behave in a manner similar to

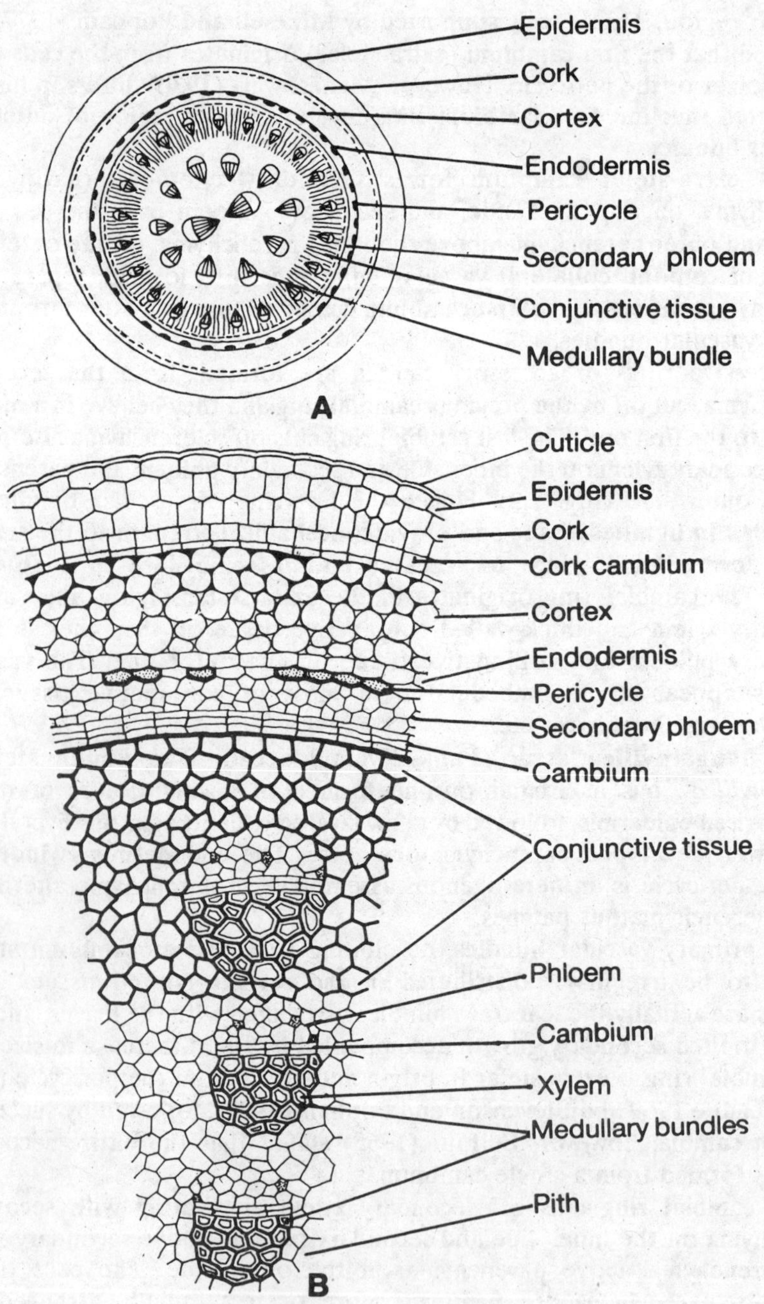

Fig. 12.3. T.S. stem of *Bougainvillea spectibilis*. A--diagrammatic and B--a cellular enlarged portion.

the first ring, as a result, concentric layers of vascular bundles are formed embedded in conjunctive tissue. This layering is conspicuous due to the prominent vessels in the xylem and the crushed cells in the oldest part of phloem (Fig. 12.3). An interesting feature is that the new cambium does not arise as a complete ring around the periphery of the vascular cylinder but in the form of strips, which join with the older cambia. Thus, an ontogenetic continuity is maintained. In some species of *Bougainvillea*, the conjunctive tissue consists entirely of sclerenchyma and cannot be demarcated from the xylem of the embedded vascular bundles, hence the secondary phloem appears in the form of islands or isolated groups apparently surrounded by conjunctive tissue. They should not be mistaken as included phloem.

(b) **Boerhaavia stem** (Fam.–Nyctaginaceae)–The young stem of *Boerhaavia* is bounded by a single layered epidermis bearing numerous multicellular hairs. The hypodermis consists of 2-3 layers of collenchyma followed by parenchymatous cortex, interspersed with few chlorophyllous cells. An indistinct endodermis and 1 or 2 layered pericycle demarcates the well developed stelar region. In *Boerhaavia*, the vascular bundles are arranged in three rings (anomalous structure)–two large, centrally placed medullary bundles, middle ring of 6-14 loosely arranged bundles and an outer ring of small 15-20 bundles. The three rings of bundles originate from the procambial strands.

Secondary growth in the two medullary bundles and in the bundles of the middle ring is limited, resulting in slight increase in size only. The intra-fascicular cambium of these bundles function normally, forming secondary xylem on the inner side and secondary phloem on the outer side, and as secondary growth progresses, the primary phloem gets crushed, forming a cap like structure of dead cells over the newly formed phloem.

However, most of the secondary growth of the stem takes place from the cambial activity of the outer ring of bundles. The intra-fascicular and interfascicular cambium join to form a cambium ring. Maheshwari (1930) observed that the ring forms secondary xylem in the intrafascicular region and lignified, conjunctive tissue in the interfascicular region on the inner side, and secondary phloem above the secondary xylem and parenchyma opposite to conjunctive tissue, on the outer side.

After the formation of secondary tissue, the cambium ceases its activity and a new cambium ring arises by the joining of secondary parenchyma cells, opposite to the conjunctive tissue, and the pericycle cells outside the phloem. This first accessory cambium behaves similarly to the vascular cambium, cutting off secondary xylem alternating with lignified conjunctive tissue on the inner side and secondary phloem opposite to secondary xylem and parenchyma above the conjunctive tissue on the outer side. This results in the formation of a fourth ring of conjoint, collateral bundles of secondary origin.

Soon this supernumerary cambium ceases to function and a new accessory cambium develops which behaves in a similar manner. The process of formation of accessory cambium rings and secondary vascular bundles may be repeated till four or more concentric rings of vascular bundles are formed. The concentric rings of vascular bundles embedded in thick conjunctive

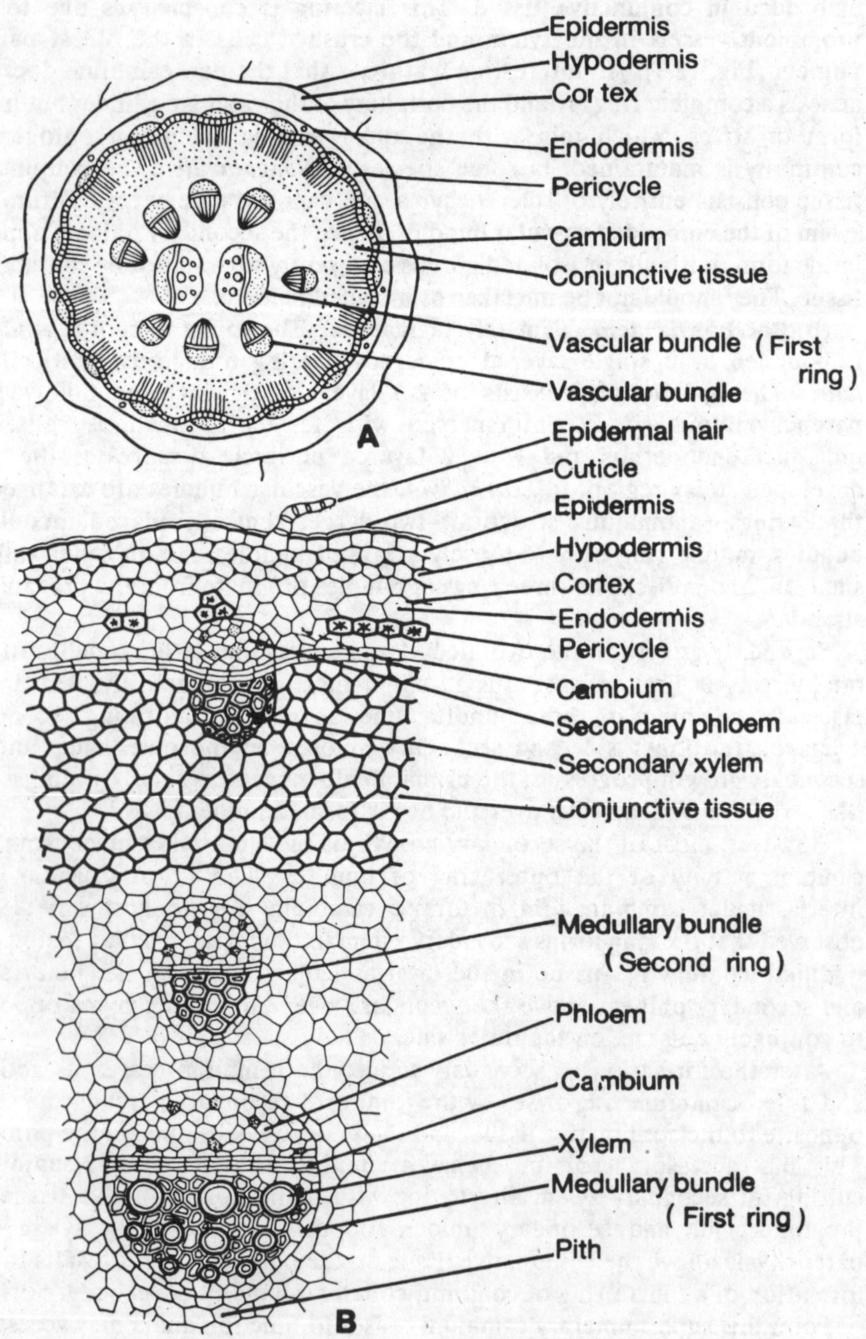

Fig. 12.4. T.S. stem of *Boerhaavia diffusa*. A--diagrammatic and B--a cellular enlarged portion.

tissue and separated by thin walled parenchymatous zone give the appearance of growth rings.

III. Primary abnormal position of cambium showing abnormal activity

In various stems such as *Thinouia*, *Serjania*, *Paullinia* etc. the cambium has unusual position showing abnormal activity, and which results in the formation of anomalous structure during secondary growth.

(a) **Thinouia ventricosa stem**–In the young condition, the cambium is not in the form of a ring but appears as a folded structure. As the secondary growth starts, the cambium separates at the folds and forms individual groups of vascular tissues. These unusually positioned cambial rings form their separate vascular tissue resulting in a peculiar lobed structure of the stem.

(b) **Serjania corrugata** and **Paullinia** (Fam.–Sapindaceae)–In these tendril climbers, specific groups of vascular tissue are constricted off from the main vascular cylinder. These constricted-off vascular tissues develop their individual cambium strips. With the initiation of secondary growth, each strip of cambium develops to form a complete and separate cambium ring, which functions normally forming secondary xylem on inner side and secondary phloem on outer side. Thus, the mature stem consists of a number of discrete vascular cylinders each of which develops its own periderm. In the mature condition, the stem seems to be made up of several smaller stems appressed to each other and giving the appearance of strands of rope.

(c) **Bauhinia langsdorffiana** (Fam.–Caesalpiniodeae)–In this plant, the cambium is in the form of strips which do not unite but remains as such. At the time of secondary growth, these strips produce secondary xylem on inner side and secondary phloem on outer side, alternately mixed with secondary parenchyma. But the secondary parenchyma is produced in excessive amounts so much so that the vascular cylinder gets broken up into numerous parts resulting in fissured or small groups of xylem.

IV. Formation of included phloem

The abnormal activity of the cambium gives rise to groups of phloem cells embedded in the secondary xylem. This is known as included phloem or interxylary phloem. Formation of included phloem occurs by two methods–

(i) **Formation and abnormal activity of arc cambium.** The extra-stelar cambium formed, behaves abnormally forming secondary xylem and sclerenchyma on the inner side whereas on the outer side it forms secondary phloem above secondary xylem and parenchyma opposite to sclerenchyma. After sometime this cambium becomes inactive below the phloem region and new arcs of cambium are formed from the parenchyma lying above the phloem. This arc cambium also behaves abnormally, forming secondary xylem alternating with sclerenchyma on the inner side and secondary phloem and parenchyma on the outer side. As such, the secondary phloem formed by the first cambium gets enclosed by the xylem and sclerenchyma cut off by the arc of cambium. This enclosed phloem is known as included phloem or interxylary phloem (Fig. 12.4) The remaining parts of the cambium at rest of the places continues its original activity and finally comes to the level of the arcs of cambium. This type of formation of included phloem is termed as concentric type and has been observed in *Achyranthes, Chenopodium, Bougainvillea, Celosia, Bosea, Pupalia, Strychnos* etc.

(a) **Achyranthes stem** (Fam.–Amaranthaceae)–The young stem has a wavy outline showing alternate ridges and furrows. The outermost layer is a single-layered, cuticularised epidermis, followed by the hypodermis which is collenchymatous beneath the ridges but chlorenchymatous beneath the furrows. The endodermis is supported by the pericycle, which is multilayered with sclerenchymatous patches on the outer side. Two medullary vascular bundles lie opposite to each other in the pith region and are decussately arranged in successive internodes. The position of medullary bundles was studied in *A. aspera* by Srivastava (1960). He observed that the two bundles may be free only in the few upper internodes but fused in the lower internodes, or may remain free throughout the length of the plant. The two fused bundles present an amphixylic structure. In other species such as *A. coynei*, the number of medullary bundles may be four which after fusion forms two bundles. The normal cambium ring is not present but the accessory cambium develops from cells of the pericycle. It behaves abnormally, forming secondary xylem alternating with conjunctive tissue on the inner side and secondary phloem above xylem and parenchyma opposite to conjunctive tissue on outer side. After sometime, the cambium strip lying below the phloem becomes inactive and new arcs of cambium develop from the parenchyma lying above it. This newly formed arc of cambium similarly cuts off secondary xylem and conjunctive tissue on the inner side and secondary phloem and parenchyma on the outer side. As a result, the secondary phloem, formed by the first cambium gets enclosed by secondary xylem and conjunctive tissue and is known as included or interxylary phloem (Fig. 12.5).

It has been observed that in certain species, the activity of cambium may change at intervals, forming secondary xylem in those regions where conjunctive tissue was being formed and conjunctive tissue where secondary xylem was being formed. Thus the secondary vascular bundles and the included phloem are found irregularly scattered.

(b) **Chenopodium stem** (Fam.–Chenopodiaceae)–The stem has a prominent ridged appearance with an outer, cuticularised single layered epidermis. The hypodermis consists of few collenchymatous cells beneath the ridges and chlorenchyma-containing cells beneath the furrows. The cortex is parenchymatous in nature followed by well defined endodermis and pericycle having patches of sclerenchyma. Vascular bundles are arranged in a ring around the parenchymatous pith. The normal cambium is not formed and secondary growth is initiated by the extra-stelar cambium originating from the pericycle cells. According to some workers this first ring of extra-stelar cambium first forms a layer of thin walled ground tissue and later the secondary vascular bundles. The primary vascular bundles, as a result of pressure from the ground tissue is pushed towards the central region and

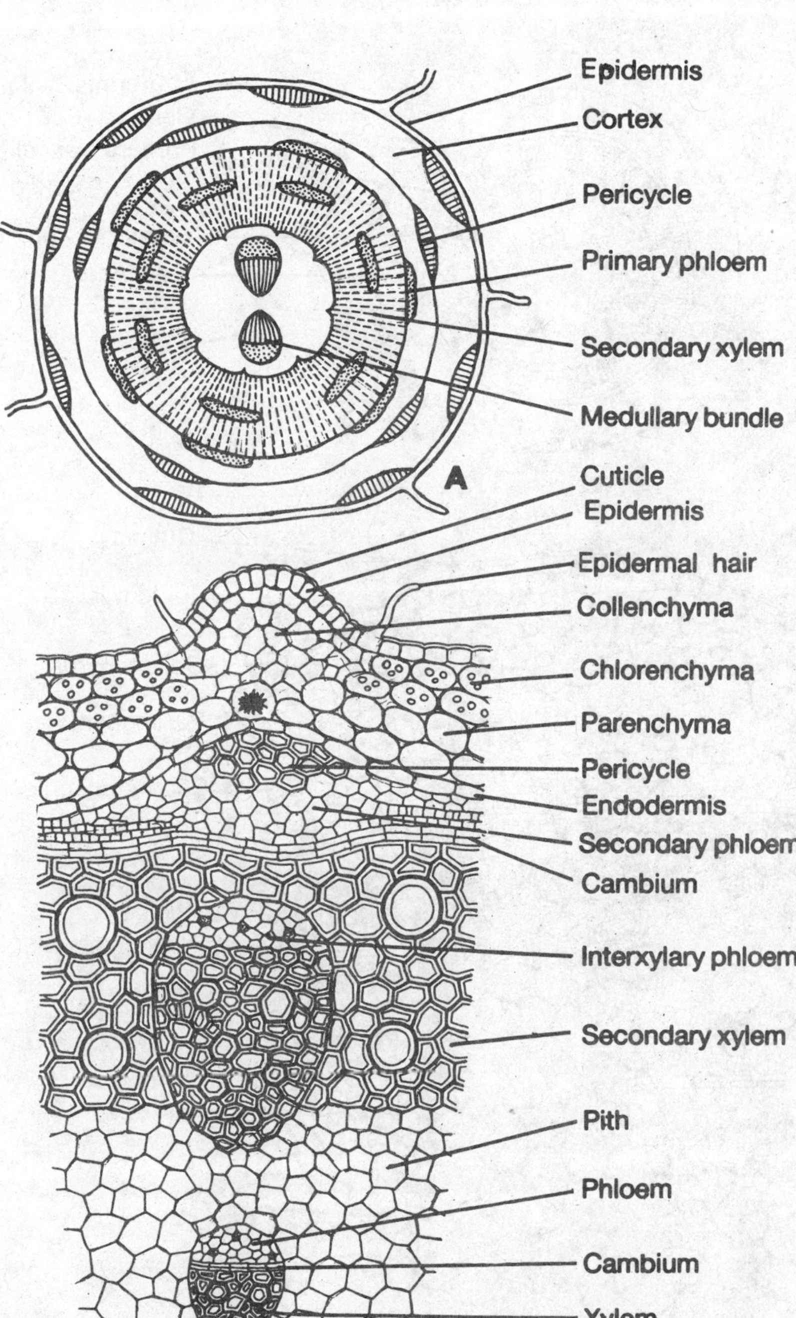

Fig. 12.5. T.S. stem of *Achyranthes*. A--diagrammatic and B--a cellular enlarged portion.

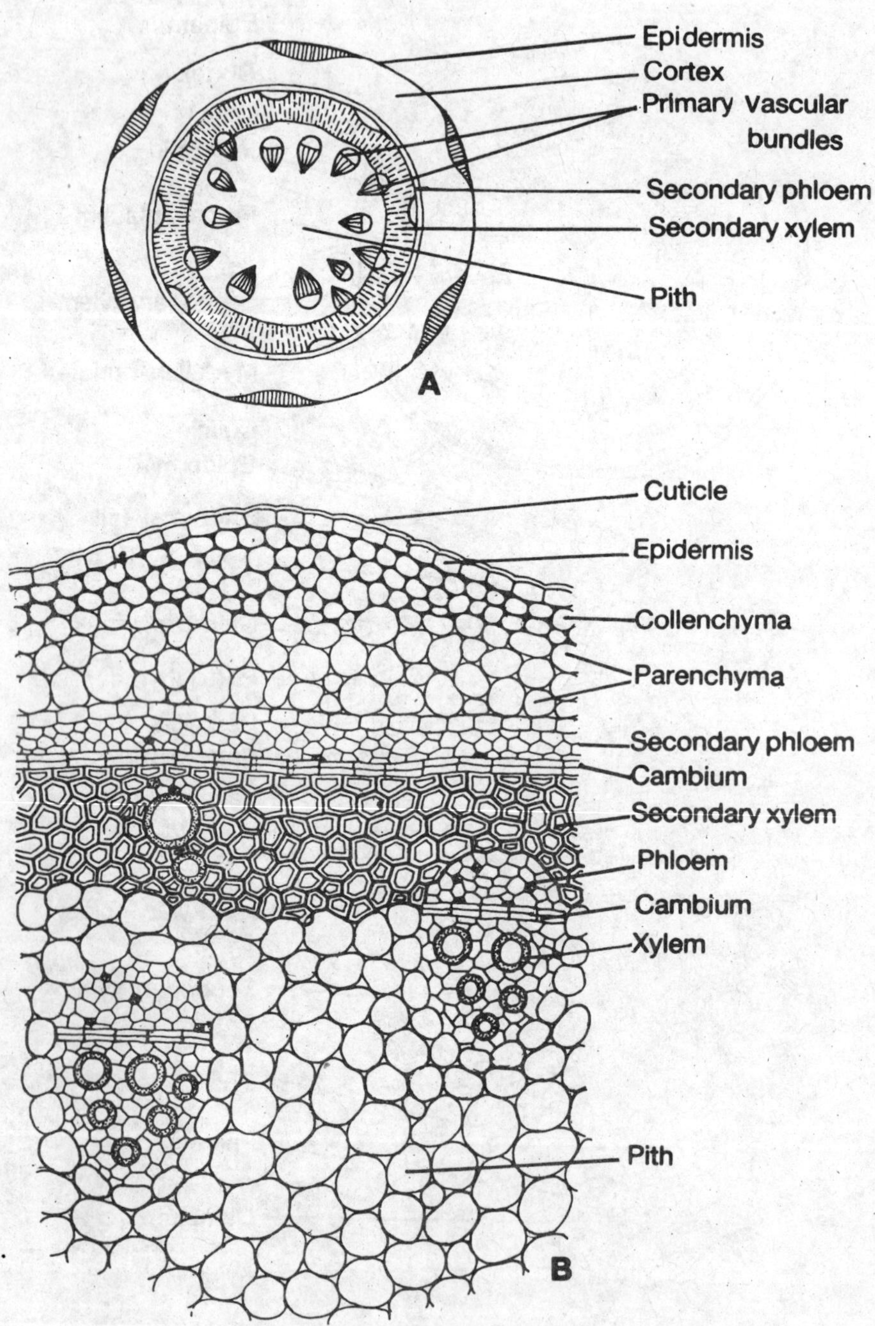

Fig. 12.6. T.S. stem of *Chenopodium*. A--diagrammatic and B--a cellular enlarged portion.

appear as medullary bundles. These primary bundles may have limited growth due to the formation of secondary tissue by their intrafascicular cambia.

The first extra-stelar cambium functions abnormally cutting off secondary xylem and conjunctive tissue on the inner side and secondary phloem and parenchyma on the outer side. The origin and function of the extra-stelar cambium is controversial. According to some workers this cambium remains active throughout the life of the plant and continuously forms secondary vascular tissue and conjunctive tissue, whereas other workers observed that the cambium is limited in its activity, hence resulting in the formation of successive arcs or rings of cambium which are responsible for forming secondary vascular tissue.

Balfour (1965) studied secondary growth in *Chenopodium* and observed that the inter- and intrafascicular arcs of cambium join to form a cambium ring, linking all the primary vascular bundles. This ring functions abnormally cutting off bands of fibres in the interfascicular region and xylem vessels and other xylem elements in the intrafascicular region towards the inner side, whereas externally the cambium forms a broad band of undifferentiated meristematic cells. This zone of undifferentiated meristematic cells increases in width, as result of which the primary phloem becomes embedded in it. Later, within the primary vascular bundle, a few layers of cambial cells, lying between the xylem and phloem become non-functional and differentiate into parenchyma or sclerenchyma fibres. The outermost cambial cells now continue their secondary growth cutting off sclerenchyma fibres and xylem vessels internally. However, few cambial cells lying within the meristematic zone just external to the xylem vessels assume meristematic activity forming a group of phloem cells surrounded by lignified tissue. This meristematic zone functions indefinitely and it gradually moves outward, leaving behind isolated groups of phloem cells embedded in the conjunctive tissue. These isolated groups of phloem cells are known as included phloem or interxylary pholem. The meristematic cells forming secondary vascular tissue are thus ontogenetically related to one another.

(c) **Bougainvillea stem** (Fam.–Nyctaginaceae)–Internal structure and formation of extra-stelar cambium and secondary vascular tissue has already been discussed previously in this chapter (see Fig. 12.3).

Bougainvillea also shows the presence of included phloem, formed by the arc cambium method which takes place followingly.

The last ring of accessory cambium forms secondary xylem and conjunctive tissue on the inner side, but on the outer side, it becomes inactive, just below the phloem region. Hence arcs of cambium develop from the parenchyma cells lying externally. These arcs or strips of cambium behave in a manner similar to the accessory cambia, with the result, that the secondary phloem formed previously gets surrounded by the newly formed xylem and conjunctive tissue. This phloem is termed as included or interxylary phloem.

(d) **Strychnos stem** (Fam.–Nyctaginaceae)–The stem has a circular outline with an outermost layer of cuticularised epidermis, followed by one or two layers of collenchymatous hypodermis. The underlying cortex is distinguishable

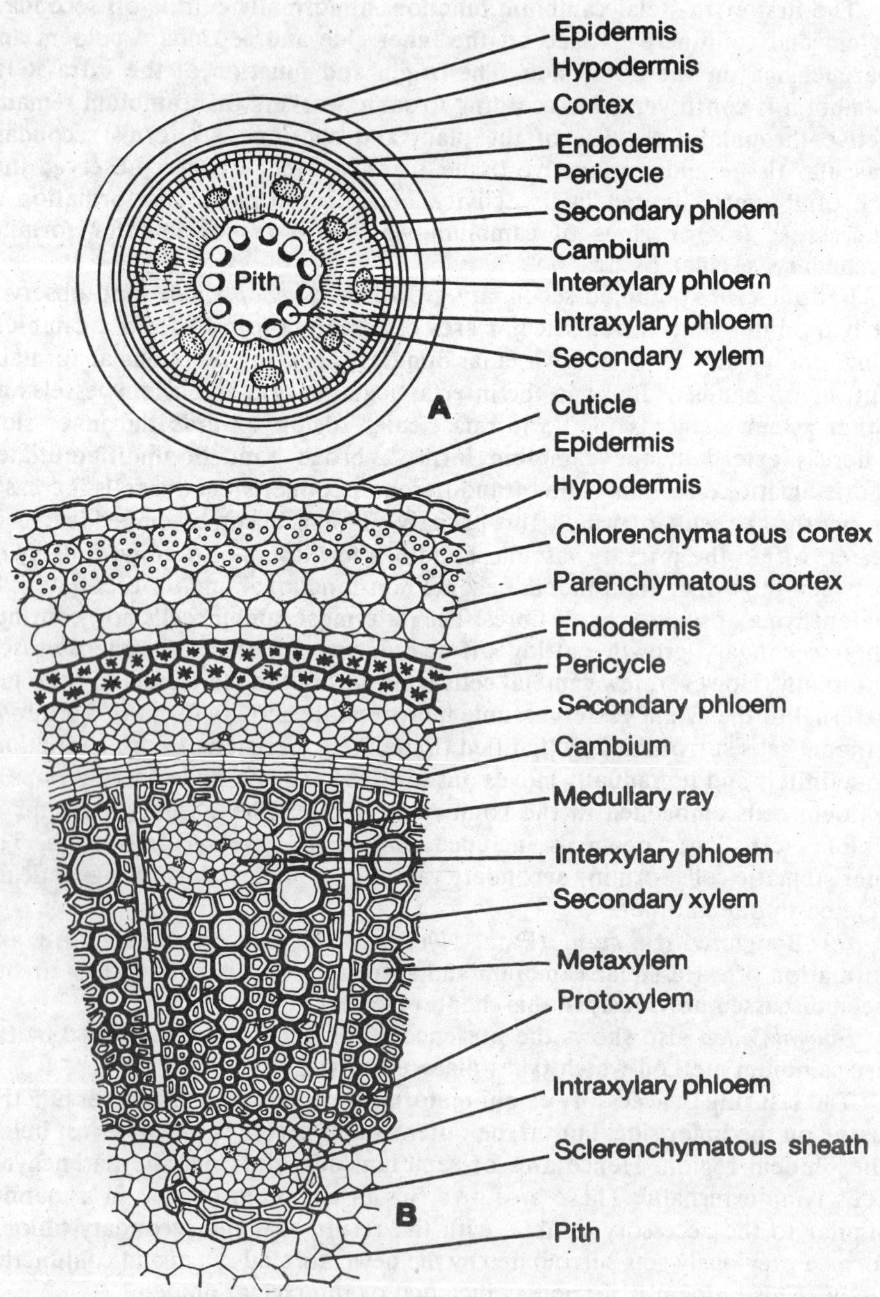

Fig. 12.7. T.S. stem of *Strychnos nux-vomica*. A–diagrammatic and B–a cellular enlarged portion.

into 2 regions, the outer composed of chlorenchymatous cells and the inner of parenchymatous cells. In the inner layers of the cortex, intercellular spaces are present. The endodermis is well defined supported by one or two layers of pericycle made of sclerenchymatous cells. The vascular bundles are arranged in a ring around the parenchymatous pith.

(ii) **Formation of included phloem by a change in the activity of the cambium.** In several families of dicots, such as Combretaceae, Malphigiaceae, Salvadoraceae, Asclepiadaceae, Acanthaceae, Gentianaceae etc. formation of included phloem occurs by a temporary change in the activity of the cambium. Under normal conditions, the accessory cambium forms secondary xylem on the inner side and secondary phloem on the outer side, but sometimes the cambium at certain regions behaves abnormally and a temporary change of activity occurs. These small strips of cambium instead of cutting off secondary xylem, now cut off thin walled parenchymatous tissue (*Leptadenia* and *Salvadora*) or secondary phloem (*Combretum* and *Entada*) on the inner side. Later, they again resume their original activity, i.e., forming secondary xylem on inner side. This deviation, from the normal function of forming secondary xylem, gives rise to the formation of phloem islands surrounded by secondary xylem and conjunctive tissue. These phloem islands are variously called included phloem or interxylary phloem (Fig. 12.8).

Included phloem has physiological significance in some xerophytic plants, since they retain and continue their function even during unfavourable conditions. These phloem islands serve to carry food to the axillary buds which on return of favourable conditions, develop into branches.

(a) **Leptadenia stem** (Fam.–Asclepiadaceae)–The young stem of *Leptadenia*, has a circular outline with an outer single layered cuticularised epidermis of compactly arranged cells. The epidermis is followed by hypodermis constituted of a few layers of collenchymatous cells. The underlying cortex is chlorenchymatous in nature followed by the layer of thin-walled endodermis. Lying beneath the endodermis, is the multilayered pericycle with patches of sclerenchyma. In the initial stage, conjoint , collateral vascular bundles are arranged in a ring around the parenchymatous pith. Secondary growth is initiated by the joining of the inter and intrafascicular cambium to form the cambium ring. This cambium functions normally at first, cutting off secondary xylem towards inner side and secondary phloem towards outer side. After sometime, the cambium at certain points, starts to behave abnormally cutting off parenchymatous cells towards inner side. This is, however, a temporary stage and soon the cambium reverts back to its original function of forming secondary xylem on inner side. This process is repeated several times resulting in the formation of a number of islands of thin walled tissue embedded in cells of secondary xylem. The cells of these islands, especially those in the centre, differentiate into sieve tubes and companion cells and the remaining cells form phloem parenchyma. Hence in the mature stem a number of phloem islands are formed.

As interesting feature of *Leptadenia* is that the stem bears three types of phloem–

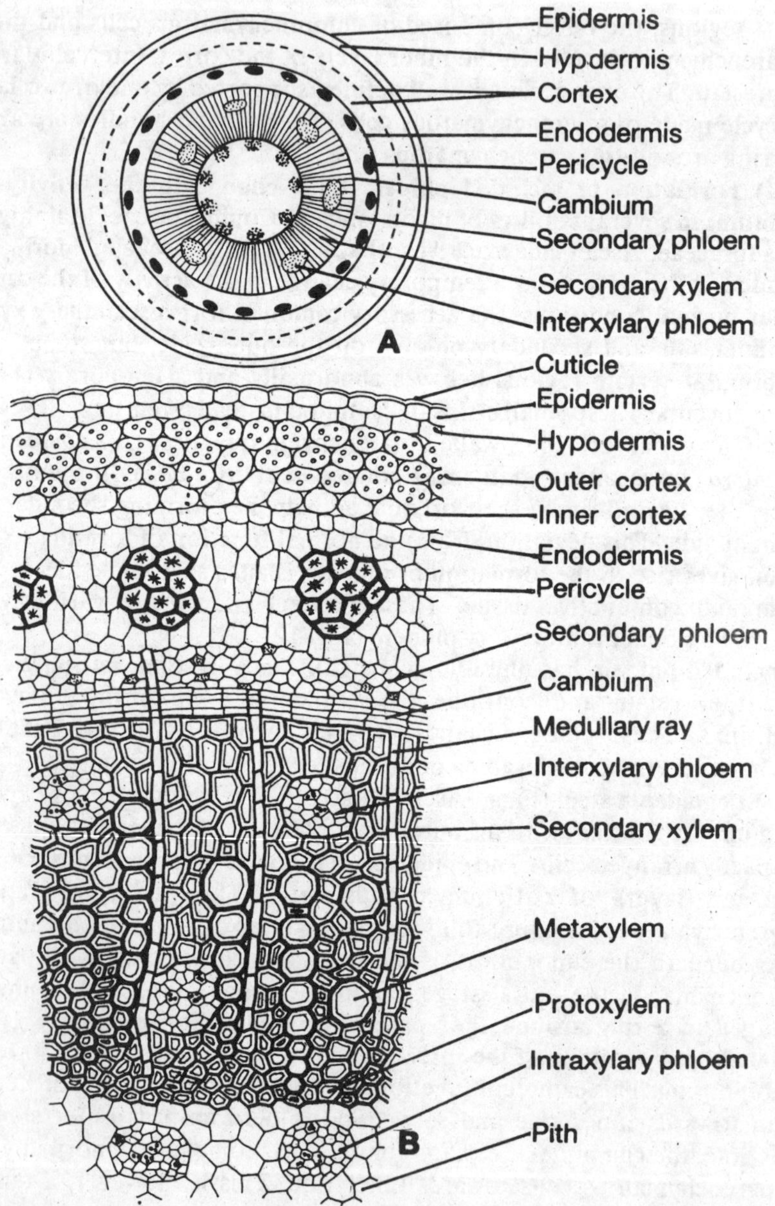

Fig. 12.8. T.S. stem of *Leptadenia spartium*. A–diagrammatic and B–an enlarged cellular portion.

(i) normal phloem (ii) interxylary phloem, and (iii) intraxylary phloem. The intraxylary phloem is present in the innermost region of the primary xylem. They may be formed by transformation of the parenchymatous cells of the pith or from xylem parenchyma (Fig. 12.8).

(b) Salvadora stem (Fam.–Salvadoraceae)–The stem is hard and has a smooth circular outline. The outermost layer is the cuticularised epidermis followed by 2-3 layers of thin walled cells of the hypodermis. The cortex lying beneath the hypodermis is constituted of parenchymatous cells having chloroplasts. The endodermis is a semi-distinct layer supported below by the heterogenous pericylcle composed of parenchyma and groups of stone cells. The vascular bundles are typically arranged in a ring. The cambium ring is formed by the union of the interfascicular and intrafascicular cambium to form a single-layered cambium ring, which, however, appears multilayered due to formation of daughter cells.

The secondary growth in *S. persica* was described by Singh (1944). The cambium functions normally cutting off secondary xylem on inner side and secondary phloem on outer side. After sometime at certain points, the cambium begins to behave abnormally cutting off parenchymatous cells instead of secondary xylem on the inner side. This is, however, a temporary change and soon the abnormally behaving cambium strip reverts back to its original function of cutting secondary xylem on inner side. This abnormal behaviour is repeated several times, resulting in the formation of a number of patches or islands of thin walled cells. The parenchymatous patches ultimately get converted to phloem parenchyma (peripheral cells) and sieve tubes and companion cells (central cells) giving rise to phloem patches termed as included phloem or interxylary phloem.

Secondary growth is initiated by the formation of the cambium ring, which at first functions normally cutting off secondary xylem on the inner side and secondary phloem on the outer side. However, after sometime small strips of cambium get detached from the main ring thus breaking up the continuity of the cambium cylinder. This breakage in continuity is, however, temporary as new complementary segments of cambium arise in the pericyle or in the peripheral region of the phloem above the non-functional segments of the cambium. These secondarly formed cambium strips join the older, dissected cambium to form a complete ring. The cambium cylinder so formed again begins its normal function of forming secondary xylem on inner side and secondary phloem on outer side. Consequently, the phloem formed by the first ring gets enclosed by the secondary xylem formed by the newly joined cambium ring. This enclosed phloem surrounded by secondary xylem is known as included phloem or interxylary phloem.

In each phloem group, a small segment of the original cambium remains embedded, which continues cutting off small amounts of secondary phloem on the outer side. As a result, the cells of the older phloem or included phloem get crushed but remain in the form of a cap on the outer face of phloem islands. An additional feature of *Strychnos* is the presence of intraxylary phloem patches just below the primary xylem and which are capped by crescent shaped masses of sclerenchymatous cells. The intraxylary phloem may be formed by the transformation of parenchymatous cells of the pith or from the xylem parenchyma (Fig.12.7).

In certain plants, such as *Combretum* and *Entada*, included phloem is formed by a novel method. In these plants, a normal cambium ring is formed

by the union of fascicular and interfascicular cambium which behaves in a normal manner, forming secondary xylem on inner side and secondary phloem on outer side. After sometime, small segments of the cambium behave

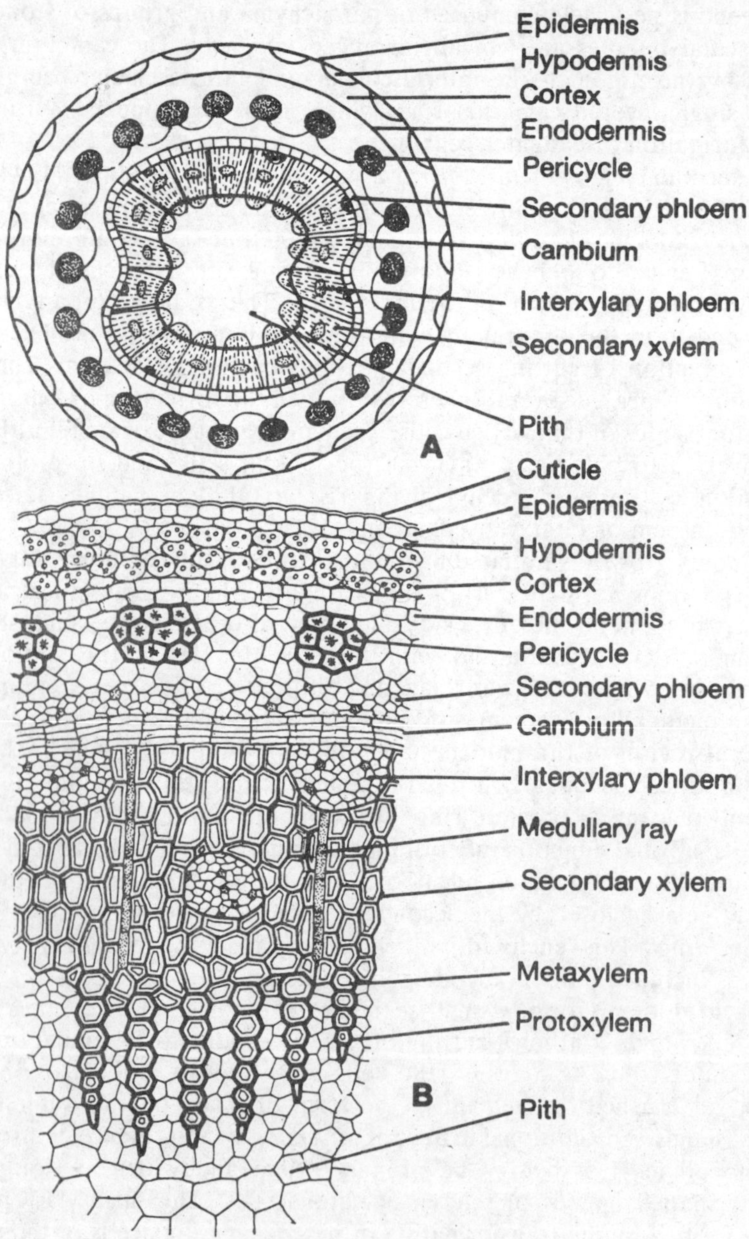

Fig. 12.9. T.S. stem of ***Salvadora persica***. A–diagrammatic and B–a cellular enlarged portion.

abnormally cutting off small groups of phloem cells on the inner side instead of secondary xylem. These segments again assume their normal behaviour of forming secondary xylem. These groups of phloem cells formed, get surrounded by secondary xylem and are termed as included phloem or interxylary phloem.

Besides these, an interesting case of abnormal secondary growth has been reported in *Adena palmata*, of family Passifloraceae (Swamy, 1975). The stem is tuberous and the cambium activity is normal giving rise to fan-shaped vascular tissue separated by parenchymatous tissue. However, after sometime, an additional inner cambium is formed on the inner side of the secondary xylem, which is active and forms medullary steles separated by parenchymatous tissue known as intermedullay stelar parenchyma.

V. Formation of interxylary cork

In several genera such as *Achillea fragmentissima, Artemesia tridentata* and *Epilobium angustifolium,* an interesting development occurs. The cambium ring cuts off secondary xylem and parenchyma towards the inner side. Soon it is observed that a layer of phellogen (cork cambium) develops in this parenchymatous zone between the secondary xylem. This layer of phellogen becomes active and cuts off cork cells which remain enclosed between two consecutive rings of secondary xylem, and is termed as *interxylary cork.* In older stages, suberization of the medullary rays may occur or certain portions of the cambium may cease its activity resulting in the splitting of the stem.

ANOMALOUS SECONDARY GROWTH IN MONOCOT STEM

In monocotyledonous plants, secondary growth is usually absent as the vascular bundles are of closed type i.e. cambium is absent. But plants like *Dracaena, Yucca, Aloe arborescens, Sansevieria* (Fam.–Liliaceae), *Agave* (Amaryllidaceae) and *Lomandra, Xanthorrhoea* and *Kingia* (Xanthorrhoeacae) exhibit secondary growth which is anomalous in origin as well as in function. *Dracaena* plant may be considered as a typical example of anomalous secondary growth.

Dracaena stem (Fam.–Liliaceae)–The young stem has a typical structure of an outer epidermis followed by sclerenchymatous hypodermis and a large number of closed, collateral vascular bundles embedded in parenchymatous ground tissue.

Secondary growth is initiated by the formation of a special meristematic zone which develops in the inner cortical region. The cortical cells lying external to the vascular bundle develop to form a multilayered secondary cambium. This cambium develops in that region of the stem which has ceased to elongate. The cells of the secondary cambium vary in shape and may be rectangular, fusiform or with one end flattened and the other narrow and pointed.

The activity of this cambium is abnormal as it cuts off a few cells on the outer side but a larger number of cells towards the inner side, i.e. towards the centre of the stem. The cells cut off towards outer side are parenchymatous in

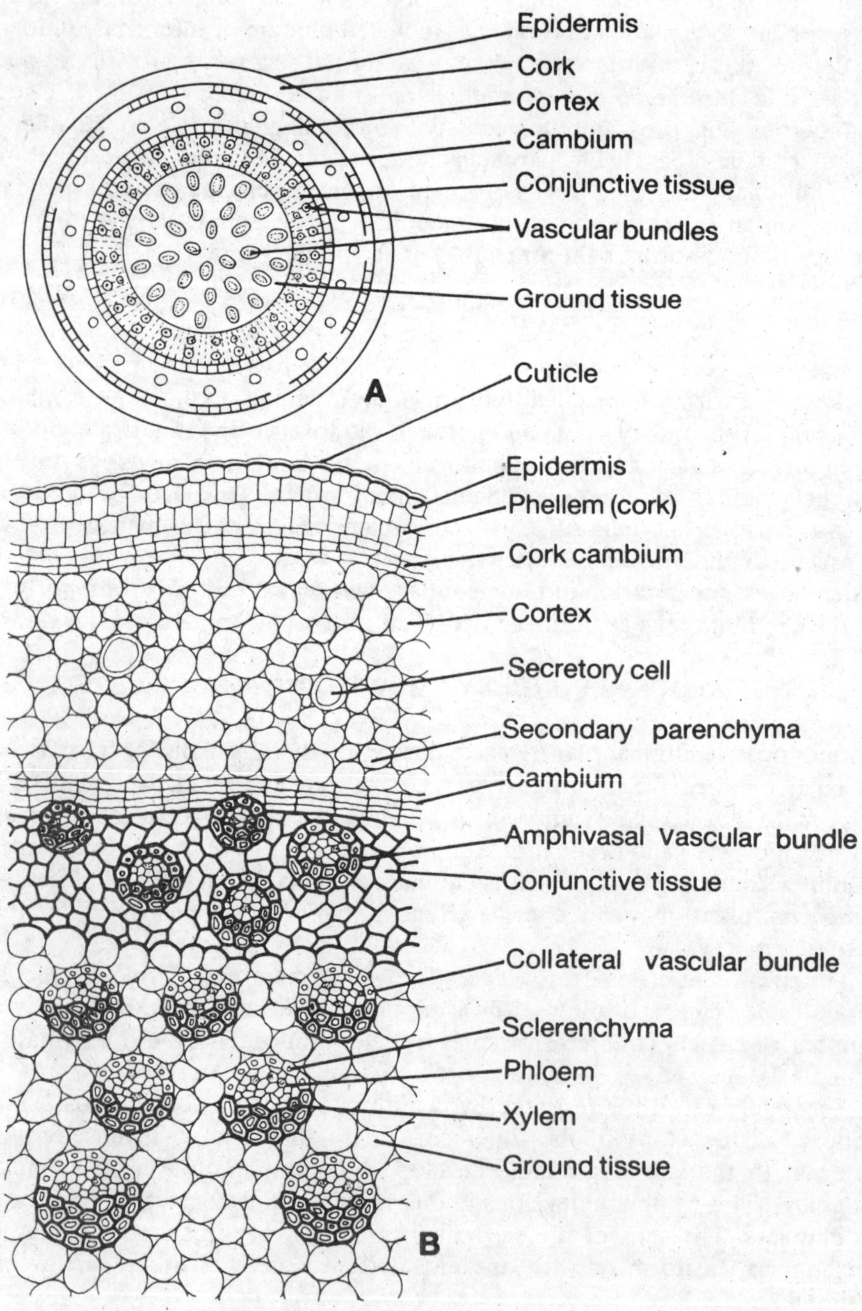

Fig. 12.10. T.S. stem of *Dracaena*. A–diagrammatic and B–a cellular enlarged portion.

nature, whereas the cells cut off towards inner side are partly parenchymatous and partly vascular in nature. The parenchymatous cells cut off on the inner side develop lignification in their cell walls and are termed as *conjunctive tissue.*

In the formation of secondary vascular bundle, usually cells from one or all the tiers of cambium participate. Sometimes, the median part of the bundle is formed from cells of one row and the lateral part from cells of the adjacent rows. However, it has been observed that secondary vascular bundle develops from a single cell, cut off by the cambial initial. This cell functions as the vascular bundle initial and divides by two anticlinal divisions to form a row of two or three cells. These cells further undergo periclinal divisions to form peripheral cells. Later divisions are irregular. The centrally situated cells of this group metamorphoses to form the phloem cells, whereas the peripheral cells differentiate into xylem element. The mature structure constituted of peripherally arranged xylem elements enclosing a central mass of phloem cells is known as amphivasal vascular bundle. The cambium then cuts of parenchyma cells on inner side which pushes the newly formed vascular bundle towards the central region.

After some time, the cambium again starts behaving abnormally, but this time the position of the vascular bundle cut off changes. The cambial initial which is to form the vascular bundle, now forms vascular elements, in these regions where it was forming parenchyma consequently the second row of vascular bundles is formed alternating in position with the first ring. Similarly, the activity and position of vascular bundles keep on changing resulting in rings of vascular bundles arranged in concentric rings by differing in position. Vascular bundles of the last ring are embedded in a mass of lignified, conjunctive tissue. The cork cambium may also develop normally and form cork cells on outer side.

In *Kingia*, development is more or less the same, except that the xylem is U-shaped, i.e. it surrounds the phloem on three sides only. The xylem of the vascular bundle is formed of tracheids and xylem parenchyma only. The vascular bundles are surrounded by a thin or thick sheath. In case of *Xanthorrhoea*, a lot of resin is secreted in the parenchyma cells cut off on the outer side, as such the vascular bundle is surrounded by a resin sheath. In *Yucca brevifolia*, cambium develops from the cells of the pericycle which assumes meristematic activity. In some monocots, development of periderm has been reported as in *Aloe* and *Cocos*.

In some monocots, such as palms, rhizomes of *Musa* and bulbs of *Tulipa* and *Galanthus nivalis*, increase in thickness of the stem takes place by a special type of meristem known as primary thickening meristem. This meristem develops in the embryonic condition only from cells lying just beneath the leaf and sheath primordia. In the embryonic stage, it appears as a flat zone. In the initial seedling stage, it is concave and appears like a cone, in the later seedling stage it becomes flattened and finally in the young plant it again assumes a concave form. The meristem initials by periclinal divisions form a meristematic layer called primary thickening meristem which contributes to the height and girth of the stem. In the palms and rhizome of *Musa*, the

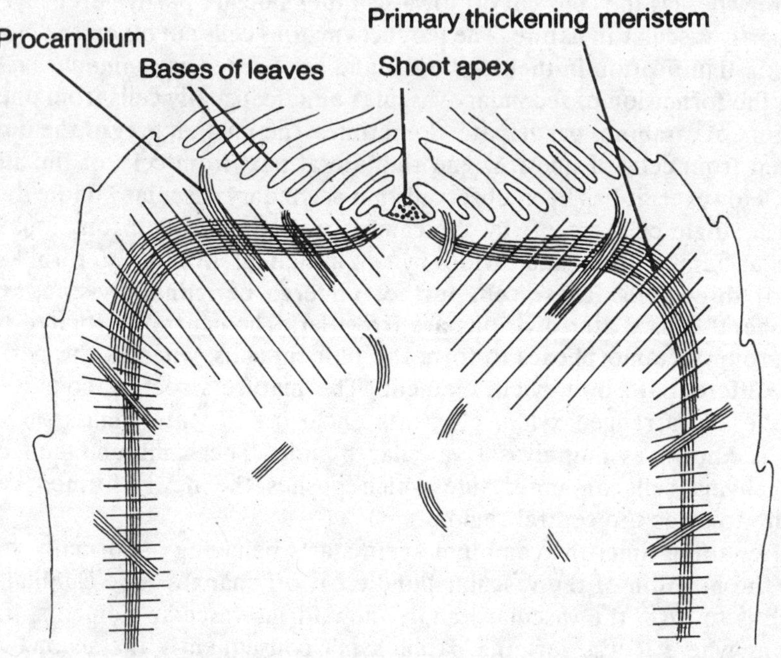

Fig. 12.11. Schematic representation of upper part of a palm plant (monocotyledon) showing meristem and growth (Redrawn after Eckradt).

procambial strands are derived usually from this meristem. The procambial strands form the vascular bundles which add to the thickness of the stem.

Another method of thickening of the stem in palms has been reported. In *Roystonea*, the parenchymatous cells in the centre of the stem as well as the undifferentiated cells of the bundle sheath constantly divide and increase in size. The intercellular spaces also grow in size resulting in the increase in thickness of the stem. This type of growth and increase in thickness of the stem is known as diffuse secondary growth.

ABNORMAL SECONDARY GROWTH IN DICOT ROOTS

In a number of dicot plants, the roots are modified to function as storage roots. Secondary growth in these roots is usually of anomalous type as an adaptation to their storage function. Members of Umbelliferae, Cruciferae, Chenopodiaceae, Compositae, Convolvulaceae, etc. are good examples of anomalous secondary growth exhibited by the roots. The aerial root of *Tinospora* shows normal nature of secondary growth but medullary rays are very conspicuous and large and store reserve food material (Fig.12.12)

Beta vulgaris root (Fam.–Chenopodiaceae)–The root has a typical dicot structure showing radial and diarch condition of stele. As it is a storage root, there occurs an increase in girth of the root to accommodate the increase in

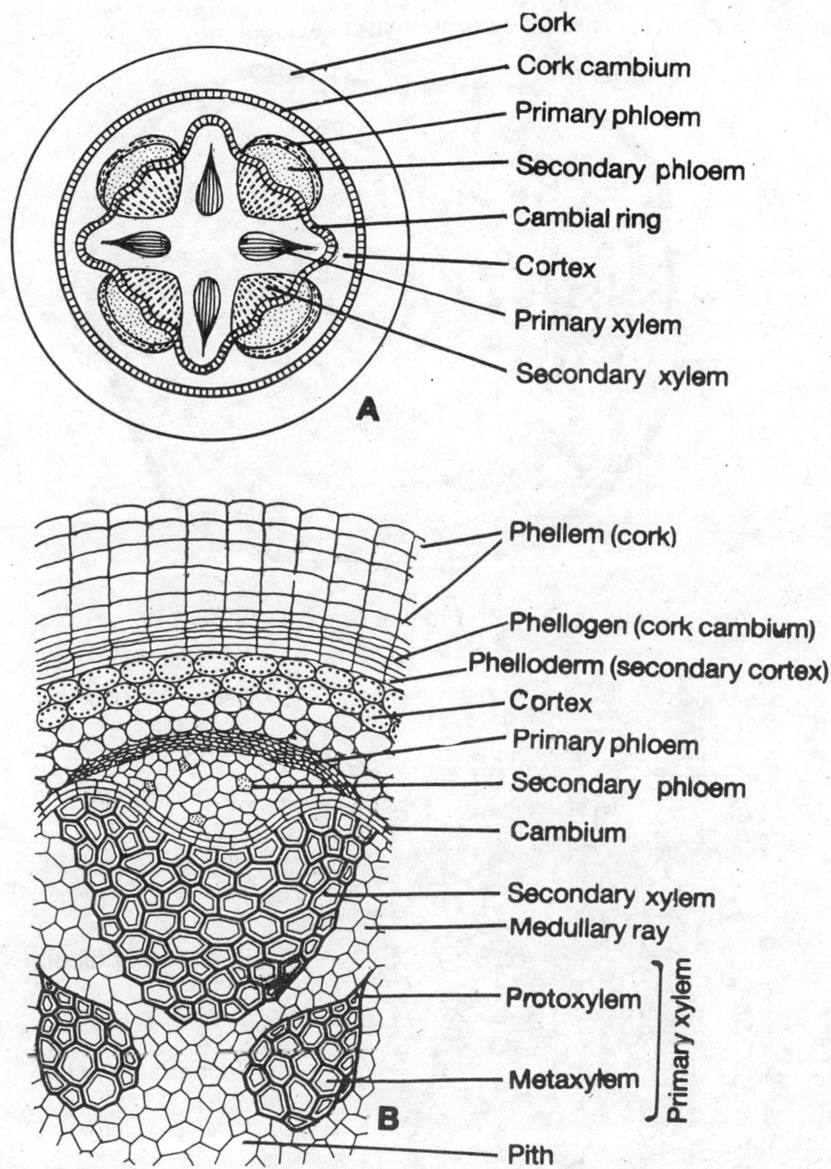

Fig. 12.12. T.S. of aerial root of *Tinospora.* A--diagrammatic and B--a cellular enlarged portion.

number of cells. Secondary growth, however, is of anomalous type due to abnormal activity as well as position of cambium (Fig. 12.13 & 12.14).

Secondary growth is initiated by the formation of the primary cambium ring, which is formed by the joining of the cells of the pericycle lying outside the protoxylem and the parenchyma cells lying between xylem and phloem.

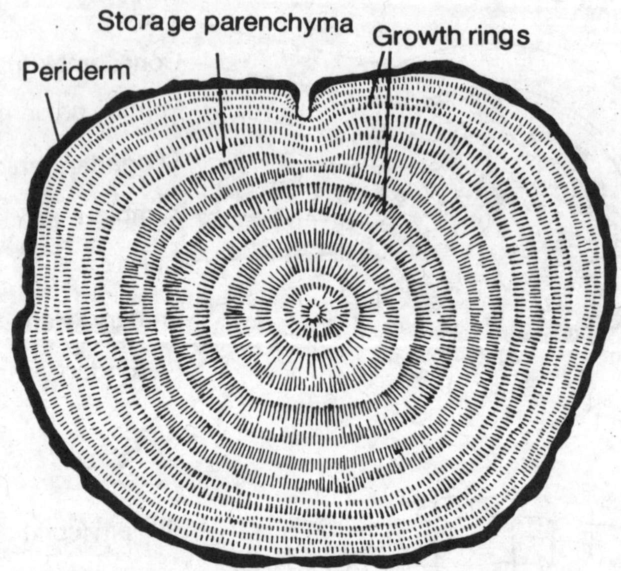

Fig. 12.13. T.S. of root of ***Beta vulgaris*** (diag.)

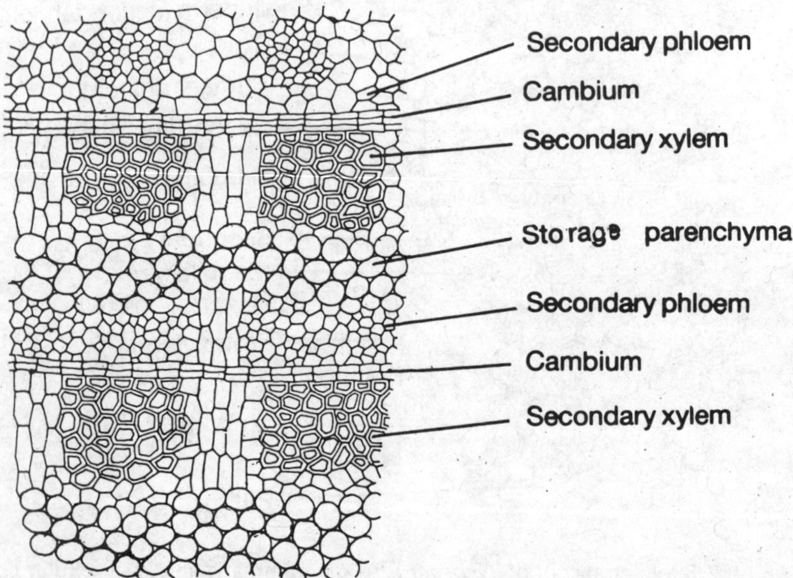

Fig. 12.14. T.S. of a portion of root of ***Beta vulgaris*** (cellular enlarged portion) showing growth rings.

This primary cambium ring exhibits abnormal activity, as it forms secondary xylem and parenchyma cells in alternate groups, towards the inner side, whereas on the outer side, it forms secondary phloem opposite to secondary xylem and parenchymatous cells opposite to inner parenchyma cells. This results in the formation of a ring of conjoint, collateral vascular bundles embedded in parenchymatous tissue. The parenchymatous cells function as storage tissue. The cambium after forming the ring of vascular bundles ceases its activity. Soon a second ring of cambium arises outside the first cambium ring.This accessory or secondary cambium develops from the phloem parenchyma cells or sometimes from the inner cells cut off by the pericycle. This secondary cambium also behaves in an abnormal manner, similar to the first cambium ring, resulting in the formation of another ring of conjoint, collateral vascular bundles arranged external to the first ring of bundles. These vascular bundles are separated by wide gaps occupied by parenchymatous cells which serve as storage tissue. This first ring of accessory cambium also ceases its function and a third cambium ring or second accessory cambium now develops outside this second ring of vascular bundles. This cambium ring develops from the cells of the pericycle and behaves in a similar manner. Towards later stages, the developing cambial ring encloses a few layers of pericyclic cells which are meristematic in nature and rapidly divide or proliferate forming layers of parenchyma cells. These cells get filled with reserve food material and function as storage tissue.

With the passage of time as more and more proliferated parenchymatous cells are formed, the vascular bundles get separated farther and farther apart. The subsequent cambial rings produce alternate layers of vascular bundles and proliferated parenchymatous cells filled with reserve food. In a transverse section, the rings of parenchyma apppear dark red (anthocyanin pigments) while those of vascular bundles appear lighter in colour. The root of *Beta vulgaris* thus increases in size by the formation of vascular bundles and proliferated parenchyma cells, rich in sugar. In mature roots, a cork cambium or phellogen develops which forms cork cells on the outer side and secondary parenchyma towards inner side.

Ipomoea batatas root (Fam.–Convolvulaceae)–The anomalous secondary growth in *Ipomoea* was described by Artschwager (1926). In young conditions, the cortex is well developed with wide intercellular spaces. A well defined endodermis demarcates the stelar region from the cortex. The stele typically shows radial and pentarch or hexarch arrangement of vascular tissue.

Secondary growth of the storage root takes place by the formation of a cambium ring which is normal in function initially, however, at later stages it begins to behave abnormally forming separate strands of secondary xylem and phloem embedded in the parenchymatous ground tissue. These strands of xylem get surrounded by individual rings of secondary cambium which originate from the parenchymatous cells of the ground tissue surrounding the xylem vessels.

This secondary cambium surrounding the xylem vessel behaves abnormally forming few elements of secondary xylem towards the centre and a few sieve tubes and laticiferous elements away from the centre. The cambium also forms considerable amounts of parenchymatous cells on both inner and outer side. These cells get filled with reserve food material (sugar) and function as storage tissue. In later stages the larger vessels often develop tyloses. From

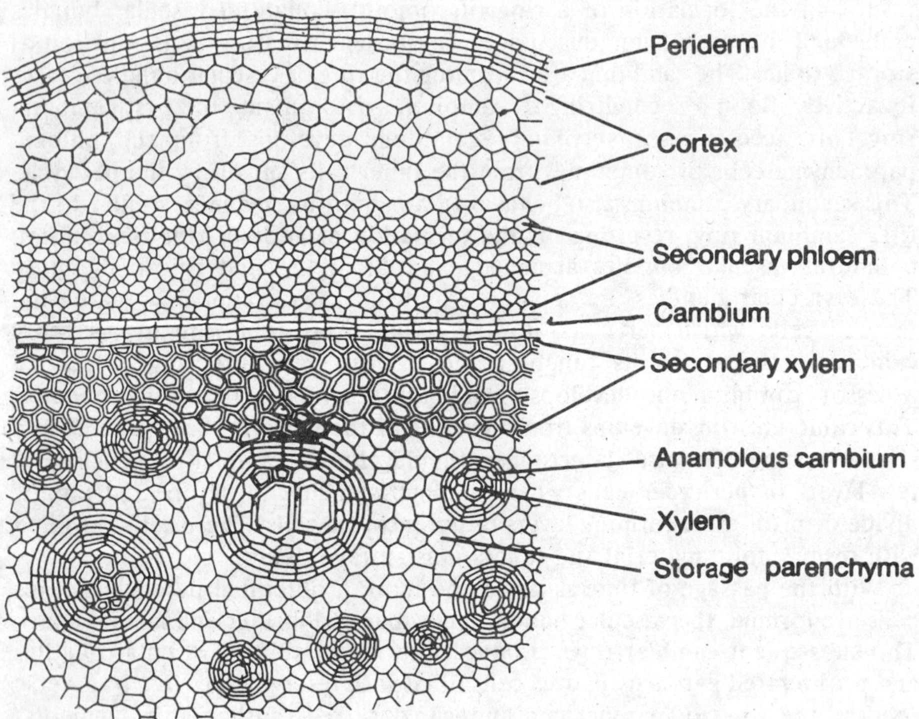

Fig. 12.15. T.S. root of *Ipomoea batatas* (an enlarged cellular portion).

the pericycle develops the cork cambium which forms the periderm.

Raphanus sativus root (Fam.–Cruciferae) The roots of radish *(R. sativus)* are fleshy, having a typical diarch primary xylem. At the beginning of secondary growth, a normal cambium is formed which cuts off secondary xylem towards inner side and secondary phloem towards outer side. The xylem cells are constituted of xylem parenchyma which repeatedly divide and proliferate in all directions thereby greatly increasing the secondary tissues. Some of these parenchymatous cells become meristematic and develop to form the secondary cambium. These cambia partly or completely surround the strands of secondary xylem scattered in the ground tissue. The activity of the secondary cambium is abnormal as it forms few xylem elements but large amounts of parenchyma both internal and external to the cambium ring. As the root matures, numerous patches of secondary xylem surrounded by secondary cambium lie scattered in the parenchymatous ground tissue. These parenchyma cells accumulate reserve food and function as storage tissue. Thus the increase in girth of the root is due to the formation of circular patches of secondary xylem and the mass of proliferating parenchymatous storage cells.

The original cambium is persistent and is visible in the peripheral region with sufficient amount of secondary phloem attached externally. The cork cambium forms a few layers of thin walled cork cells.

GENERAL QUESTIONS

1. **Discuss anomalous secondary growth found in dicot stems.**
2. **Describe the pattern of abnormal secondary growth in–**
   (i) *Leptadenia* (ii) *Boerhaavia* (iii) *Strychnos* (iv) *Salvadora.*
3. **Write notes on:–**
   (i) Interxylary phloem
   (ii) Anomalous secondary growth in *Dracaena.*
   (iii) Abnormal secondary growth in *Beta vulgaris* root.
   (iv) Interxylary cork.

CHAPTER 13

Nodal Anatomy

Nodes and internodes are distinct in the stem of Angiosperms. These may be well demarcated or indistinct as the case may be. From anatomical point of view, the place from where leaf traces or branch traces arise is referred to as **node**. The vascular cylinder shows different patterns in the regions of nodes and internodes. The distinction of node in case of monocots is very difficult due to presence of scattered vascular bundles while it is well distinct among dicotyledonous stems where vascular bundles are arranged in a ring. The vascular cyclinder is more complex in nature because of the presence and formaton of leaf and branch traces.

Leaf traces and leaf gaps

Leaf traces. In the nodal region, a portion of vascular cyclinder is diverted towards the leaf or a bud which is attached at this node. The diverted vascular tissues form the leaf trace. A leaf trace represents the lower part of the vascular supply of the leaf. A leaf trace may also be defined as a vascular bundle that connects the vascular bundle of the leaf with that of the stem. However, one or more leaf traces may be associated with each leaf and a leaf trace may traverse one or more nodes before it diverges towards the leaf.

Leaf gaps. In higher plants such as ferns and seed plants, with the divergence of the leaf traces immediately above the diverging traces, a region of parenchyma occurs in the vascular cylinder of shoot. This parenchymatous region is called leaf gap. So in a leaf gap, the parenchyma is differentiated instead of vascular tissue, in thé vascular region of the stem, for a limited distance. It is located adaxially from the diverging leaf traces. Actually, the leaf gaps are not breaks in vascular system because lateral connections between the vascular tissues occur above and below the gap.

In some ferns like *Pteridium* and *Pteris,* the leaves are so crowded that the gaps formed at the successive nodes overlap one another and that the actual vascular cyclinder appears highly dissected.

The transverse section of such rhizomes (stems) shows a circle of vascular bundles alternating with the parenchymatous leaf gaps. In certain gymnosperms, and angiosperms, the vascular system consists of several anastomosing strands so much so that leaf traces become confluent with the interfascicular areas and then the leaf gaps become uncertain and cannot be recognised.

Nodes. There are three common types of nodes found in dicotyledons: (i) **Unilacunar,** a node with a single gap and a single trace to a leaf, (ii) **Trilacunar,** a node with three gaps and three traces to a leaf (one median and two laterals), and (iii) **Multilacunar,** a node with several gaps and traces associated with a single leaf (Fig. 13.1). In most monocotyledons, leaves with sheathing leaf bases are found which receive a large number of leaf traces separately inserted around the stem. In gymnosperms, unilacunar node is common. In ferns, variable number of leaf traces enter in a leaf but they are always associated with a single gap.

Among dicotyledons, the leaf trace relationships at the nodes are thought to be of phylogenetic importance and that is why studies on nodal anatomy become valuable for systematics. Unilacunar nodes are found characteristically among members of several families such as Anonaceae, Apocyananceae, Ericaceae, Lauraceae, Solanaceae and Verbinaceae. In some cases, one to many traces arising from a single unilacunar node have been found and accordingly these may be called unilacunar one trace, unilacunar two traces, unilacunar three traces and unilacunar multitraces, respectively. Trilacunar nodes are characteristically found among members of Meliaceae, Polygonaceae, Ranunculaceae and Winteraceae. Among the members of Amentiferae (a group of monocotyledons), trilacunar nodes are found. Multilacunar nodes are found among members of Araliaceae, Chenopodiaceae and Degeneriaceae.

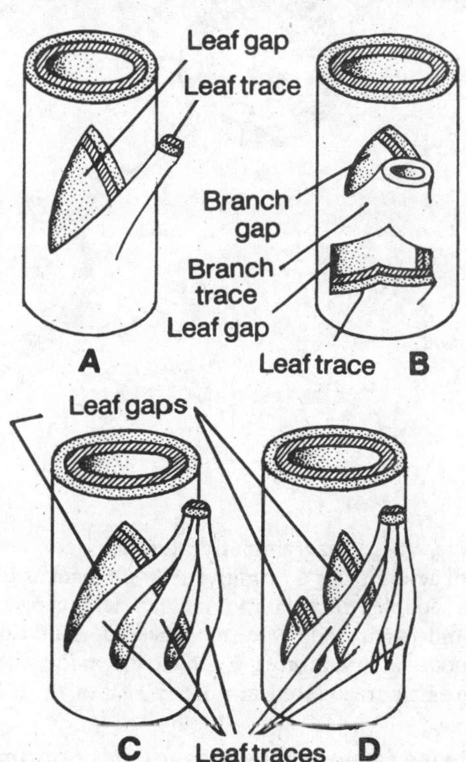

Fig. 13.1. Diagrammatic longitudinal representation of steles with different types of nodes. A–unilacunar node with a single leaf trace, B–unilacunar node with a single leaf trace but associated with a branch trace and branch gap, C–unilacunar node with three leaf traces and D–trilacunar node with three leaf traces (Redrawn after Fahn, 1967).

Sinnot (1914) proposed that trilacunar condition is the most primitive in nature from which arose unilacunar conditions by reduction process and multilacunar condition by amplification process. His proposition was widely accepted. However, Ozenda (1949) proposed that multilacunar nodes are primitive because of

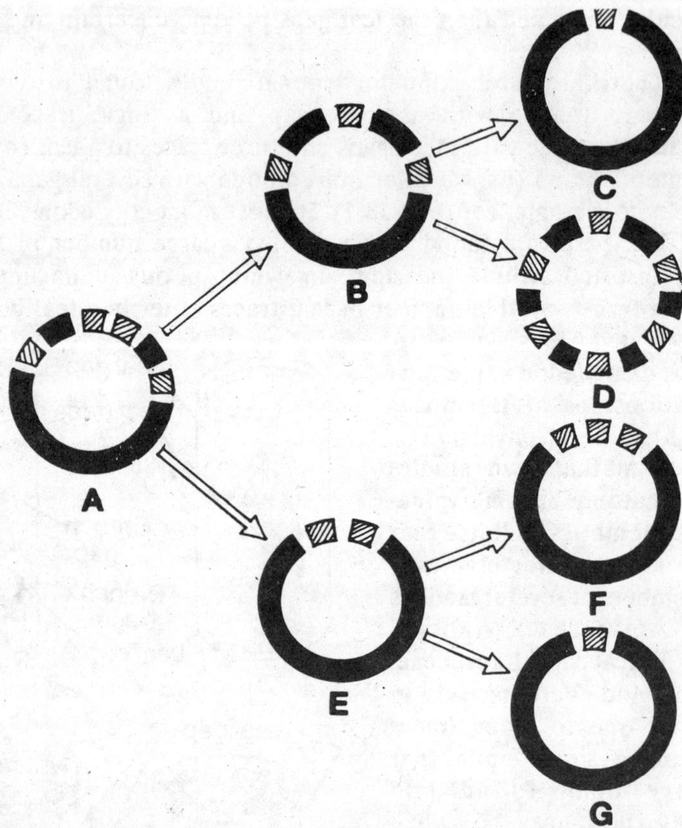

Fig. 13.2. Diagrammatic illustration of one possible course of evolution of nodal structure from a pirmitive tri-multilacunar type in dicot stems. A–trilacunar node with a double trace in the median leaf gap; B–trilacunar node with three traces, C–unilacunar node with one trace, D–multilacunar node with multitraces, E–unilacunar node with two traces, F–unilacunar node with three traces and G–unilacunar node with a single trace (Redrawn from Takhtajan, 1964).

being found among Magnoliales (a primitive family). According to him, three nodal types form a regressive series, i.e. multilacunar-trilacunar-unilacunar.

Canright (1955) and Bailey (1956) proposed that unilacunar with two distinct traces is the primitive type among angiosperms. It was supported by several other anatomists on the basis of the fact that unilacunar node with two distinct traces is characteristic not only of gymnosperms and ferns but also occur in several members of dicot families such as Labiatae, Lauraceae, Solanaceae and Verbenaceae. Pant and Mehra (1964) have not supported this concept because a vast majority of gymnosperms and angiosperms do not show the presence of two trace unilacunar node.

Takhtajan (1964) also proposed the course of evolution of nodal structure among dicotyledons. He stated that trilacunar node with two traces from the central gap is the most primitive or basic type from which arose other types of nodal structures either by reduction or by amplification (Fig. 13.2)

Branch traces and Branch gaps

The vascular supply to axillary bud or lateral branches also receive vascular traces derived from the vascular system of main axis. These traces are called branch traces or ramular traces. The vascular supply of axillary branch usually consists of two traces (two bundles) but in some plants either only one or more than two traces are found. In monocotyledons, the connection of the axillary shoot with the main stem consists of many strands.

Usually in most cases, a branch trace is found associated with the formation of a break in the vascular cylinder around and above the point of departure of the trace. This opening is called branch gap. Branch gaps are found in all siphonostelic plants (stele with pith) and not found in protostelic plants (stele without pith).

GENERAL QUESTIONS

1. Write a brief account of nodal anatomy.
2. How nodal structures have been found useful in phylogentic studies.
3. Write short notes on:
   (i) Unilacunar leaf trace (ii) Trilacunar leaf trace
   (iii) Multilacunar leaf trace.

CHAPTER 14

Anatomy of Floral Parts

Flowers are colourful attractive reproductive structures found in the most evolved plants, i.e. angiosperms. It consists of a **receptacle** on which four sets of appendages are borne. In a typical flower, the outer two sets of appendages are sterile and hence called *accessory organs* while inner two sets have fertile appendages and hence called *essential organs* (or appendages). The outer whorl of sterile appendages is called **calyx** which consists of unit structure called sepals and the inner whorl of sterile appendages is called **corolla** which consists of petals. The appendages of outer fertile whorl are collectively called **androecium** (male reproductive organs) and its unit is called microsporophyll or stamen. The appendages found in the inner fertile whorl are collectively referred to as **gynoecium** and its each unit is megasporophyll or carpel. A great range in the organisation of different floral appendages are found in nature which develops largely by reduction, fusion or modifications.

Morphological nature of flower

Flowers are specialised reproductive structures which are so distinct from other parts of the plant that they draw every one's attention. Several theories have been proposed to establish a general concept of flower or to explain commonly accepted morphological nature of flower. Following are a few important concepts of flower which have been developed by different workers from time to time.

1. **Classical concept.** Goethe (1790) proposed *theory of metamorphosis.* According to which **flower is a modified shoot** or a flower is homologous to the vegetative shoot and the lateral appendages of flower such as sepals, petals, stamens and carpels are morphologically equivalent to the foliage leaves. This theory has been most widely accepted. Thome (1963) referred to it as classical concept. He proposed that, the primitive angiosperm flower consisted of an axis with spirally arranged sterile and fertile appendages and the advanced type of flowers have evolved by modification or reduction of the various components. However, this concept has limitations, to offer a satisfactory explanations of the flowers of certain plants (taxa).

2. **Meeuse's concept.** Meeuse (1963) proposed that the flower of angiosperm

is **stachyosporous** and explained that the ovules are borne on axes or their homologies and not on the structure which are homologous of the leaves. He has visualised many similarities between *Gnetum* (gymnosperms) and certain angiospermic genera such as *Canacomyrica* of family Myrtaceae and *Sarcandra* of Chloranthaceae. Ovules in these cases are protected by three covering layers. He further explained that ovary of angiosperms have evolved from the ovule of *Gnetum*-like plants by gradually shifting of the function of the stigmatic protion of ovule from first to second or third protective layer. However, this theory fails to explain the marginal position of ovules among Ranales.

3. **Lam's concept.** Lam (1962) suggested that there is a need of two concepts to explain the concept of angiospermic flowers. He believes that there are two types of angiosperm flowers, i.e. *phyllosporous* and *stachyosporous*, both have evolved independently from ancestors like gymnosperms. In phyllosporous flowers, the ovules are borne on leaf homologies whereas in stachyosporous flowers they are borne on floral axis.

4. **Melville's concept.** Melville (1962) proposed **gonophyll concept** of flower to resolve the difficulties of foliar concept (classical concept). He proposed that the basic unit of the flower is the gonophyll, i.e. a leaf with an epiphyllous (borne on leaf) branch system. And then terms like androphyll to stamen and gynophyll to carpel and androgynophyll to both male and female organs are to be used. According to him, the gonophylls have evolved by overtopping, fusion and planation of the original dichopodial branch system. However, this theory is based on a lot of speculation.

Anatomically, the flower is a determinate stem with crowded appendages in which internodes are highly reduced or obliterated. It supports the classical concept. However, several differences between vegetative shoot and reproductive shoot (flower) are noted. The flowers show limited growth and limited activity of meristems in the formation of floral parts while growth is unlimited in a vegetative shoot. The growth period in primitive flowers with higher number of appendages and specialised flowers with limited and definite number of floral parts also show variation. In still more advanced flowers there are several specialised characters not found in vegetative shoots such as whorled arrangement of parts instead of spiral, adnation of parts of two or more different whorls, cohesion of parts within a whorl, zygomorphic instead of actinomorphic conditions and epigynous condition instead of hypogynous condition.

Anatomical features of floral parts

A typical flower is mainly comprised of sepals, petals, stamens and carpels.

Sepals. The anatomical features of a sepal resemble a leaf. Each sepal consists of epidermis, ground parenchyma and a branched vascular system. The green sepals have chlorophyllous cells but usually there is no differentiation into palisade and spongy parenchyma. Idioblasts, laticifers and tannin cells are found. Stomata and epidermal hairs may be present in epidermis. The vascular traces found in sepals are similar in origin and number. Such an organisation reveals that sepals have been derived directly from leaves and not from sterile sporophylls.

Petals. The anatomical features of a petal also show resemblance with those of leaves. Each petal consists of epidermis, ground parenchyma and more or less branched vascular system. Crystal-containing cells, laticifers, tannin cells and other idioblasts are also found. Sometimes, epidermal cells contain volatile oils which emit the characteristic fragrance of the flower. Stomata and epidermal trichomes may also be present. Cells of ground tissue contain abundant chromoplasts rich in water soluble anthocyanin.

Stamens. A stamen consists of a filament, a connective and anther. Commonly anther is dithecous or two lobed with four locules (or microsporangia). Anatomical features of the filament is relatively simple. The epidermis is cutinised and bears trichomes and stomata. A filament is uninerved, i.e. there is a single vascular bundle that runs throughout the filament and ending blindly in connective. The vascular bundle is amphicribal which remains surrounded by parenchyma.

The structure of anther is little complex. Its outermost wall is the single layered epidermis which is followed by another single layered endothecium. Cells of endothecium show hygroscopic, fibrous bands arising from the lower transverse wall and almost touching the upper transverse wall. Endothecium is seen to be well developed when anthers are mature and it helps in dehiscence of pollen grains. Next to endothecium and towards inner side are found three to four layers of cells of the wall layers. The inner most of these wall layers surrounding the microsporangium is the layer of tapetum. The tapetum is the nourishing layer and is of great physiological significance since all the nutrition which the sporogenous mass receives pass through it. The tapetal cells show various types of divisions such as endomitosis, normal mitosis and sticky type of nuclear division. The anther, has a procambial strand in the centre which later on forms the vascular bundle.

A typical dithecous anther has four microsporangia, two in each lobe. In each microsporangium, several pollen grains are formed. On the maturation of pollen, the tapetum and wall layers usually get disintegrated leaving pollen sac surrounded by endothecium and epidermis. At the time of dehiscence of anthers, the pollens are released through stomium.

Gynoecium. Gynoecium represents the female sex organ of the flower. It is constituted of one or more units known as carpels which may be free or united. Each carpel consists of an ovary, style and stigma. Of these, ovary is the important structure within which ovules are present attached on placenta.

In general, ovary and style are composed of epidermis, parenchymatous ground tissue and vascular bundles. Epidermis is cuticularised and may possess some stomata. Number of vascular bundles are usually 3-5 in each carpel.

Vascular anatomy

Puri (1952) proposed that study of vascular anatomy of flowers may help in solving several problems of morphology, homology or analogy because the fundamental vascular plan remains more or less unaltered. Such proposition thus emphasizes the importance of studies on vasculature of flower.

Anatomically, the pedicel of a flower shows close similarity with that of a stem. It consists of either a ring of vascular bundles or there is an entire vascular cylinder. The stele of pedicel when it reaches the receptacle commonly expands and then gives off vascular traces to the various floral appendages usually in the same order in which these are arranged on the receptacle. Such origin and distribution is clearly seen in simplest cases as in *Aquilegia* where vascular traces of different organs and whorls of organs arise quite independently. The appendage traces are derived from the stele of receptacle exactly as leaf traces are derived in typical stems. When the floral organs are numerous and closely placed, the gap of traces break the receptacular stele into a meshwork.

The stele of receptacle in most cases is completely used up in giving rise to the traces to the uppermost carpels but in some flowers, the stelar bundles may continue to extend above the uppermost carpels and these eventually disappear as residual tissue.

A **sepal** usually receives three traces derived from the same or different sources. Of the three, two traces are lateral and one is median. **Petals** and **stamens** receive typically one trace directly from the stele. The **carpel** normally receives three traces, one dorsal and two ventrals, each arising independently but dorsal one arises slightly ahead of stamens. In a carpel, the dorsal trace occupies the mid-rib position while ventrals give rise to the ovular traces. This type of vascular organisation is usually found in typical flowers but a considerable variation in its plan may be seen in specialised flowers.

Such a simple organisation of vascular traces in flowers can better be explained taking an example of a simple flower of *Aquilegia* of family Ranunculaceae which has been explained and studied by Eames (1931) and Tepfer (1953).

In *Aquilegia,* the vascular traces for different organs and whorls of organs arise independently, hence remain distinct bundles (Fig. 14.1 and 14.2) which fuse to form a complete vascular cylinder within the receptacle. At its first stage of branching, **sepal traces** arise first and each sepal receives three vascular traces (Fig. 14.1 and 14.2). Next to sepal traces, arise five **petal traces,** one for each petal and above the petal traces stamen traces are given off. **Stamen traces** arise from vascular cylinder in several whorls of five each (Fig. 14.1 and 14.2) but each stamen receives a single trace as in case of petals.

Next to stamen traces arise five dorsal **traces of the carpels** which occupy mid-rib positions in the five carpels (Fig. 14.1 & 14.2). Then five traces of central carpel traces arise from vascular cylinder, one pair for each carpel.

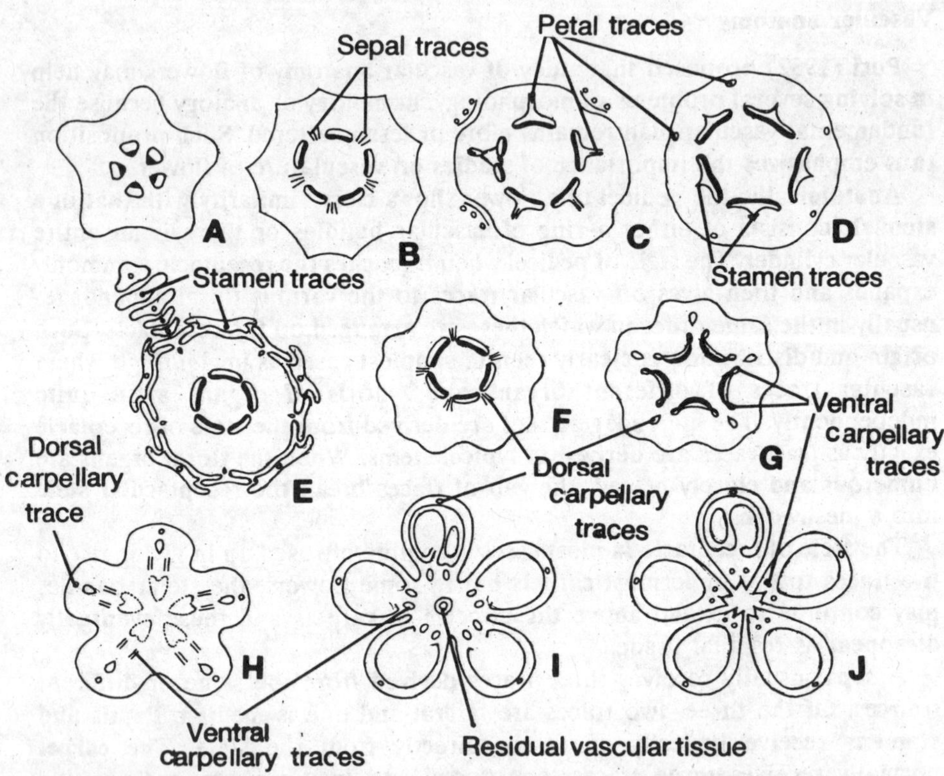

Fig. 14.1. Different cross sections of a floral bud of *Aquilegia* arranged serially from base to upwards showing vascular supply to various floral appendages. (Redrawn from Eames, 1931).

These two ventral traces position themselves on the inside of the locule. The ventral traces after giving ovular traces fade out towards the top of the locule. After giving rise to ventral carpel traces, the remaining vascular cylinder appear organised into five patches which also fade out towards the top of the receptacle (Fig. 14.2).

Modifications in vascular systems

The fundamental and simple plan of vascular system found in flowers undergoes several modifications in advanced or modified flowers. Such modifications occur by way of fusion, reduction or amplification of vascular bundles.

1. **Modification by fusion.** Fusion of vascular bundles may be of two types, i.e. cohesion or adnation. In **cohesion,** the vascular traces of the same whorl are fused with one another and in **adnation,** the vascular traces of one whorl are fused with those of another whorl.

(i) **Fusion under cohesion.** In most cases of plants with axile placentation, the two ventral carpel bundles of the same carpel, and with perietal

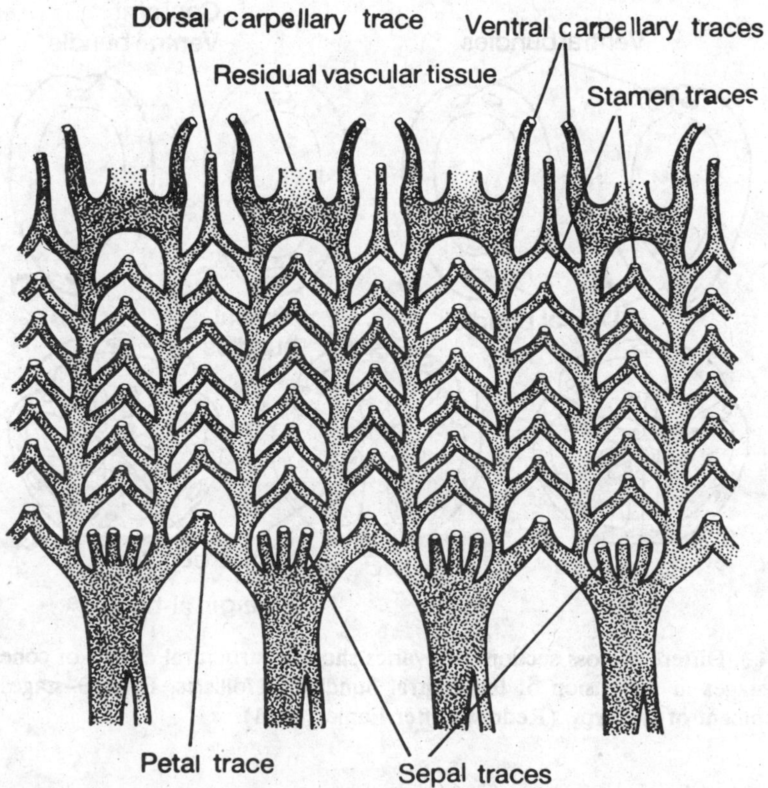

Fig. 14.2. Diagrammatic illustration of vasculature of *Aquilegia* flower being spread open (Redrawn after Tepfer, 1953).

placentation, the two ventral carpel bundles of adjacent carpels, arise as a single bundle. The fact that the bundle is double in nature is supported by the presence of bifurcated bundles in the upper region of ovary or style. So the fusion of ventral bundles of carpels, is commonly found. Similarly, such a fusion is also usually seen in case of sepals of gamosepalous calyx and petals of gamopetalous corolla. The various stages of cohesion of marginal traces of sepals can be seen in the calyx of plants belonging to Labiatae (Lamiaceae) such as in *Nepeta, Salvia, Trichostema* and *Ajuga.* In *Nepeta,* each sepal has three distinct bundles, one median and two marginal but in case of *Salvia* and *Trichostema,* the two adjacent pairs of marginal bundles fuse to the sinuses whereas other three pairs remain free (Fig. 14.3) and in *Ajuga,* all the five pairs of adjacent bundles are fused up to the sinuses (Fig. 14.3).

(ii) **Fusion under adnation.** Adnation is commonly seen in case of epipetalous stamens. The different stages of fusion can be dinstinctly seen among members of Crassulaceae in longitudinal and cross sections. All stages of fusion of organs and their respective vascular bundles may be seen among different members of Rosaceae and Ericaceae. These plants explain the

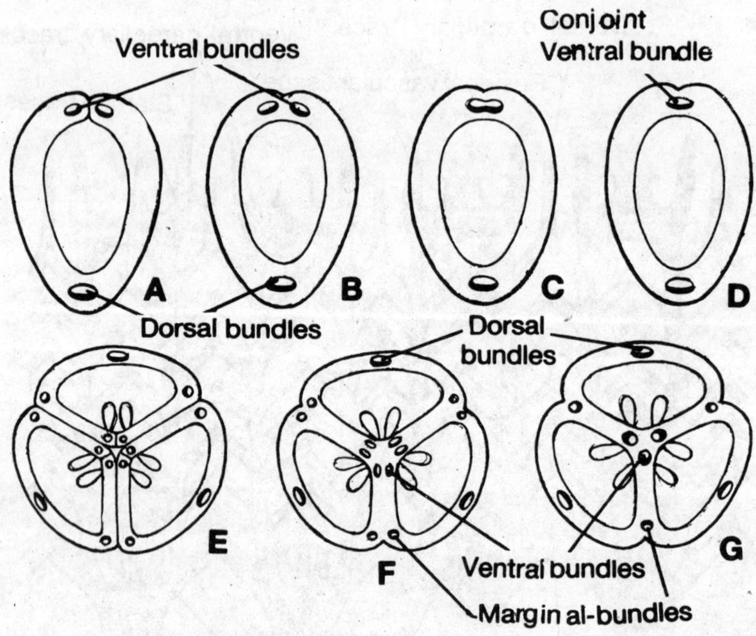

Fig. 14.3. Different cross sections of ovaries showing structural effects of cohesion. A to D–stages in the fusion of the ventral bundles of follicles, E to G–stages in the development of syncarpy (Redrawn after Eames, 1931).

origin of epigynous flowers from hypogynous ones.

2. **Modification by reduction.** A reduction in number of vascular traces is usually found in the flowers with lesser number of organs. Such reduction in number of organs may be due to loss of organs or through fusion. For example, with the reduction of corolla in *Chrysanthemum* and *Xanthium* of

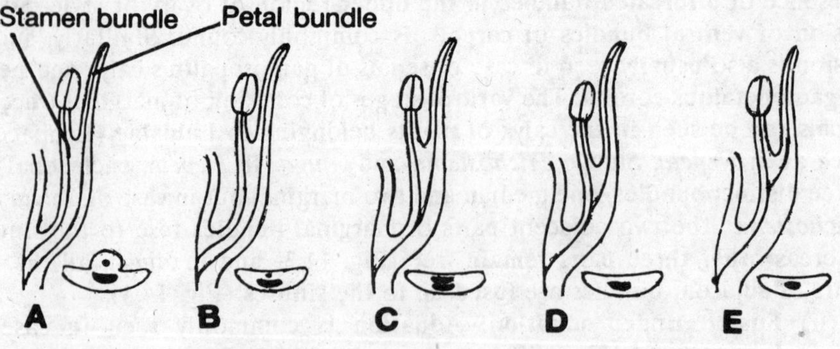

Fig. 14.4. Different stages of fusion of stamen and petal and their traces in members of *Crassulaceae*. Both L.S. and T.S. are shown (Redrawn after Eames and MacDaniels, 1947).

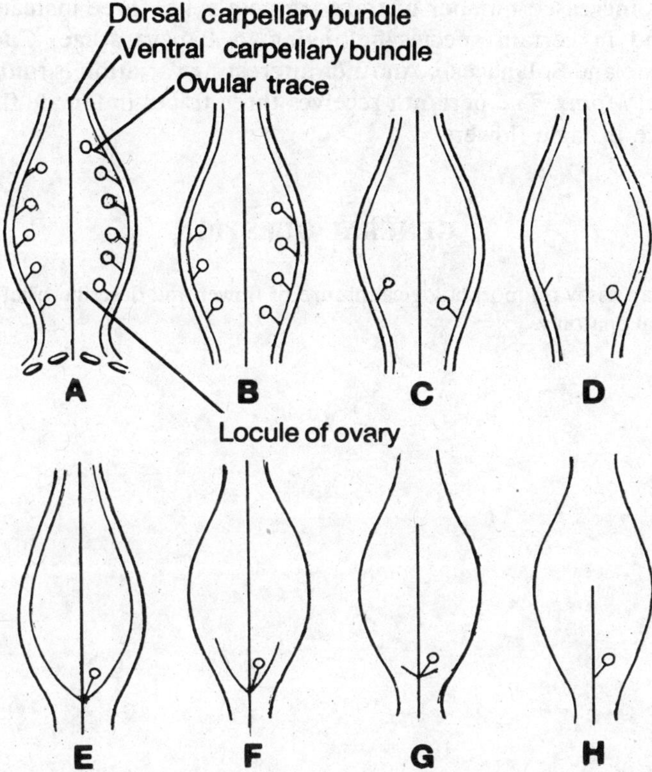

Fig. 14.5. Diagrammatic illustration. L.S. of carpel structure showing reduction of ovules and fusion and reduction of vascular supply in some Ranunculaceae and Rosaceae. A to D–follicles, A–*Helleborus*, B–*Aquilegia*, C–*Hydrastis*, D–*Waldsteinia*, E to H–achenes, E–*Geum*, F–*Ranunculus flammula*, G–*R. cymbalaria* and H–*R. aquatilis*. (Redrawn after Eames and MacDaniels, 1947).

family Compositae, the vascular supply is greatly reduced. In both these cases, the mid-rib bundle has disappeared. In *Xanthium*, the fused marginal bundles are also reduced and they never extend beyond the sinuses. The loss of marginal bundles from the sepals may be seen in some members of Primulaceae.

The reduction in ovules followed by fusion and reduction in vascular supplies is evident in the flowers showing evolutionary specialisation from a follicle with several ovules to an achene with single ovule in Ranunculaceae and Rosaceae. In Ranunculaceae, loss of ventral bundles and in Rosaceae, loss of dorsal bundles are seen. Eames and Mac Daniels (1947) proposed that the gradual loss in number of ovules and vascular supplies may be seen from *Helleborus* to *Ranunculus aquatilis* through *Aquilegia, Hydrastis, Waldsteinia, Geum, Ranunculus flammula* and *R. cymbalaria*.

3. Modification by amplification. The increase in number of vascular traces in the individual respective appendages of flower has been found in some cases. The increased number of traces in petals, i.e. three instead of one has been found in certain species belonging to Papavaraceae, Cucurbitaceae, Primulaceae and Solanaceae. Another interesting variation is found in case of perianth of *Morus*. The perianth receives three traces in female flowers and a single trace in male flowers.

GENERAL QUESTION

1. Write an essay on morphological nature of flower and describe briefly an account of floral anatomy.

CHAPTER 15

Fruit Wall and Seed Coat

FRUIT WALL

The fruit wall develops from ovary wall at maturity and then it is usually called pericarp. The **pericarp** consists of three layers, the outer *epicarp*, middle *mesocarp* and inner *endocarp*. These three layers are well distinguishable in fleshy or succulent fruits and almost indistinct in dry fruits. As such two broad categories of the fruit walls may be recognised.

(i) Dry fruit wall which mainly consists of sclerenchymatous cells, and

(ii) Succulent or fleshy fruit wall which is mainly composed of parenchymatous cells.

And accordingly two types of fruits, the *dry fruits* and the *succulent fruits* are recognised.

Dry Fruits

The dry fruit wall consists of various amounts of sclerenchyma and non-succulent parenchyma at maturity. It may be dehiscent or indehiscent in nature. The dehiscent fruit walls split open at maturity while indehiscent one remains closed.

Dehiscent fruit wall. Such dry fruits which have large number of seeds usually dehisce at maturity. This type of fruit may develop from single carpel, such as *follicle* and *legume* or from several fused (syncarpous) carpels such as *siliqua* and *capsule*. The **follicle** is a pod-like fruit which usually dehisces along the ventral suture (e.g. *Delphinium*) whereas **legume** dehisces by splitting of fruit wall along both the sutures which surrounded the fruit (e.g. *Pisum* and other members of Leguminosae).

Siliqua develops from bicarpellary and syncarpous ovary and its wall splits along both the sutures from below to upward. The **Capsule** developing from more than two carpels dehisce variously in different plants.

The **pericarp of a follicle** usually possesses a relatively simple structure and is thinner than that of siliqua and legume. It may have a narrow epicarp composed of thick-walled cells and a thin walled parenchymatous mesocarp and endocarp. The vascular supply is also simple. One median and two lateral longitudinal bundles along with several transverse branches arising from main bundle remain enclosed in sclerenchymatous sheaths. At maturity, the fruit wall splits along the line of fusion of carpel margins and dehiscence is brought about by the tension due to differential drying to parenchymatous and sclerenchymatous tissues of the pericarp.

The **pericarp of a legume** has a relatively more complicated structure than a follicle. In certain cases, the pericarp at maturity consists of a thick-walled epicarp (exocarp), a thin walled parenchymatous mesocarp and a highly sclerified endocarp. In the endocarp, several layers of sclerenchyma are found which remain arranged at an angle to the long axis of the fruit. The endocarp remains covered internally by a thin walled epidermis (Fig. 15.1). Such an orientation of sclerenchyma plays an important role in dehiscence. Their contraction is in different planes during drying of the pericarp, with the result, the fruit dehisces along two lines, one along the fusion of carpel margins and the other along median vascular bundle.

In certain legumes, such as in *Calycotome villosa*, the mesocarp is entirely collenchymatous.

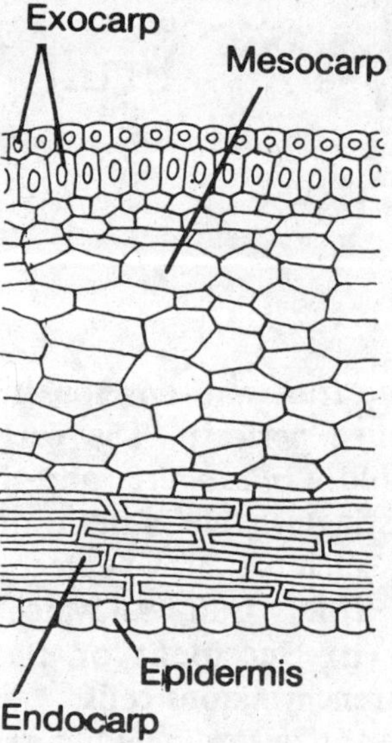

Fig. 15.1. A portion of the pericarp of a leguminous fruit (*Glycine*) (Redrawn from Monsi, 1943).

The **pericarp of capsule** has both sclerenchymatous and parenchymatous tissues with various degrees of distribution. For example, in linseed (*Linum usitatissimum*), the epicarp is sclerenchymatous whereas both mesocarp and endocarp are parenchymatous, but in tobacco (*Nicotiana tabacum*), both epicarp and mesocarp are thin walled parenchymatous and only endocarp is made up of thick walled sclerenchymatous cells. In the capsules of *Vaccaria pyramidata*, the outer wall of the epidermal cells is provided with many thick lamellae. The dehiscence of capsule is brought about either along the line of fusion of carpels or along the plane of the median bundle of each carpel.

Indehiscent fruit wall. The indehiscent fruit has only one seed because its ovary contains a single ovule. The anatomical features of fruit wall (pericarp) as well as of seed coat (testa) show resemblance.

In such fruits, the seed coat becomes more or less completely obliterated during fruit development. When pericarp and seed coat are fused, the fruit is known as a grain or *caryopsis* and if seed remains attached to the pericarp at one point only, the fruit is an *achene*. Caryopsis fruits are characteristic features of Graminae while achenes are of Ranunculaceae.

The **fruit wall of caryopsis** thus consists of the tissues of both, the pericarp and the seed coat. In *Zea mays*, the fruit wall has an epidermis covered with a cuticle. The epidermis is pitted and compressed particularly at the distal end. Epidermis is followed by one to few layers of parenchyma. Inner to this

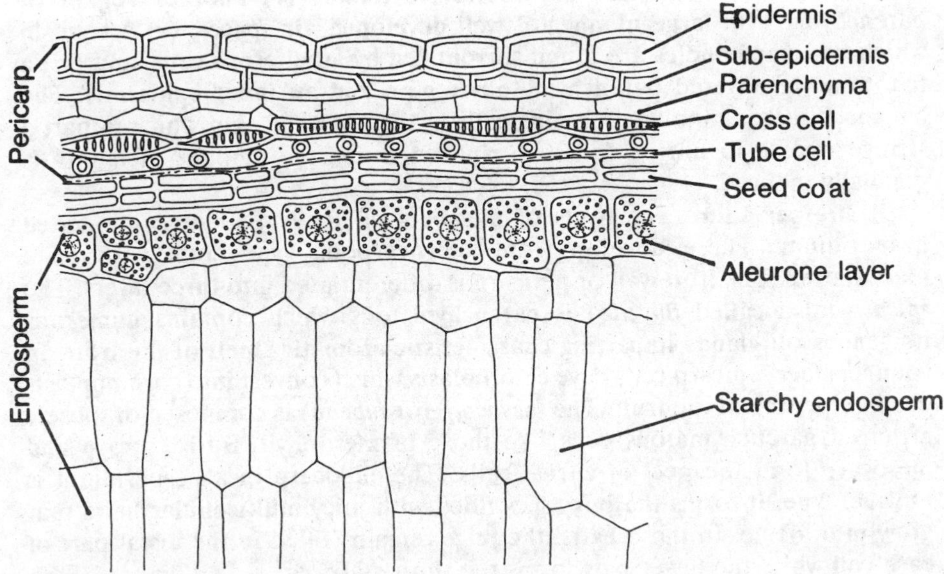

Fig. 15.2. A portion of cross section of pericarp of the caryopsis fruit (*Triticum*).

parenchymatous layer lies a layer of thick walled cross cells oriented along the long axis of grains. Inner epidermis which is the innermost layer of pericarp consists of thick-walled tube cells placed parallel to the long axis of the grains. The seed coat usually disintegrates and the remains of seed coat are found in between the endosperm and the tube cells.

In certain schizocarpic fruits like those found among members of Umbelliferae, several oil glands are characteristically found in mesocarp.

Fleshy fruits

Fleshy fruits may develop from monocarpellary or polycarpellary gynoecium. At maturity, their fruit walls do not dry up but remain indehiscent and fleshy. The anatomical features of some fleshy fruits are considered here.

(i) **Berry**. Berry fruits are commonly found among members of family Solanaceae. In tomato (*Lycopersicon lycopersicum*), the epicarp consists of a layer of polyhedral epidermal cells followed by three to four layers of sub-epidermal collenchyma. At some places, simple or glandular hairs are found developed from epidermal cells. The mesocarp consists of loosely arranged thin walled parenchymatous cells with intercellular spaces. The endocarp remains indistinguishable in a mature fruit because it disintegrates during ripening of the fruit. The pulp of mature tomato fruit consists of placental tissue which forms most of the fleshy tissue of a mature fruit.

Banana (*Musa paradisiaca* var. *sapientum*) is another good example of berry, the wall of which consists of (i) outer epidermis covered by cuticle, (ii) multilayered parenchymatous hypodermis, (iii) a broad parenchymatous region in which vascular bundles are found scattered, (iv) another region of parenchyma with large number of well developed air spaces, (v) a zone in which vascular bundles are found surrounded by laticiferous cells, and (vi) a five to seven layered parenchymatous region and an inner epidermis. The banana fruit has tannin containing cells in the pulp and skin which impart a stringent taste to unripe fruits. In ripened fruits, the amount of tannin is gradually reduced.

(ii) **Hesperidium.** The fruits of members of family Rutaceae are called hesperidium. These develop from a syncarpous gynoecium with axile placentation. The fruit wall or pericarp is differentiated into three layers. The *epicarp* (also called *flavedo*) is parenchymatous which contains numerous lysigenous oil glands imparting characteristic aromatic smell of the fruit. In younger stage, epicarp cells have chloroplasts which convert into chromoplasts during ripening of the fruit. The *mesocarp* (or *albedo*) is composed of loosely arranged parenchymatous cells like those of mesophyll. Both epicarp and mesocarp form the peel of citrus fruits. The endocarp in a hesperidium is unusual type. It forms the juicy sacs filled with juicy multicellular hairs that grow into locule. In these hairs, the juice remains filled in the distal part of each cell while the lower part forms the short stalk.

(iii) **Pepo.** The pepo develops from inferior ovary. Fruit wall is quite thick and fleshy and is formed by the pericarp (wall of the carpel) as well as the extra carpellary tissue. However, both these tissues remain indistinguishable. The epicarp and endocarp are thin and represented by outer and inner epidermis. The mesocarp is relatively quite fleshy. It consists of collenchymatous layers followed by several layers of parenchyma. Sometime, chloroplasts are found in these cells and in certain cases such as watermelon and

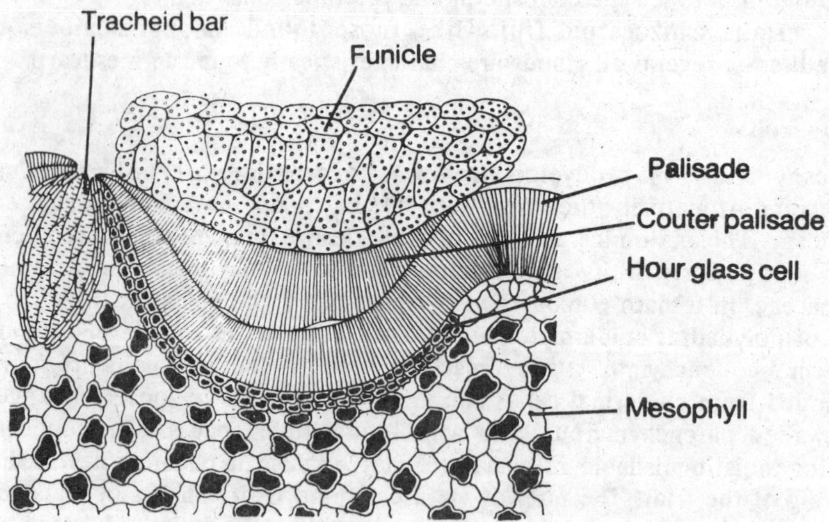

Fig. 15.3. A portion of longitudinal section of testa of a leguminous seed passing through hilum (*Phaseolus aureus*) (Redrawn from Chowdhury and Buth, 1970).

gourd, sclereids are also found. Fleshy parenchyma is followed by juicy parenchyma. The fleshy part of fruit wall has vascular bundles.

(iv) **Drupe**. Drupes develop from superior ovary. The fruit wall shows three well differentiated layers. The epicarp consists of a single layer of compactly arranged cutinised epidermal cells. Sometimes unicellular hairs are found arising from epidermal cells. The mesocarp forms the fleshy part of fruit and is made up of loose parenchymatous tissue. The endocarp is stony and consists of several layers of tangentially elongated sclereids.

SEED COAT

Seed coat develops either from one integument or from both the integuments. Prior to fertilisation, the integument consists of thin walled simple parenchymatous cells. These cells enlarge and become complex as the integument matures into the seed coat. The internal structure of mature seed coats show a great range of variation. In any case, seed coats are usually protective in nature.

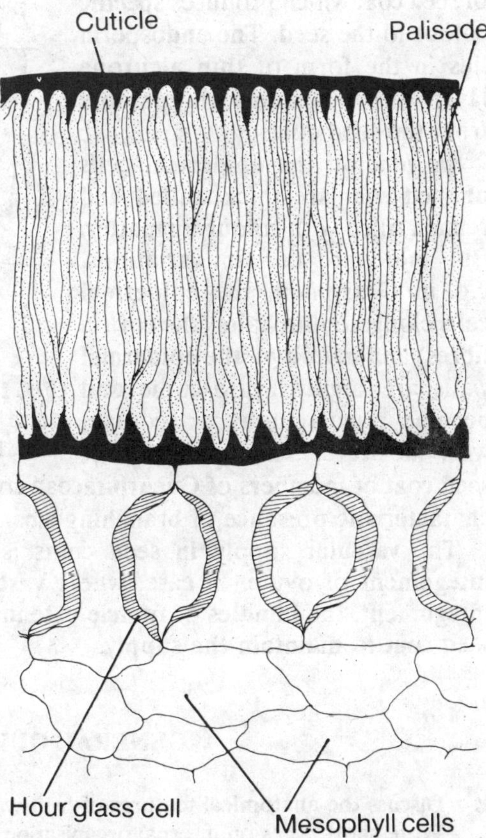

Fig. 15.4. A portion of longitudinal section of seed coat (testa) of a leguminous seed (*Vicia faba*) (Redrawn after Chowdhury and Buth, 1970).

The simplest type of seed coat is found in some epiphytes and parasites (e.g. *Orobanche*) where it consists of a single layer of cells derived from outer integument. However, the seed coats in other cases are complex in nature. Seed coats of pulses are impervious to water and gases due to deposition of thick layer of cuticle except in *Cajanus cajan* where it is very thin and smooth. The cuticle is smooth surfaced in *Cajanus cajan*, *Vicia faba*, *Dolichos biflorus* and *Phaseolus aconitifolius*, rough with dentate structures in *Lens culinaris* and *Lathyrus sativus* and papillose in *Phaseolus aureus* (Fig. 15.3). In pulses or leguminous seeds, cuticle is followed by a layer of characteristic malpighian cells, called palisade layer. It consists of closely packed palisade like cells, that elongate at right angles to the surface and differentiate into macrosclereids. Palisade cells are arranged in a single layer but in the hilum region, there are two layers, the palisade and the Couter palisade. Next to palisade is found a layer of hour glass cells (Fig. 15.4). In *Cajanus cajan* and *Dolichos lablab*, more than one layer of hour glass cells are found.

In members of Brassicaceae (Cruciferae), the seed coat is partly derived from outer integument and partly from inner integument of the óvule. Epidermis, sub-epidermis and supporting layers are derived from outer integument. The cells of epidermis are shrunken with degenerated protoplasm. Sub-epidermal layer is one to two cells in thickness, the cells of which are large but contains degenerated cytoplasm. The cells of supporting layer are arranged in single layer and palisade like pattern. The entire inner wall and lower half of the radial walls are thickened. The outer layer of inner integument gets degenerated while the cells of its inner layer forms the characteristic pigmented layer of seed coat which produces specific colour to the seed. The endosperm lies in the form of thin aleurone layer which remains inseparable from the seed coat.

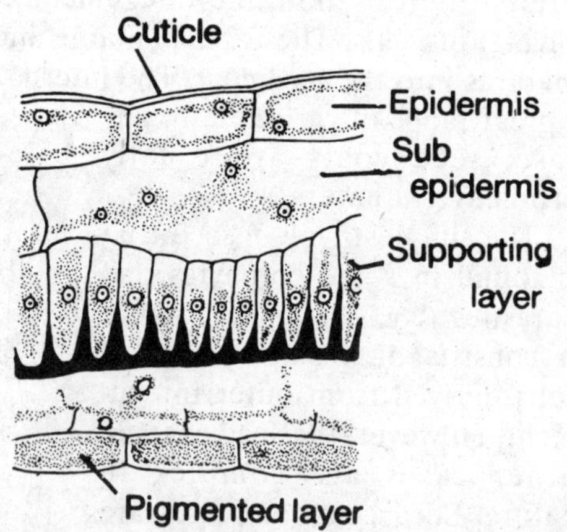

Fig. 15.5. Longitudinal section of a portion of testa of *Brassica juncea* (Redrawn after Dr. L.C. Lamba).

In general, the epidermal cells of seed coat are thick walled and remain filled with colouring matter. But there are certain variations, e.g. in *Gossypium*, the epidermal cells elongate into hairs (the cotton fibres) and in *Linum*, the epidermal walls are highly hygroscopic and become mucilaginous on contact with moisture. The epidermis of seed coat of members of Cucurbitaceae consists of highly elongated cells with characteristic presence of branching ribs.

The vascular supply in seed coats is the same as those found in the integument of ovule. In case, where vascular bundles are not found in the integument, the bundles of funicle extend and develop the vascular system of seed coat to maintain the supply.

GENERAL QUESTIONS

1. Discuss the anatomical features of fruit wall and seed coat.
2. Write short notes on internal organisation of
   (a) Pericarp (b) Seed coat.

CHAPTER 16

Ecological Anatomy

Warming (1909) classified plant communities on the basis of dependence and relation of plants to water and referred to them as hydrophytes, mesophytes, xerophytes etc. **Hydrophytes** (Gr. *hudor*-water; *phyton*-plant) are those plants that live wholly or partly submerged in water or in very wet places. **Mesophytes** (Gr. *mesos*-middle; *phyton*-plant) are those plants that live in an environment with average or optimum water supply. These are, therefore, intermediate between hydrophytes and xerophytes. **Xerophytes** (Gr. *xeros*-dry; *phyton*-plant) include those plants which live in dry places where sufficient water is not available to them. These can survive under long dry conditions and are, therefore, drought resistant plants. Between hydrophytes and mesophytes and extreme xerophytes and mesophytes, several plants showing different gradations are found. Beside these three forms, some other forms are also found in nature such as epiphytes, saprophytes, parasites, halophytes etc. **Epiphytes** grow upon other plants but they do not absorb food or minerals from them. **Saprophytes** grow upon dead organic matter while **parasites** grow with host plants (living) and obtain their food from them. **Halophytes** grow in saline soil or water.

Plants growing under different conditions become adapted structurally and physiologically to habitats in order to maintain a balance for their better growth. These adaptations are of special interest to study. Some of the anatomical features of plants belonging to these groups are discussed here.

HYDROPHYTES

Term *hydrophytes* refers to those plants which live in water or in a condition where water is available much more than their requirement. With respect to their relation to water and air, the hydrophytes are broadly divided into three groups, (i) submerged, (ii) floating, and (iii) amphibious.

I. Submerged plants

These plants grow entirely submerged in water either as attached submerged plants or submerged suspended plants. Common examples of plants belonging to this category are *Hydrilla, Vallisnaria, Zostera, Ceratophyllum, Myriophyllum,*

Elodea, Najas and *Potamogeton.* These plants show several morphological and anatomical adaptations in response to their environment.

Roots. The roots of such plants are reduced which may be either poorly branched or branching may be absent. In *Ceratophyllum* and *Utricularia* roots are absent.

Epidermis is made up of a single layer of thin-walled parenchymatous cells. The epidermal cells lack cutin or suberin and, therefore, these are able to absorb water through their general surface. The cortex is well developed and contains a large number of air cavities. Air cavitites are often separated from one another by one to many cell thick diaphragms. The air cavities have two important functions, (i) help plant in floating by increasing the buoyancy and (ii) reinforce the strength, i.e. mechanical support, through inflation with gases. The mechanical tissues like sclerenchyma are usually absent in this region.

The vascular tissues among submerged plants are least differentiated and usually consists of thin walled elements. Xylem is sometimes reduced to a central cell which remains surrounded by phloem as in case of *Elodea, Zannichellia, Potamogeton* and *Najas.* Vessels are altogether absent in several members of family Hydrocharitaceae. The reduction in xylem elements in these plants is because of the fact that they do not need any well developed system for water conduction as water is absorbed by the general surface of the body.

Stem. Stem is highly reduced in *Vallisnaria.* In other cases, it is usually very weak and flexible so that it can move according to water currents. The epidermis is single layered composed of thin walled parenchymatous cells. There is no covering of cuticle hence water and minerals are absorbed directly from the general surface. In *Ceratophyllum* and *Myriophyllum*, certain groups of specialised cells for absorption are found in epidermis and are called **Hydropoten.** Epidermal cells usually have chloroplasts and therefore capable of photosynthesis. Cortical region is often large and contain abundant and large air chambers. The air chambers are arranged symmetrically in *Myriophyllum* or scattered irregularly as in *Hydrilla* and *Elodea.* Air chambers surrounding the stele are usually small while these gradually increase in size towards periphery. Cortical cells contain chloroplasts and carry on photosynthesis. Stele is relatively less developed. Endodermis and pericycle are usually not differentiated. Xylem, usually represented by a single large cavity in the centre, remains surrounded by phloem.

Leaves. The epidermis of leaves consists of a single layer of tangentially elongated thin walled chlorophyllous cells lacking cuticle. Stomata are altogether absent, rarely non-functional and vestigeal ones may be found in some cases. The cells are capable of absorption of water and minerals and synthesis of food (photosynthesis). The mesophyll tissue remains undifferentiated but have large air chambers in continuation with stem. The air chambers are often separated by one or two celled thick diaphragms. Mesophyll cells have abundant chloroplasts. Vascular tissues mainly consists of phloem, xylem being represented by few thin walled elements. Leaves in some cases are membranous e.g. in *Elodea* these are only two cell thick except for the mid-rib region.

II. Floating plants

The floating hydrophytes may be of three types: (i) free floating, e.g. *Spirodela, Lemna, Wolffia, Wolffiella, Azolla etc.;* (ii) rooted in mud with only flowers and leaves floating on the surface of water, e.g. *Nymphaea, Nelumbium, Victoria, Limnanthemum* etc.; and (iii) roots fixed and plants float freely in water, e.g. *Pistia, Trapa, Eichhornia* etc. *Eichhornia* may be found free floating also. These plants show various adaptations in different parts.

Roots. Root system is poorly developed and usually remains unbranched. Duck weeds show various stages of disappearance of roots in floating plants, e.g. there are several roots in *Spirodela,* a single root in *Lemna* and no root in *Wolffia* and *Wolffiella.* The anatomical features of roots of floating plants show fundamental similarity with those of submerged plants. In *Lemna,* roots are without any conducting tissue.

Stem. The attached forms usually develop long horizontal stems or tubers. The epidermis consists of single layer of thin walled parenchymatous cells. Cortex contain a large number of well developed air chambers. In the rhizomes of *Nymphaea,* two to three layered collenchymatous hypodermis is present. Stellate cells are usually present in the cortex which provide mechanical support to the thin walls of air chambers. The vascular bundles are found scattered in the cortex and are present in between the air chambers (in rhizome of *Nymphaea*). Xylem consists of only tracheids and phloem is relatively better developed.

Leaves. The floating leaves are usually characterised by having a large and specialised petiole and broader lamina. Leaves are not found in some floating forms such as *Spirodela, Lemna, Wolffia* etc. Here stem takes the function of a leaf.

Petiole–Petiole is usually a well developed part of the floating leaves. In *Nymphaea,* it is very long and cylindrical while in *Eichhornia,* it is swollen and spongy. The epidermis is single layered made up of thin walled cells. Cells are compactly arranged and in most cases lack cuticle. Hypodermis is two to three layered and made up of collenchyma. The cortex is well developed and has abundant air chambers, smaller towards periphery and larger towards the centre. Trichosclereids are commonly found which provide mechanical support to the thin and delicate walls of air chambers. Vascular bundles are found scattered in between the air chambers. The xylem is poorly developed and it is usually reduced to a single large element in each bundle but phloem is normally developed.

Lamina–In Lamina, the epidermis is well developed and consists of a single layer of thin walled cells. The cells have a thick cuticle or wax on upper surface. Leaves are often amphistomatic with functional stomata only on upper side. In *Nymphaea,* the leaves are, however, epistomatic type. The mesophyll tissues are often differentiated into palisade and spongy parenchyma. The palisade cells are arranged in a single layer in *Trapa* but in several layers in *Nymphaea.* Spongy cells have large air chambers and several scattered trichosclereids (stellate cells). Air chambers provide buoyancy and trichosclereids provide mechanical support to the leaves. The vascular bundles have few xylem elements and relatively more developed phloem.

III. Amphibious plants

Amphibious plants usually grow in muddy soils or in soils saturated with water. These can withstand waterlogged conditions during rainy season or floods. Common examples of amphibious plants are *Eclipta, Lippia, Ranunculus, Sagittaria, Typha, Marsilea* etc. These plants grow under two diverse situations, the lower part of plant lives in an atmosphere of water rich conditions while upper part remains exposed to direct sun and hence live in hot conditions. Accordingly different structural adaptations are found.

Roots. Roots are well developed and have distinct root caps. The development of root hairs depends upon the situation, i.e. if growing in abundant water, no root hairs and if in well aerated and partly dried soils, several root hairs are formed. The anatomical features show the presence of both hydrophytic and mesophytic characters. The presence of aerenchyma and storage parenchyma are the hydrophytic characters and presence of mechanical tissues and well developed vascular tissues are the mesophytic characters found in them. Cortex and steles are very similar to those found in mesophytes.

Stem. Stem is usually reduced and show anatomical characters resembling both hydrophytic and mesophytic features.

Leaves. Since leaves of plants usually live in two different types of environments, heterophylly is found in several plants, e.g. *Ranunculus aquatilis, R. sceleratus, Sagittaria sagittifolia* and *Limnophylla* sp.

Epidermis is single layered. The epidermis of aerial leaves remains covered with cuticle and have functional stomata. Their mesophyll cells show differentiation into palisade and spongy parenchyma but such differentiation is not found in submerged leaves. Vascular tissues of aerial leaves is well developed. It remains surrounded by sclerenchymatous sheath in *Sagittaria* and *Typha.* In several cases, sclerenchymatous patches along the margins are found.

General characters of hydrophytes

Adaptations of hydrophytic characters can be studied dividing into two broad categories, the external and internal features.

EXTERNAL FEATURES

1. **Roots** are poorly developed, short, less branched and some times may be absent. Roots are well developed in *Nymphaea, Cyperus* and *Typha,* poorly developed in *Eichhornia, Spirodela* and *Lemna* and absent in *Wolffia, Wolffiella* and *Ceratophyllum.* In *Utricularia,* leaves take up the function of roots. In *Eichhornia* and *Pistia,* adventitious roots develop to provide buoyancy. In

Jussiaea, negatively geotropic spongy respiratory roots are found. Root hairs and root caps are completely absent.

2. **Stem** of the hydrophytes are delicate and flexible exhibiting sponginess.

3. **Leaves** are either highly dissected or narrow. In submerged plants, they are often provided with spongy covering. Emerged floating leaves possess either waxy coating (e.g. *Nymphaea*), leaf hairs (e.g. *Salvinia*) or leathery texture. Amphibious hydrophytes show heterophylly, i.e. morphologically different kinds of leaves. Submerged leaves are furcated but emerged leaves have broad lamina (e.g. *Ranunculus sceleratus*).

4. **Petioles** of the leaves are very long and delicate in the rooted plants (e.g. *Nymphaea, Sagittaria*) which help the lamina to float on water surface. In floating plants, petioles may be reduced or altogether absent. *Eichhornia* develops bulbous petiole providing buoyancy.

5. **Hydrophytes** commonly reproduce vegetatively by forming runners, offset, stolon or by fragmentation.

INTERNAL FEATURES

1. **Epidermis.** The cells of the epidermis are thin-walled and often contain chloroplasts (e.g. *Hydrilla*). Epidermis of submerged organs lacks cuticle, epidermal hairs and stomata but that of emerged organs bears waxy coating and sotmata.

2. **Cortex.** Cortex is quite well developed and parenchymatous possessing large air chambers. Few outer layers of the cortex possess chloroplasts.

3. In **leaves,** mesophyll cells are present but are not differentiated into palisade and spongy parenchyma. Spongy parenchyma with large air cavities constitute most of its part.

4. **Vascular tissues** are poorly developed. Commonly there is no distinction between xylem and phloem. Xylem is much reduced and may be represented by single element. Xylem cells are unlignified. Phloem is comparatively well developed.

5. **Mechanical tissues** are less developed. In certain cases, scattered raphides are present.

Some other characters are:

(i) Mucilaginous cells and sclereids are present.

(ii) Occasionally diaphragms form partition of air chambers.

(iii) Water is absorbed by the general surface of the submerged parts.

An amphibious plant may show hydrophytic characters in the submerged portion and xerophytic (or mesophytic) characters in the emerged portions. e.g. *Cyperus, Typha* etc.

Ecological Classification of Hydrophytes

A broad classification can be as follows:

1. **Free floating hydrophytes.** Such plants float on the surface of water and are not rooted, e.g., *Eichhornia, Lemna, Spirodela, Pistia, Salvinia, Lemnanthemum.*

2. **Floating but rooted hydrophytes.** Such floating hydrophytes remain attached to the shallow bottom of the pond, e.g. *Aponogeton, Potamogeton, Trapa* etc.

3. **Submerged floating hydrophytes.** Plants occur just below the water surface and are not attached, e.g. *Najas, Ceratophyllum.*

4. **Submerged rooted hydrophytes.** These plants occur below the surface of water and are attached by means of roots to the bottom, e.g. *Hydrilla, Utricularia, Vallisnaria.*

5. **Amphibious hydrophytes.** These plants occur in muddy places and are also called marshy plants. Such plants may even reach water surface, e.g., *Jussiaea.* Other plants are *Marsilea, Polygonum, Lippia (Phyla)* etc.

XEROPHYTES

Plants growing in dry habitats (physical dryness) or in habitats where water is unfit for absorption such as saline water (physiological dryness) are called xerophytes. Because of shortage of water, these plants develop certain characters to minimise transpiration. Xerophytes in general show following important characters.

I. External features

Roots. Root system of xerophytes is well developed, profusely branched and mechanically strong. Such plants develop a long tap root growing deep into the soil in search of water. Occasionally, roots may grow near the surface of soil to absorb the rain water.

Stem. Stems are generally hard and woody. A few xerophytes possess succulent (fleshy) stem due to abundance of mucilage which allows them to store water. Occasionally, the stem takes the function of leaf and adapts leaf-like structures to reduce the transpiring surface. When such leaf-like structure consists of one or two internodes, it is called **cladode** as found in *Asparagus* and if several internodes, it is called **phylloclade** as found in *Ruscus, Opuntia, Muehlenbeckia, Euphorbia, Casuarina* etc.

Leaves. Leaves show variation in xerophytes and accordingly these are classified into following categories.

(i) **Sclerophyllous.** Leaves are strongly cutinised and sclerified and appear strong and leathery, e.g. *Banksia.*

(ii) **Trichophyllous.** Leaves have hairy covering on the surface, e.g. *Nerium* and *Calotropis.*

(iii) **Microphyllous.** Leaves are reduced in size to avoid excessive transpiring surface, e.g. scale leaves in *Casuarina, Asparagus* and *Equisetum,* spiny in *Cactus* and needle-like in *Pinus*. In *Capparis aphylla,* leaves are not seen.

(iv) **Malacophyllous.** Leaves are fleshy due to excessive mucilage to store maximum water, e.g. *Bryophyllum, Begonia* and *Kalanchoe.*

Certain xerophytes develop caducous leaves as in *Euphorbia tirucalli* and *Capparis decidua.*

II. Anatomical features

Epidermis. Cells of epidermis are radially elongated and thick walled. The thick cutin gets deposited on the outer surface of cells. Deposition is sometimes mamillate and multilayered as in *Capparis* stem. The thickness of deposition is directly proportional to the xeric conditions. Sometimes, cuticle is supplemented by additional waxy coatings as in *Ricinus* and *Calotropis*. Occasionally, the epidermis is multilayered which in case of a leaf may be on the dorsal surface as in *Ficus* or on both the surfaces as in *Nerium*. Presence of waxy coating on epidermal surfaces and multilayered epidermis help in reducing evaporation of water through epidermis.

Leaves are **hypostomatic** and stomata are usually present in cavity, i.e. stomata are **sunken type**. The stomated cavities are often provided with a dense covering of stomatal hairs so that the air in the cavity does not come in direct contact of the wind currents. Stomata in certain desert plants such as *Capparis spinosa* and *Aristida ciliata* may sometimes get blocked due to deposition of resinous matter or wax. All these adaptations are found to reduce the loss of water during transpiration.

Epidermis of certain grass leaves have **bulliform cells** or **motor cells** which play an important role in rolling of the leaves during the period of dryness. These cells are thin walled, greatly enlarged and sensitive to turgor changes. When these are turgid, leaf remains flattened and when flaccid, the lamina rolls or minimise the exposure of transpiring surface.

Hypodermis. In most xerophytes hypodermis consists of mechanical tissues, also called sclerenchyma. There may one to many cell thick sclerenchymatous hypodermis both in stem as well as leaves. In certain cases, sclerenchyma are found in patches, e.g. in *Casuarina* stem, T-shaped sclerenchymatous hypodermis is found below the ridges. Besides providing mechnical support, sclerenchyma also protects the internal tissues from high light intensities, thus minimising the damage.

Mesophyll tissues. Mesophyll tissues are well developed and differentiated in xerophytes. These show a well-marked differentiation into well developed compactly arranged palisade tissues and loosely arranged spongy tissues. Palisade cells remain radially elongated to minimise the direct penetration of light. These are usually present towards dorsal surface and are arranged in a single layer. However, in *Nerium*, *Ficus* and *Atriplex*, the compactly arranged palisade tissues are present on both adaxial (dorsal) and abaxial (ventral) surfaces of a leaf and spongy tissues lie in between the palisade layers. In certain xerophytes, leaves are highly reduced and stem takes their function as in case of *Capparis*, *Salvadora* and *Casuarina*. In these cases, compactly arranged palisade cells are found present in the outer cortex of the stem.

The chloroplasts often show a relational distribution in palisade cells. These remain linearly arranged in a cell which has closer contact with the light. Chloroplasts can change their position in a palisade cell in response to various intensities of light.

Pericycle. In most cases, pericycle consists of either sclerenchymatous cells

Table 16.1. Comparison between three ecological groups of plants.

Hydrophytes	*Mesophytes*	*Xerophytes*
External features		
1. Roots are generally reduced, sometimes even absent. Root hairs are absent.	1. Root system is well developed. Root hairs are present.	1. Root system is very well developed, extensive and long reaching deep into the soil. These are several times larger than shoot. Root hairs are present. Roots are woody and hard.
2. Stems are highly reduced, delicate and thin, sometimes modified into rhizomes.	2. Stem is relatively developed, rigid and stout.	2. Stem is usually stout, branched, woody and sometimes underground.
3. Leaves are often thin, narrow and dissected. The emergent leaves are sometimes with large lamina, long petiole and covered with wax or hairs.	3. Leaves are large, thin, entire and usually without waxy substances.	3. Leaves are reduced, some times scaly or modified into spines, while in other cases these become leathery and thick with shining surfaces, waxy coatings or covered with hairs.
Internal features		
1. Aerenchyma are present abundantly and found in almost all vegetative parts of plant body.	1. Aerenchyma are not found.	1. Aerenchyma are not found.
2. Cuticle is entirely absent in submerged parts but a thin cuticle may be present in emergent parts.	2. Cuticle is found present in almost all green parts of plant body.	2. A thick and well developed cuticle is characteristically present all over the green parts of plant body.
3. Stomata are not found in general but may be present on the exposed upper surface of leaves.	3. Stomata are found present on both the surfaces, sometimes only on the upper surface.	3. Stomata are found fewer in number, generally confined to the lower surface of leaves. These are generally sunken type.

Contd...

Hydrophytes	*Mesophytes*	*Xerophytes*
4. Mesophyll cells are undifferentiated into palisade and spongy parenchyma.	4. Mesophyll cells are differentiated into palisade and spongy parenchyma.	4. Mesophyll cells are well differentiated and palisade cells are generally found on both sides of leaves.
5. Chloroplasts are found on outer few layers of cells, almost throughout the plant body.	5. Chloroplasts are found restricted to leaves only.	5. Chloroplasts are found in leaves and sometimes in modified stems also.
6. Epidermal cells are thin-walled.	6. Epidermal cells have thickened cell walls on outer and exposed surfaces.	6. Epidermal cells are distinctly thick-walled, almost all around the cell wall.
7. Lignified mechanical tissues are not found and vascular elements are poorely developed, sometimes even represented by thin-walled xylem elements.	7. Mechanical tissues and vascular elements are well developed.	7. Both mechanical tissues and vascular elements are very well developed, sometimes present in massive quantity.
Examples–*Hydrilla*, *Vallisnaria*, *Ceratophyllum*, *Utricularia*, *Nymphaea*, *Eichhornia*, *Potamogeton* etc.	**Examples**–*Solanum tuberosum*, *Lycopersicum*, *Nicotiana*, *Amaranthus*, *Chenopodium* etc.	**Examples**–*Capparis*, *Calotropis*, *Ruscus*, *Casuarina*, *Opuntia*, *Asparagus* etc.

or stone cells.

Vascular tissues. Vascular tissues are very well developed in xerophytes. An increased amount of vessels are found in them to allow easier conduction of water.

Laticiferous tissues. Laticiferous canals (articulated types) are found present in the stems of xerophytes such as *Calotropis, Euphorbia, Asclepias, Papaver* etc. Due to viscous nature of latex the transpiration is reduced to some extent.

SUCCULENT AND FLESHY XEROPHYTES

Succulent plants are usually not considered as true xerophytes because these retain considerable amount of water in their plant body. Common examples of succulent plants are *Bryophyllum, Bigonia, Euphorbia tirucalli, Opuntia* (cactus), *Stapelia, Ananas* etc.

The epidermal cells have waxy coatings in addition to thick cuticle. Next to epidermis are found a few layers of collenchyma followed by a photosynthetic region. The photosynthetic region consists of three to four layers of palisade cells. Inner to photosynthetic region is found a mass of water storing parenchymatous tissues. These cells remain turgid and their cytoplasm is highly vacuolated and develop succulence. Some succulent plants show enormous capacity of water storage such as *Echinocactus* and *Cereus*.

EPIPHYTES

The term epiphyte is used to refer to those plants that grow upon other plants but they do not absorb food from them. An epiphyte usually develops three kinds of roots, the clinging roots, the absorbing roots and the hanging roots. The hanging roots, are ecologically more important because these are capable of absorbing water from moisture. These are provided with an outer covering of special absorptive tissue, called velamen. Velamen usually consists of four or five layers of oblong-polygonal dead cells, the wall of which develops fibrous thickenings. The velamen acts as a kind of sponge and absorbs moisture from the surrounding air and also water trickling down the root. Velamen layers are characteristically found in case of *Vanda* and *Dendrobium* but also in several other orchids.

SAPROPHYTES

Saprophytes grow on dead and decaying organic matter and derive their food from them. Anatomical differentiation is not much important in these cases except for the presence of mycorrhiza in certain cases. Examples of saprophytic angiosperms are few and it includes *Monotrapa* (Indian Pipe) and *Neottia* (orchid).

PARASITES

The plants that grow upon other living plants and absorb their food from them are called parasites. Various degrees of parasitism are found among angiosperms. A true parasite has characteristic water and food sucking structure called haustoria. Haustoria penetrate the host tissue and establish relationship with xylem and phloem. The anatomical studies of parasites are not much important except for the presence of haustoria and their relationship.

HALOPHYTES

Halophytes grow in saline soil of water where there is abundance of salt in the soil. Typical examples of halophytes are *Basella*, *Chenopodium, Acanthus, Suaeda* etc.

Halophytes growing in marshy places near the sea shore collectively form a special vegetation called mangrove. Mangroves are commonly found in sunder bans of West Bengal. Common examples of mangroves are *Rhizophora, Sonneratia, Ceriops, Heritiera, Excoecaria* etc. Mangrove plants are characterised by having well developed stilt roots, respiratory roots or **pneumatophores and vivipary.**

Vivipary is a special kind of germination found in mangroves. The seeds begin to germinate inside the fruit while it remains attached with the parent plant. During germination, the radicle elongates considerably and swells increasing the weight. Ultimately, the seedling separates from the parent tree and falls vertically down. The radicle penetrates into the soft mud keeping the plumule and cotyledons clear above the saline water.

Anatomically, mangroves show several xerophytic characters. The epidermis is thick-walled and strongly cutinised. Sunken stomata are found in the epidermis. The hypodermis in stem consists of a few layers of sclerenchymatous cells (e.g. *Rhizophora*). Mesophyll cells are usually compactly arranged. In *Sonneratia*, mainly palisade cells are found. Mucilage cells, spicules and long stone cells (sclereids) are frequently found in between palisade cells or in cortex of stem as in *Rhizophora* and *Sonneratia.*

Plant exhibiting various ecological adaptations are categorised accordingly, and are accounted here individually for practical study point of views.

XEROPHYTES
(Examples)

A few common but important examples for anatomical studies may be broadly gouped as follows:–

1. Herbaceous stems, e.g. *Helianthus annuus, Cucurbita, Triticum* etc.
2. Woody stem, e.g. *Capparis, Salvadora, Casuarina, Bauhinia* etc.
3. Woody climbers (lianas), e.g. *Bignonia, Passiflora, Bougainvillea* etc.
4. Phylloclade, e.g. *Muehlenbeckia, Cactus, Ruscus, Euphorbia tirucalli, Casuarina* etc.

5. Cladode, e.g. *Asparagus.*
6. Dorsiventral (bi-facial) leaves, e.g. *Ficus, Nerium* etc.
7. Isobilateral (monofacial) leaves, e.g. *Zea mays.*
8. Phyllode, e.g. *Acacia auriculiformis.*

Only a few but important ones of them are described here.

CAPPARIS STEM

(*Capparis decidua*, Eng.–Capper plant, Fam.–Capparidaceae)

A transection of this common dicot stem shows following anatomical characters (Fig. 16.1)–

Epidermis. It is made up of a single layer of compactly arranged radially elongated cells. The thick multilayered cuticle forms a characteristic outer covering, punctuated by sunken stomata.

Cortex. It consists of two to three layers of closely arranged elongated cells. The cells resembling palisade cells contain abundant chloroplasts and perform the function of photosynthesis. A number of myrosin cells are found distributed among the palisade cells.

Endodermis. A single layer of compactly arranged stone cells forms the endodermis which is the innermost layer of the cortex.

Pericycle. It is a multilayered, heterogenous region made up of thick-walled stone cells alternating with patches of thin-walled cells. The cells are irregular in shape but arranged compactly.

Vascular system. Being a dicot stem, the vascular tissue is represented by a ring of conjoint, collateral, endarch and open vascular bundles, separated by narrow medullary rays. The vascular bundles consist of xylem and phloem separated by rectangular cells of the cambium. The xylem is constituted of tracheids, vessels and xylem parenchyma. The smaller protoxylem vessels are arranged towards the pith while the larger metaxylem vessels face the phloem cells. The elements of phloem, i.e. sieve tubes, companion cells and phloem parenchyma are distinctly present.

Secondary growth is initiated by the formation of a single layered cambium, which later becomes several layered. The cambium is normal in its activity cutting off secondary phloem on outer side, which forms a continuous and conspicuous region below the pericycle, and great amounts of secondary xylem, on inner side. Narrow, one to three layered medullary rays are conspicuous in the secondary xylem.

Pith. This is the central region composed of thin walled cells, which are usually polygonal in shape and compactly arranged. The pith is well developed and occupies a large area.

Features of ecological interest

Capparis stem exhibits numerous xerophytic features of ecological importance. These are:

1. The epidermis consists of thick walled, radially elongated cells. The outer surface is covered with a thick multilayered papillate cuticle.
2. Sunken stomata are present.

3. Cortex is made up of chlorophyll containing palisade cells, hence the stem performs the function of photosynthesis.

4. The cortex is reduced in size and devoid of thin-walled parenchymatous cells.

5. Endodermis is formed of stone cells.

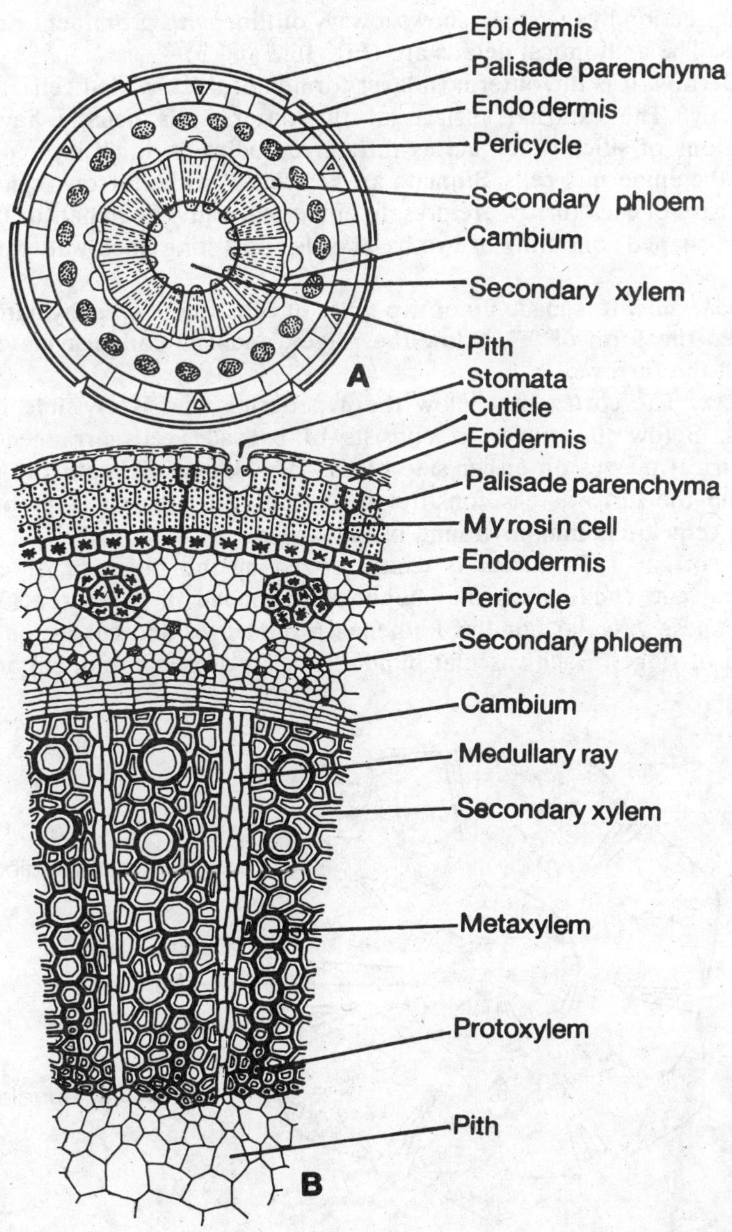

Fig. 16.1. T.S. stem of *Capparis.* A–diagrammatic and B–a cellular enlarged portion.

6. Pericycle is present in the form of patches of stone cells.
7. Secondary xylem is well developed.

CASUARINA STEM

(*Casuarina equisetifolia,* Eng.–Beefwood tree, Fam.–Casuarinaceae)

The transection of the stem shows a wavy outline with prominent ridges and furrows. The anatomical details are (Fig. 16.2 and 3)–

Epidermis. It is the outermost layer formed of thick-walled cells, arranged compactly. The external surface of the outer walls usually have heavy depositions of silica which act as refractive bodies. A thick layer of cuticle covers the epidermal cells. Stomata are restricted to the furrow regions only. At the base of each furrow are present epidermal hairs. Each hair or trichome is sickle-shaped consisting of two basal cells and a long thick-walled terminal cell.

Hypodermis. It is made up of two types of cells-the sclerenchymatous cells arranged the form of 'T', below the ridges, and few collenchymatous cells beneath the furrows.

Cortex. The cortex lies below the hypodermis and is divisible into two regions. Below the ridges, it consists of palisade cells arranged in two symmetrical masses on either side of the 'T', whereas beneath the palisade cells and the furrow, it is constituted of thin-walled parenchymatous cells. Tannin cells are commonly found in the cortex.

The cortex of *Casuarina* is characterized by the presence of conjoint, collateral and endarch vascular bundles, embedded in the parenchymatous tissue. These vascular bundles lie beneath the ridges and correspond to the number of ridges. Each vascular bundle consists of xylem, phloem and some

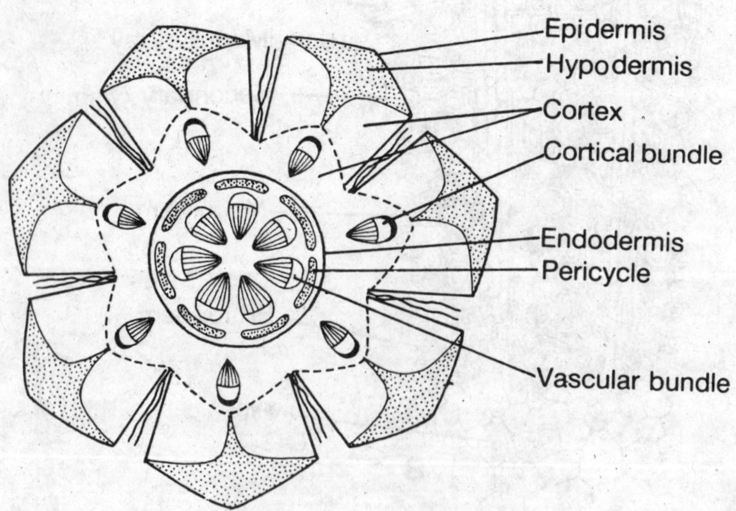

Fig. 16.2. T.S. stem of *Casuarina* (diag.).

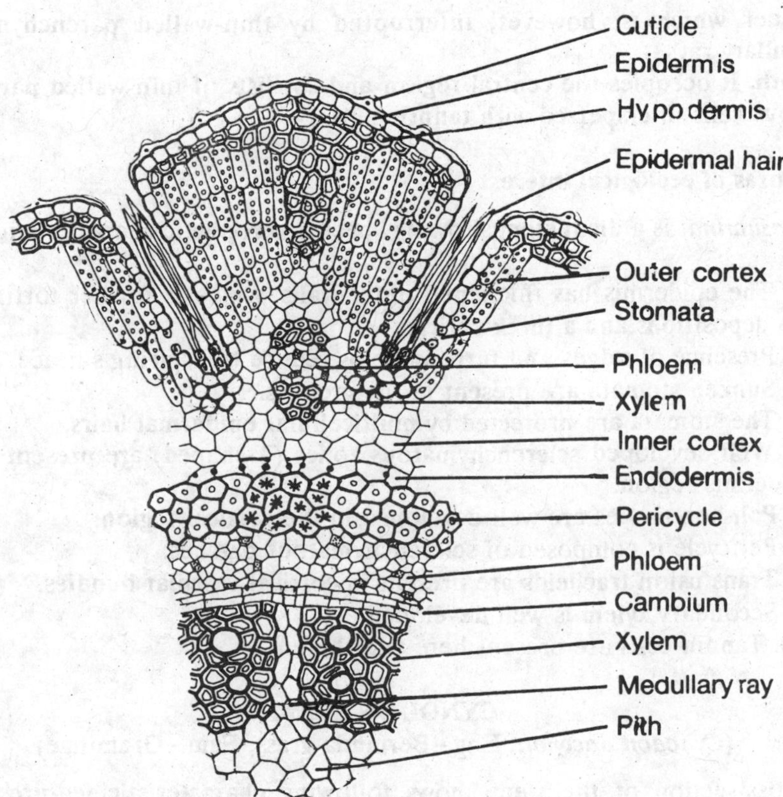

Fig. 16.3. T.S. stem of *Casuarina equisetifolia* (a cellular enlarged portion).

transfusion tracheids and is capped by a sclerenchymatous patch. These bundles are commonly referred to as cortical vascular bundles.

Endodermis. It is the innermost wavy layer made up of a uniseriate row of parenchymatous cells.

Stele. It is a polyfascicular siphonostele showing secondary growth. Before secondary growth starts, the stele is represented as a ring of conjoint, collateral, endarch and open vascular bundles present below the furrows and alternating with the cortical bundles.

Pericycle. It is present in the form of isolated patches of sclerenchymatous cells, forming a cap over each vascular bundle.

Secondary phloem. It is formed by the cambium and consists of sieve tubes, companion cells, phloem parenchyma and phloem fibres. In mature stems, it forms a continuous layer.

Cambium. In young conditions, the cambium is limited within each vascular bundle, but with the start of secondary growth, the fascicular and interfascicular cambia join to form a complete cambium ring.

Secondary xylem. It is composed of tracheids, vessels, xylem parenchyma and xylem fibres. Secondary growth results in the formation of continuous cylinder which is, however, interrupted by thin-walled parenchymatous medullary rays.

Pith. It occupies the central region and consists of thin-walled parenchymatous cells interspersed with tannin cells.

Features of ecological interest

Casuarina is a dicot stem showing specific xerophytic characteristics such as:

1. The epidermis has thickened outer walls, which is further fortified by silica depositions and a thick cuticle.
2. Presence of ridges and furrows helps reduce transpiring surface.
3. Sunken stomata are present in the grooves.
4. The stomata are protected by multicellular epidermal hairs.
5. Well developed sclerenchymatous zones (T-shaped) are present in the hypodermis region.
6. Palisade tissues are well developed in the cortical region.
7. Pericycle is composed of sclerenchymatous patches.
8. Transfusion tracheids are present in cortical vascular bundles.
9. Secondary xylem is well developed.
10. Tannin cells are present here and there.

CYNODON STEM

(*Cynodon dactylon*, Eng.–Bermuda grass, Fam.–Graminae)

A cross-section of the stem shows following characteristic features (Fig. 16.4):

Epidermis. It is the outermost single layer consisting of rectangular shaped cells arranged compactly. Stomata may be present. Externally the epidermal cells are covered by a thick cuticle.

Ground tissue. The ground tissue can be roughly divided into an outer and inner regions. The cells of the outer region are polygonal in shape, compactly arranged and may contain a few chloroplasts. The inner region of the ground tissue covers almost the entire central portion of the stem. In its outer region, few layers of sclerenchymatous cells are present, whereas the inner layers are composed of loosely arranged cells.

Stele. The stele is represented by a number of vascular bundles irregularly distributed in the inner region of the ground tissue. However, in the ventrally flattened region of the plant, these may appear arranged in two rings. The vascular bundles are oval, closed, conjoint, collateral and endarch. The outer rows of V.B. usually lie embedded in the sclerenchymatous region, whereas the vascular bundles lying in the central region are surrounded by individual sclerenchymatous sheath.

Vascular bundles are composed of phloem and xylem. Phloem consists of sieve tubes and companion cells, phloem parenchyma being absent. Xylem is generally arranged in characteristic 'V' form. It consists of circular vessels, tracheids and parenchyma. The larger elements represent metaxylem and

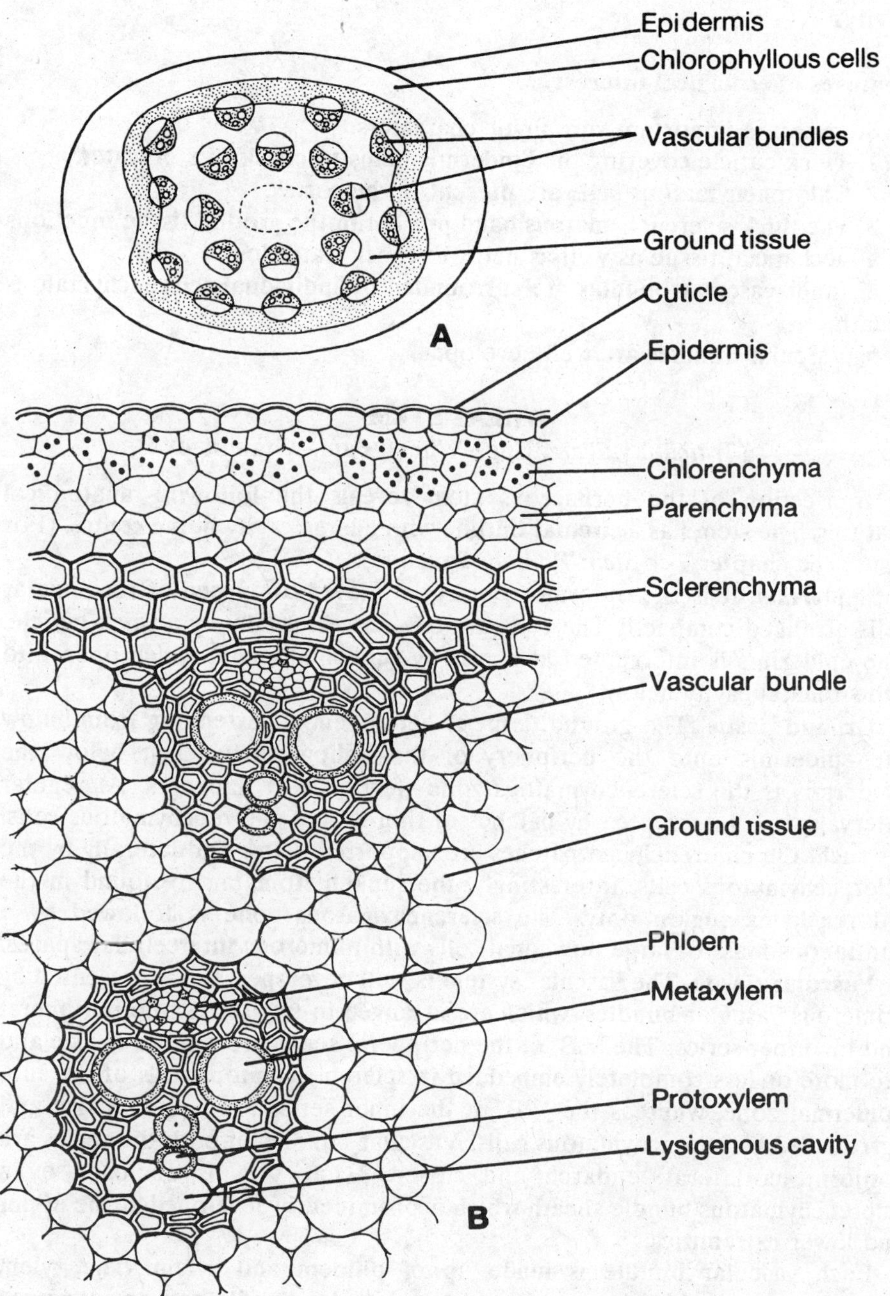

Fig. 16.4. T.S. of *Cynodon* stem. A–diagrammatic and B–an enlarged cellular portion.

smaller ones protoxylem. Just below the protoxylem lies the lysigenous cavity.

In young stems, the central region is composed of parenchymatous cells, but as the plant matures, ther central parenchyma cells degenerate to form a hollow cavity.

Features of ecological interest

Stem shows important xerophytic characters:

1. Thick cuticle covering the epidermis helps to check loss of water.
2. Chlorenchymatous cells are present in the cortex.
3. The thick sclerenchymatous band present in the ground tissue functions as a mechancial tissue as well as helps to check water loss.
4. Inner vascular bundles are surrounded by individual sclerenchymatous sheath.
5. Vascular bundles are well developed.

WHEAT STEM

(*Triticum vulgare*, Eng.–Wheat, Fam.–Graminae)

A transection of the herbaceous stem reveals the following anatomical features. The stem has a circular outline with a large cavity in the centre. (For figure see chapter *The Stem-Primary Structure)*.

Epidermis. It is the outermost, uniseriate layer composed of rectangular cells arranged compactly.The cells are covered externally by a thick cuticle. The epidermis is interrupted at places by sunken stomata, which open into sub-stomatal cavity below.

Ground tissue. The ground tissue is heterogenous, extending from below the epidermis upto the periphery of the hollow cavity. Just below the epidermis is the sclerenchymatous zone few cells in thickness. At regular intervals, it is interrupted by patches of thin-walled chlorenchymatous cells. As such, the chlorenchyma patches are supported below and laterally by the sclerenchymatous cells. Interestingly, the sunken stomata are found in the chlorenchyma region only. The sclerenchymatous zone is followed by a continuous mass of large polygonal cells with numerous intercellular spaces.

Vascular tissue. The vascular system is well developed and represented by numerous vascular bundles, which are arranged in two series–the peripheral and the inner series. The V.B. of the peripheral series are smaller in size and are more or less completely embedded in sclerenchymatous cells of the sub-epidermal zone, whereas the V.B. of the inner series are larger in size and surrounded by parenchymatous cells. Vascular bundles of both the series are conjoint, collateral, endarch and closed. Each V.B. is enclosed by a sclerenchymatous bundle sheath which is prominently developed at the upper and lower extremities.

Each vascular bundle is made up of phloem and xylem. The xylem elements occupy the lower region of the vascular bundle and are arranged more or less in Y-pattern. The larger metaxylem elements are situated towards the outer side, whereas, the smaller protoxylem elements face the inner side. The phloem consists of sieve tubes and companion cells and occurs in the peripheral region of the V.B.

A hollow cylinder is present in the centre of the stem.

Features of ecological interest

The stem exhibits following xerophytic pecularities.

1. The epidermis is covered by a thick cuticle to check loss of water by transpiration.
2. Sunken stomata are present which open into substomatal cavities.
3. The ground tissue is partly composed of sclerenchymatous cells, which give mechanical strength to the otherwise weak, herbaceous stem.
4. Patches of chlorenchymatous cells are present in between the sclerenchymatous zone which indicates that the photosynthetic activity is partly taken up by the stem, a xerophytic trait.
5. Sclerenchymatous bundle sheath encloses each V.B.
6. Vascular bundles are well developed.

BIGNONIA STEM

(*Bignonia venusta,* Fam.–Bignoniaceae)

A transection of stem of this common garden climber reveals following anatomical details (for figure see chapter *Anomalus Secondary Growth):*

Epidermis. The outline of the stem shows distinct ridges and furrows. The outermost layer is made up of a single layer of thick walled cells, covered externally by a thick cuticle.

Hypodermis. In the young condition, collenchyma is present below the ridges, whereas below the furrow 2-3 layers of chlorenchyma are present.

Cortex. The cortex consists of loosely arranged parenchymatous cells lying beneath the chlorenchymatous hypodermis. In mature stems, the cells of both the hypodermal and cortical region are replaced by cork cells.

Endodermis. The endodermis is represented by an indistinct layer.

Pericycle. It is a discontinuous layer formed of patches of sclerenchyma cells present below the ridges.

Vascular system. The vascular system is represented by a polyfascicular siphonostele, comprised of a ring of conjoint, collateral and open vascular bundles.

Abnormal secondary growth results in a peculiar arrangement. The phloem forms a continuous ring except at 4 different patches, where it projects inwards into the secondary xylem. These four projections appear as four longitudinal furrows. An interesting feature is the presence of horizontal plates of sclerenchymatous fibres (patches) in the phloem furrows. Phloem consists of sieve tubes, companion cells, phloem fibres and phloem parenchyma. Secondary xylem is well developed and consists of vessels, tracheids and xylem parenchyma. The secondary xylem appears deeply fissured due to the presence of phloem wedges. The larger metaxylem elements face the periphery whereas the smaller protoxylem are towards pith. Well developed medullary rays are present.

Pith. The central region consists of compactly arranged parenchymatous cells.

Features of ecological interest

Bignonia exhibits certain specific characters in accordance with its climbing habit.

1. Epidermis consists of thick-walled cells covered by a thick cuticle.
2. Hypodermis is well developed and formed of collenchymatous cells beneath the ridges. These cells give mechancial strength to the climber stem.
3. Presence of chlorenchymatous cells in the stem indicates that the photosynthetic function has been taken up by the stem.
4. Pericycle is present in the form of sclerenchymatous patches below the endodermis.
5. The phloem wedges are supported by horizontal bars or plates of sclerenchyma fibres which function as mechanical support.
6. Xylem is well developed consisting of thick-walled lignified cells.
7. The development of cork cells in the hypodermis and cortex region at later stage indicates the xerophytic adaptation of the climber to its climbing habit.

BOUGAINVILLEA STEM

(*Bougainvillea* sp., Fam.–Nyctaginaceae)

It is a woody climber commonly grown in gardens for ornamental purposes. A transection of the young stem exhibits following characters. (For figure see chapter *Anomalous Secondary Growth)*.

Epidermis. It is the outermost single layer composed of thin walled cells arranged compactly. The epidermis is covered externally by a thick cuticle. A characteristic feature is the presence of multicellular epidermal hairs, made up of squarish, or rectangular cells. The apical cells of the hair is rounded and slightly bulbous. Cells of the epidermis and hairs are filled with pinkish coloured sap.

Cortex. The epidermis is followed by a multilayered cortex. Just below the epidermis is present collenchymatous patches, which, however, forms a continuous layer in older stems. This is followed by a chlorenchymatous zone three to four cells in thickness. The chlorenchyma cells extend between the collenchyma patches, in young condition. In young as well as old stems, intercellular spaces are conspicious.

Endodermis. It is constituted of a distinct layer of barrel-shaped cells containing starch.

Pericycle. Beneath the endodermis lies the pericycle composed of one or two layers of cells.

Vascular tissue. The vascular system is well developed and represented by a ring of vascular bundles lying beneath the pericycle and numerous medullary bundles scattered in the pith. The bundles are conjoint, collateral, endarch and open.

The vascular bundles consist of xylem and phloem. The xylem is well developed, made up of tracheids, vessels and xylem parenchyma, whereas phloem is composed of sieve tubes, companion cells and phloem parenchyma.

In the central region are present a few large medullary bundles which are surrounded by a large number of small scattered medullary bundles. The medullary bundles are surrounded by well developed parenchymatous sheath. In older stems, as a result of secondary growth a large amount of conjuctive tissue is cut off towards inner side and which is separated by uni- or multiseriate medullary rays.

Pith. It is made up of parenchymatous cells with intercellular spaces between them.

Features of ecological interest

The stem exhibits certain anatomical features of interest.

1. Epidermis is thickly cuticularised and bears numerous multicellular hairs, which help to check water loss.

2. Cortex is composed of collenchymatous cells which give mechanical support to the climbing stem.

3. Presence of well developed chlorenchyma cells in the cortex indicates the xerophytic adaptation of the stem.

4. As a result of secondary growth extensive mass of conjuctive tissue is formed which gives mechanical strength to the climbing stem.

5. Numerous well developed medullary vascular bundles are present which also provide mechancial strength to some extent.

RUSCUS PHYLLOCLADE

(Ruscus aculeatus, Fam.-Liliaceae)

In *Ruscus*, the stem becomes modified to form a flattened, leaf like structure and is known as phylloclade. On account of flattenings it develops a dorsiventral symmetry. A transection reveals the following structues:

Epidermis. It is differentiated into upper and lower layers of epidermis, covered externally by a thick layer of cuticle. It is composed of a single layer of radially elongated cells, the continuity of which is broken by stomata.

Hypodermis. It consists of one or two layers of loosely arranged cells containing chloroplasts. The chlorenchymatous cells form more or less continuous layer below the upper epidermis but occur isolated beneath the lower epidermis.

Ground tissue. Below chlorenchymatous region is a continuous region of parenchymatous cells extending upto the lower epidermis. The cells of this region are large in size and loosely arranged. Isolated chloroplast containing cells may be present in between.

Vascular system. The vascular system is very characteristic. In the central mid-rib region, several vascular bundles aggregate and are enclosed by a common sclerenchymatous sheath. Each bundle is of concentric amphivasal

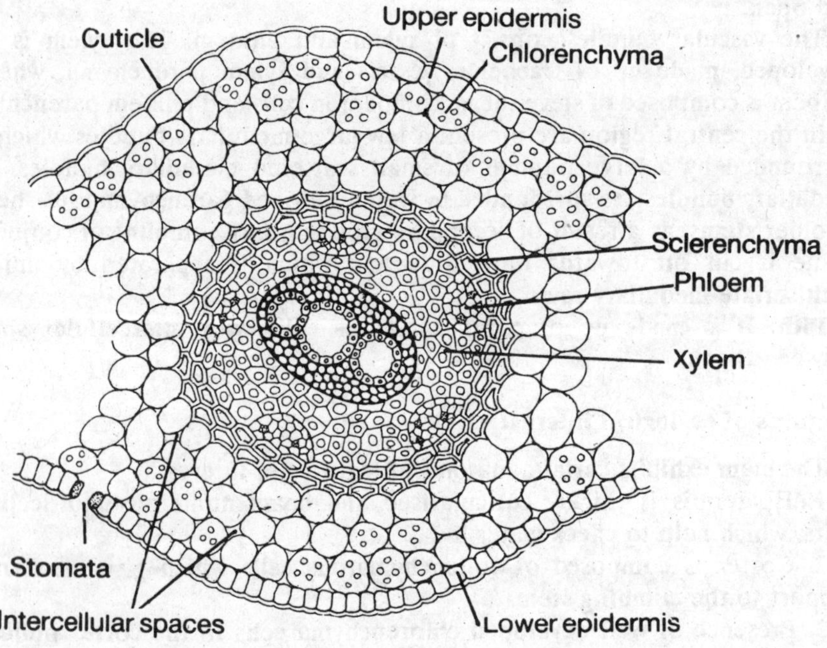

Fig. 16.5. T.S. of phylloclade of *Ruscus* (a cellular enlarged portion).

type i.e. phloem surrounded by xylem. A few parenchymatous cells are also present in between these vascular bundles.

Several vascular bundles are distributed linearly along the lamina. These are smaller in size of amphivasal type and may be enclosed by a sclerenchymatous sheath.

Features of ecological interest

The phylloclade of *Ruscus* is a modified monocot stem which exhibits features of xerophytic adaptations such as:

1. Epidermis consists of radially elongated cells.
2. A thick cuticle covers the outer walls of the epidermal cells.
3. Chlorenchymatous cells are present beneath the epidermis.
4. Thick sclerenchymatous sheath surrounding the vascular bundles is found.
5. The stem is modified morphologically and functionally into a leaf-life structure.

MUEHLENBECKIA (COCOLOBA) PHYLLOCLADE

(*Muehlenbeckia platyclada* syn. *Homocladium platycladum*, Fam.–Polygonaceae)

A transection of the phylloclade reveals anatomical features of both stem and leaf (Fig. 16.6).

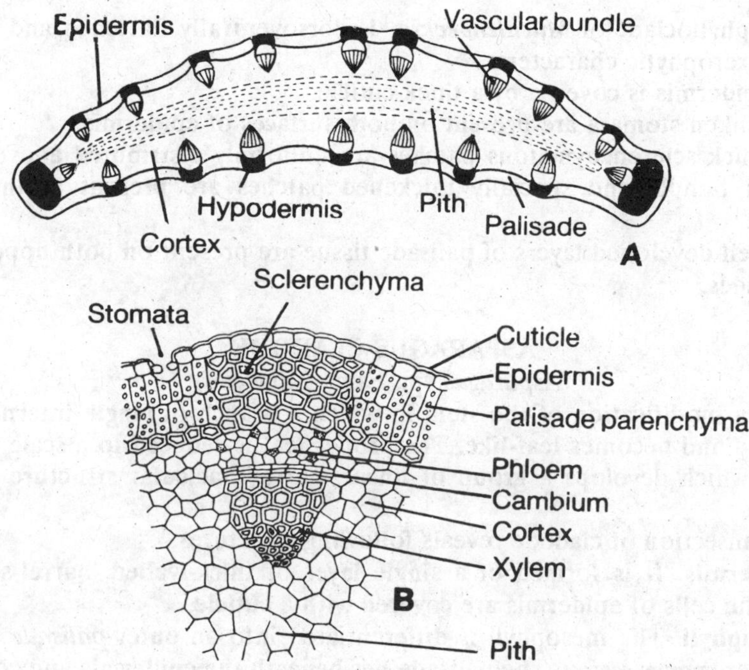

Fig. 16.6. V.S. of phylloclade of *Muehlenbeckia platycladus*. A–diagrammatic and B–a cellular enlarged portion.

Epidermis. It is the outermost layer of radially elongated cells arranged compactly in a single layer constituting upper and lower epidermis. A thick layer of cuticle covers the epidermis, the continuity of which is broken by the presence of sunken stomata.

Cortex. The cortex is represented by well developed palisade tissue, the cells of which are radially elongated, compactly arranged and contain chloroplasts. At places, above the vascular bundles are present sclerenchymatous patches, which afford rigidity and mechanical strength to the otherwise weak and flattened leafy structure.

Endodermis. An inconspicuous endodermis follow the cortex and demarcates it from the pith.

Stele. The stele exhibits polyfascicular siphonostelic condition. A number of vascular bundles are arranged in a ring and are even present, one each, at the two corners. Each vascular bundle is pear-shaped, conjoint, collateral, open and endarch. It consists of phloem and xylem. Phloem is formed of sieve tubes, companion cells and phloem parenchyma. Xylem consists of vessels, tracheids and xylem parenchyma.

Pith. The central region is occupied by a mass of compactly arranged parenchyma cells.

Features of ecological interest

The phylloclade of *Muehlenbeckia* is dorsoventrally flattened and shows certain xerophytic characters:

1. Epidermis is covered by a thick cuticle.
2. Sunken stomata are present on both surfaces of epidermis.
3. Thick sclerenchymatous patches are uniformly distributed above each vascular bundle and specially thickened patches are present at the two corners.
4. Well developed layers of palisade tissue are present on both upper and lower sides.

ASPARAGUS CLADODE

(*Asparagus* sp., Fam.–Liliaceae)

It is a modification of the stem in which a branch of single internode is flattened and becomes leaf-like. The proper leaf is reduced to a scale in the axil of which develops a group of linear, narrow, acicular structure called cladode.

A transection of cladode reveals following structures.

Epidermis. It is formed of a single layer of thick-walled, barrel-shaped cells. The cells of epidermis are covered with a cuticle.

Mesophyll. The mesophyll is differentiated into an outer *palisade region* and inner *spongy region*. The palisade lies beneath the epidermis and consists of small columnar cells containing chloroplasts. These chlorenchymatous cells are loosely arranged all around in cladode. The spongy region lies internal to palisade region. It is formed of parenchymatous cells containing chloroplasts. The cells are loosely arranged enclosing intercellular spaces.

Vascular system. The vascular system is represented by two vascular bundles situated in the centre and surrounded by a compact mass of parenchymatous cells. The vascular elements are phloem and xylem. Phloem is present on the upper side whereas xylem is present towards inner side.

Features of ecological interest

The cladode of *Asparagus* shows the following xerophytic traits:

1. The epidermis is covered externally by a thick cuticle.
2. The Palisade and spongy cells contain abundant chloroplasts, indicating that in the absence of normal leaves the photosynthetic function has been taken up by the cladode.
3. Stomata are few in number as a check against traspiration.

NERIUM LEAF

(*Nerium indicum*, Eng.–Oleander, Fam.–Apocynaceae)

A vertical section of the dicot leaf reveals that it is dorsiventrally flattened with a prominent bulge in the mid-rib region. The arrangement of tissues is as follows: (For figure see chapter on *The Leaf*).

Epidermis. The epidermis of *Nerium* leaf is highly characteristic. Both upper and lower epidermis are multilayered composed usually of three layers of compactly arranged thin walled cells. The outermost layer of both upper and lower epidermis is covered by a thick cuticle.

The lower epidermis is punctuated by the presence of sunken stomata, which are confined to the deep grooves present in the lower surface. These grooves or depressions, formed by the incurving of the epidermis are known as stomatal cavities. They are lined with numerous unicellular epidermal hairs formed by direct prolongations of the epidermal cells. These hairs surround the stomata and protect them from direct exposure to atmosphere, thus helping to reduce the rate of transpiration. The guard cells of the stomata are slightly larger and elevated above the level of epidermal cells.

Mesophyll. The mesophyll is distinguishable into well developed palisade and spongy tissue regions. However, both palisade and spongy tissues are limited to the two arms of the leaf and absent in the mid-rib region. In this region, mesophyll is represented by a few layers of collenchymatous cells, lying below the multilayered epidermis, followed by compactly arranged parenchymatous cells surrounding the vascular region.

The palisade tissue consists of 2-3 layers of compactly arranged radially elongated chlorenchymatous cells lying below the upper epidermis and only one (sometimes two) layer above the lower epidermis. The spongy tissue is composed of loosely arranged chlorenchymatous cells, enclosing large air spaces, which communicate with the such stomatal cavity. Deposits of calcium oxalate are present in the palisade tissue and are termed as **druses.**

Vascular system. A well developed vascular bundle is present in the mid-rib region. It is crescent-shaped and bicollateral with protoxylem facing the upper epidermis. Phloem is formed of sieve tubes, companion cells and phloem parenchyma. Xylem consists of vessels, tracheids and xylem parenchyma. Metaxylem lies next to the phloem and protoxylem towards the upper side.

Collateral V.B. are present in laminar region surrounded by a parenchymatous sheath.

Features of ecological interest

The leaf of *Nerium* is dorsiventrally flattened but shows an isobilateral structure. Anatomically it exhibits specific xerophytic characters, e.g.

1. A thick cuticle is present on the upper and lower epidermis.
2. The epidermis is multilayered, which helps to check excessive transpiration.
3. The stomata are sunken and found in depressions or grooves, known as stomatal cavities.
4. Stomata present in stomatal cavities are further protected by epidermal hairs, which prevent direct contact with the atmosphere.
5. Vascular bundles are well developed.
6. Calcium oxalate deposits are present in the palisade region.

FICUS LEAF

(*Ficus bengalensis,* Eng.–Banyan, Fam.–Moraceae)

The leaf of *Ficus* (dicot) shows some interesting anatomical structures. A vertical section reveals the following characters: (For figure see chapter on *The Leaf*).

Epidermis. It is constituted of two layers, the upper epidermis and lower epidermis. Both the layers of epidermis are covered by a thick cuticle. The upper epidermis is multiseriate consisting usually of three layers. The cells of the outer layer are smaller in size and divide tangentially to give rise to the two lower layers of large, more or less compactly arranged cells. At intervals, below the outer layer of the epidermis are present large cells called the **lithocytes,** which contain a cluster of calcium carbonate crystals. These bunch of grape-like clusters are named as **cystoliths** and are attached by a distinct stalk to the roof of the lithocyte. The lower epidermis is single layered, which has sunken stomata.

Mesophyll. The mesophyll is differentiated into the palisade region and the spongy region. The palisade region is made up of two or more layers of compactly arranged, elongated cells, containing chloroplasts. The continuity of the palisade region is broken above the larger vascular bundle by cellular extensions of the vascular bundles.

The spongy parenchyma region lies beneath the palisade region and extends upto the lower epidermis. It consists of loosely arranged cells enclosing small intercellular spaces. The cells of this region also contain chloroplasts.

Vascular system. There are a number of large and small vascular bundles arranged linearly. The bundles are collateral, closed and enveloped by a parenchymatous sheath. In case of the larger bundles, the bundle sheath is in contact with the epidermal layers through cellular extensions of the bundle sheath. These extensions serve in giving mechanical support to the leaf. The larger vascular bundles are characterized by the presence of sclerenchymatous patches lying next to xylem and phloem. The smaller bundles are devoid of these supporting tissues.

Each V.B. consists of phloem and xylem. Phloem is present towards lower epidermis and consists of sieve tubes, companion cells, and a little phloem parenchyma, whereas xylem faces the upper epidermis and consists of metaxylem and protoxylem vessels.

Features of ecological interest

The dorsiventrally flattened leaf shows following xerophytic characters:

1. Upper and lower epidermis are composed of thick walled cells. Upper epidermis is multiseriate in nature.
2. Both upper and lower epidermis are covered by a thick cuticle.
3. Sunken stomata are present on lower epidermis only.
4. Presence of cystolith, a deposition of calcium carbonate, in certain cells

(lithocytes) of the upper epidermis, is a highly characteristic feature of the leaf.

5. Parenchymatous cellular extensions of the bundle sheath of larger V.B. are present. The cells may sometimes possess thickenings.

6. The larger V.B. possess sclerenchyma patches next to xylem and below phloem.

ZEA MAYS LEAF

(*Zea mays,* Eng.–maize, Fam.–Graminae)

In most of the monocots, the internal structure of the leaves is more or less similar in both the upper and lower halves. Such leaves are called isobilateral (*isos*-equal, *bi*-two, *lateris*-side). (For figure see chapter *The Leaf*).

A vertical section of the leaf shows the following structures:

Epidermis. It is present on both upper and lower surfaces of the leaf. The epidermis is composed of a single layer of cells, which are more or less oval in shape. The epidermal cells of both upper and lower regions are covered by a thick cuticle. The upper epidermis is characterized by the presence of large empty and colourless cells called **bulliform cells,** present at intervals. Stomata are present on both the epidermal surfaces, so the leaf is amphistomatic type.

Mesophyll. Mesophyll tissues are undifferentiated i.e. palisade and spongy parenchyma are not distinguishable. The entire mass of mesophyll is made up of thin-walled, isodiametric cells extending from upper to lower epidermis. The cells characteristically contain abundant chloroplasts and are compactly arranged with few intercellular spaces lying below the stomata.

Vascular system. The vascular system is composed of a number of vascular bundles arranged in a parallel series. The bundles vary in size, some being small, whereas others are fairly large, occurring at regular intervals. Each bundle is collateral and closed with distinct bundle sheath. The cells of the bundle sheath are thin-walled and contain abundant starch grains. The larger vascular bundles possess a patch of sclerenchyma lying above and below them and extending upto the upper and lower epidermal layers, respectively.

Each bundle consists of xylem lying on the upper side (towards upper epidermis) and phloem on its lower side (towards lower epidermis). The larger bundles have distinct and more amount of xylem and phloem as compared to the smaller bundles. Xylem consists of a few enlarged vessels, tracheids and xylem parenchyma, whereas phloem is comprised of sieve tubes, companion cells and phloem parenchyma.

Features of ecological interest

The leaf of *Zea mays* exhibits certain xerophytic characters such as:

1. A thick cuticle is present covering the epidermal cells of both the layers.

2. Bulliform cells are present in the upper epidermis. These are motor cells which help the leaf to roll under xeric conditions, due to turgidity fluctuations of the cell. It is a special character found in plants exposed to dry conditions and helps to reduce stomatal transpiration.

3. Mesophyll region consists of compactly arranged chloroplast containing cells. This compact arrangement helps to check transpiration.

4. The larger vascular bundles are characterized by the presence of sclerenchyma patches lying above and below each bundle.

ACACIA PHYLLODE

(*Acacia auriculiformis,* Eng.–Australian Acacia, Fam.–Mimosaceae)

The petiole becomes modified to form a flattened and leaf-like structure which is known as **phyllode**. In a phyllode, both the surfaces are exposed to light, hence it has an isobilateral structure. A vertical section reveals the following anatomical details.

Epidermis. In outline, section appears flattened with a prominent, biconvex central region. The epidermis forms the outermost covering consisting of a single layer of radially elongated cells arranged compactly. It is differentiated into upper epidermis and lower epidermis. The cells are covered externally by a thick cuticle. The continuity of the epidermis is interrupted due to presence

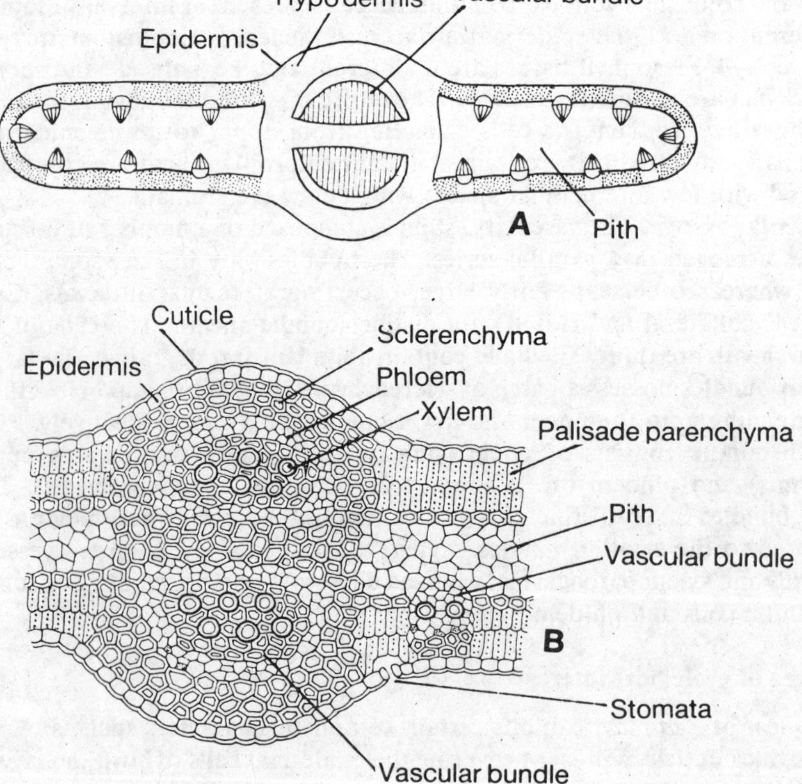

Fig. 16.7. V.S. of phyllode of *Acacia.* A–diag., B–an enlarged cellular portion.

of stomata, which are partially sunken in the lower epidermis.

Mesophyll. The mesophyll is well-developed and differentiated into palisade and spongy tissues. The palisade tissue is represented by radially elongated cells containing chlorenchyma. These palisade cells are compactly arranged in two layers below the upper as well as the lower epidermis. The continuity of the palisade region is, however, interrupted in the region of the central vascular bundle and corner V.B. In these regions, well-developed sclerenchymatous patches are present above and below the central V.B. and forming a cap above the corner bundles.

In between the upper and lower palisade layers are present thin walled parenchyma cells enclosing small intercellular spaces. This region is termed as spongy tissue and represents the flattened pith of the petiole. The cells of spongy tissue do not contain any chloroplast.

Vascular system. The vascular system is well developed being represented by a ring of vascular bundles, but appears linearly arranged due to the flattening of the petiole. In the central biconvex region, two large vascular bundles are present, which are completely covered by sclerenchymatous sheath from all sides. These V.B. possess well-developed phloem and xylem, phloem facing the periphery and xylem towards the centre. These central V.B. together give a mid-rib appearance and form a pseudomidrib.

The other V.B. of the ring are arranged in the laminar region, practically embedded in the palisade tissue. These bundles, are small pear-shaped and alternate in position at the upper and lower regions. A sclerenchymatous sheath is present like a cap over these bundles. The bundles are conjoint, collateral and closed. The presence of corner V.B. confirms the ring arrangement of V.B. (as found in stems) and justifies that the phyllode is a modified flattened petiole.

Features of ecological interest

The phyllode of *Acacia* shows certain xerophytic characters such as:

1. The epidermis is covered by a thick cuticle.
2. Partially sunken stomata are present in the lower epidermis.
3. The epidermis is formed of radially elongated cells.
4. Palisade tissues are well developed and present on both upper and lower surfaces.
5. Massive zone of sclerenchyma surrounds the bigger central bundle, and form a cap-like structure over the smaller corner bundles.

HYDROPHYTES
(Examples)

Plants growing in aquatic habitats are termed as hydrophytes. They may be broadly classified as follows:

1. Submerged hydrophytes, e.g. *Hydrilla, Elodea, Potamogeton, Myriophyllum* etc.
2. Floating hydrophytes,

(i) Free-floating forms, e.g. *Wolffia, Lemna, Ceratophyllum, Trapa, Pistia, Eichhornia* etc.

(ii) Rooted floating forms, e.g. *Nymphaea, Nelumbium, Victoria, Euryale, Trapa* etc.

3. Amphibious plants, e.g. *Sagittaria, Alisma, Typha, Radicula, Castalia, Ranunculus* etc.

A few common but important examples are discussed here.

HYDRILLA STEM

(*Hydrilla verticillata,* Fam.-Hydrocharitaceae)

A transection of the stem reveals the following anatomical details (Fig. 16.8):

Epidermis. It is the outermost layer composed of slightly radially elongated parenchymatous cells, containing chloroplasts.

Hypodermis. The sub-epidermal region consists of 1-3 layers of parenchymatous cells, more or less compactly arranged. Few chloroplasts are irregularly scattered in these cells.

Cortex. It is a distinctive layer composed of loosely arranged parenchymatous cells enclosing large intercellular spaces. These air spaces are separated by one cell thick septa called trabeculae. The inner layers of the cortex lack air spaces, and are arranged in a close manner around the vascular region.

Vascular system. The vascular system is represented by a single protostele. It is enclosed by a single layer of parenchymatous cells forming the endodermis

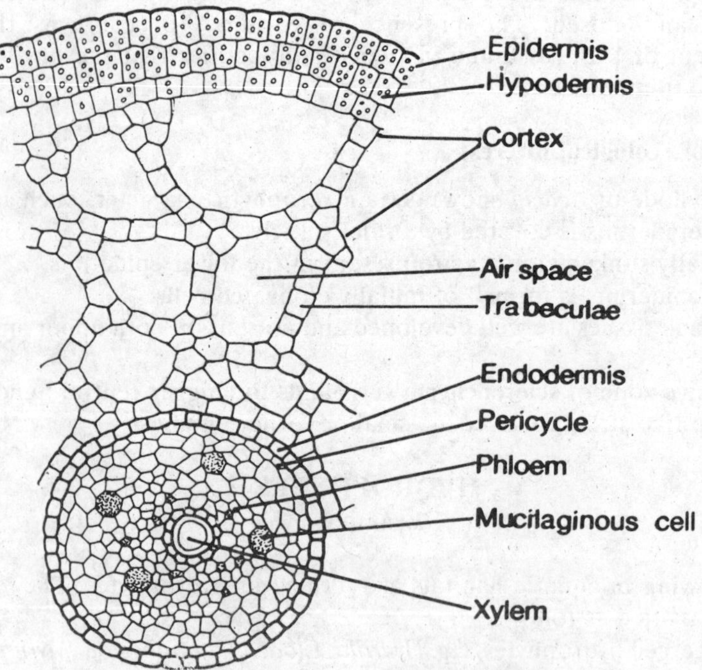

Fig. 16.8. T.S. of stem of *Hydrilla* (an enlarged cellular portion).

followed by single layered pericylce. The vascular tissue is represented by well-developed phloem consisting of sieve tubes, companion cells and phloem parenchyma. However, xylem is extremely reduced being represented by a single large central xylem cavity. Embedded in the phloem are four mucilaginous canals at four different angles.

Features of ecological interest

Hydrilla exhibits special hydrophytic characters such as:

1. Epidermis is composed of thin walled, radially elongated chlorophyllous cells.
2. Cuticle is absent.
3. Well-developed, large air spaces are present in the cortical region.
4. Mechanical tissues are absent.
5. Xylem is reduced and found in the form of a single cavity.

MYRIOPHYLLUM STEM
(*Myriophyllum* sp., Fam.–Haloragaceae)

The cross-section of the stem reveals following anatomical characters:

Epidermis. It is a single layer of thin walled cells, in which the outer cell walls are slightly thickened. Characteristically, cuticle is absent.

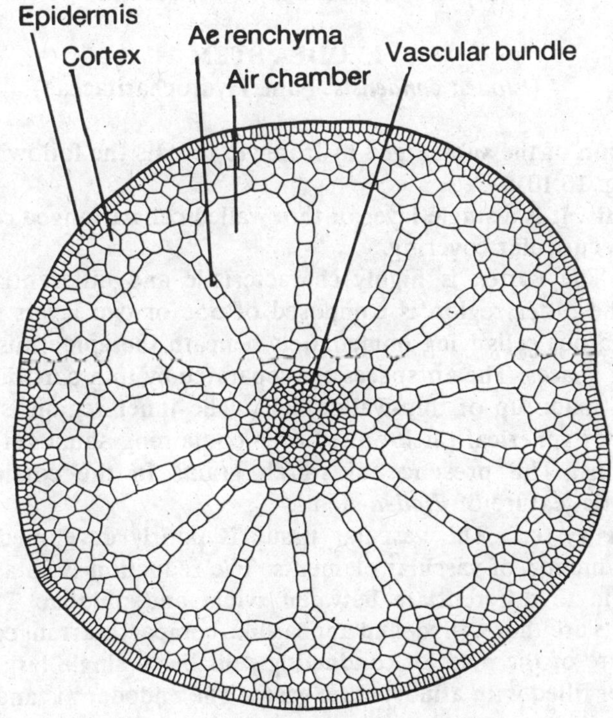

Fig. 16.9. T.S. stem of *Myriophyllum* stem.

Cortex. The cortex lies beneath the epidermis and is differentiated into two distinct zones. The outer region is composed of thin walled, compactly arranged parenchymatous cells, whereas the inner region is made up of symmetrically arranged air spaces. These air spaces are separated by thin partitions called trabeculae made up of a single layer of thin-walled cells.

Vascular system. The vascular system is poorly developed and consists of a few scattered vessels surrounded by phloem. The endodermis and pericycle are absent, and protection of vascular tissue is afforded by few parenchymatous cells surrounding them.

Features of ecological interest

Myriophyllum exhibits special characters in accordance with its submerged habit.

1. The epidermis is composed of thin walled cells.
2. Cuticle is absent.
3. The cortex has large well-developed air spaces filled with atmospheric gases. This helps in respiratory activities as well as gives buoyancy to the stem.
4. Mechanical tissues are completely absent.
5. Vascular tissue is poorly developed and represented by few scattered xylem vessels.
6. Endodermis and pericycle are completely absent.

ELODEA STEM

(*Elodea candensis*, Fam. Hydrocharitaceae)

A transection of the submerged hydrophyte reveals the following anatomical details (Fig. 16.10):

Epidermis. It is a single layer of thin walled cells arranged compactly. The cells lack a cuticular covering.

Cortex. The cortex is highly characteristic and differentiated into two regions. The outer region is composed of one or two layers of thin-walled parenchymatous cells lying immediately beneath the epidermis and followed by large air spaces. The air spaces are separated by thin partition walls called trabeculae, made up of thin-walled cells. The inner region is composed of large, oval or spherical thick-walled cells containing abundant deposition of starch grains. The presence of starch grains in the cortical cells is a characteristic feature of *Elodea* stems.

Vascular system. The vascular tissue is poorly developed and exhibits extreme reduction of vascular elements. The reduction is so acute, that it is not possible to differentiate between xylem and phloem. The xylem and phloem cells are thin walled, and appear like compactly arranged parenchyma. In the centre of the mass of conducting cells lies a single large intercellular air chamber filled with atmospheric gases. The endodermis and pericycle are not distinguishable as separate layer.

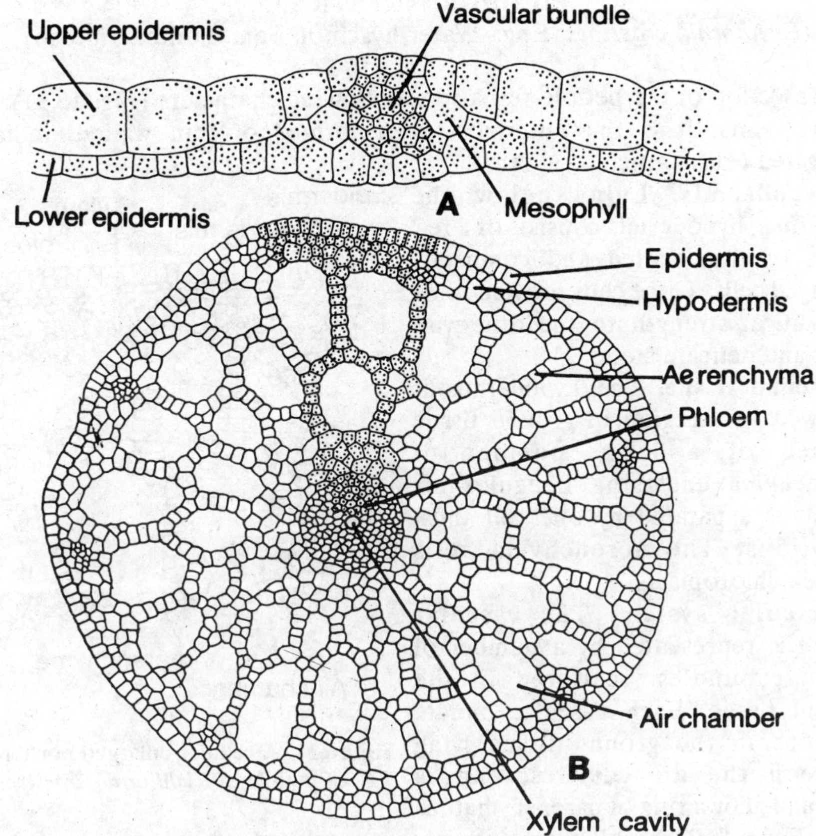

Fig. 16.10. *Elodea*. A–a cellular enlarged portion of V.S. of leaf and B–T.S. of stem.

Features of ecological interest

The plant of *Elodea* being a submerged hydrophyte exhibits certain anatomical peculiarities.

1. The epidermis is composed of thin-walled cells, lacking a cuticular covering.

2. The cortex is made up of large air spaces filled with atmospheric gases, which give buoyancy to the plant and help it to float in a submerged condition.

3. The cells of the inner cortex are thick-walled and contain abundant starch grains.

4. The most characteristic feature of the stem, is the absence of well differentiated vascular tissue. Xylem and phloem are represented by thin-walled parenchyma like cells only.

5. A single, large intercellular air chamber situated centrally in the stele is an interesting feature of the stem.

Elodea Leaf. The organisation of leaf is simple with very poorly developed vascular elements (Fig. 16.10A).

EICHHORNIA PETIOLE

(*Eichhornia crassipes,* Eng.–Water-hyacinth, Fam.-Pontederiaceae)

A transection of the petiole reveals the following characters (Fig. 16.11):

Epidermis. It is the outermost layer formed of thin walled, radially elongated cells. Cuticle is absent.

Hypodermis. Lying below the epidermis, hypodermis consists of a few layers of thin walled and compactly arranged cells. These cells provide some mechanical strength to the otherwise weak and delicate stem.

Ground tissue. The hypodermis is followed by a region of ground tissue formed of a large number of aerenchyma enclosing irregular air chambers separated by one cell thick partitions. The aerenchyma cells contain chloroplasts.

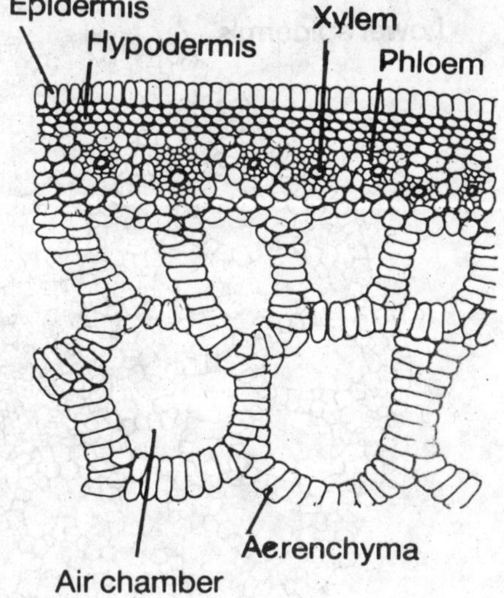

Fig. 16.11. A cellular enlarged portion of T.S. of petiole of *Eichhornia crassipes*.

Vascular system. The vascular system is represented by a number of vascular bundles, scattered in the ground tissue. Each vascular bundle develops in the ground tissue lying between the air cavities, and is surrounded by a ring of parenchymatous cells. Each V.B. is collateral and closed composed of phloem and xylem. Phloem is prominent and located in the outer region of the bundle, whereas xylem is represented by a single large element.

Features of ecological interest

Eichhornia is a free floating hydrophyte and exhibits following anatomical features:

1. Epidermis consists of thin-walled cells, lacking cuticle.
2. There is complete absence of mechanical tissue.
3. Ground tissues are composed entirely of aerenchyma which are green and enclose large air cavities.
4. Vascular bundles are poorly developed.
5. Xylem is reduced to a single element.

NYMPHAEA PETIOLE

(*Nymphaea nouchali,* Eng.–White Lotus, Fam.–Nymphaeaceae)

A transection of the petiole shows following anatomical details (Fig. 16.12):

Epidermis. It is the outermost layer of thin-walled, barrel-shaped cells. Cuticle is generally absent, but a very thin layer may sometimes be present.

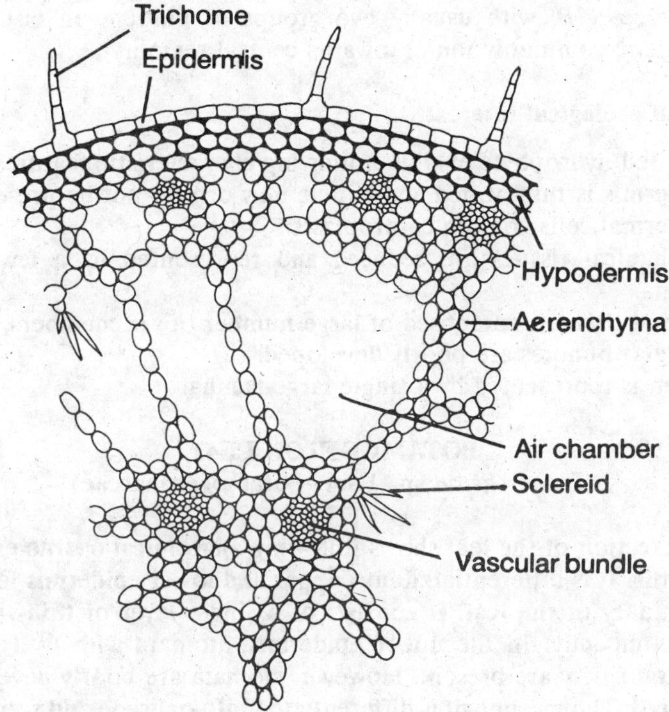

Fig. 16.12. A cellular enlarged portion of T.S. of petiole of lotus (*Nymphaea*).

Epidermal cells are characterised by the presence of chloroplasts. The cells bear a few multicelled unbranched hairs on their surface.

Hypodermis. It consists of 2-3 layers of collenchymatous cells, arranged compactly without any intercellular spaces.

Ground tissue. Below the hypodermis is the continuous region of ground tissue, characterised by the presence of large air chambers. These air chambers are separated from one another by thin-walled uniseriate or multiseriate partitions.

A characteristic feature of ground tissue is the presence of *sclereids* (trichosclereid) which protrude into the air chamber. These sclereids may be of various shapes, branched or unbranched and contain depositions of calcium oxalate crystals.

Vascular system. The vascular system is represented by a number of vascular bundles irregularly scattered in the ground tissue. They are more common in the outer compact parenchymatous region. The V.B. present in between the air chambers are surrounded by parenchymatous cells. The vascular tissue is poorly developed, xylem being represented by a single large lacunae and phloem showing normal development.

Two kinds of vascular bundles may be found:

(i) *Smaller V.B.* with only one group of phloem towards epidermis, found towards outer side.

(ii) *Larger V.B.* with usually two groups of phloem on either side of xylem, commonly found towards central region.

Features of ecological interest

The rooted hydrophyte exhibits following characteristic features:

1. Epidermis is thin-walled and cuticle may or may not be present.
2. Epidermal cells contain chloroplasts.
3. Mechanical tissues are reduced and represented by a few layers of collenchyma.
4. Ground tissue is composed of large number of air chambers.
5. Vascular bundles are poorly developed.
6. Xylem is represented by a single large lacuna.

POTAMOGETON LEAF

(*Potamogeton* sp., Fam.–Potamogetonaceae)

A vertical section of the leaf shows following anatomical characters:

Epidermis. It is differentiated into upper and lower epidermis forming the outer boundary of the leaf. It consists of a single layer of thin-walled cells arranged compactly. In the upper epidermis, stomata with well developed sub-stomatal cavity are present. However, stomata are poorly developed.

Mesophyll. The mesophyll is differentiated into palisade and spongy tissue. Palisade occurs just below the epidermis and consists of radially elongated cells, arranged compactly in three to four layers. The palisade cells contain chloroplasts.

Below the palisade, are the spongy parenchyma occupying about two thirds of the leaf area and extending upto the lower epidermis. Large air chambers

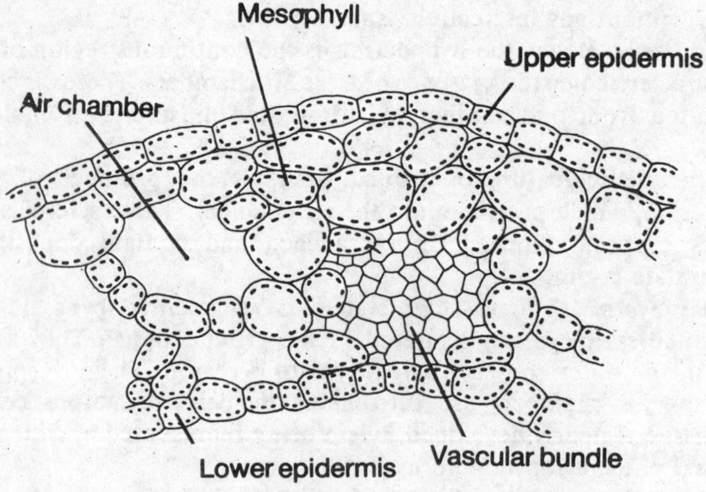

Fig. 16.13. Vertical section of a portion of the leaf of *Potamogeton*.

are present, which are separated by uniseriate partitions.

Vascular system. In the mid-rib region, a prominent vascular bundle is present, surrounded by a distinct bundle sheath. It is composed of phloem and xylem. Phloem is comparatively well developed and present towards the lower epidermis. Xylem, however, is reduced to a few elements only and faces the upper epidermis. The remaining portion is occupied by thin-walled parenchyma cells (Fig. 16.13).

Features of ecological interest

The submerged leaf of *Potamogeton* exhibits following hydrophytic characters.

1. Epidermis is composed of thin-walled cells.
2. Cuticle is absent.
3. Large number of air chambers are present.
4. Poorly developed stomata are present on upper epidermis.
5. Vascular tissues are reduced especially the xylem elements.

CERATOPHYLLUM LEAF

(*Ceratophyllum demersum,* Fam.-Ceratophyllaceae)

A transverse section through the leaf segment of *Ceratophyllum* reveals the following structures (Fig. 16.14):

Epidermis. The outline of the leaf is nearly oval, in which the epidermis forms a single layer of thin-walled cells, containing chloroplasts.

Mesophyll. The mesophyll is undifferentiated, consisting of large, thin walled cells, interrupted by a few large intercellular spaces. The cells are characterised by the presence of scattered chloroplasts.

Vascular system. The vascular system is extremely reduced and undifferen-

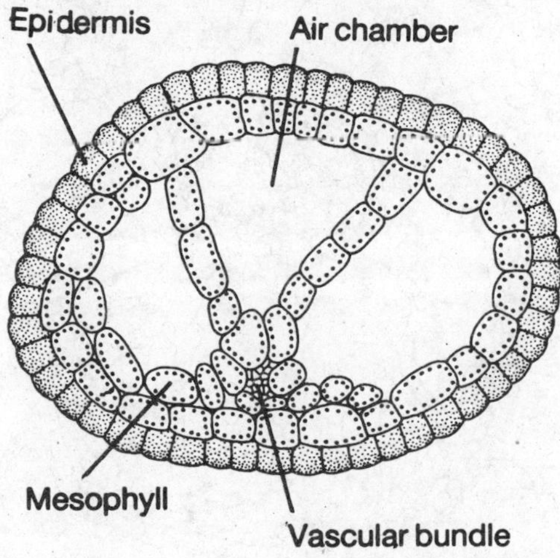

Fig. 16.14. V.S. of leaf of *Ceratophyllum.*

tiated into xylem and phloem. It is represented by a single, small vascular tract, containing a few, thin-walled elements.

Features of ecological interest

Ceratophyllum is a free floating hydrophyte which exhibits following hydrophytic features:

1. Epidermis lacks cuticle and stomata.
2. Air spaces are present in the mesophyll.
3. Mesophyll and epidermal cells contain chloroplasts.
4. There is complete absence of mechanical tissue.
5. Vascular system is extremely reduced, to a few elements only.

TRAPA LEAF

(*Trapa bispinosa,* Eng.–Water chestnut, Fam.–Trapaceae)

Vertical section of leaf of this free floating hydrophyte reveals the following structures (Fig. 16.15):

Epidermis. The epidermis is differentiated into upper and lower epidermis,

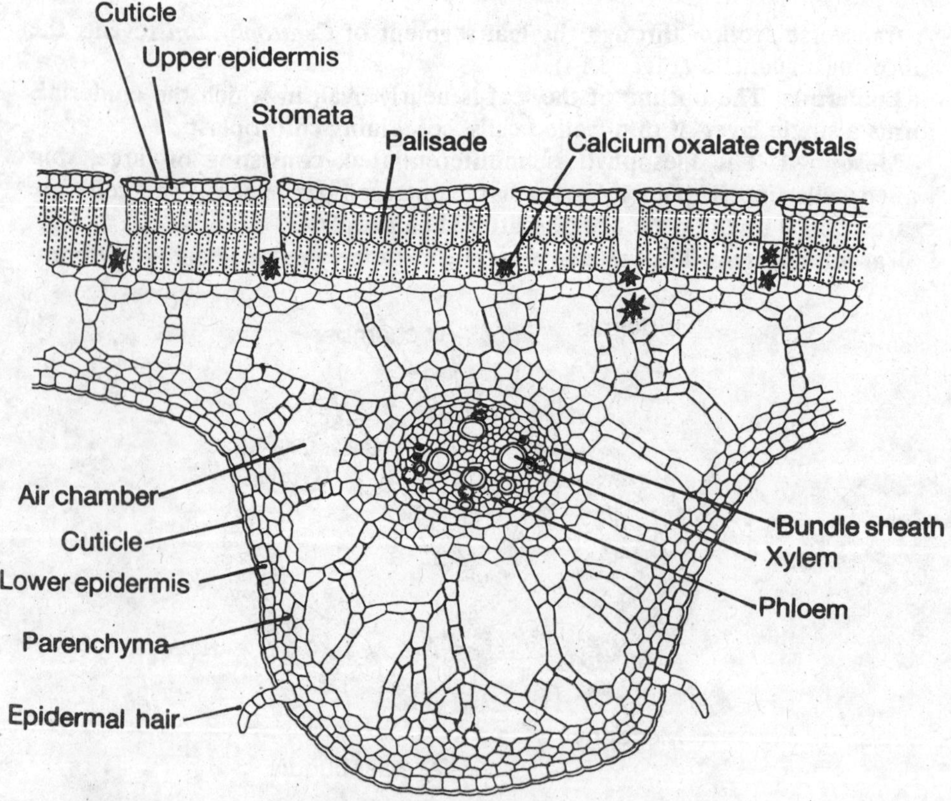

Fig. 16.15. A cellular enlarged portion of leaf of *Trapa bispinosa* (V.S.)

each consisting of a single layer of compactly arranged thin walled barrel-shaped cells. The upper epidermis is externally covered by a thin cuticle. The stomata are present only on upper epidermis, in accordance with its free floating nature.

Mesophyll. The mesophyll is well-developed and differentiated into upper palisade zone and lower spongy region. The palisade region is composed of two to three layers of rectangular shaped cells arranged compactly. The continuity of the cylinder is broken at places by sub-stomatal cavities lying at the base of the stomata. Certain cells at the floor of sub-stomatal cavities possess star-shaped sphaeraphides.

The spongy tissue is represented by one or two layers of compactly arranged parenchymatous cells, lying immediately below the palisade region, and by large air cavities, arranged regularly in a row. These air cavities are separated by uniseriate partitions.

Vascular system. The vascular system is reduced and represented by small vascular bundles. The vascular tissues are poorly developed, composed of phloem and poorly developed xylem, represented by single large cavity.

Features of ecological interest

The leaf of *Trapa* exhibits special characters which justify its free floating nature:

1. Upper epidermis is covered by a thin cuticle.
2. Stomata are present only on the upper epidermis of the leaf, which floats on the water surface.
3. Stomata open into large sub-stomatal cavities.
4. Sphaeraphides containing cells are present in the cells at the floor of sub-stomatal cavities.
5. Well-developed air chambers are present in the spongy parenchyma.
6. Vascular tissues are reduced and xylem is represented by cavity only.

NYMPHAEA LEAF

(*Nymphaea nouchali,* Eng. White lotus, Fam.–Nymphaeaceae)

A vertical section of leaf of *Nymphaea* reveals following anatomical structures:

Epidermis. It consists of upper and lower epidermis, present on both the surfaces of the leaf. It consists of a single row of compactly arranged, barrel-shaped cells. The upper epidermis is covered by a waxy cuticle, and its continuity is broken by a number of stomata. In the lower epidermis, a few slime glands are present.

Mesophyll. The mesophyll is differentiated into palisade and spongy parenchyma. The palisade tissue consists of a few layers of compactly arranged and radially elongated, tubular or columnar cells. The cells contain numerous chloroplasts arranged linearly. A few intercellular spaces may be present in between the cells (Fig. 16.16).

The spongy parenchyma lies beneath the palisade region and consists of large air chambers separated by uni- or multiseriate partitions. These partitions

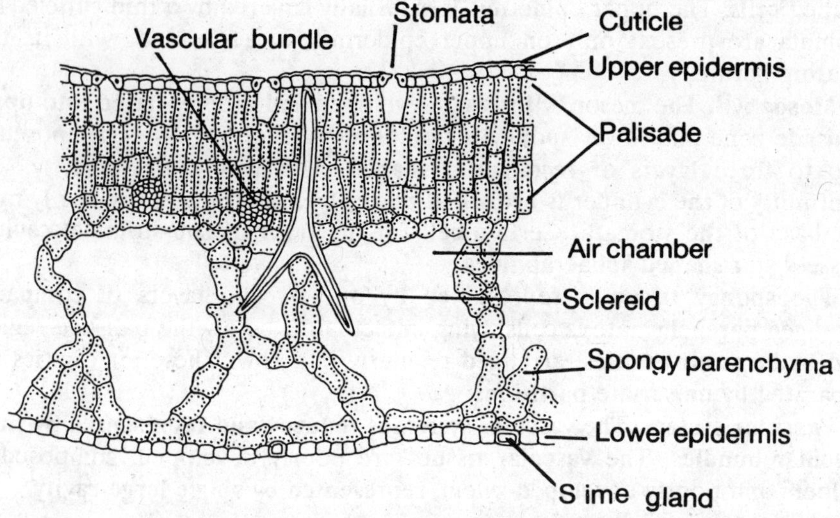

Fig. 16.16. A cellular enlarged portion of V.S. of leaf of lotus (*Nymphaea*).

are formed of thin-walled cells containing chloroplasts. In the palisade as wcll as spongy region, trichosclereids of different shapes are present.

Vascular system. The vascular system is poorly developed and represented by small vascular bundles embedded in the palisade region, all along the lamina region and about 3 or 4 V.B. occurring together in the central region. The vascular bundles are surrounded by parenchymatous bundle sheath. Each V.B. is collateral and closed. In the mid-rib region, double, mixed or fused vascular bundles occur, which are usually inversely oriented. The vascular bundles consist of poorly developed xylem and comparatively larger phloem. In the laminar region, the xylem of the V.B. remains oriented towards upper epidermis.

Features of ecological interest

A vertical section of the leaf exhibits special hydrophytic characters as follows:

1. Upper epidermis contains stomata, whereas lower epidermis lacks stomata. This justifies that leaf is floating with exposed upper surface.
2. Large number of air chambers are present.
3. Chlorophyll containing palisade tissues are well-developed, and present in several layers below the upper epidermis.
4. There is a complete absence of mechanical tissue, only trichosclereids provide support.
5. Vascular bundles are reduced and xylem is represented by one or two elements only.

TYPHA LEAF

(*Typha angustata,* Fam.–Typhaceae)

A vertical section of *Typha* exhibits following anatomical characteristics.

Epidermis. It is the outermost layer composed of barrel-shaped cells. The cells are covered externally by thick cuticle. At places, the upper as well as lower epidermis are interrupted by simple stomata (Fig. 16.17).

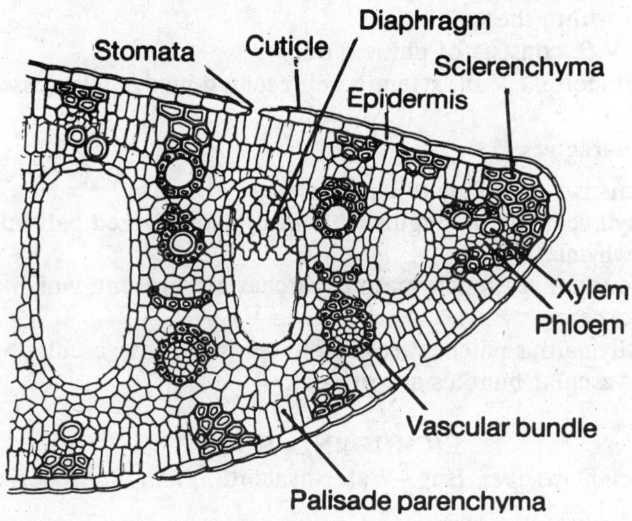

Fig. 16.17. V.S. of leaf of *Typha* (a cellular enlarged portion).

Mesophyll. The mesophyll is differentiated into palisade region and spongy region. The palisade tissue is represented by columnar cells arranged in three to many layers. The cells contain chloroplasts. At places, the palisade tissue is interrupted by sclerenchymatous patches and sub-epidermal vascular bundles. These sclerenchymatous patches are prominent at the two corners. The spongy tissue is traversed by large intercellular spaces, each separated by many layered partition wall. At times, thin-walled diaphragm cells are present in these air chambers.

Vascular system. The vascular system is represented by two types of vascular bundles–(i) sub-epidermal vascular bundles and (ii) vascular bundles of partition wall. The sub-epidermal vascular bundles are present below the sclerenchymatous patches and are of collateral and closed type. They are composed of xylem and phloem. Xylem consists of vessels, tracheids and xylem fibres. The phloem is present in the form of patches.

The vascular bundles of partition wall are present in each partition wall of the air space. Usually the middle vascular bundles are the phloem bundles i.e. these consist of phloem cells only and each one is surrounded by thick sclerenchymatous sheath. The other vascular bundles are collateral, closed and there may be sclerenchymatous patches below these vascular bundles.

Features of ecological interest

Being an amphibious plant, *Typha* exhibits both xerophytic and hydrophytic characters. These are:

Hydrophytic characters

1. Stomata are simple and hydrophytic type.
2. Large number of air chambers are present. Some may show the existence of diaphragm within them.
3. Certain V.B. consists of phloem only.
4. In sub-epidermal V.B., xylem is represented by a single vessel only.

Xerophytic characters

1. Epidermis is covered by thick cuticle.
2. Mesophyll cells are distinguishable into many layered palisade cells and spongy parenchyma.
3. Sub-epidermal sclerenchymatous patches are present, which are prominent at the corners.
4. Sclerenchymatous patches are present below certain vascular bundles.
5. Several vascular bundles are present.

EICHHORNIA ROOT

(*Eichhornia crassipes,* Eng.–Water-hyacinth, Fam.–Pontederiaceae)

A transection of the root reveals following features (Fig. 16.18):

Epidermis. It is the outermost layer which is composed of simple barrel-shaped cells. Cuticle is absent. Root hairs are absent.

Cortex. The cortex is well-developed and differentiated into three distinct zones. Outer cortex consists of compactly arranged parenchymatous cells. Middle cortex is well-developed and covers the major portion of the root. It is composed of large number of conspicuous air chambers, which are separated from each other by uniseriate trabeculae. Occasionally, the trabeculae branch and join each other forming an anastomosing system. The inner cortex consists of 3-4 layers of loosely arranged parenchymatous cells, enclosing intercellular spaces. The cells are nearly spherical in shape and regularly arranged in concentric layers.

Vascular system. The stele is enclosed by a single layered endodermis and a pericycle. The vascular bundles are of radial type and more than six in number. The xylem shows exarch condition, metaxylem towards the centre and protoxylem facing the periphery. Phloem is well-developed and present in between the xylem groups.

Pith. The central portion of the stele is occupied by compactly arranged parenchymatous cells and is known as pith.

Features of ecological interest

The root of *Eichhornia* exhibits the following hydrophytic characters:

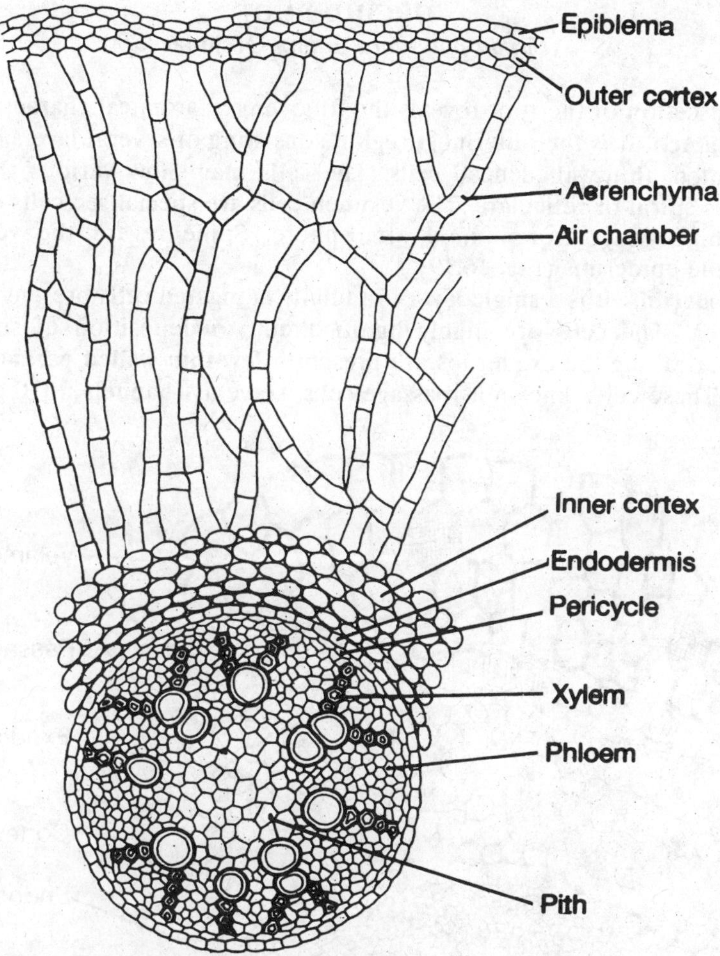

Fig. 16.18. A cellular enlarged portion of T.S. root of water hyacinth (*Eichhornia crassipes*).

1. Epidermis is thin-walled, lacking cuticle and root hairs.
2. Cortex is well developed and consists of large number of air chambers.
3. Mechanical tissues are absent.
4. Xylem elements are reduced.

EPIPHYTES
(Examples)

Plants which grow on other plants for mechanical support only but are nutritionally independent are known as epiphytes. Their roots are hygroscopic and specially modified to absorb water from the humid atmosphere. Some examples are *Dendrobium*, *Vanda*, *Cymbidium* etc.

ORCHID ROOT

(*Vanda parviflora*, Fam.–Orchidaceae)

A transection of the root reveals the following anatomical characteristics:

Velamen. It is the outermost region consisting of several layers of radially elongated, thin-walled dead cells. The cells may show various thickenings such as spiral or reticulate. The velamen cells are specialised cells capable of absorbing water from humid atmosphere. Some regard the velamen as multiple epidermis (Fig. 16.19).

Exodermis. It is a single layer of radially elongated cells present below the velamen. The cells are highly lignified due to deposition of suberin. At intervals along the exodermis are present a few thin walled parenchymatous cells. These cells, known as passage cells, serve as channels for the entrance

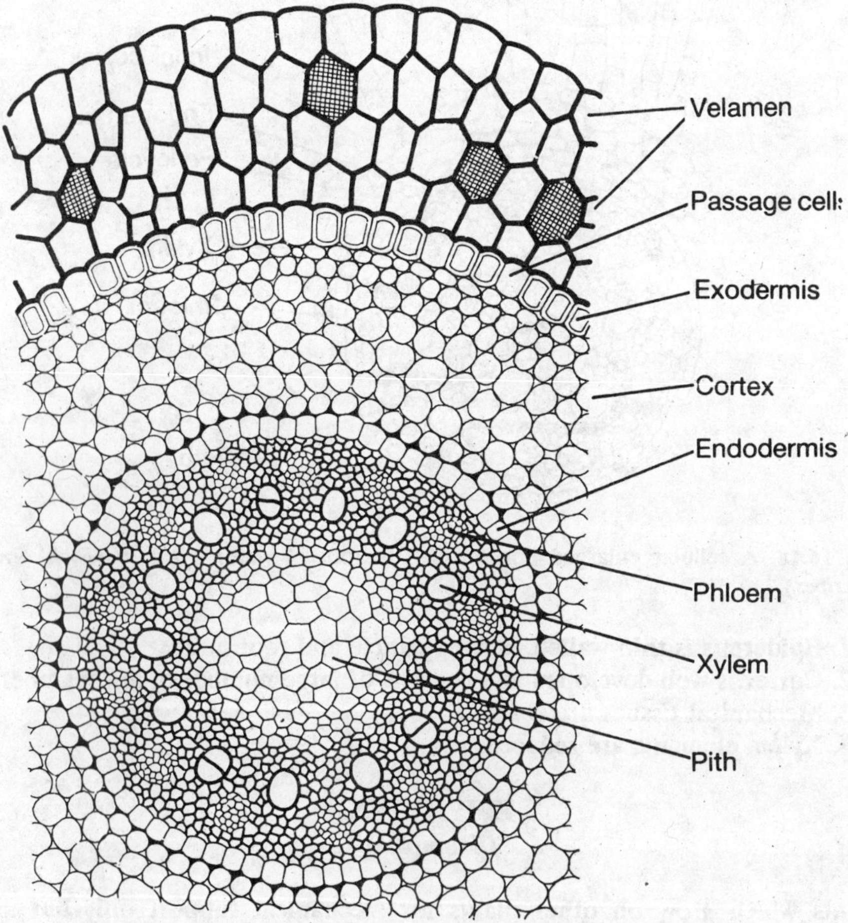

Fig. 16.19. A cellular enlarged portion of T.S. of root of an orchid, ***Vanda.***

of water from the velamen into the vascular region.

Cortex. It is a well-developed region composed of loosely arranged, irregular-shaped cells. A few air chambers and secretory sacs are also present in this region. In young conditions, some of the cells contain chloroplasts.

Endodermis. It is single layered composed of lignified barrel-shaped cells, interrupted at places by few thin-walled passage cells.

Pericycle. It lies beneath the endodermis forming a single layer of thin-walled cells.

Vascular system. The vascular system is well-developed, represented by radial vascular bundles, numbering more than six. The vascular elements consist of phloem and xylem and are surrounded by lignified, conjunctive tissue. Phloem consists mainly of sieve tubes and companion cells, whereas xylem is composed of vessels and tracheids showing exarch condition. The large metaxylem elements show reticulate thickenings and are situated towards the centre, whereas the smaller protoxylem elements are annularly or spirally thickened lying towards periphery. Conjunctive tissue composed of lignified parenchymatous cells lies between pith and pericycle. The xylem and phloem are embedded in these tissues.

Pith. In the central region well-developed pith is present. It consists of spherical parenchymatous cells with intercellular spaces. The pith cells may develop thickenings at later stage.

Features of ecological interest

The epiphytic root of *Vanda* (orchid) exhibits certain special characters:

1. Presence of specialised water absorbing tissue, the *velamen*, is the characteristic feature of the root.

2. The thick-walled cells of exodermis provide mechanical support to the aerial, pendant roots.

3. The passage cells are present in the exodermis which facilitate water conduction from the velamen to the xylem cells.

4. Secretory cells present in the cortex help the root to tide over dry conditions.

5. Presence of air chambers in the cortex gives lightness to the hanging roots.

6. Presence of chloroplasts in the cortical cells, during young conditions, suggests their probable role in photosynthesis.

7. Well-developed, thick-walled conjunctive tissues are present in between vascular elements. These cells provide mechanical strength to the otherwise soft and delicate roots.

HALOPHYTES
(Examples)

Plants growing in salty marshes or near salt lakes and sea shores are known as halophytes. The soil in such places is said to be physiologically dry, since the water is not available to the plants due to high osmotic concentration. Common examples are *Rhizophora, Avicennia, Sonneratia* etc.

RHIZOPHORA ROOT (PNEUMATOPHORE)

(*Rhizophora mucronata,* Fam.–Rhizophoraceae)

Rhizophora grows in mangrove swamps and is characterized by negatively geotropic roots. These roots grow above the ground and are known as breathing roots or pneumatophores. A transection of the root shows following features (Fig. 16.20):

Cork. The outer 4-6 layers consist of compactly arranged, suberised cells. The cork cells are dead cells and are formed by the activity of the cork cambium. The continuity of the cork layer is broken at places by lenticels which facilitate gaseous exchange, specially absorption of oxygen for respiratory purpose. Underlying the cork cells, is present one or two layers of

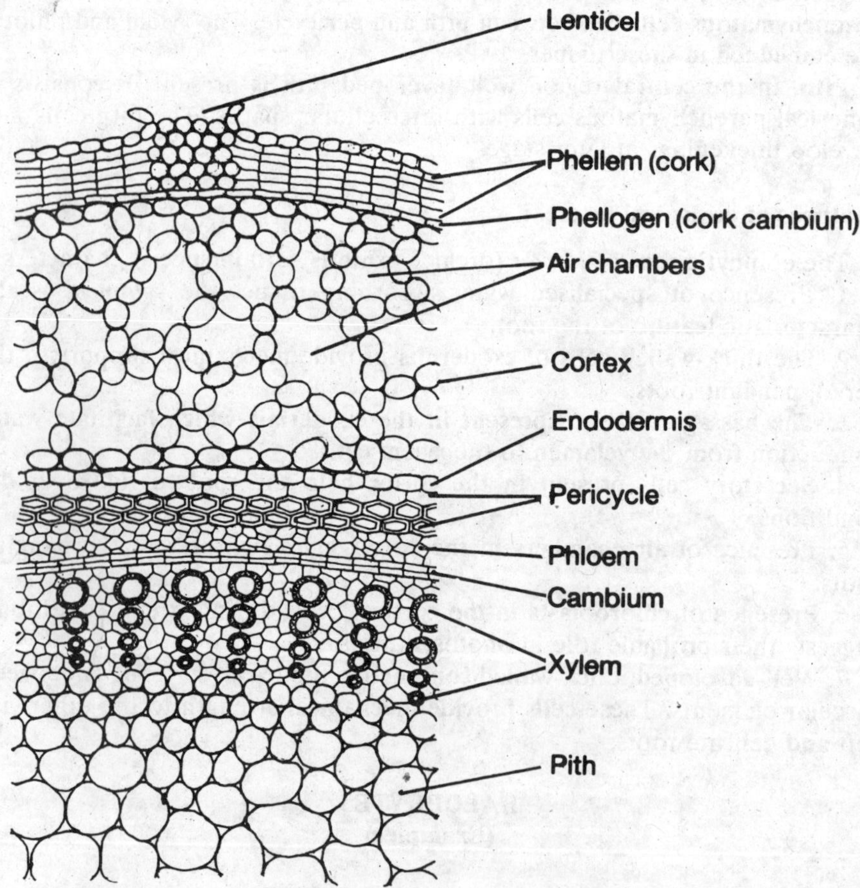

Fig. 16.20. A cellular enlarged portion of T.S. of *Pneumatophore* (respiratory root of a mangrove).

cork cambium. Cork remains absent in young roots.

Cortex. It is well-developed consisting of large number of air chambers, separated by partition walls. In the terrestrial part of the root, the cortex is composed of two types of cells–

(i) radially elongated cells connected with each other; and

(ii) vertically elongated cells having circular lumina.

Endodermis. It is well developed, prominent layer composed of barrel-shaped cells arranged in a single layer.

Pericycle. The pericycle is multilayered and lies beneath the endodermis. It consists of an outer single layer of parenchymatous cells, followed by two layers of sclerenchymatous cells.

Vascular system. The vascular system is conjoint, collateral and open. It consists of a continuous layer of phloem, lying towards the pericycle, followed by a layer of cambium and finally a continuous layer of xylem. Metaxylem elements are large and faces the periphery, whereas the small protoxylem elements face the pith region, so root shows presence of endarch type of xylem.

Pith. The pith is made up of a large number of rounded cells, which are loosely arranged enclosing intercellular spaces.

Features of ecological interest

The pneumatophores present an interesting case of root anatomy.

1. Well-developed cork region is present which communicates with the outer atmosphere through lenticels.

2. The cortex is composed of large air chambers which help it to remain above the water surface.

3. Radially elongated thickened cells in the cortex are found which give mechanical strength to the roots.

4. Sclerenchymatous pericycle also aids in giving mechanical strength.

5. Pith composed of loosely arranged parenchymatous cells, enclosing intercellular spaces, helps in keeping the roots afloat.

GENERAL QUESTIONS

1. Discuss the role of environment on the anatomy of plants.
2. "Hydrophytes are well adapted to their environment", justify the statement giving suitable examples.
3. Discuss, with suitable examples, the various morphological and anatomical pecularities exhibited by xerophytes.
4. What are the characteristic ecological features of the following plants.

   1. Phylloclade of *Muehlenbeckia*
   2. Pneumatophore of *Rhizophora*
   3. *Ficus* leaf
   4. *Vanda* root
   5. *Elodea* stem
   6. *Nymphaea* petiole
   7. Petiole of *Eichhornia*
   8. *Typha* leaf
   9. *Casuarina* stem
   10. *Acacia* phyllode.

CHAPTER 17

Systematic Plant Anatomy

Anatomical features of vegetative as well as reproductive organs have been studied to help in studies on systematics, particularly for resolving controversial taxonomic problems. In fact, such studies have been successfully employed for the solution of taxonomic problems and to the elucidation of phylogenetic relationships. In several cases, anatomical studies have helped in identifying the fragmentary material which lacked flowers and fruits in the herbarium specimens.

Several internal features of the plants have been found useful and important in systematic studies. A few are discussed here after dividing them into two broad categories, the anatomy of vegetative organs and that of flowers and seeds (reproductive organs). Of vegetative organs, wood anatomy in relation to taxonomy has attracted wide attention.

VEGETATIVE ORGANS

Trichomes

Trichomes have been frequently used for systematic comparisons of closely linked taxa. They have proved most useful to taxonomists at generic and specific levels. Heintzelman and Howard (1944) supported the segregation of two genera, *Ottoschulzia* and *Poraqueiba* of family Icacinaceae on the basis of evidence from trichomes. Controversy about the link between Typhaceae and Sparganiaceae could be resolved on basis of presence of sessile or glandular trichomes. Similarly, Inamdar (1967) confirms the position of *Nyctanthes* within family Oleaceae on the basis of structure and ontogeny of trichomes.

Stomata

Stomata have been studied since long back, even in 19th century. Vesque (1889) classified stomata into four categories which he named after the families in which these were found. This system of classifying categories of stomata are followed even now-a-days. Von Cotthem (1970) proposed a detailed account of classification of stomata based on the number and position of subsidiaries. Krishnamurthy and Kannabiran (1970) have successfully used structure of stomata coupled with subsidiary epidermal features in the identification of fragmentary materials in the pharmacognostical studies.

Pant and Mehra (1965) and Bir and his coworkers (1971) supported the division of family Rubiaceae into three subfamilies, Rubioideae, Cinchonoideae and Guettaroideae. Kothari and Sah (1974, 75) while studying taxonomic and phylogenetic significance of stomata in the Papilionaceae, proposed that tribes like Sophoreae, Podalyrieae and Gensteae should be considered more primitive due to presence of anomoperigenous and aniso-mesoperigenous stomata than Dalbergieae and Phaseoleae. Similarly, the ontogenetic studies on stomata of Caricaceae and Cucurbitaceae suggest close relationship between them due to presence of aniso-mesoperigenous and aperigenous stomata in both these families.

Epidermis

Characters of epidermal cells like shape, pattern of thickness, nature of sculpturing on their walls, inclusions of epidermal cells have been found useful in taxonomic studies. The importance of papillate epidermal cells, shape of silicified and suberised cells and their relationships have been found valuable in diagnostic studies in Graminae. Distribution of silica bodies among members of Cyperaceae provides a good taxonomic criteria. Metcalf (1968) reported that wedge-shaped silica bodies are restricted to *Mesomelaena*, *Neesenbeckia*, *Ptilianthelium* and *Scirpodendron*; bridge-shaped bodies to *Mapania*; spherical and hemispherical warty bodies to *Acriulars* and *Scleria*; spherical and hemispherical echinulate to *Rhyncospora* and hemispherical smooth bodies to *Lophoschoenus*. Similarly, Jain and Singh (1974) have been able to distinguish Himalayan species of *Quercus* (oak) on the basis of epidermal characters.

Leaf anatomy

Though leaves show almost an obstinate organisation of internal tissues, yet several characters like structure of mesophyll, bundle sheath, presence and absence of hydathodes, foliar nectaries and oil glands, stomatal crypt, suberisation and position of abscission layer have been found valuable for taxonomic studies. In grasses, the general leaf anatomy has been found a very valuable character useful in grass systematics. Govindarajalu (1969) proposed a key to distinguish various species of *Cyperus* on the basis of leaf anatomy.

Occasionally, the distribution of mechanical tissues has been found quite useful. Ayensu (1974) has shown that *Vellozia* and *Barbacenia* of Velloziaceae can be distinguished from each other by the form of sclerenchyma in the leaves.

A considerable amount of studies on leaf anatomy in relation to taxonomy have been made by Wunderlich (1950), Hagerup (1953), Fahn (1954), Tomlinson (1959), Cutler (1965) and Metcalf (1968).

Nodal anatomy

Distribution of different types of nodes, such as unilacunar, trilacunar, multilacunar and unilacunar two traces and number of gaps have been found characteristic to a single family or sometimes even for one order. Such characters, therefore, may be valuable. For example, unilacunar node is found in the entire order of Centrospermales. However, a comparative study of nodal anatomy may also be valuable for showing relationships of genera or even species.

Bailey and Howard (1941) proposed that on the basis of nodal structure, the subfamily Icacinoideae of family Icacinaceae may be divided into two different sections, one section characterised by having trilacunar nodes and other by having unilacunar nodes.

Stem anatomy

In many cases, the primary internal structures of stem has been found very useful in taxonomic studies. For this purpose characters like distribution and patterns of thickening of collenchyma, transformation in cortex, distribution of fibres, variations in endodermis, characteristics of pith, distribution of vascular bundles etc. have been considered. Some examples are considered here. Variation in endodermal cells may provide systematic criterion in families like Piperaceae, Compositae and Lamiaceae. Carlquist (1961) proposed that species of *Dubantia* and *Fitchia* may be distinguished on the basis of anatomical differences in pith.

Vascular organisation in the stem has been found taxonomically useful in several cases. For example, the presence of bicollateral vascular bundles arranged in two rings forms a characteristic feature of the stem of Cucurbitaceae. Similarly, the presence of accessory cortical or medullary bundles are found restricted to some families like Nyctaginaceae, Amarantaceae, and Chenopodiaceae and these prove valuable for identification purposes. Ayensu (1970) proposed that two morphologically similar species of *Dioscorea*, *D. cayenensis* and *D. rotundata* may be distinguished on the basis of arrangement of vascular bundles in the stem. Similarly, different species of *Potamogeton* may be distinguished on the basis of vascular bundles.

Sclereids

Variation in sclereids in structure and distribution have been found useful for some taxonomic studies. These have been studied from this point of view among families like Connaraceae, Limoniaceae, Nymphaeaceae, Oleaceae, Theaceae etc. and among genera *Camellia*, *Capparis*, *Jasminum*, *Rhizophora*, *Cyrtandra* etc.

Cambium

The nature of cambium such as storied or non-storied and its composition like homogenous or heterogenous ray initials have been found useful in phylogenetic and taxonomic studies. Ghouse and Yunus (1974) and Ghouse and Hasmi (1977) have used characteristic of development of ray initials, their frequency, ability to organise into different fusiform ray initial units etc. for taxonomical application. Ghouse and Yunus (1974) distinguished the species of *Dalbergia* of African origin from those of Indian species, on the basis of cambial structure. The African species have uniseriate ray initials while Indian species have multiseriate ray initial units.

Wood anatomy

Wood anatomy has been found invaluable in solving some taxonomic problems. Carlquist (1961) stated that the well established trends of xylem evolution should be taken into account by any systematist interested in formulating relationships or evolutionary status of angiosperm taxa. The various features which have been used are as follows:

(i) **Vessels**–Features like distribution, pattern, diameter and frequency, perforation and thickening and presence of tyloses have been found useful in taxonomic and phylogenetic studies. Vessels found solitary are considered as primitive than those aggregated in groups, such as pore multiples, pore clusters and pore chains. Similarly, non-porous wood is more primitive in which vessels are not found (as in gymnosperms) than porous wood (as in angiosperms) and diffuse porous wood is more primitive than ring porous wood.

(ii) **Wood parenchyma**–Distribution of wood parenchyma has been found very useful in diagnostic studies. It may be of two types–

(A) **Apotracheal type** in which case the parenchyma are distributed without any specific relation to vessels. These are not in contact with vessels. These may be of three different types, (a) **diffuse apotracheal type** (parenchyma cells occur singly among the fibres and tracheids), (b) **banded apotracheal type** (parenchyma cells occur in bands) and (c) **terminal apotrachcal type** (bands of parenchyma cells are confined to the ends of growth rings, as in *Michelia* and *Acer*). Diffuse types are found in *Adina cordifolia* and *Dillenis* while banded type in *Pterygota* and *Lophopetalum*.

(B) **Paratracheal type** in which case parenchyma cells are distributed in close association with vessels. It may be of following types–(a) **abaxial type** (parenchyma cells found in contact with abaxial surfaces of vessels, the surface away from the centre of vessels) and (b) **vasicentric types** (parenchyma cells completely surround the vessel). If the vasicentric type of parenchyma forms the wing-like appendages, it is called *aliform vasicentric type* (as in some leguminous plants) and if it forms a ring-like extension around many vessels which coalesce and forms irregular bands, it is called *confluent aliform vasicentric type*. Simple vasicentric types forming a fairly uniform sheath around vessels are found in *Acacia catechu* while aliform type in *Albizzia lebbek* and confluent type in *Ongelnia oojeinensis*.

(iii) **Rays**–Features like ray abundance, ray dimensions, width and cellular composition, degree of wall thickness of ray cells, pitting in ray parenchyma

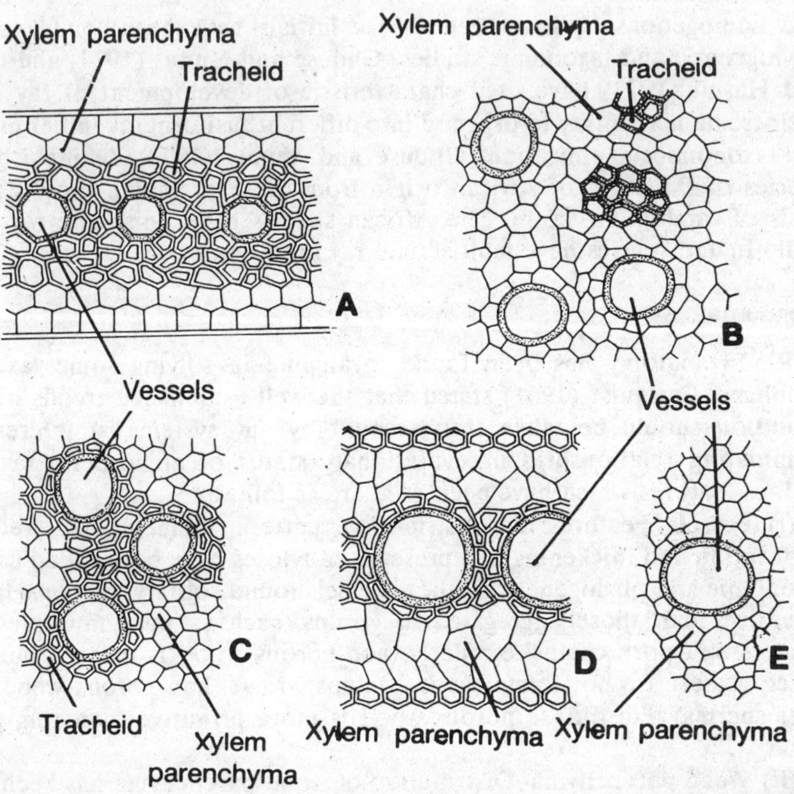

Fig. 17.1. Different types of xylem parenchyma. A–terminal apotracheal type, B–paratracheal-Vasicentric confluent aliform type, C–diffuse apotracheal type, D–banded apotracheal type, and E–paratracheal type.

and homogenous and heterogenous nature have been variously used for taxonomic and phylogenetic studies. It is believed that heterogenous rays are primitive than homogenous rays.

(iv) Storied structure–The term refers to the arrangement of cells or tissues in horizontal series as seen in tangential section. Storied structures are considered advanced over the ones with non-stratified cells.

Thus, wood anatomy has been found quite useful in taxonomic studies. It is further supported by the following key which can be used for the identification of common Indian woods. It has been prepared by the hard works of Indian plant anatomists.

KEY TO IDENTIFICATION OF SOME COMMON WOODS

I. *Wood non-porous or without vessels*

1. Resin canals absent	Fir (*Abies pindrow*) (wood white, odourless, indistinct sap wood and heart wood)

2. Resin canals present
   (i) Resin canals in long tangential bands — Deodar (*Cedrus deodara*) (wood yellowish, light, with strong pungent odour)
   (ii) Resin canals scattered
      (A) Wood white and odourless — Spruce (*Picea smithiana*)
      (B) Wood yellowish to pale reddish with strong resinous odour - Chir (*Pinus roxburghii*)

II. *Wood porous or with vessels:*

1. **Wood ring porous** to semi-ring porous.
   (i) Rays distinct into very broad and very fine types, pores in long radial, oblique or flame-like groups - Indian oak (*Quercus* spp.)
   (ii) Rays one type, not prominent, and pores not arranged in long radial, oblique or flame like groups.
      (A) Wood with ripple marks, water extract gives stong fluorescence - Bijasal (*Pterocarpus marsupium*)
      (B) Ripple marks absent, no fluorescence in water extract, wood smells like leather. - Teak (*Tectona grandis*)

2. **Wood diffuse porous**
   (i) Gum ducts vertical - Sal (*Shorea robusta*) (wood coarse textured, vessels large and heavily plugged with tyloses)
   (ii) Vertical gum ducts absent
      (A) Ripple marks present - Sissoo (*Dalbergia sissoo*) (wood golden brown, gummy deposits found in vessels)
      (B) Ripple marks absent
         (a) Pores in long radial chains or oblique groups - Ebony (*Dispyros melanoxylon*)

(b) Pores not in radial chains or oblique groups; pores large and visible to the naked eyes and parenchyma usually distinct to the eyes.

(i) Parenchyma vasicentric to aliform

* Wood light coloured, soft to moderately hard -Mango (*Mangifera indica*)
* Wood dark coloured and very hard -Laurel (*Terminalia tomentosa*)

(ii) Parenchyma paratracheal -Babul (*Acacia nilotica*)

(iii) Parenchyma apotracheal, diffuse and indistinct -Semul (*Bombax ceiba*)

Wood anatomy has also played an important role in solving certain taxonomic problems. A few of them are mentioned here.

1. Engler and Prantl (1889) proposed amentiferae as least specialised group but Hutchinson (1959), Cronquist (1968) and Takhtajan (1969) consider Mangoliales as most primitive group. The characteristic anatomical features of certain members of Magnoliales such as presence of some vesseless genera, some with long but narrower vessels, oblique end walls, wood diffuse porous type, pores distributed singly and heterogenous rays support Hutchinson's view. At the same time, Amentiferae have specialised wood. They have ring porous wood, simple perforation plates, short and round vessels with opposite or alternate pittings and homogenous rays.

2. Hutchinson (1959) placed Annonaceae in a separate order Annonales whereas Takhtajan (1969) placed it under order Magnoliales. Purkayastha (1980) studied wood anatomy of the family and found that Annonaceae have much advanced type of wood than Magnoliales and thus he supported Hutchinson in this respect.

3. The anatomical studies made on members of Parietales reveal that Parietales of Engler and Prantl (1889) is a heterogenous assemblage and the division of the order into Parietales and Guttiferales by Wettstein is justified.

4. Schumann (1897) divided Rubiaceae into two subfamilies, Cinchonoideae and Rubiodeae but Bremekamp (1966) divided it into eight subfamilies. Knock-Noorman (1969) studied the wood anatomy of Surinam species of the family and Purkayastha (1980) studied their Indian species and found that divisions proposed by Bremekamp (1966) are more in conformity with structure of wood.

5. Studies of Purkayastha (1980) on calcium oxalate crystals found in wood of family Sapotaceae reveal that the two genera of the family, *Manilkara* and *Mimusops* cannot be morphologically distinguished but they can be separated on the basis of chambered crystalliferous strands of calcium oxalate.

FLOWERS

Studies on floral anatomy have also proved useful in problems related to taxonomic and phylogenetic studies. A flower has typical vascular ground plan, i.e. sepals receive three distinct vascular traces, petals and stamens receive a single trace and carpels three traces each. During the course of evolution and specialization of flower, the vascular plan has undergone considerable modifications through reduction, amplification, cohesion and adnation of vascular bundles. Such modifications have been found useful to the systematics. A few examples are taken into account here.

The separation of *Paeonia* from family Ranunculaceae and placing it in a separate family of its own Paeoniaceae by modern systematics have been supported by anatomical features of flowers. Other members of Ranunculaceae receive three traces in each sepal and one trace in each petal but in case of *Paeonia*, they receive few to several traces. Stamen traces in other members of Ranunculaceae are derived independently but in *Paeonia*, stamens are supplied by few large trunk bundles. Similarly, each carpel in *Paeonia* receives several traces as against three traces in each carpel of other members of Ranunculaceae. These anatomical differences justify the separation of *Paeonia* from Ranunculaceae.

Some other good examples where floral anatomy has played important role in solving taxonomic problems may be the cases of *Trapa* and *Ruppia* and *Zannichelia*.

Engler and Diels (1938) included *Trapa* in the family Trapaceae after separating it from Onagraceae. This has been supported by vascular plan of flowers which differs in *Trapa* and other members of Onagraceae. Furthermore, the inferior ovary of *Trapa* has been considered to be receptacular due to down turning of the receptacular bundles, whereas in other genera of the family Onagraceae, the ovary has been regarded as appendicular in the fusion of the bundles present in the same radii.

Ruppia and *Zannichelia* were separated from the family Potamogetonaceae and were placed by Hutchinson (1959) under separate families, Ruppiaceae and Zannicheliaceae. The studies on floral vasculature were carried out by Singh (1965). He proposed that such a treatment by Hutchinson should not be supported. The difference between different floral characters of *Potamogeton*, *Ruppia* and *Zannichelia* show stages in reduction and these should be retained in the same family.

Role of floral anatomy in phylogenetic problems has also been found valuable and hence advocated, e.g. the fact that families like Menispermaceae, Calycanthaceae and Annonaceae have originated from Ranunculaceae is supported by the organisation of floral vasculature. Similarly, the general vascular plan of flower and the course of placental bundles particularly in Caryophyllaceae and the Polemoniaceae suggest that the latter have been derived from the Caryophyllaceous stock (Singh and coworkers, 1987). Studies on floral anatomy of family Solanaceae and Scrophulariaceae suggest that the two families show great affinity in the presence of anatomically parietal placentation and in the uniformity of vasculature of the gynoecium.

SEED ANATOMY

Seed is defined as a ripened ovule. The presence of seed is the distinguishing feature of phanerogams from rest of the plants. The several exomorphic characters of seeds have been used in identification and classification of angiosperms. However, in recent past, diagnostic anatomical features of seeds have also been looked upon by different workers such as Martin (1946), Misra (1970), Bouman (1974), Corner (1976), Singh and Dathan (1980) etc. Martin (1946) after studying the various features of the seed anatomy of more than twelve hundred seeds proposed a classification of seeds on the basis of size, relation of embryo and endosperm, position of embryo etc. Corner (1976) classified bitegmic dicotyledonous seeds on the basis of place of differentiation of the mechanical layer and recoginsed two broad classes of seeds, the *testal* and the *tegmic* types. In the testal type, the mechanical layer differentiates in the outer integument while in case of tegmic type, it differentiates in the inner integument. Each type has been further divided into three subtypes being referred to as exo, meso- and endo-testal or tegmic type.

The application of seed anatomy in solving taxonomic and phylogenetic problems is still in beginning stages. Such studies can also prove helpful in crime detection, particularly related with adulteration of commercial and agricultural crops.

GENERAL QUESTIONS

1. Write a brief account on systematic plant anatomy.
2. Discuss the role of wood anatomy in relation to taxonomy.
3. "Wood anatomy has been found invaluable in solving some of taxonomic problems". Discuss.
4. Write short notes on (a) Floral anatomy and taxonomy. (b) Seed anatomy and taxonomy.
5. Discuss how anatomical studies have proved useful in solving phylogenetic problems of angiosperms?

Selected Readings

Addicot, F.T. and R.S. Lynch 1955. Physiology of abscission. *Ann. Rev. Pl. Physiol.* **6**: 211-238.

Ahmad, Z., M.I.H. Khan and H. Begum 1969. Periderm in some perennial plants. In K.A. Chowdhury (Ed.) *Recent Advances in the Anatomy of Tropical Seed Plants.* pp. 75-83. Hindustan Publishing Corp., Delhi.

Airy Shaw, H.K. 1973. *A Dictionary of the Flowering Plants and Ferns.* University Press, Cambridge.

Bailey, I.W. 1923. The cambium and its derivative tissues. IV. The increase in girth of the cambium. *Amer. J. Bot.* **10** : 499-509.

Bailey, I.W. 1944. The development of vessels in angiosperms and its significance in morphological research. *Amer. J. Bot.* **31** : 421-428.

Bailey, I.W. 1953. Evolution of tracheary tissue of land plants. *Amer. J. Bot.* **40** : 4-8.

Bailey, I.W. 1956. Nodal anatomy in retrospect. *J. Arnold Arbor.* **37** : 269-287.

Balfour, E. 1965. Anomalous secondary thickening in Chenopodiaceae, Nyctaginaceae and Amaranthaceae. *Phytomorphology* **15** : 111-122.

Barnard, C. 1957. Floral histogenesis in the monocotyledons. I. The Gramineae. *Aust. J. Bot.* **5** : 1-20.

Behl, H.M. and B. Tiagi 1977. Developmental anatomy of seed and fruit in *Cicer arietinum* Linn. *J. Indian bot. Soc.* **56** : 313-321.

Bir, B., T. Rajagopal and N. Ramayya 1971. Studies on the structural and developmental variation and distribution of stomata in the Rubiaceae. *Bot. J. Linn. Soc.* **64** : 295-310.

Boke, N.H. 1940. Histogenesis and morphology of the phyllode in certain species of *Acacia. Amer. J. Bot.* **27** : 73-90.

Brazier, J.D. and G.L. Franklin 1961. Identification of hardwoods–a microscopic key. *Forest Prad. Res. Bull.* **46.**

Brown, W.H. 1935. *The Plant Kingdom.* Ginn & Co., Boston.

Brown, W.V. 1958. Leaf anatomy in grass systematics. *Bot. Gaz.* **119** : 170-178.

Brown, W.V. 1959. Grass leaf anatomy–its use in systematics. In D.L. Bailey (Ed.) *Recent Advances in Botany,* Vol. I, pp. 105-108, Univ. of Toronto Press, Toronto.

Buvat, R. 1952. Structure, evolution et fonctionnement du méristème apical du quelques dicotyledones. *Annls. Sci. nat. (Bot.) Ser. II.* **13** : 199-300

Buvat, R. 1955. Le méristème apical de la tige. *Année Biol.* **31** : 596-656.

Calvin, M. 1959. From microstructure to macrostructure and function in the photochemical apparatus. In *The Photochemical Apparatus–Its Structure and Function. Brookhaven Symp. Biol.* **11** : 160.

Carlquist, S. 1958. Structure and ontogeny of glandular trichomes of Madinae

(Compositae). *Amer. J. Bot.* **45** : 675-682.

Carlquist, S. 1961. *Comparative Plant Anatomy.* Holt, Rinehart and Winston, New York.

Clowes. F.A.L. 1961a. *Apical Meristems.* Blackwell, Oxford.

Clowes, F.A.L. 1961b. Cellular organization in root apical meristems. In D.L. Bailey (Ed) *Recent Advances in Botany.* Vol. I. pp. 791-794. University of Toronto Press, Toronto.

Clowes, F.A.L. 1965. Meristems and the effect of radiation on cells. *Endeavour* **24** : 8-12.

Corner, E.J.H. 1976. *The Seeds of Dicotyledons.* University Press, Cambridge.

Cronquist, A. 1968. *The Evolution and Classification of Flowering Plants.* Houghton Miffun, Boston.

Cronshaw, J. 1971. Support and protection in plants. In V.G. Dethier *et.al.* (Eds.) *Topics in the Study of Life.* Pp. 127-138. Harper & Row, Publishers, New York.

Cronshaw, J. 1974. Phloem differentiation and development. In A.E. Robards (Ed.) *Dynamic Aspects of Plant Ultra-structure.* pp. 391-413. McGraw-Hill Book Co., London.

Cutter, E.G. 1959. On a theory of phyllotaxis and histogenesis. *Biol. Rev.* **34** : 243-263.

Cutter, E.G. 1965. Recent experimental studies of the shoot apex and shoot morphogenesis. *Bot. Rev.* **31** : 7-113.

Cutter, E.G. 1971, 1978. *Plant Anatomy: Experiment and Interpretation.* Edward Arnold (Pubs.) Ltd., London.

Dycus, A.M. and L. Kundson 1957. The role of the velamen of the aerial roots of Orchids. *Bot. Gaz.* **119** : 78-87.

Eames, A.J. 1931. The vascular anatomy of the flower with refutation of the theory of carpel polymorphism. *Amer. J. Bot.* **18** : 147-188.

Eames, A.J. 1953. Floral anatomy as an aid in generic limitation. *Chronica Bot.* **14** : 126-132.

Eames, A.J. 1961. *Morphology of the Angiosperms.* McGraw-Hill Book Co., Inc., New York.

Eames, A.J. and L.H. Mac Daniels 1947. *An Introduction to Plant Anatomy.* McGraw-Hill Book Co., Inc., New York.

Esau, K. 1953, 1965. *Plant Anatomy.* John Wiley & Sons. Inc., New York.

Esau, K. 1977. *Anatomy of Seed Plants.* John Wiley & Sons, Inc., New York.

Esau, K. and V. I. Cheadle 1969. Secondary growth in *Bougainvillea. Ann. Bot. (N.S.)* **33** : 807-819.

Esau, K., V.I. Cheadle and E.M. Gifford Jr. 1953. Comparative structure and possible trends of specialization of the phloem. *Amer. J. Bot.* **40** : 9-18.

Evert, R.F., J.D. Davis, C.M. Tucker and F.J. Alfieri 1970. On the occurrence of nuclei in mature sieve elements. *Planta* **95** : 281-296.

Fahn, A. 1953. The topography of the nectary in the flower and its phylogenetic trend. *Phytomorphology* **3** : 424-426.

Fahn, A. 1967. *Plant Anatomy.* Pergamon Press, Oxford.

Fahn, A. 1975. Secretory cells. In H.Y. Mohan Ram *et. al.* (Eds) *Form, Structure and Function in Plants.* pp. 313-316. Sarita Prakashan, Meerut.

Fahn, A. 1979. *Secretory Tissues in Plants.* Academic Press, London.

Foster, A.S. 1936. Leaf differentiation in angiosperms. *Bot. Rev.* **2** : 349-372.

Foster, A.S. 1938. Structure and growth of the shoot apex in *Ginkgo biloba. Bull. Torrey bot. Club* **65** : 531-556.

Foster, A.S. 1939. Problems of structure, growth and evolution in the shoot apex of seed plants. *Bot. Rev.* **5** : 454-470.

Fuller, H.J. and O. Tippo 1954. *College Botany*. Holt, Rinehart and Winston, Inc., New York.

Gawadi, A.G. and G.S. Avery, Jr. 1950. Leaf abscission and the so-called 'Abscission layer'. *Amer. J. Bot.* **37** : 172-180.

Gifford, E.M., Jr. and G.E. Corson, Jr. 1971. The shoot apex in seed plants. *Bot. Rev.* **37** : 143-229.

Gunning, B.E.S. and M.W. Steer 1975. *Plant Cell Biology–An Ultrastructural Approach.* Edward Arnold (Publs.) Ltd., London.

Haberlandt, G. 1914. *Physiological Plant Anatomy*. Macmillan & Co., London.

Hall, M.A. 1976. *Plant Structure, Function and Adaptation.* The Macmillan Press Ltd., London.

Hatch, M.D. and C.R. Slack 1966. Photosynthesis by sugarcane leaves. A new carboxylation reaction and the pathway of sugar formation. *Biochem. J.* **101**: 103-111.

Hickey, L.J. 1973. Classification of the architecture of dicotyledonous leaves. *Amer. J. Bot.* **60** : 17-33.

Hickey, L.J. and J.A. Wolfe 1975. The basis of angiosperm phylogeny: Vegetative morphology. *Ann. Miss. bot. Gdn.* **62** : 538-589.

Holman, R.M. and W.W. Robbins 1934. *A Textbook of General Botany*. John Wiley & Sons, Inc., New York.

Hooke, R. 1665. Of the shematisme or texture of cork, and of the cells and pores of some other such forthy bodies. *Micrographia,* London.

Jane, F.W. 1962. *The Structure of Wood.* A. & C. Black, London.

Kribs, D.A. 1937. Salient lines of structural specialization in the wood parenchyma of dicotyledons. *Bull. Theory bot. Club* **64** : 177-188.

Krishnamurthy, K.H. and B. Kannabiran 1970. Histomorphology of foliar epidermis and pharmacognosy in Asclepiadaceae. *J. Indian bot. Soc.* **49** : 105-113.

Lang, W.H. 1930. *Strasburger's Text-book of Botany.* 6th Ed. Macmillan & Co. Ltd., London.

Maheshwari, P. 1930. Contributions to the morphology of *Boerhaavia diffusa. J. Indian bot. Soc.* **9** : 42-61.

Martin, A.C. 1946. The comparative internal morphology of seeds. *Am. Midl. Nat.* **36** : 513-660.

Meidner, H. and T.A. Mansfield 1968. *Physiology of Stomata.* McGraw-Hill Book Co., Inc., New York.

Melville, R. 1962. A new theory of the angiosperm flower I. The gynoecium. *Kew Bull.* **16** : 1-50.

Melville, R. 1963. A new theory of the angiosperm flower. II. The androecium. *Kew Bull.* **17** : 1-66.

Metcalfe, C.R. 1959. A vista in plant anatomy. In W.B. Turrill (Ed.) *Vistas in Botany.* pp. 76-99. Pergamon Press, London.

Metcalfe, C.R. 1963. Comparative anatomy as a modern botanical discipline. In R.D. Preston (Ed.) *Advances in Botanical Research.* Vol. I. pp. 101-147. Academic Press, London.

Metcalfe, C.R. 1968. Current development in systematic plant anatomy. In V.H. Heywood (Ed.) *Modern Methods in Plant Taxonomy.* pp. 45-57, Academic Press, London.

Metcalfe, C.R. and L. Chalk 1950. *Anatomy of the Dicotyledons.* 2 Vols. Clarendon Press, Oxford.

Mühlethaler, K. 1961. Plant cell walls. In J. Brachet & A.E. Mirsky (Eds) *The Cell.* Vol. II, pp. 85-134. Academic Press, New York.

Paliwal, G.S. 1972. Stomata. In T.M. Varghese & R.K. Grover (Eds.) *Vistas in Plant*

Sciences. Vol. 2, pp. 23-48. Searchmates Publications, Sangrur.

Pant, D.D. 1965. On the ontogeny of stomata and other homologous structures. *Plant Sci. Ser. (Allahabad)* **1** : 1-24.

Pant, D.D. and B. Mehra 1964. Nodal anatomy in retrospect. *Phytomorphology* **14** : 384-387.

Park, R.B. and J. Biggins 1964. Quantasomes: size and composition. *Science* **114** : 1009-1011.

Parthasarathy, M.V. 1975. Sieve-element structure. In M.H. Zimmermann and J.A. Milburn (Eds.) *Transport in Plants. I. Phloem Transport.* pp. 3-38. Springer-Verlag, Berlin.

Pate, J.S. and B.E.S. Gunning 1972. Transfer cells. *Ann. Rev. Pl. Physiol.* **23** : 173-196.

Philipson, W.R., J.M. Ward and B.G. Butterfield 1971. *The Vascular Cambium : Its Development and Activity.* Chapman and Hall, London.

Puri, V. 1952, Floral anatomy in relation to taxonomy. *Agra Univ. J. Res.* (*Sci*) **1**: 15-35.

Puri, V. 1958. Floral anatomy and taxonomy. *Indian bot. Soc. Mem.* **1** : 15-18.

Puri, V. 1962. Floral anatomy in relation to taxonomy. *Bull. bot. Surv. India* **4**: 161-165.

Purkayastha, S.K. 1980. Wood anatomy in relation to taxonomy. In P.K.K. Nair (Ed.) *Glimpses in Plant Research.* Vol. V. *Modern Trends in Plant Taxonomy.* Pp. 192-198. Vikas Publishing House Pvt. Ltd., New Delhi.

Ray, P.M. 1963. *The Living Plant.* Holt, Rinehart & Winston, Inc., U.S.A.

Sattler, R. 1965. Flower. In *McGraw-Hill Yearbook Science and Technology.* McGraw Hill Book Co., New York.

Sattler, R. 1973. *Organogenesis of Flowers. A Photographic Text-Atlas.* Univ. of Toronto Press, Toronto.

Scott, D.H. and G. Brebner 1894. On the secondary tissues in certain monocotyledons. *Ann. Bot.* **7** : 21-62.

Singh, B. 1943. The origin and distribution of inter- and intra-xylary phloem in *Leptadenia. Proc. Indian Acad. Sci., Sect. B* **18** : 14-19.

Stant, M.Y. 1952. The shoot apex of some monocotyledons. I. *Ann. Bot.* (*N.S.*) **16** : 115-128.

Stebbins, G.L. and G.S. Khush 1961. Variation in the organization of the stomatal complex in the leaf epidermis of Monocotyledons and its bearing on their phylogeny. *Amer. J. Bot.* **48** : 51-59.

Takhtajan, A. 1964. *Foundations of the Evolutionary Morphology of Angiosperms.* (in Russian). Moscow.

Takhtajan, A. 1969. *Flowering Plants–Origin and Dispersal.* Oliver & Boyd Ltd., Edinburgh.

Tepfer, S.S. 1953. Floral anatomy and ontogeny in *Aquilegia formosa* var. *truncata* and *Ranunculus repens. Univ. Calif. Publ. Bot.* **25** : 513-648.

Thoday, D. 1939. The interpretation of plant structure. *Adv. Sci.* **1** : 84-104.

Tippo, O. 1946. The role of wood anatomy in phylogeny. *Am. Midl. Nat.* **36** : 362-372.

Tomlinson, P.B. 1959. Structure and distribution of sclereids in the leaves of palms. *New Phytol.* **58**: 253-266.

Tomlinson, P.B. 1961. *Anatomy of Monocotyledons.* **II.** *Palmae.* Clarendon Press, Oxford.

Torrey, J.G. 1963. Cellular patterns in developing roots. *Symp. Soc. Exp. Biol.* **17**: 285-314.

Torrey, J.G. 1968. *Development in Flowering Plants.* 3rd ed. Macmillan & Co., New York.

Troughton, J. and L.A. Donaldson 1972. *Probing Plant Structure.* Chapman & Hall, London.

Van Cotthemn, W.R.J. 1970. A classification of stomatal types. *Bot. J. Linn. Soc.* **63** : 235-246.

Van Fleet, D.S. 1950. A comparison of histochemical and anatomical characteristics of the hypodermis with the endodermis in vascular plants. *Amer. J. Bot.* **37** : 721-725.

Wardlaw, C.W. 1957a. The reactivity of the apical meristem as ascertained by cytological and other techniques. *New Phytol.* **56** : 221-229.

Wardlaw, C.W. 1957b. On the organization and reactivity of the shoot apex in vascular plants. *Amer. J. Bot.* **44** : 176-185.

Wardlaw, C.W. 1965. *Organization and Evolution in Plants.* Longmans Green & Co., London.

Wardlaw, C.W. 1968. *Morphogenesis in Plants.* Methuen & Co., London.

Wareing, P.F. 1951. Growth studies in Woody species. IV. The initiation of cambial activity in ring-porous species. *Physiol. Plant.* **4** : 546-562.

Warming, E. 1909. *Ecology of Plants. An Intoduction to the Study of Plant Communities.* Clarendon Press, Oxford.

PART TWO

EMBRYOLOGY AND PALYNOLOGY

Reproductive Organs

Pollination and Fertilization

Embryogenesis

Apomixis and Polyembryony

Embryology in relation to Taxonomy

Experimental Embryology

Palynology

CHAPTER 1

Introduction

Goethe (1790) proposed that flower is a modified shoot or a flower is homologous to the vegetative shoot and the lateral appendages of flower, such as sepals, petals, stamens and carpels are morphologically equivalent to the foliage leaves. Calyx and corolla are sterile appendages whereas androecium and gynoecium are fertile appendages.

The different parts of flower are as follows-

(i) **Floral axis.** The flower is borne on an axis which is differentiated into the *pedicel* and the *thalamus*. The pedicel may be elongated or reduced and expands above into a receptacle or thalamus. On the thalamus are borne four distinct whorls arranged one above the other. The lower two whorls are known as the *non-essential* or *accessory whorls*, whereas, the upper two are known as *essential* or *reproductive whorls*.

(ii) **Calyx.** It is the lower most whorl or the first accessory whorl which consists of a number of leafy structures, the *sepals*. The sepals may be fused (gamosepalous) or free (polysepalous). These are usually green but may be variously coloured or petaloid as in *Salvia*. Their main function is to protect the floral buds, but when they are petaloid these attract insects and help in pollination. In some cases, these develop spurs (as in *Delphinium*) and store nectar, or help in fruit dispersal (as in several members of Asteraceae).

(iii) **Corolla.** The second whorl lying above the sepals is known as corolla. It consists of brightly coloured foliaceous units called *petals*. Their main function is to attract insects or birds for pollination. However, in bud condition, the petals enclose the immature reproductive units and provide protection from environ-

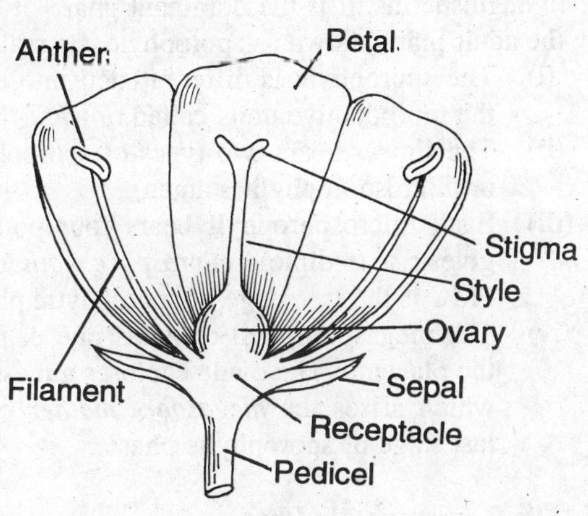

Fig. 1.1. Different parts of a flower.

mental hazards. Petals may be free (polypetalous) or fused (gamopetalous). In Monochlamydeae group, the sepals and petals are indistinguishable and are collectively called *perianth* and their individual units are called *tepals*.

(iv) **Androecium**. It is the first reproductive whorl lying adjacent to the corolla. Androecium comprises the male reproductive units called *stamens* or *microsporophylls*. Each stamen consists of a short or elongated stalk the *filament* which terminates above into an expanded lobed head, the *anther*. The anther may consist of two lobes (dithecous or bicelled) or a single lobe (monothecous or unicelled). In dithecous anthers, the two lobes are connected by a ring of tissue known as *connective*. A typical (dithecous) anther encloses four microsporangia (two in each lobe), each gives rise to a large number of pollen grains or microspores.

(v) **Gynoecium.** It is the second and innermost essential or reproductive whorl, occupying the central position. Gynoecium represents the female reproductive organ and is made up of one or more functional units called *carpels* or *megasporophylls*. A single carpel (also called pistil) consists of a swollen base known as *ovary*, which elongates above into a filliform *style* and terminates into a capitate, lobed or flattened head, the *stigma*. The ovary is the most important part and bears one or numerous, small rounded bodies, the *ovules*. Within each ovule is present a single oval sac called *embryo sac* (female gametophyte) which contains the fertile *egg* (female gamete).

LIFE CYCLE OF AN ANGIOSPERM

The life cycle of an angiosperm shows a distinct alternation of generation, following one after the other. The two phases are, the *sporophytic phase* (asexual or diploid phase) and the *gametophytic phase* (sexual or haploid phase). The salient features of life cycle are as follows:—

I. The Sporophytic Phase

In angiosperms, it is the dominant phase of longer duration and is represented by the adult plant known as sporophyte. Its main features are—

(i) The sporophyte is differentiated into root, stem and leaves and it bears the reproductive units called *flowers*.

(ii) The flower bears two types of sporophylls, the megasporophyll (carpel) or microsporophyll (stamen).

(iii) Each microsporophyll bears four pollen sacs (microsporangia), which gives rise to diploid *microspore mother cell* or pollen mother cell (PMC). PMC is the last stage of sporophytic phase.

(iv) The megasporophyll contains one or more ovules attached variously to the placenta. The ovule encloses a mass of tissue known as *nucellus* from which arises the *megaspore mother cell* (MMC). MMC represents the last stage of sporophytic phase.

II. The Gametophytic Phase

The gametophytic phase is the nutritionally dependent phase of short duration.

It is represented by male or *microgametophyte* and the female or *megagametophyte*. Its important features are—

(i) Each microspore mother cell undergoes meiotic division to produce a large number of microspores called pollen grains.

(ii) The pollen grains germinate to form the male or microgametophytes. Microgametophyte is extremely reduced and consists of a pollen tube containing the tube nucleus and two male nuclei. The two, non-motile male nuclei represent the male gametes and are the last structures of the gametophytic phase.

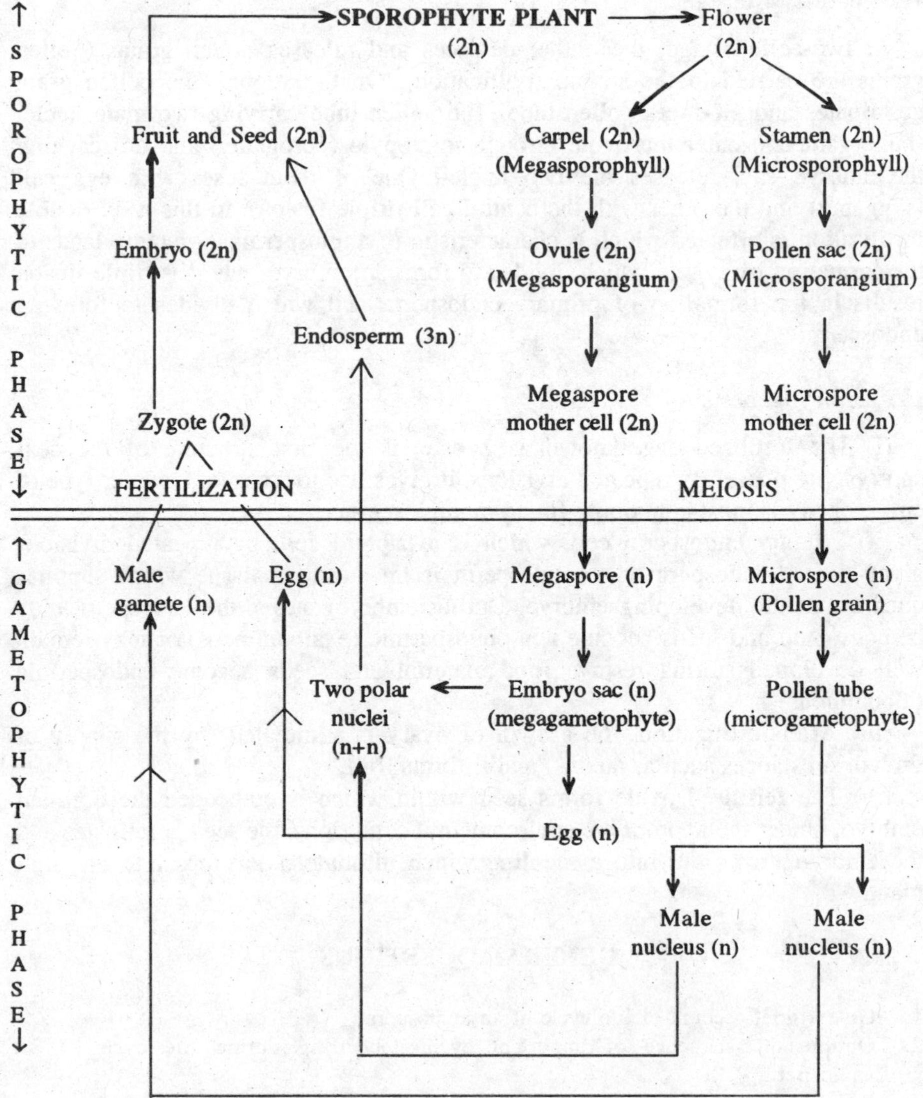

Fig. 1.2. Schematic representation of the life-cycle of an angiosperm.

(iii) The megaspore mother cell after meiotic division forms four megaspores. In most angiosperms, polygonum type of embryo sac is found which is monosporic type, i.e. it develops from a single megaspore and the remaining three megaspores lying towards micropyle are degenerated.

(iv) A mature embryo sac is seven celled and eight nucleate, i.e. three antipodal cells, one egg cell, two synergid cells and a central cell. Central cell is formed by fusion of two polar nuclei. The egg cell represents the female gamete and with its formation, the gametophytic phase ends.

III. Fertilization

At two-celled stage, the anther dehisces and releases pollen grains. Pollen grains are carried to the stigma (pollination). On the stigma, the pollen grain germinates and produces pollen tube. The pollen tube carrying two male nuclei (male gametes) enter the ovule through micropyle (porogamy) and on reaching the embryo sac, releases the two nuclei. One of them fuses with egg cell (syngamy) and the other with the central cell (triple fusion). In this way, double fertilisation is affected which is characteristic of angiosperms. Syngamy leads to the formation of zygote which divides to form an embryo, whereas triple fusion results in the formation of primary endosperm cell which divides to form an endosperm.

IV. Embryogenesis

(i) The fertilised egg, known as zygote, is the first structure of the next sporophytic phase. By repeated divisions, it gives rise to embryo. Crucifer type of embryogeny is most commonly found in angiosperms.

(ii) Primary endosperm cell, which is a triploid cell, by repeated divisions gives rise to endosperm. The endosperm is the nutritive tissue which supplies nutrition to the developing embryo. During embryogenesis, the endosperm may be consumed and seeds become non-endospermic (exalbuminous) or may remain well developed storing reserve food material and seeds become endospermic (albuminous).

(iii) After fertilization, the growth of ovary is stimulated (by the supply of growth substances such as auxins) and it forms fruit.

(iv) The fertilised ovule forms seed within which is embedded the dormant embryo. Under the appropriate environmental conditions, the seed germinates and the embryo grows out into a seedling which ultimately develops into an adult plant.

GENERAL QUESTIONS

1. Give a brief account of life cycle of an angiosperm.
2. Outline the sequence of important events occurring during life cycle of an angiosperm.

CHAPTER 2

Microsporophyll

The stamen (or microsporophyll) is the male reproductive organ of the plant. A single stamen typically consists of an elongated stalk,—the *filament* and at the tip of which is present—the *anther* (or microsporangium). A typical anther is a bilobed structure, the lobes of which are connected by a structure called *connective.* The anther is usually bilobed (also called bicelled, dithecous or tetrasporangiate) structure, each lobe of which comprises two pollen sacs or microsporangia. In mature and dehiscing anthers, the two microsporangia of each lobe appear as a single mass due to breakdown of the partition tissue.

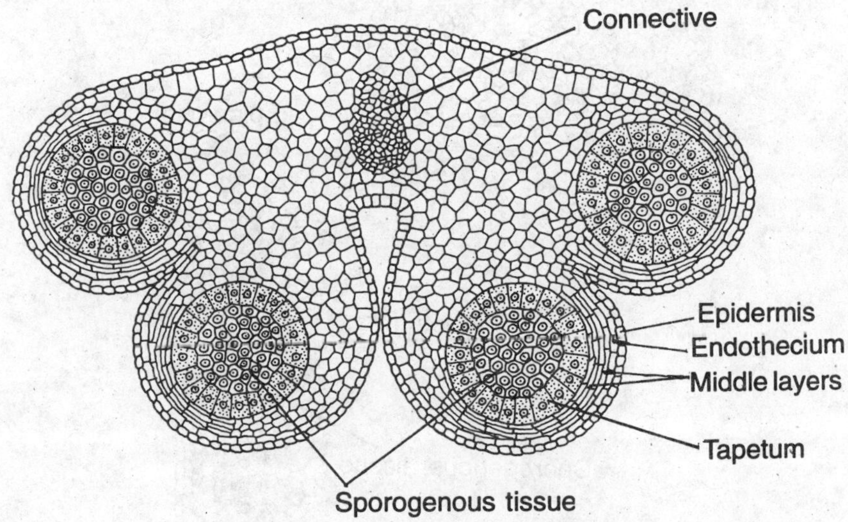

Fig. 2.1. Transection of a young anther.

In some plants like *Moringa, Wolffia* and members of family Malvaceae, the anthers are monothecous, i.e. they have a single lobe with two microsporangia. These may also be called unicelled or bisporangiate. In *Arceuthobium* (a parasitic mycelial dicot), there is a single microsporangium per anther.

Development of Microsporangium

The young developing anther consists of a mass of homogenous, meristematic cells bounded by a single layered epidermis. As the anther develops, it assumes a four lobed appearance, each lobe giving rise to a microsporangium. Development of microsporangium takes place as follows:

(i) In the first stage of development, hypodermal cells in each lobe assume large size, elongate radially and the nucleus becomes conspicuous. These highly prominent cells constitute the *archesporium* and show considerable variations.

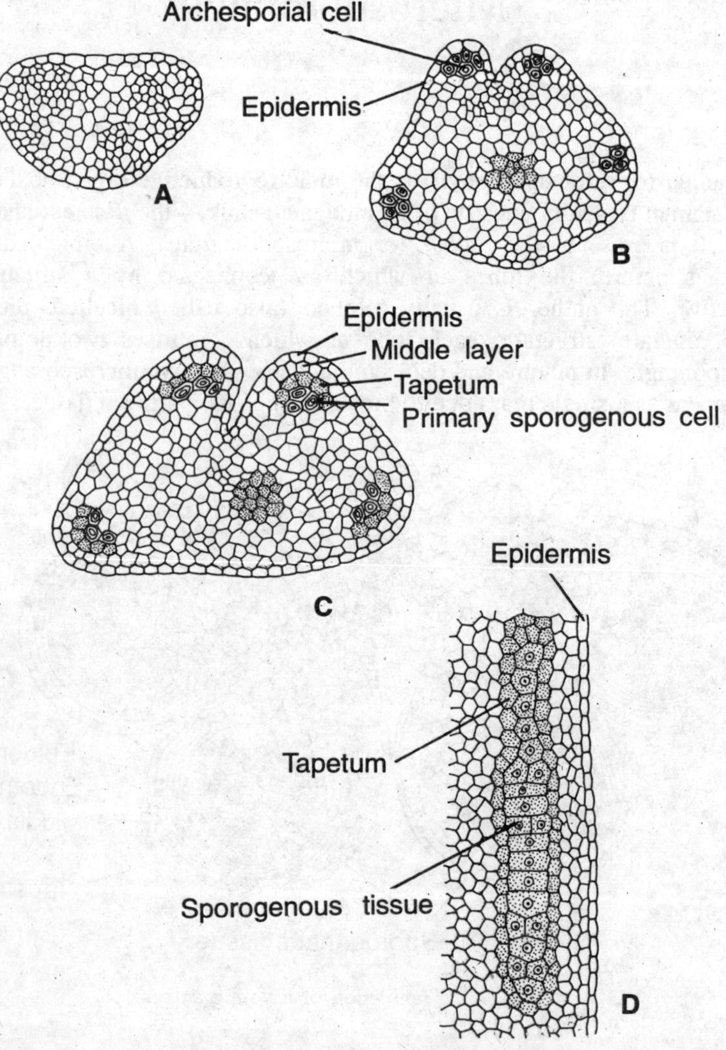

Fig 2.2. Development of microsporangium. A to C—different stages in the development of anther (shown in T.S.) and D—a longisection of anther showing the origin of inner portion of tapetum from the cells adjacent to sporogenous tissue.

In a cross section, a single archesporial cell may be seen, as in *Boerhaavia* and *Dionaea,* or a plate of cells as in *Urginea* and *Ophiopogon.* In the longitudinal plane, the row of archesporial cells may be one cell thick as in *Enalus,* or two cell thick as in *Boerhaavia* or several celled in thickness as in *Urginea.*

(ii) The archesporial cells divide by a periclinal division, cutting off an outer *primary parietal layer* and an inner *primary sporogenous layer.*

(iii) The cells of primary parietal layer divide by a number of anticlinal and periclinal divisions to form 2 to 5, concentric-layered *anther wall.*

(iv) The primary sporogenous cells may function directly as the spore mother cells or may undergo one or two mitotic divisions to form a large number of spore mother cells (also called Pollen mother cells or PMC).

The ontogeny of the microsporangium may be explained as follows:

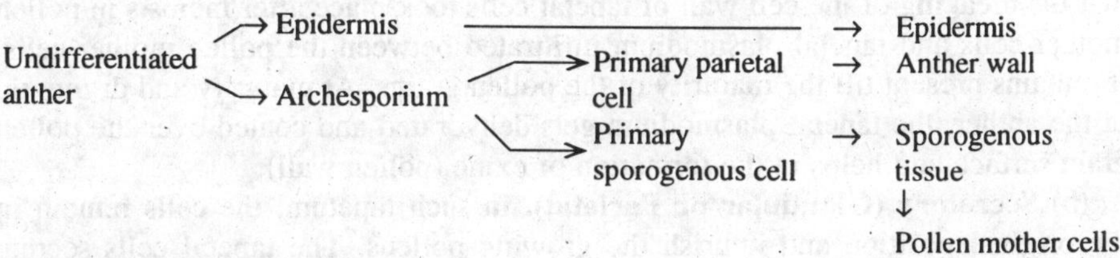

Structure of anther wall

The anther wall is derived from the parietal layer and consists of—

(i) an epidermis,
(ii) endothecium,
(iii) middle layers, and
(iv) tapetum.

(i) **Epidermis.** It is the outermost layer, made up of flattened and greatly stretched cells. In *Arceuthobium*, the epidermal cells develop fibrous thickenings and is known as *exothecium*. The epidermis provides protection to the developing anther.

(ii) **Endothecium.** The single layered endothecium consists of radially elongated cells, which develop fibrous bands, arising from the inner tangential wall and terminating near the outer wall of each cell. The endothecium is not a continuous layer all around microsporangium, but its cells remain thin-walled and form a strip of thin walled cells along the line of dehiscence, called *stomium* (fig. 2.5) in each anther lobe. The fibrous bands, the differential expansion of the outer and inner walls and the hygroscopic nature of the endothecial cells, all combine to bring about dehiscene of the anther. Spores are released through stomium.

(iii) **Middle layers.** Middle layers consist of two or three layers of flattened cells, which are of short duration. In some plants, the cells are persistent and function as storage centre for starch, which is used during pollen development.

(iv) **Tapetum.** It is the inner most layer composed of a single layer of cells with dense cytoplasm and prominent nuclei. In most angiosperms, the tapetum is of dual origin, since the outer part is formed by the parietal layer and the inner portion is derived from the connective tissue. In some plants like *Alectra* *thomsoni,* **dimorphic tapetum** is found, i.e. tapetal cells of different origin with different morophology.

On the basis of behaviour, two types of tapetum are known.

(a) **Amoeboid (Invasive or Periplasmodial).** The cell wall of such tapetal cells ruptures and the protoplast moves out into the sporangial mass. The protoplasts of tapetal cells intrude between the pollen mother cells and developing pollen grains. After intrusion, the tapetal protoplasts fuse with each other and form a mass of tapetal periplasmodium. Eames (1961) suggested that this kind of tapetum is of primitive type and found mostly in lower monocots and in some lower dicots.

Mepham and Lane (1969) studied the periplasmodial tapetum in *Tradescantia* and found it to be a highly organised and functional structure. They discovered that the breaking of the cell wall of tapetal cells took place after meiosis in pollen mother cells and tapetal plasmodium infiltrated between the pollen mother cells. It remains present till the maturity of the pollen grains. At maturity and drying up of the anther, the tapetal plasmodium gets dehydrated and coated over the pollen grain surface and helps in the formation of exine (pollen wall).

(b) **Secretory (Glandular or Parietal).** In such tapetum, the cells remain in the original position and nourish the growing pollens. The tapetal cells secrete nourishing substances into the microsporangium (anther sac) throughout the pollen development and degenerate only after pollen maturity. This type of tapetum is commonly found in angiosperms.

An interesting feature of secretory type of tapetum is the presence of a number of spherical bodies, the *pro-Ubisch bodies*, along with mitochondria, plastids and dictyosomes. The pro-Ubisch bodies are of lipid nature and make their appearance during sporogenesis. As cell division proceeds, the number of pro-Ubisch bodies increases and a marked shifting of position towards the region nearest to the anther cavity, is seen. At the tetrad stage of spores, these increase further in number and get surrounded by a layer of ribosomes. Eventually, the pro-Ubisch bodies are extruded from tapetal cells and pass through the tapetal cell membrane into the space between membrane and the spore cell wall. Over here, these are rapidly coated with sporopollenin and are known as *Ubisch bodies.* Ubisch bodies are spheroidal in shape. Functionally, these play a decisive role in organising exine pattern as clearly evidenced in case of *Mirabilis jalapa* and *Silene pendula*.

Tapetal membrane. In both types of tapetum, an additional membrane called tapetal membrane is formed. In the secretory type of tapetum, the Ubisch bodies are arranged on this membrane. The functional significance of this layer is not very clear.

Cell division in tapetal cells. A significant feature of tapetum is the nuclear behaviour. It may be of following types.

(a) *Endomitosis.* In this type of cell division, duplicaton of the chromosomes and the separation of the chromatids occur without disorganisation of nuclear membrane. Since spindle formation does not take place in this process, a large polyploid nucleus is formed, e.g. *Cucurbita pepo*.

(b) *Formation of restitution nuclei.* In this type of cell division, chromosome duplication and chromatid separation occur normally till the early anaphase stage, but later the two sets of chromosomes rejoin within a common nuclear membrane. The nucleus containing double set of chromosomes is known as restitution nucleus.

(c) *Multinucleate condition.* In some cases, e.g. *Mimusops elengi* and *Iodina rhombifolia,* during mitotic division, nuclear division is not followed by cell wall formation, hence tapetal cell becomes multinucleate.

(d) *Polyteny.* Polyteny is the condition in which the number of chromonemata in each chromosome increases. It has also been reported in certain tapetal cells.

Functions of tapetum. The different functions of the tapetum may be summarised as follows:

1. The primary use of tapetum is to provide nourishment to the developing pollen grains.

2. During the post-meiotic stage, the tapetum plays an important role in the formation of pollen wall.

3. The amoeboid type of tapetum is responsible for producing callase enzyme which causes degradation of callose wall of the microspore tetrads. Callase is essential for normal differentiation of spore tetrad.

4. The tapetum is responsible for transferring two important substances, (i) pollen kit and (ii) tryphine. Pollen kit is a mixture of carotenoids and hydrophobic liquids, whereas tryphine consists of hydrophilic substances and proteins.

5. The pollen wall contains specific proteins, partly contributed by the tapetum. When the pollen grains are moistened, these proteins are released and are the cause of hay fever and pollen allergy.

6. The proteins present in the pollen wall have great biological significance, i.e. in the recognition of compatible pistils. If the pollen falls on an incompatible stigma, these proteins produce callose plugs to inhibit the formation of pollen tube.

Sporogenous tissue

The primary sporogenous layer forms the sporogenous tissue and gives rise to the microspore mother cells or pollen mother cells (PMC). The sporogenous tissue may function directly as the microspore mother cells or undergo a series of mitotic divisions before sporogenesis. In either case, the cells of the last mitotic division function as microspore mother cells. The PMC are polygonal in shape and closely packed, but as the anther increases in size, these become loosely arranged and rounded in shape. All spore mother cells have the potential of forming pollen grains, however, some of them disintegrate and act as nourishing tissue for the developing pollen grains.

Microsporogenesis. Each PMC (2n) undergoes meiotic division to form a group of four microspores (n). (fig. 2.3). The four microspores so formed are collectively called pollen tetrad. During the meiosis in PMC, two patterns of cell wall formation are found.

(i) ***Successive type***. In successive type of cytokinesis, each nuclear division is successively followed by cell wall formation. The first nuclear division of meiosis is accompanied by cell wall formation, which is of centrifugal type, i.e. a cell plate is formed in the centre and then extends laterally, resulting in the formation of two haploid cells, i.e. a dyad. The two cells of the dyad undergo the second meiotic division (meiosis II), which is followed by wall formation and results in a tetrad of haploid cells, arranged in an *isobilateral* manner. This type of cytokinesis is commonly found in monocots.

(ii) ***Simultaneous type***. In this kind of cytokinesis, meiosis I is not followed by cell wall formation. The diploid nucleus of PMC completes both parts of meiotic division (meiosis I and II) resulting in four haploid nuclei lying in a common mass of cytoplasm of PMC. Cell wall formation now occurs simultaneously in between the four nuclei, resulting in the formation of four haploid cells (microspores). The wall formation is of centripetal type, i.e. wall formation starts from the lateral walls and meet in the centre of the cell. This kind of cytokinesis may lead to the arrangement of microspore tetrads in various ways, and it is common among dicots.

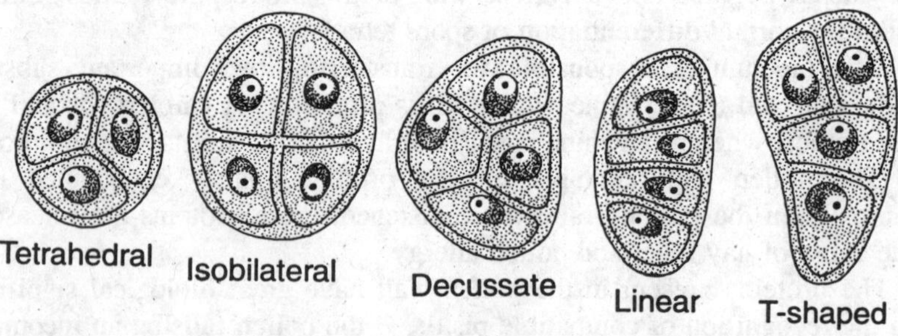

Fig. 2.3. Different types of arrangement of microspores or pollen grains (Redrawn, Johri and Bhatnagar, 1955).

Microspore tetrads. The four spores of a tetrad are completely separated from each other, as well as from other spore tetrads of the same sporangium. However, cytoplasmic connections may exist even in the tetrad stage. The arrangement of spores in the tetrads may show variation, such as tetrahedral, isobilateral, decussate, linear and T-shaped (fig. 2.3). In *Aristolochia elegans*, all five types of arrangement have been observed.

Abnormal types of microspore arrangement

After meiotic division,the microspores usually separate from one another to form independent pollen grains. However, in *Drosera* and *Drimys*, the four microspores of a tetrad remain together and form *compound pollen grains*. In many membes of Asclepiadaceae and Orchidaceae, all the pollen grains of a pollen sac (microsporangium) remain united in a single mass, known as ***pollinium***. A ***pollinium*** consists of a sac-like structure containing a mass of pollen grains. The anterior end of pollinium is attached to an elongated arm-like structure, the *translator* or *caudicle*, which in turn is attached to a spherical, knob-like *corpusculum* (fig. 2.4). Pollinia may be studied in genera like *Calotropis, Asclepias, Coelogyne, Pergularia* etc. An interesting case has been observed in *Hyphaene*, where more than four spores occur in a tetrad, leading to a condition called *polyspory*.

In *Hydrilla, Cryptostegia, Typha* and *Juncus*, the microspores remain in groups of four while in members of Mimosoidae, these are found in groups of 8 to 64. In some orchids like *Neottia*, the groups remain enclosed in a frothy structure called *massullae*.

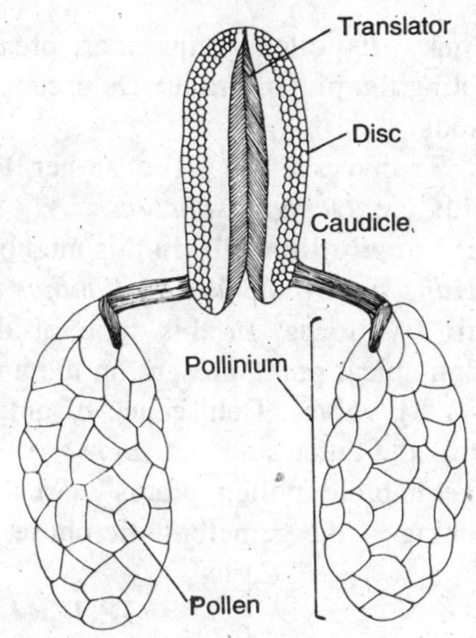

Fig. 2.4. Paired pollinia of *Asclepias*.

Dehiscence of microporangium (anther)

The fully developed microsporangium consists of a mass of haploid pollen tetrads enclosed within the sporangial wall. On maturation of the anther, the sporangial wall consists of the epidermis and the endothecium, wheras the middle layers and the tapetum disintegrate. The partition wall between the two anther lobes of one side disintegrates and the two pollen sacs appear as one (fig. 2.5).

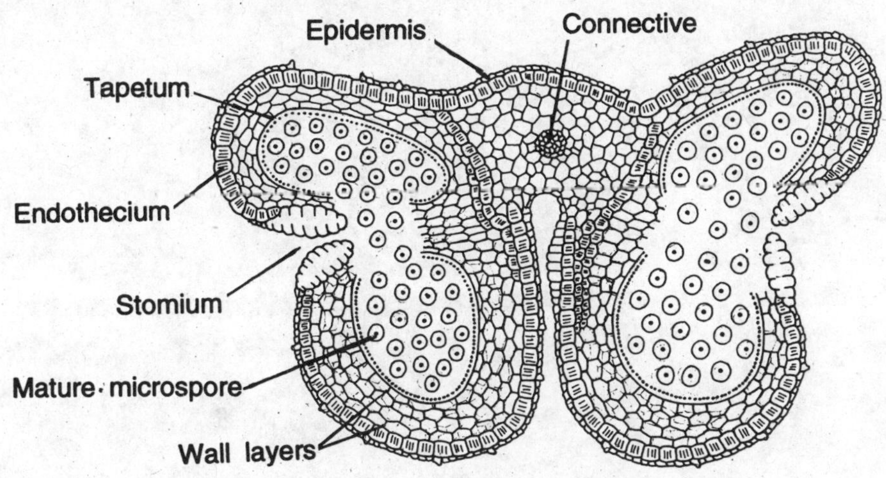

Fig. 2.5. Transection of a mature anther showing dehiscence.

The mass of pollen grains exert pressure on the anther wall which bursts open liberating the pollen grains. Dehiscence of anther may take place by the following methods:

(i) **Transverse slit**. The anther lobe dehisce by transverse slits, along the breadth, e.g. *Ocimum sanctum*.

(ii) **Longitudinal slit.** In this method, the anther lobes split by longitudinal slits, e.g. *Hibiscus, Gossypium, Helianthus* etc.

(iii) **By pores.** In this type of dehiscence, the pollen grains are liberated through apical pores present on the anther lobes, e.g. *Solanum*.

(iv) **By valves.** Dehiscence of anther lobes is affected by valves, which open towards the outer side, e.g. *Berberis.*

The haploid pollen grains marks the end of the sporophytic phase and the beginning of the gametophytic phase. Microspore is the first gametophytic cell.

GENERAL QUESTIONS

1. Describe the development of microsporangium in angiosperms and throw light on its origin.
2. Outline the various events taking place in the microsporangium leading to the process of fertilization.
3. Briefly discuss about microsporogenesis in a typical angiosperm.
4. Write short notes on—
   (i) Tapetum (ii) Pollinium (iii) Sporopollenin (iv) Dehiscence of anthers.

CHAPTER 3

Microgametophyte
(Male gametophyte)

The male gametophytic generation begins with the formation of microspores. The microspores are released in the anther locule by the breakdown of the common callose wall of the spore tetrad. However, the microspores start synthesising their individual callose walls prior to their release. The microspores after release from the tetrads are known as pollen grains and represent the first cell of the gametophyte.

Microgametogenesis

Microgametogenesis involves several dividing steps, starting from pollen germination and culminating in the formation of male gametes. Pollen germination starts while it is within the pollen sac. After being released from the spore tetrad, the pollen grain absorbs locular fluid and expands considerably. Vacuoles make their appearance and the cytoplasm gets limited to a thin film lining the cell wall. The nucleus shifts from the centre to thc pcriphery and prepares to divide mitotically. Nuclear division is asymmetric and gives rise to two unequal cells, the larger *vegetative cell* and the smaller *generative cell.*

However, before mitotic division starts in a pollen, two important changes are observed—

(i) *Migration of nucleus*–The nucleus shifts from the central to a peripheral position, which is in a definite direction and marks the position of the generative cell. This position is always constant and is genetically controlled in each species. However, some variations have been found, as in *Spiranthes*, the generative cell is cut off towards the distal pole (opposite to the tetrad centre), and in *Drimys*, it is cut off on the proximal pole (towards the centre of the tetrad).

(ii) *Polarisation of cell organelles*–The cytoplasm becomes highly vacuolated, resulting in the displacement of the cell organelles. Polarisation of the cell organelles occur and most of the plastids, mitochondria and lipid bodies migrate towards the region which is to form the vegetative cell.

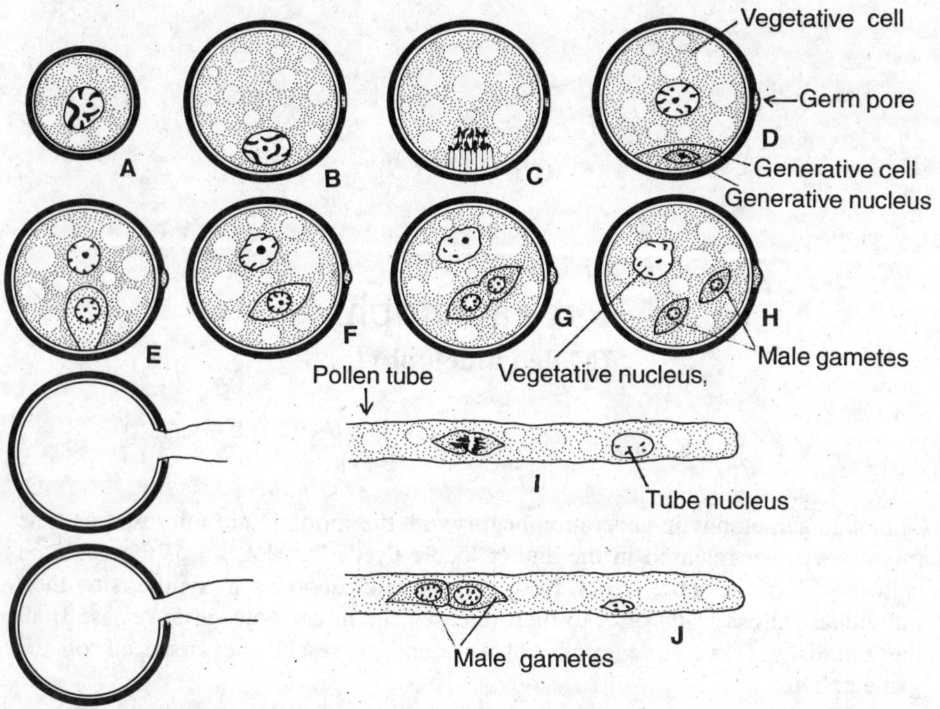

Fig. 3.1. Development of the male gametophyte in angiosperms (Adapted from Maheshwari, 1950)

Formation of vegetative and generative cells. Nuclear division is followed by cytokinesis resulting in the formation of a large vegetative cell and a small spindle-shaped generative cell. The partition wall between the two cells finally disappears and the generative nucleus lies within the cytoplasm of the large vegetative cell.

Vegetative cell. The vegetative cell is large and naked and occupies a major part of the pollen grain. After division of the pollen nucleus, the vegetative nucleus continues to grow. The cell organelles like plastids and mitochondria increase in size as well as in number. The RNA and protein contents increase considerably followed by a decrease in vacuole size, which soon disappears. In several plants, the vegetative cell contains stacked, rough ER which are released during pollen germination. Rough ER probably contains storage products, necessary for pollen germination and growth of the tube. The storage products are usually reserve RNA and proteins, as these are required during pollen germination. The nucleus also shows considerable changes. The nuclear membrane becomes highly convoluted and the nucleolus becomes prominent. However, in mature stages, the nucleolus disappears.

Generative cell. In the early stage of pollen division, the vegetative and the generative cell are separated by two plasma membranes. Later, wall formation starts between the cell membrane of the generative cell and the intine of the pollen wall until the two ends of the growing wall meet and fuse and the generative cell is pinched off.

The nature of the wall of the generative cell has been variously interpreted. Some workers report the presence of a definite cell wall, others state that only an electron transport layer is present between the two cell membranes, whereas others suggest that there is no wall around the generative cell. However, the absence of true cellulosic cell wall is confirmed. Echlin (1972) reported the presence of callose wall in majority of plants, while Owen and Westmuckett (1983) reported the presence of B-1, 4-glucans along with callose deposition.

The generative cell is small in size and spherical in shape, at the time of detachment from the pollen wall. During pollen development, it changes shape, becomes elongated and appears vermiform. The elongated shape of the generative cell facilitates its movement into the pollen tube.

The cytoplasm of the generative cell is hyaline, highly reduced and contains the cell organelles like mitochondria, ribosomes, ER, microtubules and dictyosomes. Initially when the generative cell is spherical in shape, the microtubules are scattered randomly in the cytoplasm, but as the cell assumes elongated shape, the microtubules become oriented with their long axes parallel to the long axis of the cell. Hence, these axial microtubules play an important role in maintaining the elongated shape of the generative cell. Eventually, the generative cell looses contact with the pollen wall and shifts into the vegetative cell. The pollen grain, thus, becomes two-celled.

The function of the generative cell is to give rise to the male cells, i.e. the male gametes or sperms. The generative nucleus divides by mitosis followed by cytokinesis, and a wall separates the two cells. The partition wall, as well as the surrounding wall of the sperm disappear and the naked sperms are released. An important feature of the sperm formation is that their separation is caused by degradation of the partition wall and not by the furrowing of the binucleate cell. The cytoplasm of male cell generally contains the organelles present in the generative cell.

Nuclear division in the generative cell may takc placc under two conditions-

(i) while the pollen grains are confined within the anther, i.e. pollens are shed at the three-celled stage, as in *Beta* and *Hordeum*, or

(ii) after the pollen has been shed from the anther, i.e. at the two-celled stage. In the two celled pollen grain, the generative cell may divide, after the pollen has alighted on the stigma (e.g. *Holoptelea integrifolia*) or in the pollen tube just before it reaches the embryo sac. The latter case is the most common method. In rare cases, the generative cell divides after the pollen tube has reached the embryo sac (e.g. *Euphorbia terracina*).

However, in majority of plants, the generative nucleus divides before the pollen has shed, i.e. at the three celled stage. The division of the generative nucleus at an early stage is an indication of the suppression of the male gametophyte-a characteristic feature of vascular plants.

Differences in the behaviour of vegetative and generative cells. The two cells, vegetative and generative, though arising from a single parent cell exhibit different behavioural patterns, which are based on nuclear and cytoplasmic differences between them. The differences are as follows:

Generative cell	*Vegetative cell*
1. After pollen mitosis, DNA content increases 2C-level.	1. After pollen mitosis, DNA content increases to 1C-level.
2. Lysine rich histone is associated to DNA.	2. It is absent.
3. RNA and protein synthesis are practically absent.	3. RNA and protein synthesis are excessively high.
4. The generative cell is metabolically inactive.	4. The vegetative cell is metabolically active and supplies the required metabolites to the generative nucleus through cytoplasmic connections.

Internal organisation of mature pollen grains

Brewbaker (1957) reported that about seventy per cent of the angiosperms (e.g. Solanaceae and Rosaceae) possess bicellular pollen grains whereas thirty per cent (e.g. Cruciferae, Compositae, Caryophyllaceae and Graminae) exhibit tri-cellular pollen grains. Both types of pollen grains have a single vegetative cell and a single generative cell. The generative cell later produces two male gametes. The two male (sperm) cells lie in the cytoplasm of the vegetative cell enclosed by their individual cell membranes as well as an internal membrane of the vegetative cell.

The work done on three-celled pollen grains of *Brassica* spp., *Spinacea oleracea* and *Zea mays* reveals that the sperms exhibit remarkable dimorphism. In *Plumbago zeylanica*, one gamete is larger in size with long slender projections and contains nearly all the mitochondria, while the other gamete is small and contains all the plastids. The two sperms of the three-celled pollen grain are usually linked by plasmodesmata, and also, one of the sperms remains joined by cytoplasmic connections to the vegetative nucleus to form a three-celled *male germ unit*.

The formation of male germ unit is of universal occurrence in all angiosperms. The male germ unit travels through the pollen tube as one unit and on reaching the ovule, the vegetative nucleus separates from the sperms, followed by separation of the two sperms.

Unusual types of pollen development

In majority of angiosperms, the pollen mother cell divides by meiotic division forming four haploid daughter nuclei, each of which then functions as the functional pollen grain. However, deviations from this normal type of development have been observed in many species. Some of them are discussed here.

1. **Pollen development in Cyperaceae.** In members of Cyperaceae, PMC

divides meiotically to form four haploid nuclei, out of which a single functional nucleus remains in the centre and the three non-functional nuclei are pushed to one side of the cell. The functional nucleus produces the vegetative as well as the generative cell, whereas the non-functional nuclei eventually degenerate.

2. **Pollen grain develops embryo sac (Ne'mec phenonenon).** A single pollen grain usually contains two or three nuclei. However, in some members of Liliaceae, pollen grains with more than three nuclei have been observed. These abnormal types of male gametophytes lead to the formation of female gametophytes or embryo sac-like structures. Embryo sac-like pollen grains had been first observed in the petaloid anthers of *Hyacinthus orientalis* by Nemec.

In *Hyacinthus*, some of the microspores enlarge, escape from the spore wall and form sac-like structures. The nucleus of these sac-like structures divides by three mitotic divisions forming eight nuclei which organise themselves like an eight-nucleate embryo sac. An interesting feature of these pollen embryo sacs is that if they are placed along with normal pollen grains, on an agar medium, the pollen tubes formed by normal pollen grains showed chemotropic response towards the pollen embryo sacs.

3. **Pollen grains develop into sporophytes.** Guha and Maheshwari (1966) working on *Datura innoxia* succeeded in developing sporophyte from pollen grain. It has been suggested that the pollen grains have the potential of adopting any one of the three developmental procedures, e.g.

(i) They form male gametes under normal conditions.
(ii) They form female gametophytes under special conditions.
(iii) They grow into apogamous sporophytes under appropriate nutrient culture medium.

4. **Pollen grain sterility (Male sterility).** In nature, some individuals in a crop remain unable to produce fertile pollen. This condition is known as male sterility. Male sterility has been found in maize, wheat, sorghum, barley, cucumber, carrots, tomatoes and onion. Male sterility may be due to a recessive gene in a chromosome, an organelle other than nucleus (cytoplasmic male sterility) or due to an interaction of both nuclear and cytoplasmic genes. Plant breeders have successfully used male sterility for the production of hybrids especially in maize and sunflower.

GENERAL QUESTIONS

1. Describe the germination of microspore (pollen grain) and the formation of male gametophyte.
2. Describe briefly the development and germination of pollen grains in angiosperms.
3. Write short notes on—
   (i) Nature of pollen tube
   (ii) Generative cell
   (iii) Ne'mec phenomenon
   (iv) Vegetative cell

CHAPTER 4

The Megasporophyll

The female reproductive whorl of a flower is known as the *gynoecium*, which consists of unit structure called *carpel* (or pistil). A carpel normally consists of three morphologically and functionally distinct parts, the *stigma*, the *style* and the *ovary*. Stigma is the apical knob-like structure on which pollen grains are deposited after pollination. Style is the long tubular part that joins stigma with ovary. Ovary is the basal swollen and fertile part of the carpel. The ovary may be unilocular (as in Compositae and Cucurbitaceae), bilocular (as in Solanaceae), trilocular (as in Liliaceae) or pleurilocular (as in Malvaceae). The ovules (or integumented megasporangia) are found in the ovary part of the carpel.

The Ovule

The ovule or megasporangium arises from the placenta and the arrangement of ovules in an ovary is called placentation, e.g. marginal (Leguminosae), parietal (Brassicaceae), axile (Solanaceae, Malvaceae, Liliaceae), basal (Compositae), superficial (Nymphaeaceae) etc.

Structure of ovule

The ovule is a small oval structure attached to the placenta by a short stalk or *funicle* (funiculus). It comprises a central mass of tissue called *nucellus* which remains surrounded by one or two layers of protective cells, called *integument*. (fig. 4.1.) The point of attachment of funicle to the ovule body is termed as *hilum*. In case when ovule is inverted, its body

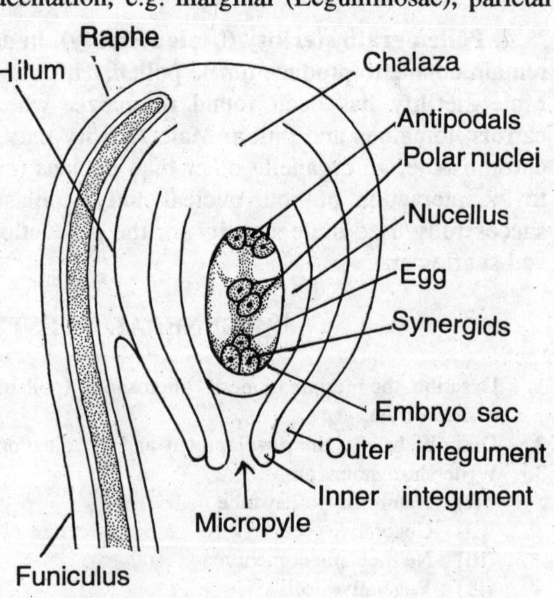

Fig. 4.1. L.S. of the ovule (anatropous type).

fuses with the funicle forming a ridge-like structure called *raphe*. The raphe extends forwards and fuses with the nucellus and the integuments to form the *chalaza* (fig. 4.1).

Integuments. Nucellus remains surrounded by the integument except in the micropylar region. Ovules in which integuments are present are known as *tegmic* type, whereas ovules lacking integuments are called *ategmic* type as found in *Loranthus* and *Santalum*. An ovule, if surrounded by one integument, is called *unitegmic* type and if by two integuments, is *bitegmic* type. Unitegmic ovules are generally found in members of gamopetalae (sympetalae) as in Compositae while bitegmic type in polypetalae and monocots. In certain cases, a third integument or *aril* develops as a funicular outgrowth from the base of ovule. In Litchi, the aril is fleshy and forms the edible part of the fruit. In castor (*Ricinus*), the cells of the integument near the micropyle divide and enlarge to form *caruncle*. Caruncle is the micropylar outgrowth of integuments, but some times considered as the third integument. In *Asphodelus* (a member of Liliaceae), the third integument is called *arm*.

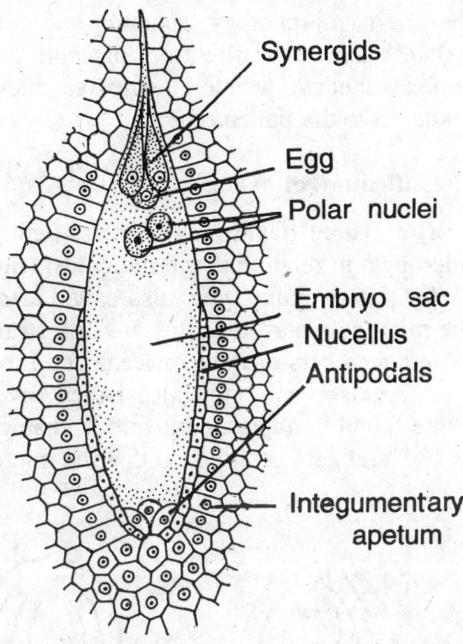

Fig. 4.2. Embryo sac with degenerating nucellar epidermis and the integumentary tapetum has developed from the inner layer of integument.

In members of Sympetalae, possessing unitegmic ovules, the nucellus degenerates at an early stage, and the inner integument becomes highly specialised to form a single or multilayered **endothelium**. The endothelium consists of radially elongated cells rich in cytoplasm, protcins, RNA, carbohydrates, ascorbic acid etc. Functionally, endothelium is similar to tapetum i.e. it provides nourishment to the embryo sac, hence sometimes also called *integumentary tapetum*. A significant feature of endothelium is that the cells exhibit *polyploidy* which may be as high as 32 n as found in *Pedicularis*.

Micropyle. The integuments surround the nucellus except in the terminal (apex) region, where an opening is left, known as micropyle. If both the integuments are involved, the opening formed by the outer integument is called *exostome* while that formed by the inner integument is called *endostome*. Within the micropyle are deposited synergid-synthesised chemotropic agents, which stimulate the entry of the pollen tube.

Obturator. In certain ovules, obturators are found. These are associated with directing the growth of the pollen tube towards the micropyle. These may be multicellular with glandular hairs as in *Tetragonia*. A considerable variation in the origin, morphology, anatomy and extent of development of the obturator is found. These may arise from placenta or funiculus or from both and produce a surface exudate which is nutritive and provides a mechanical and chemical guidance to the pollen tube.

Classification of ovules

The mature ovule may be classified into six types based on the position of micropyle in relation to the funiculus. (fig. 4.3.)

1. **Orthotropous type** (also called *atropous*). The mature ovule is straight and the micropyle and funiculus are in a straight line and above the hilum, as found among members of Polygonaceae and Piperaceae.

2. **Anatropous type** (also called *inverted*). It is the most common type of ovule found in angiosperms and is characteristic of sympetalae. The ovule bends at 180° and gets inverted, so that the micropyle comes to lie close to the funicle.

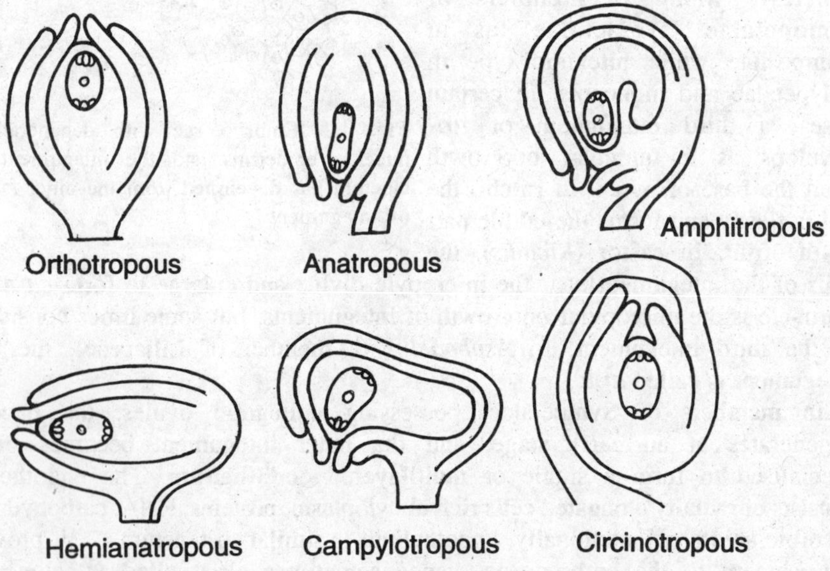

Fig. 4.3. Different types of ovules.

3. **Campylotropous type.** In this type of ovule, the curvature is incomplete, and the body of the ovule lies at right angle to the funiculus, and is somewhat curved, as in Capparidaceae and Leguminosae.

4. **Amphitropous type.** When the curvature of the body of the ovule becomes more than that of campylotropous and the embryo sac assumes a horse-shoe shape, the ovule is said to be amphitropous type. It is commonly found in Alismaceae and Butaceae.

5. **Hemianatropous type** (also called *hemitropous*). It is a modified type of

anatropous ovule in which the body of ovule lies at right angle to the funicle and micropyle and chalaza lie in a straight line. The micropyle is not found near the hilum, e.g. *Ranunculus* and in members of Primulaceae.

6. **Circinotropous type.** It is a special type of ovule formed due to unilateral growth. Initially, the nucellar mass is in line with the funicle and appears orthotropous, then unilateral growth of ovule takes place and it gets inverted, thus becoming anatropous. The curvature continues till the ovule takes a complete 360° turn and the micropyle again faces upwards, e.g. *Opuntia* (Cactaceae).

Development of ovule

The ovules develop from the inner surface of the ovary which is lined with a special ring of tissue, the *placenta.* The ovule primordium appears as a small mass of homogenous tissue on the surface of the placenta. It grows rapidly and assumes a conical structure with a rounded tip. This mound-like structure is the future nucellus. It increase in size by repeated divisions and projects out. As the development continues, the integument primordia originate close to the base of nucellar tissue and grow upwards forming the integuments. In bitegmic ovules, the inner integument grows faster than the outer integument. Although integuments are initiated after the nucellus, they develop faster and completely cover the nucellus, except in the micropylar region.

Megasporogenesis

The various steps leading to the formation of the megaspore is known as *megasporogenesis.* Formation of megaspore takes place as follows—

(i) A single cell in the hypodermal region of the nucellus functions as the *archesporial initial.* The cell is differentiated during the early stage of the development. The archesporial initial is recognised by its larger size, prominent nucleus and dense cytoplasm. Further development of archesporial initial depends upon the type of ovule.

(a) **Tenuinucellate type.** In this type of ovules, the archesporial cell directly functions as the *megaspore mother cell,* so that the sporogenous cell is hypodermal is nature and is surrounded by single layered nucellus (parietal cells are not formed), e.g. *Elytraria acaulis.* (fig. 4.4).

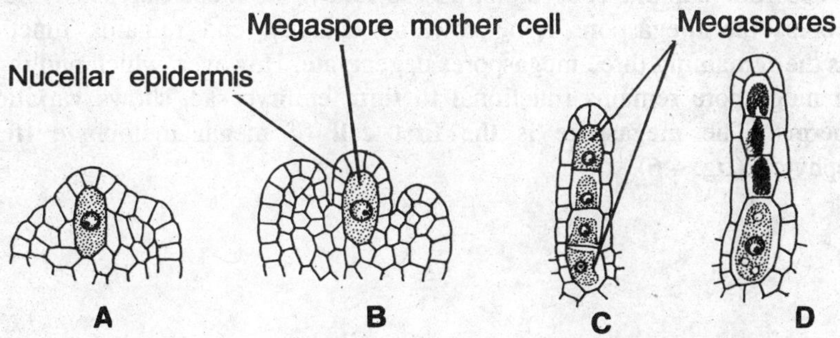

Fig. 4.4. Megasporogenesis in a tenuinucellate ovule. A and B-megaspore mother cell stage and C and D-megaspore tetrad stage. (Redrawn, Johri and Singh, 1959).

(b) **Crassinucellate type.** In these ovules, the archesporial cell divides by periclinal division forming outer *primary parietal cell* and inner *primary sporogenous cell*. The parietal cell may remain undivided or undergo periclinal and anticlinal divisions forming several parietal cells, thereby pushing the sporogenous cell deep into the nucellar tissue. The sporogenous cell is thus sub-hypodermal in nature, e.g. *Myriophyllum intermedium*. (fig.4.5).

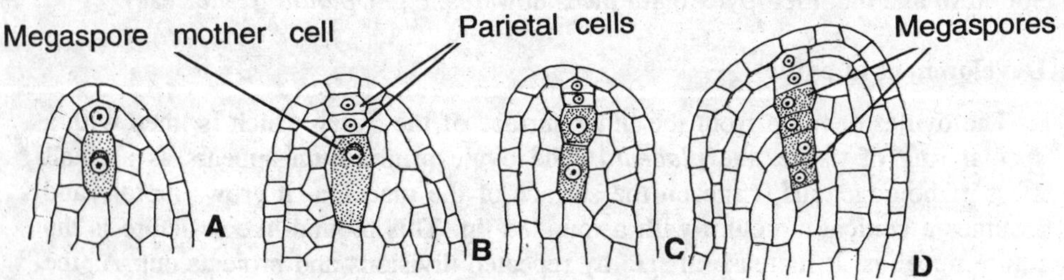

Fig. 4.5. Megasporogenesis in crassinucellate ovule.

(c) **Pseudo-crassinucellate type.** In this type of ovules, the sporogenous cell becomes sub-hypodermal due to divisions in nucellar epidermis, e.g. *Nigella damascena*.

Davis (1966) reported tenuinucellate ovules from 105 families, crassinucellate ovules from 179 families and pseudo-crassinucellate ovules from 11 families.

Single-celled archesporial initial is most common, but in Crossomataceae and Paeoniaceae, it consists of a group of cells, whereas in *Sedum*, the archesporium is multicellular in nature.

(ii) The primary sporogenous cell functions directly as the megaspore mother cell, or megasporocyte, but in members of Sympetalae, the primary archesporial cell functions as the megasporocyte. In either case, the megaspore mother cell (2n) divides by reduction division (meiosis I) forming a dyad, which again divides (meiosis II) to form a linear tetrad (n). In some forms, e.g. *Rumex* and *Rheum*, T-shaped tetrads have been reported, whereas, isobilateral and tetrahedral tetrads have been reported from Crassulaceae and Hydrocharitaceae, respectively.

(iii) The four haploid cells of tetrad are known as megaspores. Of the four megaspores, the megaspore lying towards chalazal end remains functional, whereas the remaining three megaspores degenerate. However, which and in what form, a megaspore remains functional to form embryo sac, shows variation in angiosperms. The megaspore is the first cell of megagametophyte (female gametophyte). (fig. 4.6).

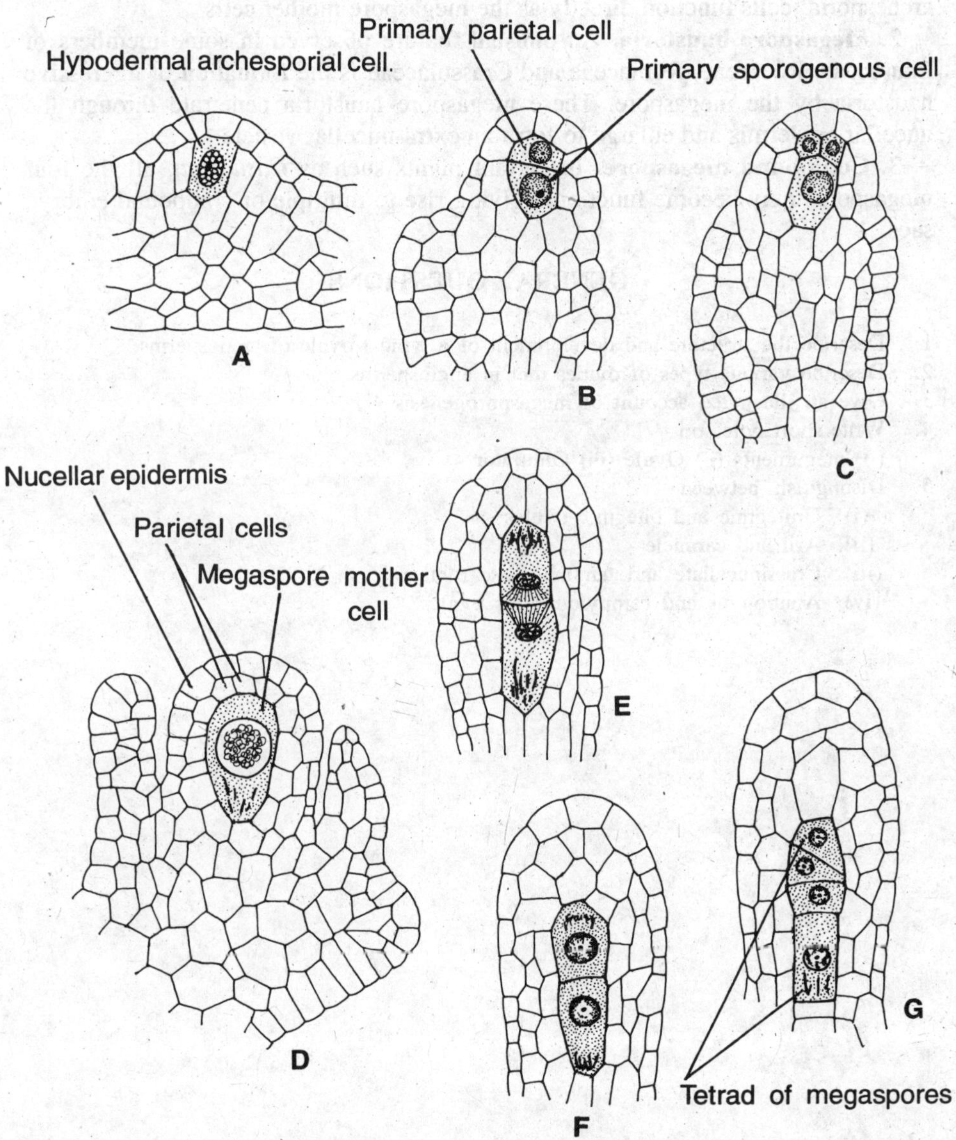

Fig 4.6. Megasporogenesis in the ovule showing the development of megaspores (Redrawn Maheshwari, 1950).

Special types of Megasporogenesis

In certain members, megasporogenesis shows variation from normal types and gives rise to abnormal structures, These are-

1. **Reduced ovules.** In certain members of Ranunculaceae, e.g. *Adonis* and *Clematis*, the ovary is usually uniovulate. Each carpel has a single fertile ovule and 2-4 sterile ovules. In *Dendrophthae,* there is no ovule formation and the archesporial cells function directly as the megaspore mother cells.

2. **Megaspore haustoria.** An unusual feature observed in some members of Lauraceae, Rosaceae, Rubiaceae and Crassulaceae is the formation of aggressive haustoria by the megaspore. These megaspore haustoria penetrate through the nucellar epidermis and enlarge to form an extra-nucellar vesicle.

3. **Compound megaspore.** In certain plants such as *Casuarina*, all the four megaspores may become functional giving rise to multiple or compound embryo sacs.

GENERAL QUESTIONS

1. Describe the structure and development of a typical ovule of angiosperms.
2. Describe various types of ovules met in angiosperms.
3. Give an illustrated account of megasporogenesis.
4. Write short notes on—
   (i) Integuments (ii) Ovule (iii) Obturator
5. Distinguish between—
   (i) Unitegmic and bitegmic ovules
   (ii) Aril and caruncle
   (iii) Crassinucellate and tenuinucellate ovules
   (iv) Anatropous and campylotropous ovules.

CHAPTER 5

Megagametophyte
(Female Gametophyte)

The haploid megaspore marks the beginning of the female gametophytic generation. In angiosperms, the female gametophyte is represented by an extremely reduced sac-like structure-the *embryo sac*, which remains embedded in the sporophytic tissue. It is totally dependent upon the surrounding sporophytic tissues for its nourishment.

Development of embryo sac

(*Polygonum* type)

Of the four megaspores formed by meiotic division in MMC, the megaspore lying towards chalazal end forms embryo sac. Its developmental stages are as follows:

(i) The functional megaspore begins to elongate along the micropylar-chalazal axis. In the cytoplasm, several small vaculoes make their appearance, which later fuse to form a single large vacuole.

(ii) The nucleus of megaspore divides mitotically forming two daughter nuclei. The spindle of dividing nucleus is oriented along the long axis of the cell, so that the two nuclei, face the two opposite poles. The nuclear division is not followed by wall formation, however, the two nuclei are pushed towards the opposite poles by an expanding central vacuole.

(iii) Both nuclei now divide twice giving rise to four nuclei at each pole. The eight nuclei so formed, however, remain embedded in a common cytoplasm.

(iv) The embryo sac now elongates considerably and undergoes cellular organisation.

(v) At the micropylar end of the embryo sac are present four nuclei, which get organised into 3+1 arrangement, in which three cells form the *egg apparatus* and the fourth is left free in the cytoplasm of central cell to form the *upper polar nucleus.*

(vi) The *egg apparatus* consists of one egg and two synergids. The major portion of egg apparatus is surrounded by the central cell, only its micropylar end is attached to the wall of the embryo sac. The three cells are almost of equal length, however, the egg cell extends a little more into the central cell.

(vii) At the chalazal end, three nuclei out of four, join to form the three *antipodal cells* and the fourth nucleus functions as the *lower polar nucleus*. Later, the lower polar nucleus comes to lie alongside the upper polar nucleus.

This pattern of embryo sac development is known as **Polygonum type** and is of common occurrence amongst majority of angiosperms (about 81 per cent).

Structure of Mature Embryo-sac

The mature embryo-sac is an ellipsoidal structure with tapering ends and possessing a seven-celled, eight-nucleate organisation, i.e. 3-celled egg apparatus, 3 antipodal cells and a binucleate central cell (secondary nucleus). The seven cells of the female gametophyte are formed within wall of the parent megaspore. The various constituents of the female gametophyte are briefly discussed. (fig. 3.1).

Embryo-sac wall. The wall of embryo-sac presents interesting features. It is considered to be multilayered and pecto-cellulosic in nature. The innermost layer is represented by the wall of functional megaspore, whereas, the outer layer comprises crushed cells of nucellus. In *Stipa,* the gametophyte wall is made up of three parts. The innermost portion is contributed by the gametophyte, the middle portion by the degenerated megaspore and the outer portion by the nucellar cells.

Egg apparatus. The egg apparatus is present at the micropylar end of the embryo-sac and consists of two synergids and one egg cell. The synergids are elongated structures lying at the micropylar end of the embryo-sac and partly enclosing the egg. The cytoplasm of synergids is rich in different organelles. An interesting feature of cytoplasm is its polarized nature, i.e., the chalazal end is occupied by vacuoles, whereas cytoplasm and a prominent nucleus are present in the micropylar half.

Each synergid is associated at its micropylar end with a filiform apparatus. The filiform apparatus is a mass of finger-like projections of the synergid wall consisting of tightly packed cellulosic microfibrils enclosed by a non-fibrillar sheath.

The life span of synergids is of short duration. In embryo-sacs possessing two synergids, one degenerates before the pollen tube enters the embryo-sac, whereas other, known as persistent synergid, degenerates after the pollen tube has discharged its male gametes into the embryo-sac. *Synergid haustoria* have been observed in a number of plants, such as *Cotula, Sedum, Cortaderia* etc. In *Quinchamalium,* the tips of large synergids elongate, break through the embryo-sac and grow upwards into the style and function as synergid haustoria. These haustoria absorb the nutrients from the neighbouring tissues and transfer it to the synergids.

Functionally, synergids are highly active metabolically and perform three important functions.

(i) These secrete chemotropically active substances which help in directing the growth of pollen tube.

(ii) The degenerating synergids form the site for the discharge of pollen tube contents.

(iii) The filiform apparatus of the synergids help in the absorption and conduction of nutrients from the nucellus into the female gametophyte.

The *egg cell* lies slightly above the synergids, enclosed by a pecto-cellulosic cell wall. In the early state of development, the egg shows polarisation of cellular contents. The micropylar end of the egg cell is occupied by a large vacuole, whereas the chalazal end is occupied by the organelles. The mature egg also contains large amount of starch which is utilized during fertilisation and early embryogeny. In *Nicotiana rustica,* the egg contains two RNA-positive bodies that are involved in enzyme synthesis required during early embryogeny. In *Plumbago capensis,* finger-like projections arise from the micropylar end of the egg. In rare cases, such as in *Cortaderia jubata,* the egg develops into an haustorium.

Antipodals. The antipodals consist of three cells present at the chalazal end of the embryo-sac. They do not undergo any special change and usually degenerate before or immediately after fertilization. However, some variations in their structural and cytological features do take place, such as in grasses, mitotic divisions result in a large number of antipodals, as high as 300 in *Sasa paniculata.* Sometimes, karyokinesis is not accompanied by cytokinesis resulting in multinucleate condition, as has been observed in *Zea mays, Stackhousia, Tagetes* etc.

In many plants, such as *Grindelia, Haplopappus, Argemone, Quinchamalium* etc. *antipodal haustoria* have been reported.

The antipodal cells carry out three important functions.

(i) These are nutritive in function, nourishing the embryo-sac. The antipodal cells have increased DNA content (through multinucleate condition, polyteny or endopolyploidy) and are somewhat similar to glandular anther tapetum and integumentary tapetum.

(ii) These contain a large amount of starch, lipid and proteins which are consumed by the developing embryo and endosperm.

(iii) Certain substances are produced and secreted by the antipodals, which control the growth and development of endosperm.

Central cell. The central cell lies in the centre of embryo-sac. It is the largest cell in the embryo-sac and forms the endosperm. Its cell wall shows considerable variation. It is usually thickest in the region adjacent to the nucellus and thins towards the chalazal end of the egg apparatus. The central cell remains connected by plasmodesmatal connections to the egg, synergids and antipodals.

The cytoplasm is rich in organelles performing intense metabolic activities. A large central vacuole containing reservoir of amino acids, inorganic salts and sugars is present and which is utilized during fertilization and endosperm development.

The central cell possesses two nuclei called *polar nuclei.* The polar nuclei are either centrally suspended by cytoplasmic strands or are embedded in the cytoplasm near the egg apparatus. The polar nuclei fuse to form the *secondary nucleus* and participate in double fertilization.

The central cell draws its nourishment from the nucellus or integument through cell wall projections. In *Aquilegia vulgaris,* these haustorial projections also secrete enzymes that cause digestion of nucellar tissue.

Types of Embryo-sacs

In majority of angiosperms, the embryo-sac develops from only one of the four megaspores formed by meiotic division of megaspore mother cell. However, formation of female gametophyte from two or all four megaspores has been observed in many species. As such, embryo-sacs have been classified into monosporic, bisporic or tetrasporic types depending upon whether one, two or four megaspores are involved in the formation of an embryo-sac. The different types of embryo-sacs are as follows:

1. Monosporic embryo-sacs (fig. 5.1)

This type of embryo-sac is derived from one of the four megaspores. It is characterised by genetically identical nuclei, as it is derived through mitosis of single nucleus. Monosporic embryo-sacs are of two types.

(i) **Polygonum type.** It is the most common type and was first observed in *Polygonum*. The embryo-sac develops from a single chalazal megaspore and is eight nucleate, comprising three antipodals, a binucleate central cell and a three-celled egg apparatus (i.e. 3+2+3).

(ii) **Oenothera type.** In this type, the embryo-sac develops from a single micropylar megaspore and is four nucleate, organised into a three celled egg apparatus and an uninucleate central cell (i.e. 3+1). This type of embryo-sac is found among members of family Onagraceae.

2. Bisporic embryo-sacs (fig 5.1)

In plants bearing bisporic embryo sacs, the megaspore mother cell undergoes first meiotic division which is followed by wall formation, giving rise to haploid dyad cells. One of the dyad cells is the functional cell, whereas the other dyad cell, facing the micropylar end degenerates. The functional cell divides by the second meiotic division, which is not followed by wall formation, hence a binucleate functional megaspore is formed. Both megaspore nuclei participate in embryo-sac formation and divide by two mitotic divisions forming eight nuclei. The eight nuclei are organised into three-celled egg apparatus, a binucleate central cell and three-celled antipodals. A bisporic embryo-sac is constituted of two genetically different nuclei, hence of the eight nuclei formed, four are of one type and the other four of a different type. Bisporic embryo-sacs are of two types.

(i) **Allium type.** This type of embryo-sac was first observed in *Allium fistulosum* by Strasburger (1879) and is found in all species of *Allium*. The embryo-sac develops from the chalazal dyad cell. Distribution of nuclei follows the general pattern, i.e. 3+2+3.

(ii) **Endymion type.** This type of embryo-sac develops from the micropylar dyad cell. In *Schisandra chinensis*, deviation from normal bisporic has been observed and the mature embryo sac is bisporic but four-nucleate.

3. Tetrasporic embryo-sacs (fig. 5.1)

In such cases, the megaspore mother cell (2n) divides by meiotic division and

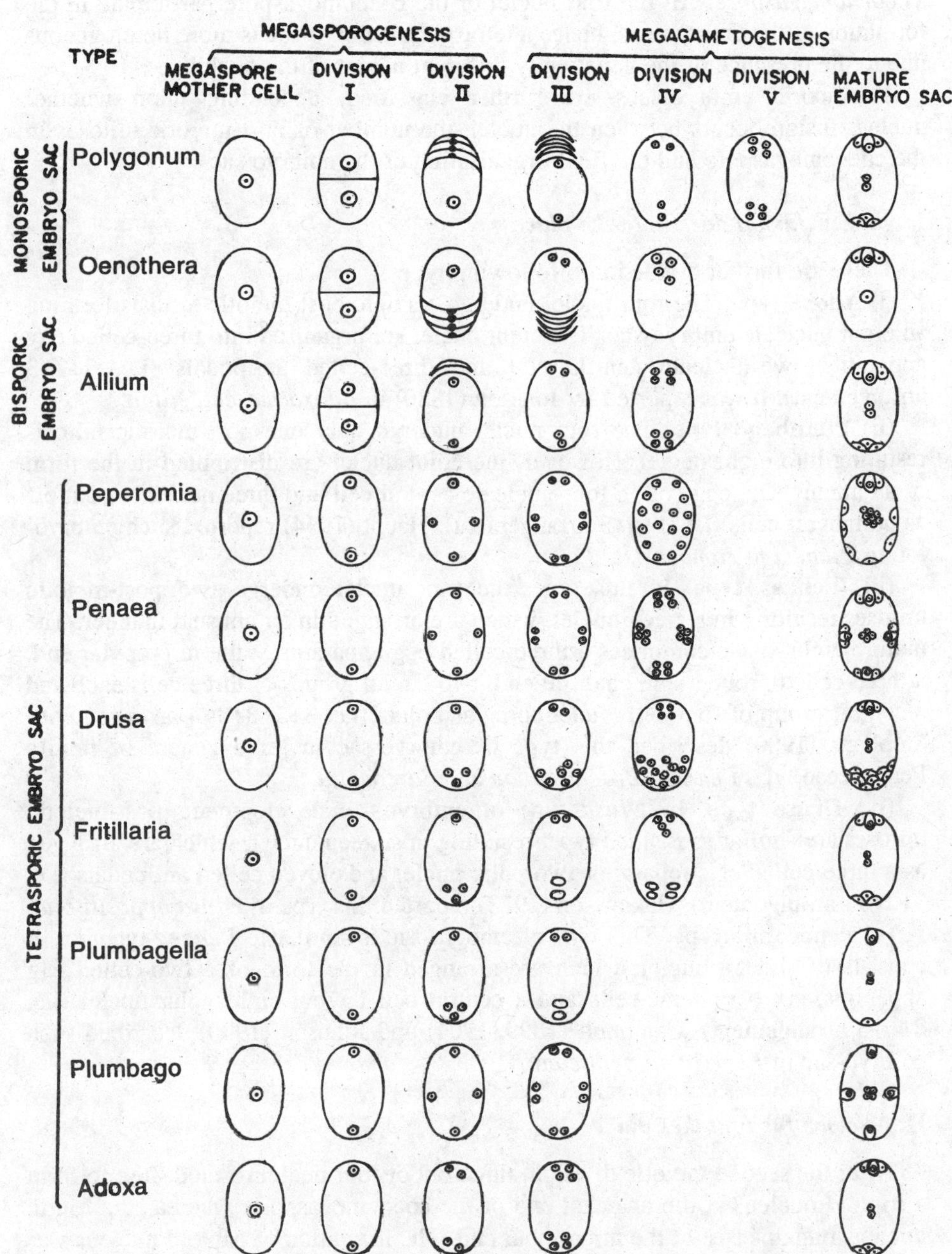

Fig. 5.1. Important types of embryo-sac development met within the angiosperms (After Maheshwari, 1950).

neither of the two divisions is followed by wall formation, so that the four haploid nuclei remain in a common cytoplasm. Such a four-nucleate megaspore is known as *coenomegaspore*. All the four nuclei of the coenomegaspore participate in the formation of the embryo-sac, hence a tetrasporic embryo-sac is more heterogenous due to the presence of the genetically different nature of the nuclei.

Tetrasporic embryo-sacs are further classified, depending upon whether nuclear fusion occurs between the nuclei, the number of post-meiotic mitoses in the coenomegaspore and the final organisation of the embryo sac.

I. *Nuclear fusion does not take place*

These are further divided into following types.

(i) **Adoxa type.** The four nuclei undergo a single post-meiotic mitosis forming an eight nucleate embryo-sac. The eight nuclei are organised into three-celled egg apparatus, two-nucleate central cell and three-celled antipodals (i.e. 3+2+3 arrangement). It was reported by Jonsson (1880) in *Adoxa moschatellina.*

(ii) **Plumbago type.** The four nuclei undergo only one post-meiotic mitosis resulting into eight nuclei. However, the eight nuclei are distributed in the form of an uninucleate egg cell, a four-nucleate central cell and three nuclei are cut off as peripheral cells (i.e. 1+4+3 arrangement). Haupt (1934) reported such embryo-sac in *Plumbago campensis.*

(iii) **Penaea type.** In this type, the four nuclei undergo two post-meiotic mitoses resulting in sixteen nuclei, which are arranged in an unusual manner. The mature embryo-sac comprises a three-celled egg apparatus at the micropylar end, a three-celled group at the chalazal end, two lateral groups of three cells each and a central group of four cells functioning as polars (i.e. 3+3+3+3+4 arrangement). Stephans (1900) described this type of embryo-sac in three genera of family Penaeaceae, viz. *Penaea, Brachysiphon* and *Sarcocolla.*

(iv) **Drusa type**. In *Drusa* type of embryo-sac development, post-meiotic mitoses are similar to *Penaea* type, resulting in sixteen nuclei, which are arranged as a three-celled egg apparatus, two polar nuclei and eleven celled antipodals (i.e. 3+2+11 arrangement). Hakansson (1923) reported this type in *Drusa oppositifolia.*

(v) **Peperomia type.** This type of embryo-sac is similar to *Penaea* type, i.e. it consists of sixteen nuclei, which are arranged in the form of a two-celled egg apparatus, six peripheral cells and a central cell having eight polar nuclei (i.e. 2+6+8 arrangement). Campbell (1899, 1901) and Johnson (1900) described such embryo-sac in *Peperomia pellucida.*

II. *Nuclear fusion takes place*

After the second meiotic division, three out of four nuclei formed, fuse to form a triploid nucleus at the chalazal end of the coenomegaspore, whereas the fourth nucleus remains free at the micropylar end. The formation of triploid nucleus was reported by Bambacioni, hence also called Bambacioni effect. This type of embryo-sac shows two variations-

(i) **Fritillaria type.** Both the nuclei, i.e. one haploid and one triploid, undergo two divisions giving rise to four nuclei at each pole. The eight-nucleate embryo-sac consists of a three-celled haploid egg apparatus, three-celled triploid antipodals and a central cell with two polar nuclei, one haploid and one triploid (i.e. 3+3+2 arrangement). Treub and Mellink (1880) reported such embryo-sac in various species of *Lilium*.

(ii) **Plumbagella type.** In this type of embryo-sac, the two nuclei (one haploid and one triploid) divide once to form four nuclei, i.e. two haploid and two triploid nuclei. These four nuclei are arranged in two groups of two nuclei each. One group is present at the micropylar end consisting of one haploid egg and one haploid upper polar nucleus, the other group present at chalazal end consists of one triploid antipodal cell and one triploid lower polar nucleus. Hence the polar nuclei comprises one haploid and one triploid set. This type of embryo-sac was first observed by Fagerlind (1938) in *Plumbagella micrantha.*

SPECIAL TYPE OF TETRASPORIC EMBRYO-SAC

This special type of embryo-sac was discovered by Martinoli (1939) in *Chrysanthemum cinerariaefolium.* The coenomegaspore contains four nuclei arranged in 1+2+1 arrangement, i.e. one at the chalazal end, one at the micropylar end and two in the centre of the cell. Further development may occur by two methods-

1. ***Two central nuclei do not fuse, but remain closely associated***—The two nuclei at the micropylar and chalazal ends undergo two divisions forming four nuclei at each end. The embryo-sac now consists of ten nuclei. The four nuclei at the chalazal end form the antipodals, while out of four nuclei at micropylar end, three form the egg apparatus and the fourth nucleus associates with the central cell to form the polar nuclei (i.e. 4+3+3 arrangement).

2. ***Two central nuclei fuse to form a diploid nucleus***—The coenomegaspore now contains three nuclei, each of which undergoes two mitotic divisions forming twelve nuclei arranged in three groups of four each. At the chalazal end, the four nuclei along with three nuclei (2n) from the central cell form the antipodals (7), whereas at the micropylar end, three nuclei form the egg apparatus and the fourth nucleus associates with the diploid nucleus of the middle group to form the central cell. The embryo-sac consists of seven antipodals, three egg apparatus and two central cells (i.e. 7+3+2 arrangement). The additional three diploid nuclei present in antipodal region are called *additional antipodals*.

Haustorial Nature of Embryo-sacs

In majority of plants, the embryo-sac derives its nourishment from the surrounding nucellar tissue. However, in members of Santalaceae and Loranthaceae, haustorial behaviour of the embryo-sac is of common occurrence. The synergids and antipodals may give out haustoria, or micropylar and chalazal extensions may grow out into the surrounding sporophytic tissues and function as haustoria. These extensions are referred to as *micropylar caeca* or *antipodal caeca.* In extreme cases, the entire embryo-sac may grow beyond the ovular tissue and get lodged directly on the placenta, e.g. *Santalum* and *Quinchamalium* etc. In members of Loranthaceae, the embryo-sacs grow upward into the style, upto various heights. The largest embryo-sac has been observed in *Moquiniella rubra,* where the embryo-sac tip grows upto the style, touches the stigma and then curves back and grows downwards into the style. In such case, fertilization may occur either in the basal region or at a certain distance above in the stylar region: (The haustoria developing from antipodals and synergids have been described earlier).

Evolutionary Significance

Schnarf (1939) proposed that monosporic eight-nucleate (*Polygonum* type) embryo-sac is most primitive type, from which all other types have been evolved. This idea has been widely supported and it also gets support from female gametophytes of Pteridophyta and Gymnosperms which are of monosporic type.

The possibility of the origin of all other types from the monosporic eight-nucleate embryo-sac made this view more acceptable. The derivation of *Oenothera type* of embryo-sac presents no difficulty as this would require only two divisions to take place between the functioning megaspore stage and the differentiation of egg apparatus. The absence of one division only results in the formation of four nucleate embryo-sac found in *Oenothera.*

Both bisporic type of embryo-sacs, i.e. *Allium* and *Endymion types* can similarly be evolved. In these cases, the first meiotic part of division is accompanied with wall formation, but no wall formation takes place after second divison. As a result two dyad cells are formed, of which only one remains functional. The two megaspore nuclei of functional dyad would then require to undergo two divisions only to form the eight-nucleate stage.

He further proposed that failure of wall formation both in first and second divisions of meiosis in megaspore mother cell results in the derivation of tetrasporic types of embryo-sacs. The four megaspore nuclei lying in the same membrane show variable arrangements. For example, in *Peperomia* and *Adoxa* types, the arrangement is 2+2 (two nuclei at the micropylar part and two nuclei at chalazal part of the embryo-sac). In *Plumbagella, Drusa* and *Fritillaria types,* the arrangement is 1+3 (one at the micropylar and three at the chalazal end). Two divisions in them give rise to *Drusa type*. The fusion of three chalazal nuclei during first mitosis results in a secondary four-nucleate stage and this gives rise to *Plumbagella* and *Fritillaria types*. One more mitotic division in these nuclei gives rise to *Fritillaria type* while its absence to *Plumbagella type*.

Accepting these propositions, it becomes evident to accept that the monosporic eight-nucleate embryo-sac is the primitive and all other types have evolved from it. The *Allium* and *Adoxa types* are believed to have evolved as a result of reduction in *Polygonum type*, while *Oenothera type* is reduced *Polygonum type, Plumbago* is reduced *Penaea type* and *Polygonum type* is a reduced *Fritillaria type*.

GENERAL QUESTIONS

1. Describe the development of female gametophyte in angiosperms.
2. Describe briefly about different types of embryo-sac met within angiosperms.
3. With the help of diagrams only, trace the development of megaspore till the stage of fertilisation.
4. Write short notes on-
   (i) Filiform apparatus
   (ii) Egg apparatus
   (iii) Tetrasporic embryo-sac
   (iv) Central cell
5. Briefly describe the development of *Polygonum* type of embryo-sac.

CHAPTER 6

Pollination

The term pollination refers to the transfer and deposition of pollen on the stigma of the flower. It may be of two types.

(i) *Natural pollination,* which occurs naturally in the plants, and

(ii) *Induced pollination,* which is carried out by artificial means, to produce plants with better desirable characters. Plant breeders commonly refer the term *hybridization* to the process.

Only natural pollination is described here. It is mainly of two types.

I. Self pollination

II. Cross pollination

SELF POLLINATION

It involves the transfer of pollen grains from the anther to the stigma of the same flower (it is called AUTOGAMY) or another flower of the same plant (it is called GEITONOGAMY). Self pollination regularly occurs in groundnut (*Arachis hypogea*), tobacco (*Nicotiana*), barley (*Hordeum vulgare*), oat (*Avena sativa*), wheat (*Triticum*), pea (*Pisum*) etc.

Factors Favouring Self Pollination
(Contrivances for self pollination)

Self pollination is promoted by certain floral adaptations. These are-

1. **Homogamy.** It refers to maturation of the sex organs at the same time (simultaneous maturation of sex organs). Self pollination is possible only in those flowers which show homogamy. It is commonly found in *Mirabilis jalapa, Argemone mexicana, Tagetes indica, Vinca rosea* etc. These flowers develop likewise structural adaptations, e.g. (i) in *Mirabilis*, the filaments of stamens and the style get spirally coiled, (ii) in *Argemone*, pollen grains shed on the hollow of the petals and get in contact with the stigma when the flower closes and (iii) in *Vinca*, the anthers are situated near the narrow throat of the corolla tube and when stigma emerges out, it receives pollen.

2. **Cleistogamy.** In certain plants, flowers do not open at the time of fertilization and do not expose their sex organs. This phenomenon is known as

cleistogamy. In cleistogamous flowers, there are three characters to favour self pollination, e.g.

(i) flowers are underground,

(ii) flowers are bisexual and

(iii) flowers never open.

Cleistogamous flowers are found in *Commelina bengalensis* and some members of family Violaceae, Balsaminaceae and Polygonaceae. However, in *Commelina bengalensis*, both cleistogamous and chasmogamous flowers are found. Cleistogamous flowers are found underground, whereas chasmogamous flowers are aerial and show cross pollination.

Fig. 6.1. A plant of *Commelina bengalensis*.

CROSS POLLINATION

The transfer of pollen grains from the anther of a flower to the stigma of another flower is known as cross pollination or *allogamy*. Term *xenogamy* is also used but it refers to pollination between flowers of different genetic constitution.

Factors Favouring Cross Pollination

(Contrivances for cross pollination)

Cross pollination is favoured by certain adaptations. A few of them are as follows.

1. **Dicliny or Unisexuality.** Plants bearing unisexual flowers possess only one kind of sex organ, i.e. either male or female. As such cross pollination is the only possible method in such plants, as in papaya (*Carica papaya*), date palm (*Phoenix sylvestris*), mulberry (*Morus alba*) etc.

2. **Dichogamy.** It refers to maturation of sex organs at different time periods, hence self pollination becomes impossible. Dichogamous flowers may show two conditions.

(i) ***Protandrous.*** When anthers mature before the stigma, the condition is called protandrous. Hence when anthers dehisce, the pollens are carried to another flower by pollinating agents. It has been observed in members of Labiatae, Malvaceae and Rubiaceae.

(ii) ***Protogynous.*** When female sex organ, i.e. stigma matures before the dehiscence of anthers, it is called protogynous. Such stigma receives pollen from other sources, thereby bringing about cross pollination, as in *Magnolia, Michaelia* and *Adhatoda*.

3. **Heteromorphy.** The phenomenon of bearing two types of morphologically

dissimilar flowers is known as *floral heteromorphism*. The heteromorphic flowers bearing two different kinds of stamens and styles are called *dimorphic flowers*. The most common example of dimorphism is the primrose (*Primula vulgaris*) which shows distyly. It bears two types of flowers.

(a) ***Long-styled or pin-eyed.*** The flowers bear carpels with long style and long papillate stigma but short stamens and small pollen grains.

(b) ***Short-styled or thrum-eyed.*** The flowers bear carpels with short style, small stigma but long stamens and large pollen grains. Some species of *Oxalis, Linum* and *Lythrum* show trimorphism (fig. 6.2). The difference in the length of the stamens and style prevents self pollination, hence cross pollination occurs in such flowers.

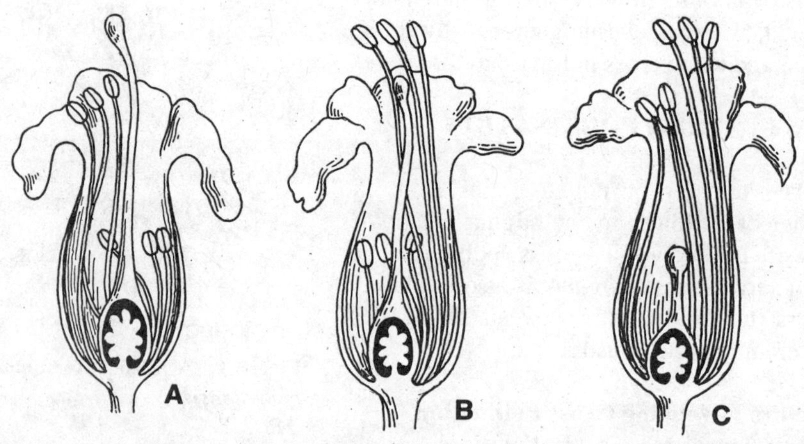

Fig. 6.2. Trimorphism in flowers of *Lythrum salicaria*. The three types of flowers with three different positions of anthers and stigmas so that there is cross pollination involving three heights.

4. **Herkogamy.** The word *herkos* means barrier. In certain flowers, morphological barriers develop which makes self pollination impossible, e.g. in flowers of *Calotropis, Asclepias, Pergularia* (family Asclepiadaceae), *Coelogyne* and *Zeuxine* (family Orchidaceae). In these cases, pollen grains remain aggregated together in the form of pollinia and cannot dehisce. The pollinia are dispersed by visiting insects. The pollinia has a sticky glandular disc or *corpusculum* which gets attached to the leg of the insect and the two connectives get coiled around the leg. When the insect carrying pollinia visits another flower, cross pollination is carried out. This kind of pollination is also called *clip mechanism* or *translator mechanism*.

The presence of extrorse anthers in a flower is another example of herkogamy as found in *Gloriosa*.

5. **Self sterility.** Flowers in which pollens are incapable of causing fertilization are called *self-sterile* or *self incompatible*. Such plants which are unable to produce fertile pollens are called *male steriles*. Male sterility has been recorded in

maize, wheat, sorghum, barley, cucumber, carrots, tomatoes, onions, sunflower etc. Male sterility may be due to

(i) a recessive gene in a chromosome

(ii) an organelle other than nucleus (i.e. cytoplasmic male sterility), or

(iii) an interaction of both nuclear and cytoplasmic genes.

Plant breeders have considerably used male sterility for the production of hybrids in maize and sunflower.

Agents of Cross Pollination

In angiosperms, pollens are immotile and thus, have to be carried to the stigma by external pollinating agents. Depending upon their nature, the agents may be—

(i) Biotic, e.g. insects, birds, snails etc. or

(ii) Abiotic, e.g. wind and water.

Some important examples are discussed here.

BIOTIC AGENTS

1. **Entomophily.** It refers to the pollination brought about by insects. It is by far the most effective and popular method. Entomophilous flowers develop certain adaptations. These have bright, attractive petals and produce odours to attract insects. The colour and odours may be of different kinds. Flowers of *Colocasia, Arum* and *Rafflesia* attract carrion flies by producing smell like rotten decaying meat. The pollen grains of insect pollinated flowers are rough and sticky and produced in fewer numbers. The stigmas are expanded or lobed and rough so that they can readily receive the pollen. A large number of such flowers produce nectar. Bees are the chief visitors of such flowers and in turn bring about pollination. This is in accordance with the principle *quid pro quo* (i.e. some thing for some thing), in which the bees take nectar from the flowers and in return, pollinate the flowers.

Many plants are pollinated by butterflies. Moths which generally visit flowers at night are active pollinators and for them, fragrance is the most important attractant.

Salvia, Yucca, Ficus and *Calotropis* are other important insect pollinated plants.

2. **Ornithophily.** Pollination carried out by birds is called ornithophily. Such flowers have usually brightly coloured petals and produce nectar to attract birds. *Bignonia capreolata* is pollinated by small humming birds, *Strelitzia* is pollinated by honey birds and *Aloe* by female sun birds.

Some flowers show variations in shape, as in red silk cotton (*Salmalia*), coral tree (*Erythrina*) and bottle brush (*Callistemon*). *Salmalia* is pollinated by crows and myenas. These birds obtain staple food from flowers in the form of nectar.

3. **Cheiropterophily.** Pollination by bats is called cheiropterophily. Bats are nocturnal animals and the flowers they visit are large and have a strong scent. Bats move swiftly and transport pollens over long distances, sometimes upto 30 km. Such flowers possess dull colours and emit strong musty odour which attracts the bats. They also produce large quantities of nectar, which is lapped by the bats

and in the process, pollination is affected. Sausage tree (*Kigelia pinnata*) is bat pollinated. Some of these flowers produce a large number of prominent stamens, e.g. baobab tree (*Adansonia*) has some 1500-2000 stamens. *Bauhinia megalandra* and *Eperua falcata* are also bat pollinated.

4. **Malacophily.** Pollination by snails and slugs is called malacophily. It is not a common method, rather it usually takes place in subterranean runners and creepers which grow closely adhering to the soil. The flowers are small and inconspicuous and emit an earthy odour to attract snails and slugs.

ABIOTIC AGENTS

1. **Anemophily.** Pollination by wind is called anemophily. Most grasses and cereals like wheat and maize, bhang, coconut, date-plam etc. are wind pollinated. Such flowers are small, inconspicuous, lacking fragrance and nectar. The perianth lobes are reduced, or some times absent. These produce large quantities of pollens. A maize plant produces about 18,500,000 pollens whereas *Mercurialis annua* produces about 13,52,000,000 pollens and a bhang (*Cannabis*) plant produces over 5,00,000 pollens.

The pollen grains are small, smooth and dry. These plants in majority bear unisexual flowers.

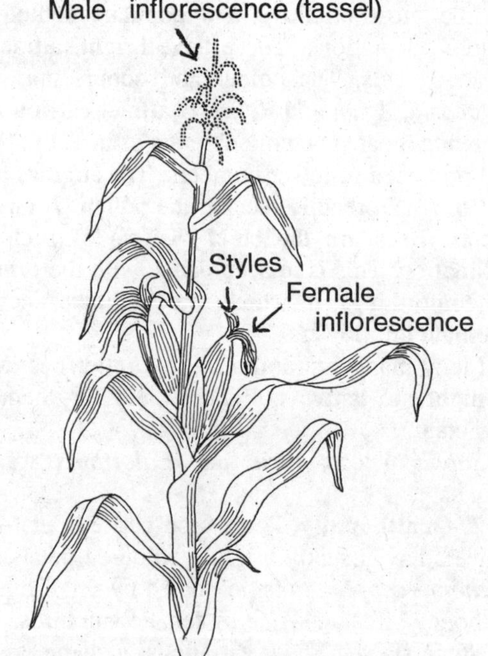

Fig. 6.3. A maize plant (monoecious plant).

In *male flowers*, the filament of each stamen elongates beyond the perianth and exposes the anthers to the wind. Anthers are versatile so that they can swing in all directions and dispose their pollens, as in most grasses. In *female flowers*, stigmas may be large and feathery as in grasses, or brush-like as in *Typha* or have entire stigmatic surface receptive as in *Quercus.* In Maize, the female flower bears a very long (longest in plants) and silky style to catch pollen.

One of the finest examples of anemophily is the phenomenon of *sulphur rain* or *sulphur shower* in the hills. Enormous quantities of yellow, powdery, winged pollen grains from pine forests blow down the hill sides giving the appearance of yellow, sulphur rain.

2. **Hydrophily.** Pollination by water is called hydrophily. It may be of two types-

(i) ***Ephydrophily.*** It refers to pollination occurring on the water surface as in *Elodea, Hydrilla* and *Vallisnaria.*

(ii) ***Hyphydrophily.*** It refers to pollination occurring beneath the water surface as in *Najas, Zostera, Ceratophyllum* etc. *Zostera marina* exhibits a high degree of floral adaptation to facilitate pollination. The pollen grain is long, upto 250 μm, and needle-like and stigma is also long and exposed. The pollen floats on water surface and on coming in contact with the stigma coils around it and gets attached to the stigmatic surface, where pollen germination proceeds.

A summary of pollination and its agencies may be enumerated as follows.

POLLINATION

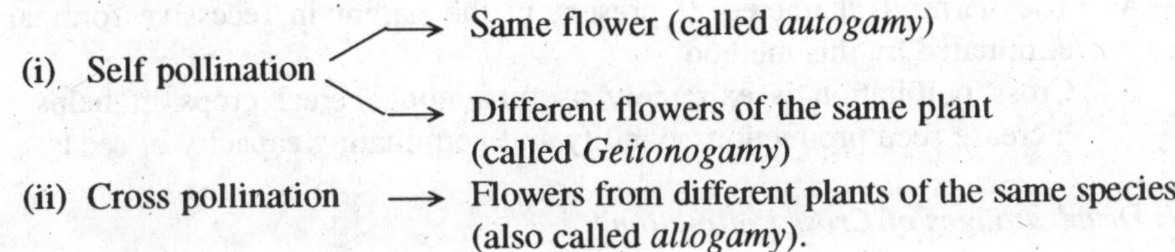

AGENCIES OF POLLINATION

(i) Biotic-
1. Entomophily (pollination by insects)
2. Ornithophily (pollination by birds)
3. Cheiropterophily (pollination by bats)
4. Malacophily (pollination by snails)

(ii) Abiotic-
1. Anemophily (pollination by wind)
2. Hydrophily (pollination by water).

ADVANTAGES AND DISADVANTAGES OF POLLINATION

Some important points are as follows:

I. *Advantages of Self Pollination*

1. Fertilization is fully ensured in such bisexual flowers where stamens and carpels mature at the same time.
2. Parental characters are preserved indefinitely.
3. There is no wastage of pollen grains because fertilization process may be successfully carried out with small number of pollen grains.
4. The flowers need not to be showy and large.
5. Fragrance and nectar production is not required.
6. By these economy, a plant saves considerable nutrient material which would otherwise be used in the production of pollen, nectar, scent and large petals.

II. *Disadvantages of Self Pollination*

1. Continued self-pollination generation after generation may lead to weakening of the offspring.

2. Seeds produced through this method are poor in quality and give rise to weak offsprings.
3. The defective and weaker characters of plants cannot be eliminated.
4. New varieties cannot be produced by self-pollination. Thus, there is a little chance for improvement in the next generation.

III. *Advantages of Cross Pollination*

1. The offsprings are healthier.
2. A large number of viable seeds are produced.
3. New varieties may be produced by cross pollinating two different varieties of the same species or even two species.
4. The harmful characters if present in the parent in recessive form are eliminated by this method.
5. Cross pollination is extremely useful among cereal crops. It helps to increase seed production, viability and germinating capacity of seeds.

IV. *Disadvantages of Cross Pollination*

1. The pollination is not always certain because some pollinating agent is always required which may or may not be available at the proper time.
2. A large number of pollens are to be produced to ensure chances of pollination which means a lot of wastage of pollens.
3. The process is uneconomical for the plant because the flowers have to be large, coloured, scented and have to produce nectar for attracting pollinating agents.
4. If in a crossing, a diseased plant gets involved, it will lead to weak and diseased offsprings.
5. A significant disadvantage is the possible appearance of harmful characters, in dominant form, in the offsprings.

SOME SPECIAL EXAMPLES OF POLLINATION

1. Salvia (Garden sage) (Fig. 6.4)

It is an example of insect pollinated (entomphilous) plant. The flowers exhibit various morphological adaptations and contraptions to trap the insects and ensure pollination. Flowers have bilipped corolla and dimorphic stamens functioning on *lever mechanism*. One half of each anther is fertile and the other half is sterile and both the fertile and the sterile lobes are situated on the lower corolla lip forming a platform for the insects. When a bee alights on the lower lip, it pushes the sterile lobe and the upper fertile lobe automatically comes down and sheds its pollen on the back of the bee. The pollen-carrying bee now visits another flower, and the hanging stigma rubs against the pollen-laden back of the bee, and thus pollination is affected.

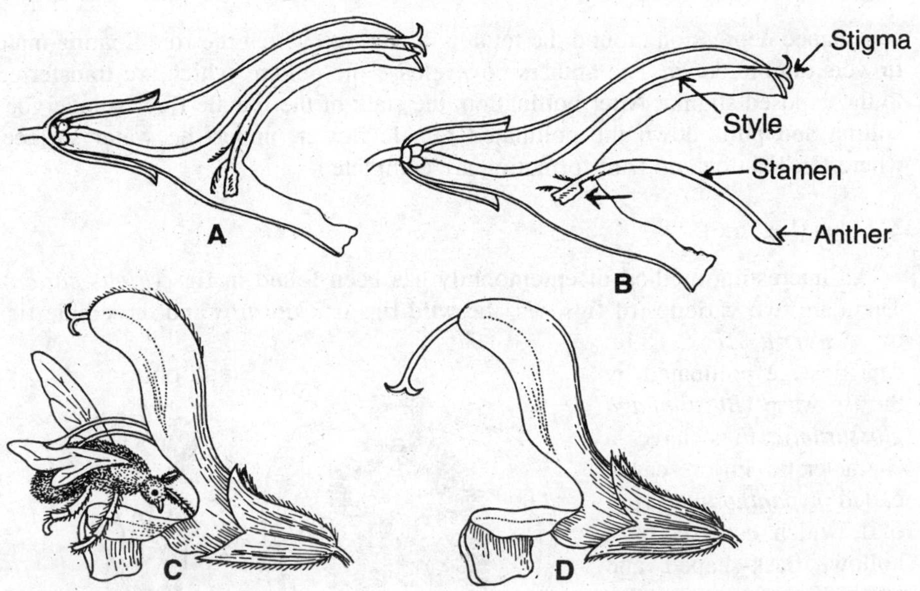

Fig. 6.4. Pollination in *Salvia*. A-L.S. of flower showing position of pistil and stamen, B-L.S. of flower showing movement of stamen when pressed in the direction of the arrow, C-the bee (pollinating agent) being dusted with pollens, and D-a flower with mature stigma (hanging stigma) ready to contact the pollen-laden back of bee.

2. Vallisnaria (Ribbon weed) (Fig. 6.5)

It is a submerged aquatic plant, but exhibits a highly interesting example of ephydrophily. Plants are dioecious bearing male and female flowers under water. The male flower is represented by three tepals (perianth) and two stamens. The stamens have elongated filaments which push the anther lobe beyond the perianth. On attaining maturity, the male flowers get detached from the plant and float on the water surface. The female flowers are also borne under water on slender stalks. As the female flowers approach maturity their stalks begin to elongate, thereby pushing the flowers upwards to the water surface. On reaching the water surface, a change in position of the flowers from vertical to horizontal takes place. The horizontally placed flowers exert pressure, which results in a small,

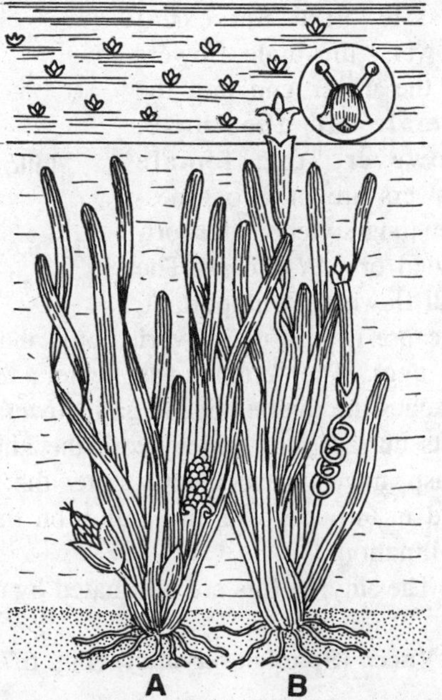

Fig. 6.5. Pollination in *Vallisnaria spiralis*. A-a submerged male plant and B-a submerged female plant. In inset (circle) a male flower with two erect stamens is shown.

cup-shaped depression around the female flower into which the free floating male flowers tumble down. The anthers now release the pollens which are transferred to the exposed stigma. After pollination, the stalk of the female flower undergoes coiling and pulls down the pollinated female flower, under the water surface, where fertilization and fruit formation are completed.

3. Ficus (Fig. 6.6)

An interesting method of entomophily has been found in figs (*Ficus carica*). There are two varieties of figs, *viz.* the wild figs or *Caprifigs* and the edible figs or *Smyrna figs*. The caprifigs are pollinated by the fig wasp (*Blastophaga glossurum*). Figs have a characterstic inflorescence called *hypanthodium* (fig. 6.6), which consists of a hollow, flask-shaped and succulent receptacle, enclosing a cavity, which opens outside by an apical pore. The inner surface of the hollow receptacle bears sessile and unisexual flowers, the male flowers at the upper end and the female at the base. However, the female flowers are of two types- the long-styled and short-styled or *gall flower*. The gall flowers are important, because it is in their ovaries only that the wasp, *Blastophaga glossurum* can lay its eggs. The female wasp enters the hypanthodium through the apical pore, reaches the ovaries of the gall flowers, deposits its eggs and while emerging out, gets dusted with pollen from the anthers of male flowers. The pollen-carrying wasp may now accidently enter the hypanthodium bearing long-styled flowers, and in turn deposits the pollen on the stigma of these flowers, thereby causing pollination.

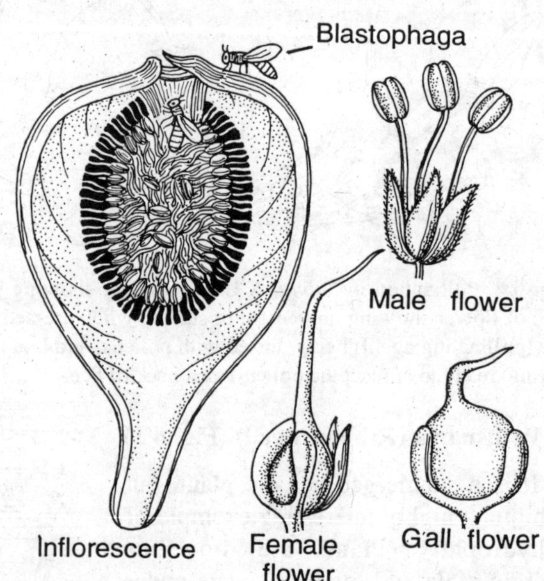

Fig. 6.6. L.S. of fig inflorescence with male, female and gall flowers shown separately.

The smyrna figs are pollinated by caprifig wasp in a similar manner.

4. Yucca (Spanish bayonet) (Fig. 6.7)

Yucca represents a remarkable and striking example of total interdependence between a plant and an insect. It is a xerophytic plant bearing white pendulous flowers, borne on a long scape. In *Yucca* flowers, the styles are hollow, with a longitudinal canal, at the lower end of which the stigmatic surface is located so that pollination is not possible unless the pollen reaches the bottom of this tube. And also within the flower, the stigma extends beyond the anthers, hence self pollination is not possible. The only insect capable of pollinating *Yucca* flower is yucca moth (*Tageticula*). The flowers open at night and exude a heavy fragrance, which attracts the moth. The female moth visits the *Yucca* flowers at night, and collects the pollen in the form of balls, held in specially constructed maxillary palps. The moth then visits another flower, penetrates ovary and deposits its fertilized eggs amongst the ovules, after which it reaches the stigma and presses the pollen ball into the stylar canal. Thus, pollination is brought about. The relationship between the two is a classical example of obligate symbiosis, in which both flower and insect help each other.

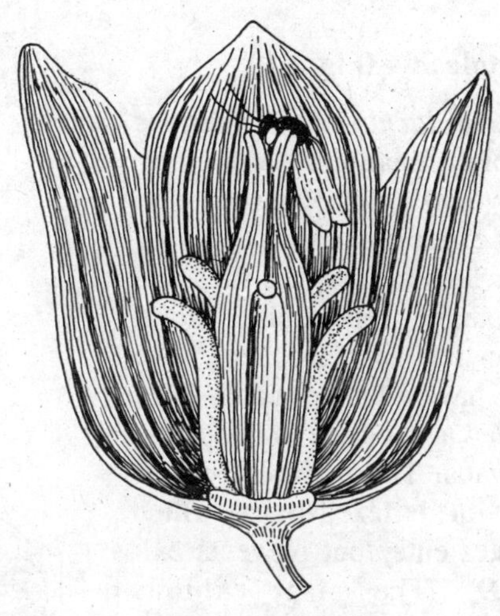

Fig. 6.7. Flower of *Yucca* with insect.

The important points of *Yucca* and moth relationship are these- (i) that *Yucca* flowers can be pollinated in nature only by *Yucca*-moths, (ii) that the larvae of moths can derive nutrition only from *Yucca* ovules and from no other kind of food, and (iii) that these moths can complete their life cycles only with the aid of each other. Thus, it is an amazing example of parallel evolution of plants and insects.

5. Ophrys (an orchid) (Fig. 6.8)

Ophrys speculum, an orchid, exhibits an unique mode of entomophily, by a hairy wasp (*Colpa aruea*). The wasp lives in subterranean burrows but mates above the ground, however, the male wasp emerges about four weeks before the female wasp. The time of emergence of the male wasp coincides with the opening of the orchid flowers. Interestingly, the orchid flowers not only resemble the female wasp, but also possess the same odour. The male wasp, being inexperienced, considers the *Ophrys* flowers to be the female

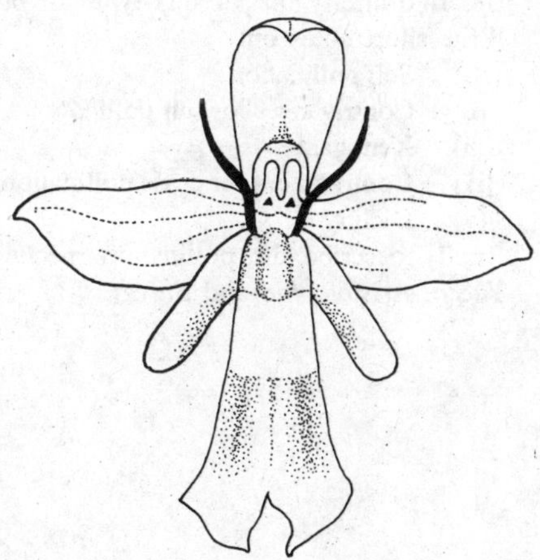

Fig. 6.8. A flower of *Ophrys*.

wasp and enters it for copulation. It repeats the pseudo-copulation act with a number of flowers and thereby causes pollination in a very effective manner.

6. Aristolochia (Fig. 6.9)

In *Aristolochia,* entomophilous mode of pollination is ensured by the *fly-trap mechanism.* The flowers exude a pungent, decaying odour which attracts the flies and also exhibit remarkable floral modification to trap them. The flowers have a long perianth tube, expanded at the base to house the sex organs. The perianth tube is lined with deflexed hairs, which facilitates entry but prevents exit (fig. 6.9). The plant exhibits protogyny, i.e. the pistil matures before the anthers mature. On maturity, the foetid smell of the flower attracts the small flies which slip down the deflexed hairs and fall on the stigma. If they are carrying pollens, some of these fall on the stigma, thus effecting pollination. Later, when anthers mature, the flies get dusted with pollens, meanwhile the hairs dry up, and the tube bends backwards, making way for the exit of flies.

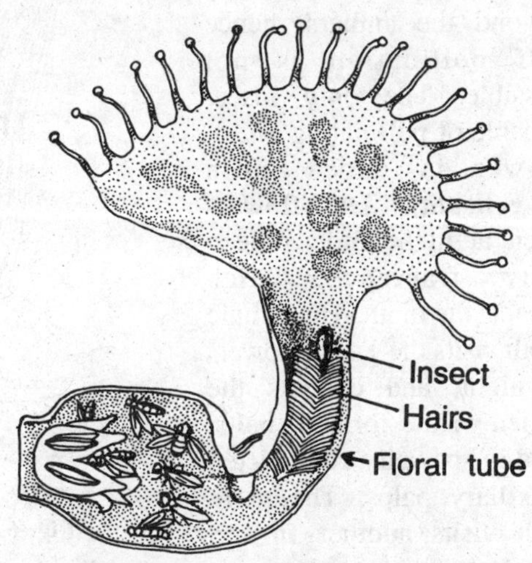

Fig. 6.9. A flower of *Aristolochia.*

GENERAL QUESTIONS

1. Describe briefly the various types of pollination met in angiosperms.
2. Write short notes on:
   - (i) Self pollination
   - (ii) Contrivances for self pollination
   - (iii) Xenogamy
   - (iv) Contrivances for cross pollination
   - (v) Dicliny
   - (vi) Herkogamy
   - (vii) Cleistogamy
   - (viii) Geitonogamy
   - (ix) Anemophily
   - (x) Entomophily
3. Briefly describe the pollination mechanism found in *Ficus, Vallisnaria, Zea mays, Yucca, Aristolochia* and *Salvia.*

CHAPTER 7

Fertilization

The fusion of male gamete with the female gamete is known as fertilization. In majority of angiosperms, the female gamete being located within the ovary is quite far away from the stigma, the site of deposition of male gamete. Hence to affect fertilization, it is essential that the male gamete be transported from the stigma, via the style to the ovary. Thus, the process of fertilization comprises two important steps-

(i) Germination of pollen and the formation of pollen tube.
(ii) Fusion of male gamete with the female gamete (the egg).

A pollen grain usually germinates to form only one pollen tube, however, polysiphonous pollens are found in members of family Malvaceae and Cucurbitaceae. In members of Amentiferae, branched pollen tubes are found.

Growth of pollen tube

Pollen grains are resting structures. The first essential requirement for pollen germination is the high relative humidity. The stigma possesses a specialised surface for catching and retaining the pollen. However, its main function is to provide water and essential nutrients, in the form of exudates, necessary for pollen germination. These exudates are secreted by the epidermal stigmatic papillae and accumulate in the schizogenous cavities present beneath the epidermis of the stigma. The stigmas which secrete exudates are called *wet stigmas*, as in *Petunia,* whereas stigmas which do not secrete exudates are called *dry stigmas* as in *Gossypium*. The exudate is highly viscous and adhesive with a high refractive index and appear as droplets on the surface of stigma. Chemically, it consists of lipids, phenols, free sugars, amino acids, peptides and proteins. Functionally, it helps in retention of the pollen grain and aids in germination.

A fully developed pollen tube has a compartmentalised appearance due to the presence of transverse callose plugs, which originate as a ring on the inner walls, grow inwards towards the centre and finally seal the tube. These plugs prevent the backflow of cytoplasm and nuclei.

Pathway of pollen tube

The pollen grain adheres to the stigmatic papilla, where it imbibes water, swells up, ruptures along the germ pores and the intine comes out in the form of a pollen tube. The pollen tube traverses the surface of the stigmatic papilla or passes through the cellulose-pectic layer of the papilla walls and finally reaches the base of the papilla, from where it migrates through the intercellular spaces of the stigmatic tissue. Further growth of the pollen tube depends upon the structural nature of the style. Structurally, there are three types of styles-

(i) ***Solid style***—This type of style is found mainly in dicots and is characterised by the presence of a central core of elongated, highly specialised cells, called transmitting tissue. The pollen tube makes its way through the pectin-rich wall layer of the transmitting tissue or through the intercellular spaces, by the enzymatic activity of the tip.

(ii) ***Hollow style***—Such style is commonly found in monocots and is characterised by a canal running continuously from the stigma to the base of style. The canal may contain a secretional fluid (as in *Amaryllis*) or may remain dry (as in *Crocus*). The secreted fluid contains a nutritive fluid for the growing pollen tubes.

(iii) ***Semi-solid style***—This type of style is found in *Artabotrys* and in members of family Cactaceae. Structurally, it is of intermediate type between solid and hollow style and the transmitting tissue occurs only on one side of the stylar canal.

Entry of Pollen tube into the ovule

The pollen tube after traversing the style reaches the ovary to enter into the ovule and then into embryo sac. There can be three methods of pollen tube entry.

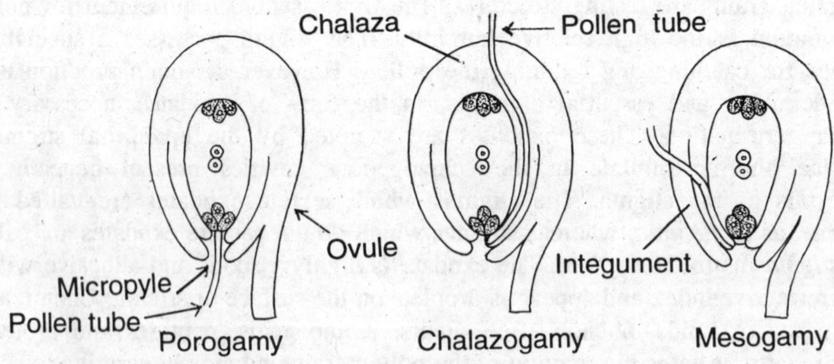

Fig. 7.1. Mode of pollen tube entry into the ovule.

(i) ***Porogamy***. In majority of plants, the pollen tube enters the ovule through the micropyle. This mode of entrance is called porogamy and the fertilization process as *porogamic fertilization.* Micropylar entrance of pollen tube and its growth towards the embryo sac is considered as a chemotropic movement. Growth of pollen tube may be facilitated by special organ called obturator which develops in the form of swelling, from the placenta and functions like a bridge for the pollen tube. Sometimes, the distal cells of integument dissolve *in situ* and secrete a mucilaginous secretion to facilitate the movement of pollen tube (as found in *Paspalum orbiculare*).

(ii) ***Chalazogamy.*** During fertilization, when pollen tube enters through the chalazal end to reach the ovule, the process is called chalazogamy, as in case of *Casuarina*.

(iii) ***Mesogamy.*** When pollen tube enters through the integuments as in *Cucurbita*, the process is called mesogamy and the mode of fertilization as mesogamic fertilization.

However, the mode of entry of pollen tube has got no phylogenetic significance. It may show variations. In case of *Brassica oleracea*, both porogamy and mesogamy have been reported. In certain cases, a distinct micropyle is absent and the embryo sac elongates greatly so as to meet the pollen tube at a certain distance up in the style, as in some members of Loranthaceae.

Penetration of Pollen tube into the embryo sac

Whatever may be the mode of entry of pollen tube, it grows through the nucellar tissue and penetrates into the embryo sac through the micropylar end. The penetration of pollen tube into the embryo sac may take place by three ways-

(i) between the wall of embryo sac and one of the synergids,
(ii) between the egg and one of the synergids, or
(iii) directly into one of the synergids.

Electron microscopic studies have shown that the pollen tube enters at the apical region of the filiform apparatus, grows through it and reaches the cytoplasm of the synergid. The pollen tube discharges its contents into the synergid, but does not grow beyond it, into the embryo sac. Usually, the penetrated cell is destroyed due to the entrance of the pollen tube, whereas, the other synergid degenerates after some time.

Discharge of Pollen tube contents

The process of discharge takes place in seconds and includes two sperms and the vegetative nucleus or its remains and the cytoplasm. The cytoplasm of the pollen tube does not mix with the cytoplasm of the synergid and both remain as separate entities. Prior to discharge, the tip of the pollen tube swells up and some times bifurcates into two short branches. Very rarely, the pollen tube tip does not rupture, but sub-terminal openings are formed, through which the contents are released. (Fig. 7.2.B)

The two sperms are true cells, each bound by a plasma membrane, and possessing the usual cytoplasmic organelles. Motility of the sperms is associated with the microtubules present within the cytoplasm.

Passage of sperms

The pollen tube discharges the two sperms into the synergid, in the form of intact cells. It was generally accepted that only the male nuclei migrate out leaving cytoplasm behind. Recent studies have revealed that male cytoplasm is also involved in fertilization. Synergids are believed to function in three ways—(i) it secretes some substances which exercise a chemotactic influence over the pollen tubes and directs their growth, (ii) the degenerating synergids form the seat for pollen tube discharge in the embryo sac, and (iii) the synergids have nutritive role. However, the nutritive role of synergids is doubted.

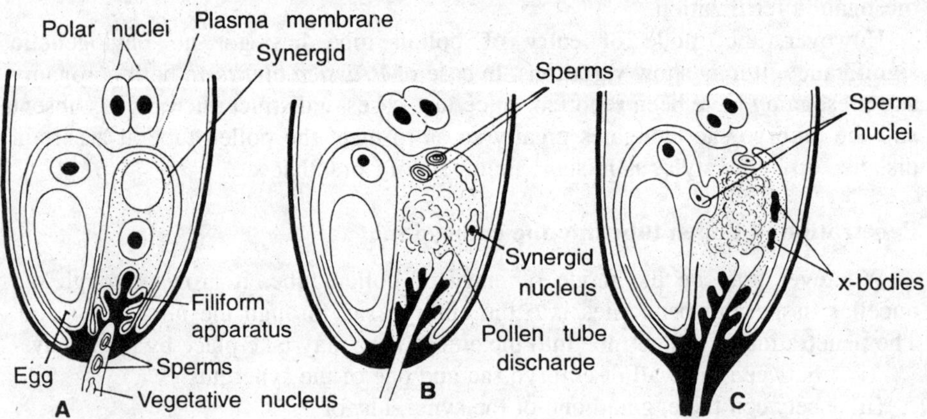

Fig. 7.2. Sperm transfer in embryo sac. A-entrance of pollen tube, B-discharge of pollen tube and C-proposed method for gamete transfer.

Fusion of gametes (Double fertilization)

After the pollen tube has discharged its contents into the embryo sac, one of the male gametes fuses with the egg (syngamy) and the other fuses with the secondary nucleus (triple fusion). The phenomenon of double fertilization is characteristic to angiosperms only and it was discovered by Nawaschin (1898) in *Lilium* and *Fritillaria*.

Syngamy. The process of fusion of a male gamete with the female gamete is called syngamy. The male nucleus which fuses with the egg is larger than that which fuses with the secondary nucleus. Grassimova and Navashina (1960) recognised three types of syngamy.

(i) ***Premitotic***. The sperm nucleus, on coming in contact with the egg nucleus fuses immediately to form the diploid zygote, which then undergoes division, e.g. Graminae and Compositae.

(ii) ***Post-mitotic***. The sperm nucleus approaches the egg nucleus and remains in contact with it for a short period. Fusion takes place only after both the nuclei have entered into mitotic division as in *Lilium* and *Fritillaria*.

(iii) ***Intermediate***. The sperm nucleus first completes its mitotic division and then fuses with the egg nucleus. However, the contents of the two nuclei show incomplete mixing, as in *Impatiens*.

Triple fusion. In majority of plants, a single polar nucleus (n) from the micropylar end and a single polar nucleus (n) from the chalazal end migrate towards the centre of the embryo-sac and fuse to form the diploid secondary nucleus. During the process of double fertilization, one male nucleus (n) fuses with the secondary nucleus (2n) to form the triploid primary endosperm nucleus (3n). Such a fusion of three nuclei is called *triple fusion*.

Abnormal features

Polyspermy. In certain unusual cases, more than two sperms are released into the embryo-sac. This phenomenon is called *polyspermy*. Normally, only one pollen tube enters the embryo-sac and releases two sperms. But, in rare cases, more than one pollen tube may enter the embryo-sac or the single pollen tube may release more than two sperms. Polyspermy may result in fertilization of egg nucleus by more than one sperm nucleus, to form a polyploid zygote; or the supernumery sperms may fuse with other cells of the embryo sac, such as synergid or antipodals to form multiple embryos (polyembryony). Rarely, the embryo-sac receives more than one pollen tube with several sperms released into the embryo-sac. One of the sperms from one pollen tube fuses with the egg nucleus, whereas, one sperm from other pollen tube fuses with the secondary nucleus. This abnormality is called *heterofertilization*.

Persistent and branched pollen tubes. In majority of angiosperms, the pollen tube after discharging its contents, collapses, starts degenerating and after fertilization completely disappears. But in some cases, the pollen tube may persist for as long as three weeks till the time of embryo development. In *Cucurbita* and some members of Onagraceae, the pollen tube after entering the embryo-sac, begins branching in the apical region. These branches grow between and around the inner and outer integuments and penetrate the nucellus and other ovular tissues, behaving in a haustorial fashion i.e. *haustorial pollen tube*. However, some workers are of the opinion that the branches simply ramify between the ovular tissue and do not have any haustorial function. As such, the haustorial behaviour of the branches of the pollen tube is uncertain.

GENERAL QUESTIONS

1. Describe the process of fertilization in angiosperms.
2. Enumerate the various events from the time of pollen germination on the stigma leading to fertilization.
3. Write an illustrated account of the process of fertilization and embryogenesis in a dicotyledonous plant.
4. What is double fertilization in angiosperm and mention its significance.
5. Write short notes on
   (i) Porogamy, mesogamy and chalazogamy
   (ii) Double fertilization
   (iii) Triple fusion
   (iv) Syngamy
   (v) Polyspermy

CHAPTER 8

Sexual Incompatibility

In natural conditions, a wide variety of pollen grains alight on the receptive stigma, however, not all of them germinate nor bring about fertilization. If a carpel, bearing functional gametes is pollinated with viable and fertile pollens, yet fails to produce seeds, the two individuals (male and female) are said to be *incompatible* and the phenomenon is termed as *sexual incompatibility*. Sexual incompatibility may exist between members of different species, i.e. *interspecific* or between members of same species, i.e. *intraspecific*. The mystery underlying sexual incompatibility have been to some extent unravelled on physiological and biochemical facts and which suggest that the basic mechanism controlling both inter-and intraspecific incompatibility is more or less similar.

Intraspecific or Self Incompatibility

A large number of angiospermic plants have the characteristic to be fertilised successfully, only by the pollen of other plants. Perhaps the most effective and natural method which prevents self pollination is self-incompatibility. Based on morphology of mating types, self incompatibility may be of two types—

(i) **Homomorphic.** In a species, when all the mating types are morphologically similar, it is said to be of homomorphic type. Plants of this type are identified by breeding test only.

(ii) **Heteromorphic.** Heteromorphic type refers to the conditions when two or three distinctly different morphological mating types are present within a species and may be identified without breeding tests. The mating types usually differ in the structure and position of the sex organs, i.e. stigma and anthers. Heteromorphic incompatibility may be regulated by one gene with two alleles as in *Primula* or by two genes with two alleles as in *Lythrum*.

Genic Control of Self Incompatibility

Incompatibility is a gene controlled physiological process. Depending upon whether the genic control is exercised by the male gametophyte or the sporophyte, self incompatibility may be of two types—

(i) **Gametophytic self incompatibility (GSI).** The genotype of the male gametophyte determines the incompatibility process, e.g. *Trifolium* and members of Graminae, Liliaceae and Solanaceae.

East and Mangelsdorf proposed the *opposition-s-alleles* hypothesis, according to which the incompatibility reactions are controlled by single *S* gene having several alleles. If pollen grains possess the *S-allele* common to any one of the two alleles present in the gametic cells of female, the fertilisation will not be effected between the two. This may be illustrated as follows—

Sporophyte S_1S_2 alleles ⟶ Microsporogenesis [S_1, S_2 POLLEN] ⟶ Pollination ⟶ S_1, S_2 OVARY ⟶ Incompatible No fertilization

Sporophyte S_2S_3 ⟶ Microsporogenesis [S_2, S_3 POLLEN] ⟶ Pollination ⟶ S_1, S_2 OVARY ⟶ 50% fertilization with S_3 pollens

Sporophyte S_3S_4 ⟶ Microsporogenesis [S_3 — Functional, S_4 — Functional POLLEN] ⟶ Pollination ⟶ S_1, S_2 OVARY ⟶ 100% fertilization

(ii) Sporophytic self incompatibility (SSI). The process of incompatibility is controlled by the genotype of the sporophytic tissue. In such a system, all pollens behave in a similar manner irrespective of the *S-allele* they carry, and will be non-functional on a S_1S_2 plant. This may be explained as follows—

Sporophyte S_1S_2 ⟶ Micro-sporogenesis ⟶ S_1 or S_2, S_2 or S_3, S_1 or S_3, S_1 or S_4 POLLENS ⟶ Pollination ⟶ S_1S_2 PISTIL ⟶ *No Fertilization*

S_3S_4, S_4S_5, S_3S_5 POLLENS ⟶ Pollination ⟶ S_1S_2 PISTIL ⟶ ***Fertilization***

Physiology and Biochemistry of Sexual Incompatibility

The physiological and biochemical aspects of incompatibility are recognised by two reactions:

(i) ***Recognition reaction.*** Recognition reaction occurs at the molecular level, in the sense that the carpel recognises whether a pollen grain is compatible and should be accepted or is incompatible and should be rejected.

(ii) ***Rejection reaction.*** The rejection reaction sets in with pollination. It may occur on the stigma and is expressed by inhibiting pollen germination or it may occur in the style in the form of inhibition of pollen tube growth.

Physiological and biochemical studies on inter and intra-specific incompatibility have revealed the role played by pollen wall and the importance of pollen proteins in the pollen-stigma interactions. Green (1894) was the first to observe that the pollen grains on hydration released some hydrolytic enzymes, which played a significant role in the pollen-stigma interaction. His discovery was confirmed by various scientists. However, Lewis and his coworkers are credited with the discovery that inter and intra-specific incompatibility is associated with recognition proteins held extra-cellularly in the pollen wall. These sporophytically and gametophytically derived proteins, held in the pollen wall are released on hydration of the pollen.

Stigma-surface inhibition. Various experiments on these lines have revealed that the recognition and rejection reactions occur on the stigma, also, if the pollen which fails to germinate on the stigma, is introduced directly into the style, it causes normal fertilisation. This clearly suggests that stigma is the barrier of pollen germination.

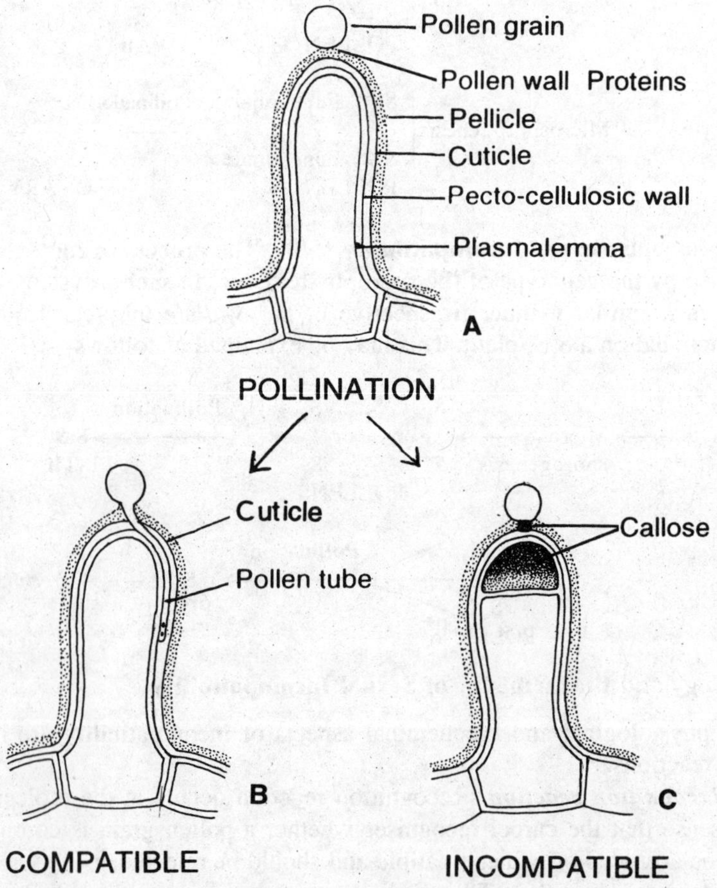

Fig. 8.1. Figure showing pollen stigma interaction in Cruciferae and Compositae. The growth of incompatible pollen is inhibited and it can not cross the deposition of callose (after Shivanna, 1977).

In members of Cruciferae, such as *Brassica* and *Arabis*, the stigma is covered with a cuticle. If the pollen can penetrate the cuticle, it is termed as *compatible pollination*, whereas if the pollen fails to grow through the cuticular layer, it is referred to as *incompatible pollination*. The compatible stigma activates the enzyme cutinase which is responsible for penetrating the cuticle. However, electron microscopic studies do not support this differential behaviour of pollen.

In the light of recent researches, incompatible pollination due to *stigma-surface-inhbition* is based on following features—

(i) The rejection-reaction exhibited by incompatible stigmas lack copious exudates, i.e. remain dry. However, these stigmas possess an extremely active, hydrated layer, the *pellicle* over the cuticle.

(ii) The pellicle comprises of lipid containing a variety of proteins which possess esterase activity.

(iii) The pollen, upon falling on the stigma, stimulates exudation of water by the papillate cells. The hydrated condition of stigma in turn stimulate the outflow of exine-bound proteins.

(iv) The released exine proteins are received by the pellicle, where the pellicle proteins and the exine proteins interact resulting in the incompatible reaction.

(v) An interesting feature of the stigma-rejection reaction to an incompatible pollen is the appearance of the callose plugs just below the pollen contact point. However, the rejection reaction of the stigma is absolutely localised and in no way inhibits the germination of neighbouring pollens.

Stylar inhibition. In gametophytic self incompatibility (GSI) systems, the pollen germinates, followed by growth of pollen tube to two third distance into the style, and then the rejection reaction sets in, stopping further growth of pollen tube. Incompatibility due to stylar inhibition is based on the following characteristics—

(i) The cells of the style secrete S-allele specific proteins which induce incompatibility reactions.

(ii) In *Nicotiana alata*, a high activity of peroxidase-10 has been observed in the styles with incompatible pollen tubes. Based on this observation, the probable involvement of peroxidase-10 in rejection reaction of incompatible pollen tubes, has been suggested.

(iii) In *Petunia hybrida*, the degeneration of the tube cytoplasm and the thickening of the walls of the pollen tube signifies the incompatibility reaction.

(iv) In *Lilium longiflorum*, the tip of the pollen tube with incompatible styles show different structure and are unable to absorb the secretions of the pistil, hence exhibit retarded growth.

(v) The styles of self incompatible species, such as *Lycopersicon peruvianum* exhibit high RNAase activity. It has been proposed that this RNAase of the style enter the pollen tube cytoplasm. This results in checking the growth of the pollen tube.

(vi) The pollen tube in an incompatible style may show various types of abnormal behaviour, such as branching of the tip of the pollen tube, development of callose plugs or restricted growth of pollen tube.

Methods to Remove Sexual Incompatibility

Sexual incompatibility may be removed by regulating the mechanisms which govern it. Some of these methods are—

1. **Pollination methods.** Modifications in the pollination patterns have been found useful in overcoming incompatibility. Pollination changes which have been proved effective are—

(i) ***Bud pollination.*** Intraspecific incompatibility may be overcome by pollination at an early bud stage, as in *Petunia axillaris.*

(ii) ***Intra-ovarian pollination.*** Pollens having stigmatic or stylar incompatibility, can be pollinated by introduction of pollen suspension directly into the ovary. This process has been proved successful in *Argemone mexicana, A. ochroleuca, Papaver somniferum* and *P. rhoeas.*

(iii) ***Mixed pollination.*** The stigma is pollinated by a mixture of incompatible and compatible pollens. The compatible pollen is known as *Mentor pollen* or *Recognition pollen.* Interspecific incompatibility has been considerably controlled by this method in a number of plants, such as *Cucumis, Nicotiana, Cosmos, Petunia* etc.

Stettler and Ager (1984) have proposed a mechanism to explain mentor pollen effects as follows—

Table 8.1. Control of incompatibility by mentor pollens

Site of mentor effect	*Proposed mechanism*
At Stigmatic surface	(i) Mentor pollen provides recognition proteins which permit incompatible pollen to germinate. (ii) Mentor pollen provides P-factor which interacts with S-factor from stigma to render it accessible to incompatible pollen.
In Style	Mentor pollen provides a pollen growth promoting susbtance which permits incompatible pollen to sustain tube growth.
In distal tissue of Pistil	After pollen tube penetration, mentor pollen produce substances which promote growth of ovules, ovary and other fruit tissues

(iv) ***Stub-pollination.*** In certain plants, such as *Ipomoea trichocarpa,* the stigma is the site of incompatibility. If stigma and a small part of style are removed and pollens are dusted on the cut surface (also called stub), a fairly successful pollen germination is achieved.

(v) ***Test tube pollination.*** It was first reported by Kanta and coworkers in *Papaver somniferum.* In this process, stigma, style and ovary are removed and pollens are dusted on the exposed ovules. The pollinated ovules are then cultured in an appropriate nutrient medium resulting in the formation of seeds.

2. ***Irradiation method.*** Sexual incompatibility may be overcome by irradiation, which is supposed to induce cytological changes in the pollen or remove the inhibitory substances in the style.

3. ***Modification of the stigma.*** Self incompatibility in *Brassica campestris* has been reported to be removed by applying NaCl to the stigmatic surface, 10-15 minutes before pollination. NaCl application modified the stigmatic responses and increased adhesion and germination of pollen. It also helped to reduce callose formation, thus facilitating growth of pollen tube.

4. ***Heat treatment of style.*** In certain plants, self-incompatibility may be reduced by supplying high temperature, upto 50°C. In *Lilium longiflorum,* pre-pollination heat treatment of the style for six minutes at 50°C, has been found to successfully remove self-incompatibility.

5. ***Parasexual hybridisation.*** A revolutionary approach for removing sexual incompatibility is the process of parasexual hybridisation. The process involves the protoplasts of somatic cells and is executed by three steps—

(i) *Isolation of protoplasts.* The protoplasts of somatic cells are bound by cellulose cell wall and cemented to neighbouring cells by pectin. The adjacent cells are first separated by treatment with pectinase enzyme and then the cell walls are digested by cellulase enzyme, thereby releasing the protoplast.

(ii) *Fusion of isolated protoplasts.* The naked and isolated protoplasts are made to fuse in the presence of fusogens, such as sodium nitrate and poly- ethylene glycol (PEG). In recent years, electrically stimulated protoplasmic fusion is carried out.

(iii) *Culture of protoplast.* The fused protoplast is now grown in a culture medium enriched with nutrients. Carlson and coworkers (1972) produced first parasexual hybrid between *Nicotiana glallca* and *N. langsdorfii.*

Now using the techniques of protoplast fusion, several somatic hybrids have been raised, e.g. *Brassica campestris* x *B.oleracea, Datura innoxia* x *D. candida, Solanum tuberosum* x *Lycopersicon esculentum, Datura innoxia* x *Atropa belladona* etc. The process of parasexual hybridisation and successful regeneration of crop plants has been reported in alfalfa, citrus, clover, maize, mustard, potato, rice and sandal wood.

GENERAL QUESTIONS

1. Describe briefly the various aspects of sexual incompatibility.
2. Write short notes on-
   (i) Self incompatibility
   (ii) Physiology and biochemistry of sexual incompatibility
   (iii) Methods to remove sexual incompatibility.

CHAPTER 9

Endosperm

Endosperm is a nutritive tissue for the developing embryo in phanerogams (gymnosperms and angiosperms). In *Gymnosperms*, the endosperm is haploid and formed before fertilization, whereas in *Angiosperms*, it is triploid and develops after fertilization.

The endosperm in angiosperms is generally a product of three fused nuclei (i.e. one male nucleus and two polar nuclei).

TYPES OF ENDOSPERMS

In general, the endosperms have been divided into three categories on the basis of pattern of formation, (i) Nuclear endosperm, (ii) Cellular endosperm, and (iii) Helobial endosperm. Davis (1966) reported that out of 228 families of angiosperms, nuclear type of endosperm is found in 116 families, cellular type in 72 families and helobial in 17 families. In the *nuclear type*, the first and subsequent divisions of primary endosperm nucleus are free nuclear type without any immediate cell wall formation. In *Cellular type*, the first and subsequent nuclear divisions are accompanied by wall formation whereas in *Helobial type*, the first nuclear division is followed by wall formation and the subsequent divisions are of free nuclear type.

Nuclear Endosperm

Nuclear endosperm is characterised by free nuclear divisions of the primary endosperm nucleus. The nuclei may remain either free or may become separated by wall formation at a later stage. Usually first few divisions of nuclei are synchronous, i.e. simultaneous, but in later stages, the nuclei are observed in different stages of division. After the formation of a few thousands free nuclei, these are pushed more and more towards the periphery of the embryo sac, due to more and more enlargement of the vacuole. Most of the nuclei become aggregated either towards micropylar end or towards chalazal end of the embryo sac. An interesting condition is found in *Musa errans*, in which some of these nuclei form isolated groups and are called *nodules*. These nodules develop distinct cytoplasmic wall and moving to the centre of the sac, develop as separate endosperm mass.

After free nuclear divisions, cell wall formation begins. Normally, the cell wall formation starts from periphery towards centre or from apex towards base. Rarely, it may start from chalazal end and proceed towards micropylar end, as found in *Elatine*. The entire embryo sac becomes filled with cells or the entire endosperm may remain in the free nuclear state excepting one or two layers of cells towards

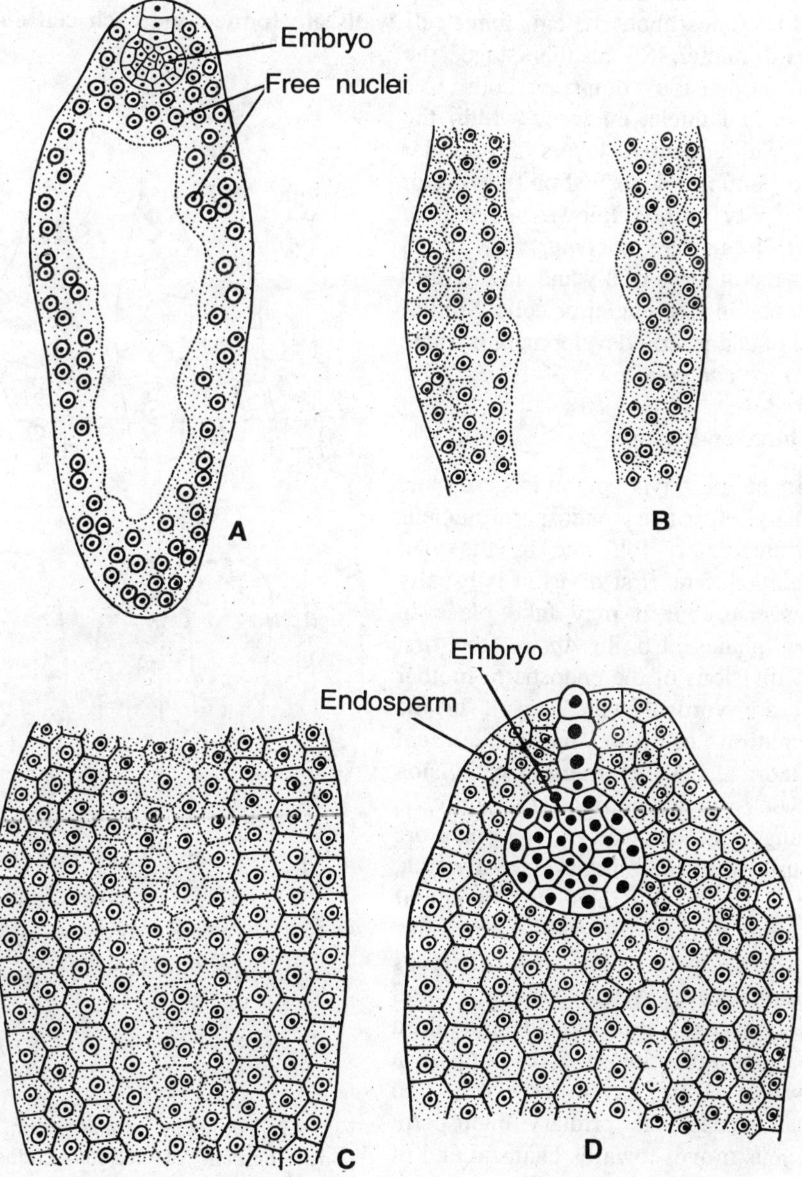

Fig. 9.1. Formation of nuclear endosperm.

periphery. In *Phaseolus*, cell formation may be restricted only to the micropylar part of the sac around the embryo. Cell formation is, however, never found in *Oxyspora* and *Limnanthes*.

Coconut (*Cocos nucifera*) is an interesting example of nuclear endosperm. Initially, it undergoes a number of free nuclear divisions and by the time fruit becomes 5 cm. long, clear water containing free nuclei accumulates in the embryo sac. The fluid is now called *liquid syncytium* or milk of endosperm. When fruit becomes about 10 cm. long, cell walls are formed and each cell encloses several nuclei. So at this stage, the fluid suspension comprises both, free nuclei and nuclei enclosed within the cell. Subsequently, layers of cellular endosperm are deposited on the wall of the cavity and is known as *coconut meat*. In mature coconuts, the liquid endosperm turns milky and now it does not contain free nuclei or cells. Similar type of endosperm development is found in *Areca catechu*.

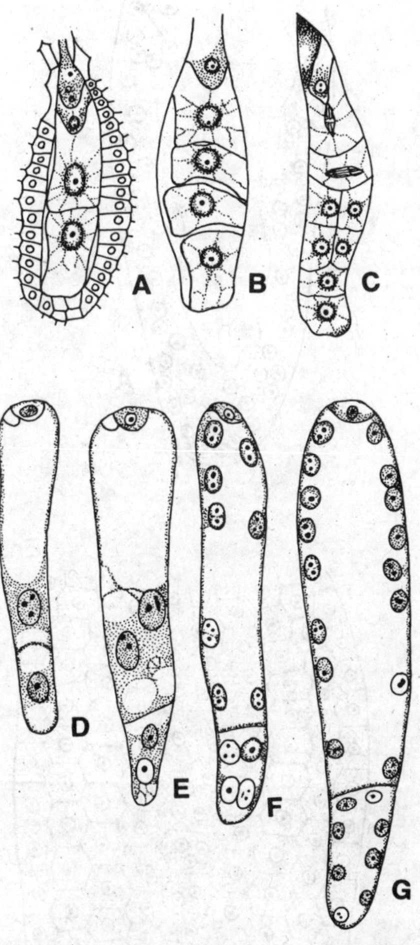

Fig. 9.2. Formation of endosperm. A to C-development of cellular endosperm and D to G-development of helobial endosperm.

Cellular endosperm

In cellular type of endosperm, the division of primary endosperm nucleus is immediately followed by the wall formation. The first division is usually transverse, but it may take place in other planes also. In *Adoxa*, the first two divisions of the endosperm mother cell are vertical which result in the formation of four large identical cylindrical cells. A transverse division is then formed resulting into eight cells arranged into two rows. After one more (fourth division) transverse division, the further divisions are irregular forming a mass of endosperm cells. Among members of families Annonaceae, Aristolochaceae and Boraginaceae, the second and third divisions are transverse resulting in a row of four or more cells. In Acanthaceae, the primary endosperm nucleus moves towards chalazal end of embryo sac and then divides to form two cells of unequal size. The lower small cell is called chalazal chamber while the bigger one is called micropylar chamber which divides and forms a linear row of three cells. The central cell develops into the endosperm proper and remaining cells form the micropylar and chalazal haustoria.

In members of Loranthaceae, a composite endosperm is formed by fusion of the endosperms of different embryo sacs. Since ovule is not found in these plants, the embryo sacs of an ovary come closer to each other and during the course of development, the endosperms of different embryo sacs fuse and form a composite structure.

Cellular type of endosperm formation is found chiefly in dicotyledons, saprophytes and parasites (*Cuscuta* remains an exception).

Helobial endosperm

It is found commonly among monocotyledons. Out of 17 families showing helobial type, 14 belong to monocots. It is of intermediate type between the other two kinds of endosperms. In helobial type, the first division is followed by a wall formation dividing embryo sac into two chambers, the micropylar and the chalazal. The micropylar chamber is usually larger than the chalazal chamber. Several free nuclear divisions in micropylar chamber and only few divisions in chalazal chamber take place.

Stenar (1928) described the endosperm development in *Eremurus* as an example of a typical helobial endosperm. After the formation of two chambers, free nuclear divisions are found in both chambers, but more rapidly in micropylar chamber. When there are four nuclei in chalazal, the micropylar has eight and when there are eight nuclei in chalazal, the micropylar has sixteen and so on. In older ovules, the chalazal chamber begins to show degeneration. When cell formation takes place in the micropylar chamber, the chalazal chamber is almost crushed, showing only a few disorganised nuclei. Thus, the cellular tissue, developed with the nuclei of micropylar chamber, represents the endosperm.

Ruminate endosperm

Sometimes ruminate endosperm is considered as a special case. Periaswamy (1962) described ruminate type of endosperm as one which shows some degree of irregularity and unevenness in its contour. The rumination of endosperm begins at a late stage of development and earlier it may be cellular, nuclear or helobial type. It is not a fourth type of endosperm. It has been found in Annonaceae, Myristicaceae, Rubiaceae, Aristolochaceae, Vitaceae and Palmae where endosperm is mostly rigid and furrowed. There may be two main reasons of rumination:

(i) due to the irregular development of the inner surface of the seed coat, or
(ii) the outer tissue of the nucellus and integuments may invaginate into the endosperm bringing about the ruminate condition.

In *Passiflora*, unequal radial elongation of cells of any one layer of the seed coat brings about irregularity on the inner surface of the seed coat. In *Cocoloba urifera* and *Diospyros chloroxylon*, the irregularities which arise on the inner surface of the seed coat directly affect the surface of endosperm and make it ruminate. In *Degenaria*, the growth of endosperm remains slow and in the meantime, the seed enlarges to its maximum size. It gives rise to a ruminate nucellus. Later, endosperm increases rapidly in size and shows an unequal peripheral activity. The developing endosperm fits itself within the seed coat and ruminates. In *Areca catechu*, the endosperm is ruminate in appearance due to the infolding of the dark inner seed coat into the light coloured endosperm.

Cytology of endosperm

Normally, the endosperm in angiosperms is formed after triple fusion and, therefore, it is triploid in nature. However, certain variations have also been observed. The endosperm in *Oenothera* is diploid, in *Pepromia*, it is multiploid, and in *Balanophora*, it is haploid. In *Petunia*, it is mosaic type consisting of two different colourations. In its formation, the diploid secondary nucleus divides twice mitotically before fertilization resulting into four diploid nuclei. Of the four, only one fuses with the male nucleus and produces triploid tissue. The remaining three divide further and form a mass of diploid tissue. Thus, the endosperm in *Petunia*, consists partly of triploid cells and partly of diploid cells. In Orchidaceae and Podostemonaceae, the endosperm formation is suppressed.

Endosperm haustoria

In certain angiosperms, the endosperm forms haustoria to obtain nutrition from various tissues of the ovule. The absorbed food is transported to the endosperm tissue to be utilized by developing embryo. Such haustoria have been found in all the three types of endosperm. Some examples are discussed here.

IN FREE NUCLEAR ENDOSPERMS

In **Grevillea robusta,** free nuclear type of endosperm is found. Later, its upper part becomes cellular while its lower part develops into tubular and

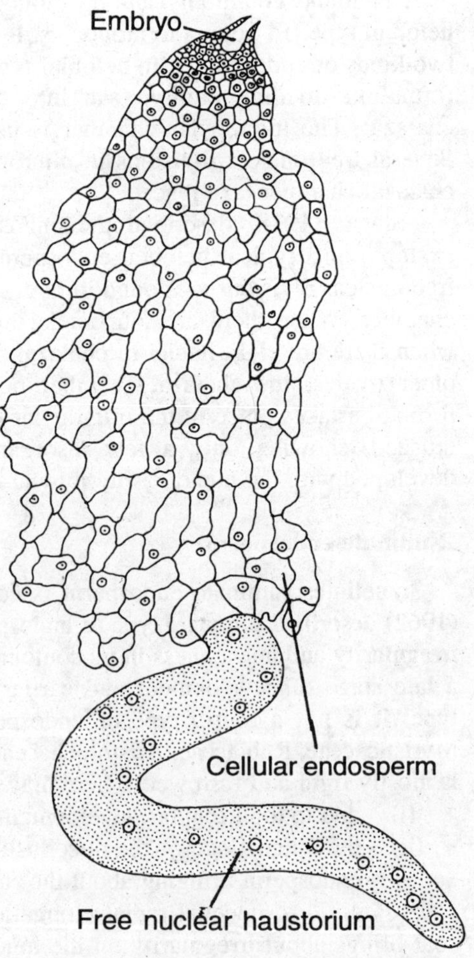

Fig. 9.3. The endosperm haustoria in *Grevillea robusta*.

coiled structure called vermiform appendage which remains non-cellular and multinucleate (fig. 9.3). This appendage functions like haustorium which penetrates chalazal tissues and consumes it. After absorbing the chalazal tissue, the vermiform appendage divides into several multinucleate partitions giving rise to secondary endosperm.

In **Macadamia ternifolia**, the endosperm is nuclear type. Cell wall formation takes place only in the upper part whereas the lower part with few nuclei remains coenocytic and develops several lobed haustorial structure penetrating the nutritive tissue. Such haustorial lobes, after consuming food materials from the basal part of the ovule, become functionless.

In **Lomatia**, while upper part of endosperm becomes cellular, the lower part remains coenocytic and haustorial, absorbing food materials from chalazal tissue. Several unicellular projections arise from this haustorial part for additional absorption of food material.

IN CELLULAR ENDOSPERMS

In **Impatiens roylei**, the development of endosperm is cellular type, and in this case aggressive micropylar haustorium is formed. A profusely branched haustorium develops from the uppermost cell of micropylar chamber. (fig. 9.4.A)

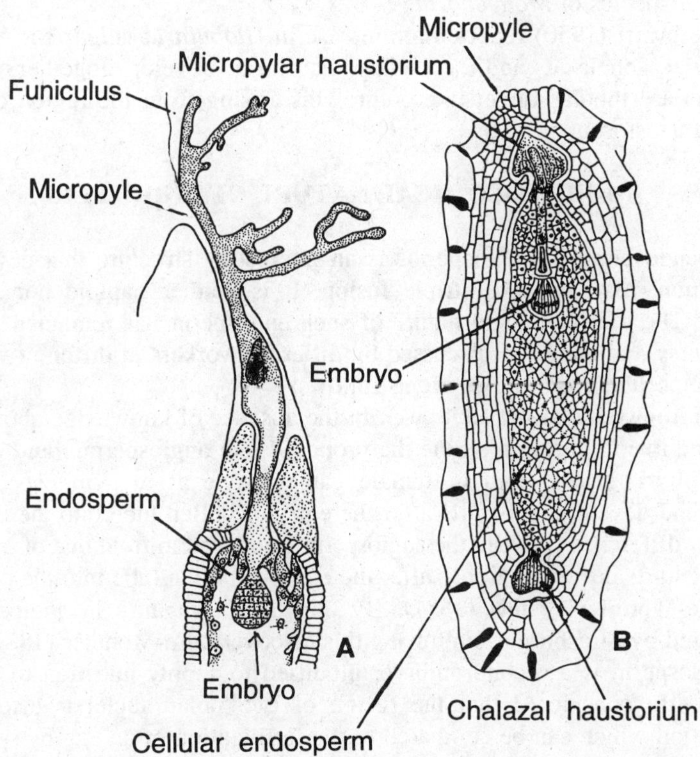

Fig. 9.4. Endosperm haustoria. A-hypha-like branched haustorium in *Impatiens roylei*, and B-both micropylar and chalazal haustoria shown in L.S. of ovule of *Lobelia amoena*.

In **Nemophila,** the primary endosperm nucleus divides into two equal cells, of which the lower cell functions as a chalazal haustorium. Some times, it may form a long lateral branch which penetrates into the funiculus and may reach even upto the placenta. In case of *N.aurita,* large micropylar and chalazal haustoria are developed. In *Lobelia amoena,* both micropylar and chalazal haustoria are found (fig. 9.4B).

In **Ruellia** and some other members of Acanthaceae, both micropylar and chalazal haustoria develop from the endosperm. After the first nuclear division of primary endosperm nucleus, micropylar and chalazal cells are formed. The chalazal cell gives rise to haustorium that remains coenocytic. The micropylar cell divides transversely and its upper cell gives rise to micropylar haustoria. This endosperm is partially cellular and partially free nuclear.

IN HELOBIAL ENDOSPERM

In **Monochoria,** the endosperm is of helobial type. The chalazal chamber contains several nuclei as compared to micropylar chamber due to unequal number of nuclear divisions. The two tubular outgrowths develop from two lateral sides of micropylar chamber which grow in the direction of chalazal tissue, penetrate into it to absorb food materials. Such lateral haustoria have been found in several species of *Monochoria.*

Maheshwari (1950) reported a rare case in *Globularia vulgaris* in which case, micropylar, chalazal and secondary haustoria develop together. Secondary haustoria are tubular finger-like outgrowths arising from the region close to the chalazal haustorium.

MORPHOLOGICAL NATURE OF ENDOSPERM

The endosperm in angiosperms is an interesting structure that develops after fertilisation, as a result of triple fusion. It is neither haploid nor diploid but triploid. The morphological nature of such endosperm has remained a matter of controversy and has been discussed by different workers in different ways. Some of the important propositions are as follows:

1. **Hofmeister's view (1858, 61).** In the absence of knowledge about syngamy and triple fusion during his time, he proposed that angiospermic **endosperm is a gametophytic tissue** (part of female gametophyte as in gymnosperms) whose growth and development starts after the entry of pollen tube into the embryo sac. The only difference between the angiospermic endosperm from that of gymnosperm is that it starts developing only after the entry of pollen tube into the embryo-sac.

2. **Le-Monnier's view (1887).** By this time, syngamy in angiosperms was discovered by Strasburger. Following this discovery, Le-Monnier (1887) proposed that endosperm is a second embryo, modified to supply nutrition to the zygotic embryo. He considered that the fusion of two polar nuclei is also an act of fertilisation which can be compared to the fertilization of an egg by sperm. Thus,

according to him, the **endosperm is a second embryo**, i.e. diploid and a sporophyte. The occurrence of double fertilization was not known till then.

3. **Nawaschin's view (1898).** He discovered double fertilisation or triple fusion. He reported that triple fusion is an act of **true fertilization** in which *two* polar nuclei fuse with *one* male nucleus of a pollen tube.

4. **Sargant's view (1900).** She supported Nawaschin (1898) and further clarified the act of sexual fusion as consisting of the union of one male element coming from pollen tube and one female element coming from micropylar pole as the upper polar nucleus (the sister cell of the egg). She considered polar nucleus akin to egg nucleus. However, due to fusion of one more polar nucleus coming from chalazal end, the chromosomal balance is disturbed, and the second embryo is, thus, disfigured from the very beginning. This results in the formation of a **form-less mass of tissue**, called endosperm, which does not interfere with the development of first embryo.

5. **Strasburger's view (1900).** He proposed that triple fusion is not a true act of fertilization, rather it imparts a stimulus for the growth of reserve food material in the form of endosperm. He believed that, fusion of secondary nucleus with a male nucleus is not a real fertilization rather a vegetative fertilization. Thus, he considered **endosperm as a reduced female gametophyte** or a belated gametophytic tissue.

6. **Thomas and Ne'meck's view (1910).** They believed that by the fusion of polar nuclei with male nucleus, a tissue is formed which might be more suitable for the nourishment of the embryo. This nutritive tissue arises from the same mixed stock than the tissue which would have been formed by mother stock only. The triple fusion, thus, gives rise to a tissue, physiologically more harmonious with the embryo. They proposed that the hybridity of the endosperm is a systematic procedure of making its composition suitable for the use of developing plant.

7. **Modern view.** The views discussed earlier were based on the fact that the endosperm is always a product of triple fusion. Recent studies have revealed that it is not a regular feature. Variations are there in nature.

In *Oenothera type* of embryo-sac, there is a single polar nucleus. Such one polar nucleus is found in other members of family Onagraceae. Fusion of one polar nucleus with one male nucleus results in the formation of a **diploid endosperm**. This diploid endosperm has the same genetic constitution as that of true embryo, but still the two structures are very different.

In *Ditepalanthus*, the primary endosperm nucleus is **tetraploid** formed by the fusion of three polar nuclei and one male nucleus. Likewise, in *Fritillaria* and *Plumbagella*, the primary endosperm nucleus is **pentaploid** formed by fusion of one triploid lower polar nucleus, and one haploid upper polar nucleus with one male nucleus. The endosperms of *Penaea* and *Plumbago* are also pentaploid. A still higher number of ploidy in endosperm nucleus has also been found. In *Acalypha indica*, the endosperm is **nonaploid (9n)**, while *Peperomia hispidula*, the endosperm is 15n.

Such examples showing variations in the formation of endosperm nucleus indicate that there is no uniformity in the origin of the endosperm. But one character is uniform that whatever may be the number of fusing nuclei, the result is always the formation of a form-less mass of tissue functioning as a source of nutrition to the zygote and the developing embryo.

Studies have been further extended on the lines of experimental morphogenesis using tissue culture techniques. The aim of such studies have been to evaluate whether endosperm tissue still retain the power of forming complete plant parts or such potential has been lost. Johri and his coworkers for the first time demonstrated convincingly that the endosperm of *Exocarpus cupressiformis* form the organs. Buds arising from the peripheral cells of endosperm possessed distinct shoot apices with well differentiated vascular system. Since then organogenesis has been successfully induced in the endosperm cultures of several species. Indian botanists have worked considerably on these lines (Rangaswamy and Rao (1963), Satsangi and Mohan Ram (1965), Johri and Shrivastava (1973), Johri and Bhojwani (1977) etc.). In endosperm cultures of *Croton bonplandianum* only shoots were formed whereas from the endosperm of *Putranjiva roxburghii* and *Jatropha panduraefolia*, both roots and shoots were differentiated. This shows the nature of endosperm which is so unbalanced in nature that it forms a form-less nutritive tissue only.

At present, the most accepted view regarding the morphological nature of the endosperm is that it is neither a gametophyte nor a sporophyte. It is altogether a new structure of complex morphological nature, characteristic to angiosperms. It is considered as an undifferentiated tissue of polyploidous nature, the composition of which has been adjusted according to the nutritive need of the developing embryo.

GENERAL QUESTIONS

1. What is endosperm and how is it formed? Describe its function and morphological nature.
2. Describe about origin, structure and significance of endosperm in angiosperms.
3. Write short notes on—
   (i) Nuclear endosperm
   (ii) Cellular endosperm
   (iii) Helobial endosperm
   (iv) Endosperm haustoria
   (v) Morphological nature of endosperm

CHAPTER 10

Embryo

The egg cell of the embryo sac is fertilized by the male nucleus to form a zygote. The zygote undergoes a brief period of rest, which differs from species to species. However, in *Oryza* and *Crepis*, the zygote begins to divide a few hours after fertilization. During the resting stage, the zygote undergoes several changes such as reduction in vacuolar size, increase in cytoplasm and aggregation of ribosomes to form polyribosomes. The various steps from zygote leading to the formation of embryo are called *embryogeny*. The embryogeny found in dicots and monocots are discussed separately. However, the early stages of embryogeny show similarities in both dicot and monocot embryos. The plane of first division of the zygote is almost always transverse, forming a small apical cell and a large basal cell. The two-celled structure is known as *pro-embryo*.

DICOTYLEDONOUS EMBRYO

On the basis of further development of pro-embryo, dicotyledonous embryos have been classified into five main types (Maheshwari, 1950).

I. The terminal cell divides longitudinally

(i) The basal cell contributes only a little towards embryo formation— ***Crucifer type***

(ii) The basal and terminal cell, both, participate in the formation of embryo— ***Asterad type***

II. The terminal cell divides transversely

(i) The basal cell contributes a little towards embryo formation. It divides to form the suspensor— ***Solanad type***

(ii) The basal cell does not divide and if suspensor is formed, it develops from the terminal cell— ***Caryophyllad type***

(iii) The basal cell as well as terminal cell participate in the formation of the embryo— ***Chenopodiad type***

Development of Embryo

Hanstein and Famintzin (1870) studied the embryo development in *Capsella bursa-pastoris* (Shephard's purse), which represents the typical dicot and crucifer type of embryogenesis. It may be summarised as follows:

The zygote divides by a transverse division forming two cells, an upper or *terminal cell* (ca) and a lower or basal cell (cb). The basal cell further divides transversely forming two cells, whereas terminal cell divides by a vertical division to form two longitudinal cells (fig. 10.1). Thus, a four celled *pro-embryo* is formed. The terminal cell again divides by a second vertical division, at right angle to the first one, resulting in four cells, and referred to as the *quadrant stage*. A transverse division now takes place in the four cells, giving rise to eight cells, arranged in two tiers of four cells each. It is known as the *octant stage* (fig. 10.1).

In the octant stage, the upper tier of four cells is known as the *epibasal* or *anterior octants*, whereas the lower tier is referred to as *hypobasal* or *posterior octants*. The eight cells of the octant stage now divide by a periclinal division resulting in a sixteen-celled stage, arranged as eight outer cells and eight inner cells. The *outer* eight cells form the *dermatogen*, which further divide anticlinally to form the *epidermis*. The *inner* eight cells divide by vertical and transverse divisions to differentiate an *external* layer of *periblem*, lying below the dermatogen, and a *central* region of *plerome*. The periblem develops to form cortex, whereas plerome forms the stele.

As these developments are taking place, the *two cells* of the *basal cell* divide by four to six transverse divisions to form a long six to ten celled *suspensor*. By the time, the embryo is sixteen celled and assumes a globular shape, the suspensor is fully formed and helps to push the growing embryo, deep into the nourishing tissue of the endosperm. The *uppermost* cell of the suspensor (facing the micropyle) enlarges to form a vesicular cell, the *haustorium*. The haustorium is normally unbranched, but may branch in certain cases like in *Sedum aere*. The *lowermost* cell of the suspensor (lying adjacent to proper embryo) is known as *hypophysis*, which partly contributes to embryo structure. The hypophysis now divides transversely forming two cells, which further divide by two vertical divisions at right angles to each other resulting in eight cells, arranged in two tiers of four cells each. The *upper tier* of four cells (adjacent to the suspensor cells) function as initials of *root cap* and *root epidermis*, whereas *lower tier* of four cells (facing the embryo) function as initials of *root cortex*. These cells further divide and develop to form the root cap, root epidermis and cortex of radicle, respectively.

Further divisions in the embryo, especially in the region where the cotyledons are to develop, results is the formation of heart-shaped embryo. The hypocotyl and cotyledons by further transverse divisions elongate rapidly and enormously, so much so that the space within the ovule gets limited for the expanding embryo. As a result, a marked curvature in the growth of cotyledons is observed, which now curves and gets placed within the ovule, usually parallel to the axis of the embryo. The mature embryo appears as a horse-shoe shaped structure in a longitudinal section, with a terminal stem tip bordered by two laterally placed cotyledons.

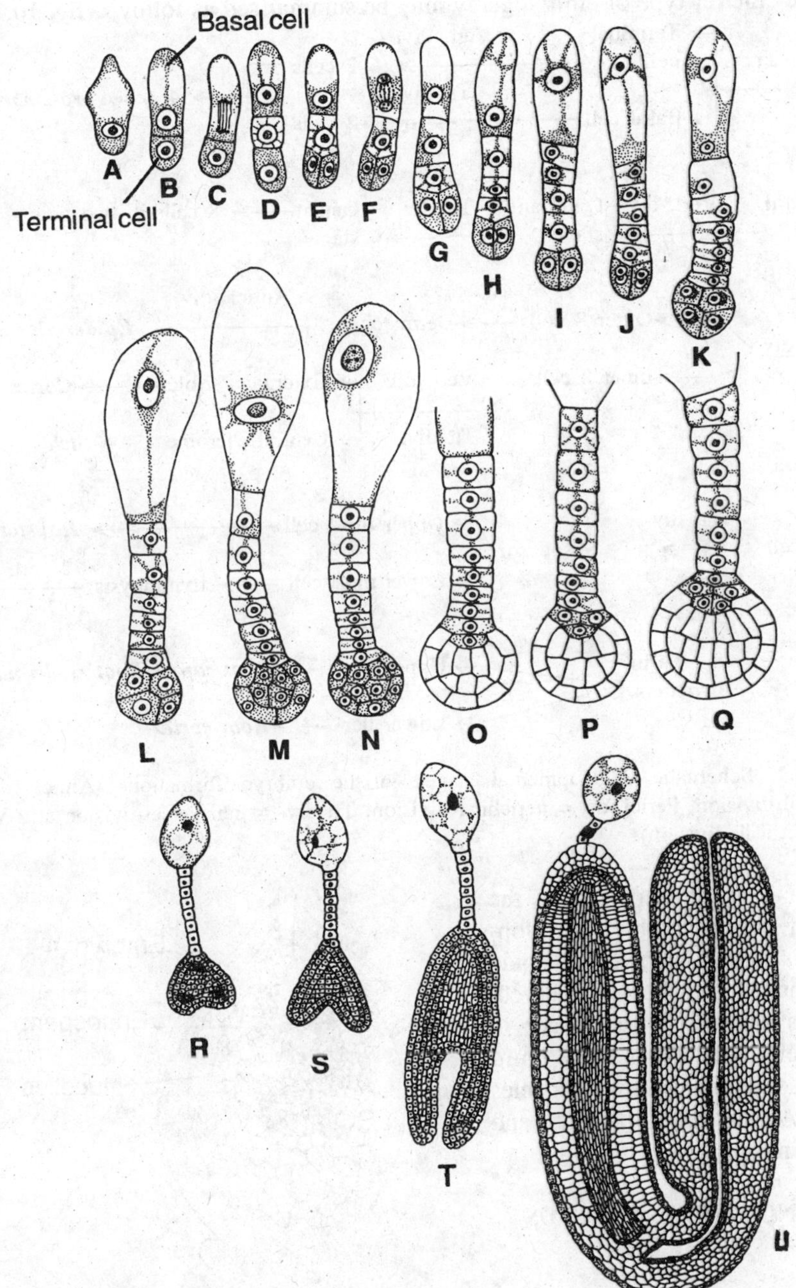

Fig. 10.1. Embryo development in *Capsella bursa-pastoris*.

The crucifer type of embryogemy may be summarised as follows (fig. 10.2)

Zygote —Tr. div.→
- → Terminal cell —Vert. div.→ 2 cells
- → Basal cell —Tr. div.→ 2 cells

2 cells + 2 cells → 4 celled-***pro-embryo***

Terminal cell —Vert. div.→ Quadrant stage —Tr. div.→ Octant stage → 8 cells in two tiers

Pericl. div
- → Outer 8 cells → Dermatogen —Anticl. div.→ ***Epidermis***
- → Inner 8 cells —Vert. div. / Tr. div.→
  - → External Periblem → ***Cortex***
  - → Central Plerome → ***Stele***

Basal cell —Tr. div.→ Suspensor
- → Upper most cell → ***Haustorium***
- → Lower most cell → hypophysis —2 vert. div.→

8 cells in two tiers
- → Upper tier → ***Root cap & Root epidermis***
- → Lower tier → ***Root cortex***

Fig 10.2. Schematic developmental stages in the embryo formation. (Anticl. div.= anticlinal division, Pericl. div.= periclinal division, Tr. div. = transverse division and Vert. div.= vertical division).

In majority of dicotyledons, the crucifer type of embryo development is found. However, some variations have been observed, such as absence of suspensor in members of Trapaceae and Orchidaceae, or suspensor becomes elongated as in Rubiaceae and Leguminosae.

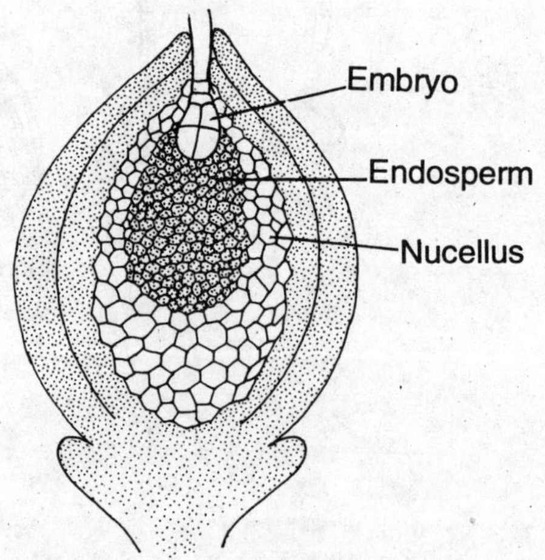

Fig. 10.3. General plan of development of embryo and endosperm.

MONOCOTYLEDONOUS EMBRYO

Embryo development in monocots is essentially similar to that of dicots, during the early stages, but the mature embryo differs considerably.

Embryo development of *Sagittaria sagittifolia* has been considered to be a typical monocot type and is commonly observed in primitive families.

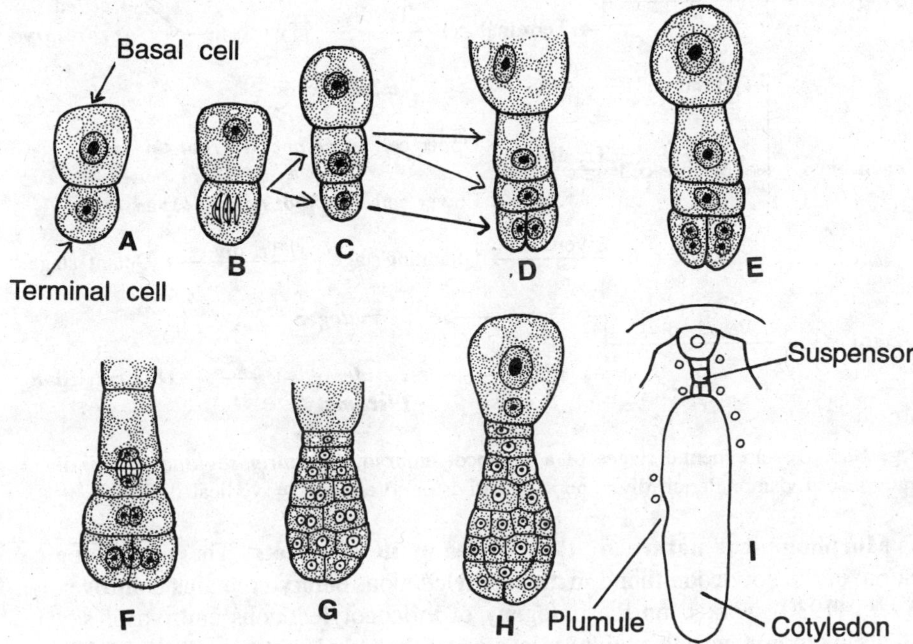

Fig. 10.4. Embryo development in *Sagittaria* (monocotyledon).

The zygote divides by a transverse division, giving rise to a terminal cell (ca) and a basal cell (cb). The ***basal cell*** (*lying towards the micropylar end*) is larger in size and does not divide any further, but simply enlarges to form a prominent vesicular cell. The ***terminal cell*** divides transversely forming two cells (t_1 and t_2), resulting in a three celled *pro-embryo* (fig. 10.4). The *lowermost cell* (t_1) divides by two vertical divisions at right angles to each other giving rise to four cells (*quadrant stage*). Each of the four cells further divides by a transverse wall to form eight cells (*octant stage*) arranged in two tiers of four cells each. The eight cells next divide by periclinal divisions to differentiate eight outer and eight inner cells. The outer cells give rise to the *dermatogen*, whereas, inner cells develop to form the *periderm* and the *plerome*.

The ***middle cell*** (t_2) of the three-celled pro-embryo divides transversely to form two cells. Its upper cell by repeated divisions develops to form the *hypocotyl*, *root tip* and a short, 3 to 6-celled *suspensor*. The lower cell after multiple divisions gives rise to the lateral *shoot apex*.

The eight cells of the octant stage, after differentiating the periderm, plerome and dermatogen, divide further and develop to form a single cotyledon. Later, the shoot apex originates from a lateral position, situated between the base of the cotyledon and the embryo axis. Embryo development in monocots may be summarised as follows (fig. 10.5)—

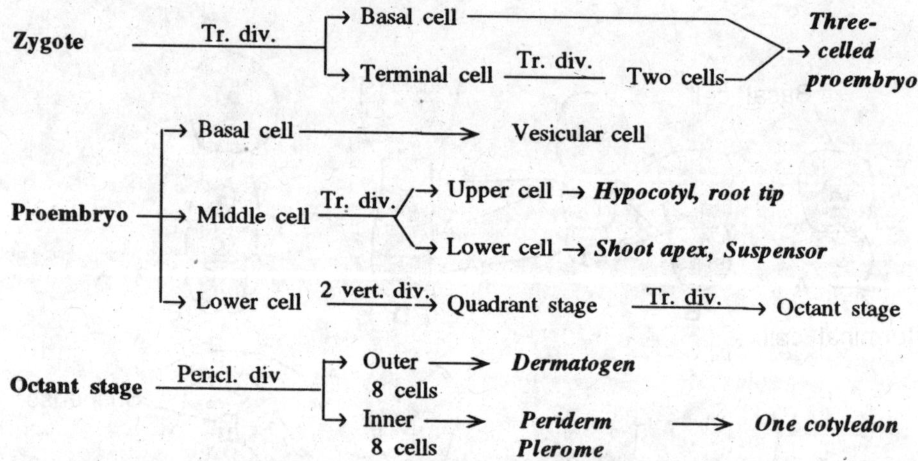

Fig. 10.5. Developmental stages of a monocot embryo, *Sagittaria sagittifolia* (Tr. div. = transverse division, Pericl. div.= periclinal division, Vert. div. = vertical division).

Morphological nature of the cotyledon in monocots. The morphological nature of the cotyledon found in monocotyledonous embryo remains controversial. Avery (1930) worked on the ontogeny of monocotyledonous embryo of several plants including wheat, maize, rice etc. He believed that scutellum is the true cotyledon, coleoptile is the second leaf and the elongated space between the scutellum and the coleoptile, i.e. mesocotyl represents the first internode. Arber (1934) proposed that the scutellum and the coleoptile together represent the cotyledon. The coleoptile functions as the cotyledonary sheath. However, several other workers believe that coleoptile represents the true cotyledon and the scutellum is the lateral outgrowth of the young axis.

Differences between dicotyledonous and monocotyledonous embryos

Not only the number of cotyledons in two embryos but also their position make the important differences between the two. In dicotyledonous embryo, there are two cotyledons and these are lateral in position, whereas in monocotyledonous embryo, only one cotyledon is found which occupies terminal position. Wardlaw (1955) remarked that *in the dicotyledonous embryo, the plumule is typically distal and is situated symmetrically between two equivalent cotyledons; in the monocotyledonous embryo, the shoot apex occupies a lateral position in the some what cylindrical embryo and the cotyledon is terminal.* Swamy *et al.* (1972) working on the ontogeny of monocotyledonous embryo established that the epicotyl is truly a terminal structure. They established that from one and the same terminal tier the epicotyl and cotyledon arise. The apparent lateral position of the epicotyl is due to early growth of cotyledon and slow growth of the epicotyl after its initiation.

Lakshmanan (1972) proposed that the number of cells of the terminal quadrant of the two embryos is quite different. In dicots, the two diagonally opposite cells of the terminal quadrant give rise to the two cotyledons whereas in monocots, the single cotyledon might be formed by all the four cells, or three cells or two adjacent cells of the terminal quadrant.

In monocotyledonous embryo, the *plumule* is considered to be lateral which remains surrounded by coleoptile, and the *radicle* remains surrounded by coleorrhiza. In a dicotyledonous embryo, the plumule is terminal and both coleoptile and coleorrhiza are considered to be absent.

Variations in the structure of suspensor

An unique, ephimeral structure arising from the radicular end of the pro-embryo is the suspensor. During the early stage of embryo development, the suspensor grows faster than embryo, however, in the mature embryo, only the remnants of suspensor are seen. The primary function of the suspensor is to provide a nutrition-rich environment for the embryo which is accomplished by pushing the embryo deep down into the tissue of the endosperm. In members of family Loranthaceae, the suspensor is exceptionally long to enable it to bring the embryo down from the style into the endosperm.

The occurrence of suspensor haustoria in some members of Rubiaceae has been reported. Lloyd (1902) believed that the function of the suspensor in these forms is therefore not only to bring the embryo into favourable position with relation to the food supply in a mechanical sense, but to act as a temporary embryonic root. Now, the presence of suspensor haustoria is known in several plants. These cells have a structure some what similar to *transfer cells*, which are involved in the transportation of elements across the cell membrane. As such, it is presumed that these suspensor haustoria are also involved in the absorption of nutrients from the endosperm and transferring them to the embryo.

The suspensor haustoria show structural and functional variation in different genera of angiosperms. In majority of angiosperms, such as in *Capsella,* the suspensor is long, thread-like and terminated by a globular basal cell, which is haustorial in nature. The cells of haustoria show wall ingrowths which are actively involved in the absorption of nutrients from the endosperm and transferring them to the embryo.

In *Dicraea stylosa* (Podostemaceae), the basal cell enlarges and sends out 4-5 thin-walled haustorial branches which grow between the two integuments of the ovule.

In *Sedum* (Crassulaceae), the apical cell of the two-celled pro-embryo forms the embryo as well as a four-celled suspensor. The basal cell elongates and sends out haustorial branches, like fungal hyphae, which penetrates through the integuments, upto the raphe.

In *Myriophyllum* (Haloragaceae), the two-celled proembryo consists of a small terminal cell and a large basal cell. The basal cell divides longitudinally forming two daughter cells which expand laterally forming wing-shaped haustoria.

In members of Orchidaceae, different types of suspensors have been studied. They may be single-celled, uniseriate filament of 5-10 cells, resemble a bunch of grapes or form a mass of 6-10 tubular cells.

Various *in vitro* studies of embryo growth, with or without suspensor, have shown that suspensor definitely stimulated the growth of the embryo, especially at the heart-shaped stage.

GENERAL QUESTIONS

1. Describe the development of a typical dicotyledonous embryo.
2. With suitable examples describe the structure and development of monocotyledonous embryo.
3. Enumerate briefly the post-fertilization changes found in dicotyledonous embryogeny.
4. Write short notes on—
   (i) Crucifer type of embryogenesis
   (ii) Monocotyledonous embryo
   (iii) Differences between dicotyledonous and monocotyledonous embryos
   (iv) Suspensor

CHAPTER 11

Apomixis

In general two types of plants are recognised, (i) those which have normal sexual cycle and involve two important steps in life cycle, i.e. *meiosis* and *syngamy* (fertilisation). These are called *amphimictic* type of plants, (ii) those in which the usual method of sexual reproduction is replaced by a kind of asexual reproduction. These are called *apomictic* plants. The term apomixis was given by Winkler (1908, 1934). He defined apomixis as the substitution of the usual sexual reproduction by a form of reproduction which does not involve meiosis and syngamy. He used the term to signify the production of a sporophyte from a gametophyte without any gametic fusion. It interrupts the normal alteration between a reduced gametophyte and an unreduced sporophyte.

Maheshwari (1950) divided apomixis into two types.

(i) *Recurrent apomixis*—which includes vegetative propagation and agamospermy,

(ii) *Non-recurrent apomixis*—where haploid embryo sac, developed after meiosis, develops into embryo without fertilization. Heslop-Harrison (1972) proposed that non-recurrent type should not be included within apomixis because it involves meiosis.

I. RECURRENT APOMIXIS

Two main categories of recurrent apomixis are now recognised.

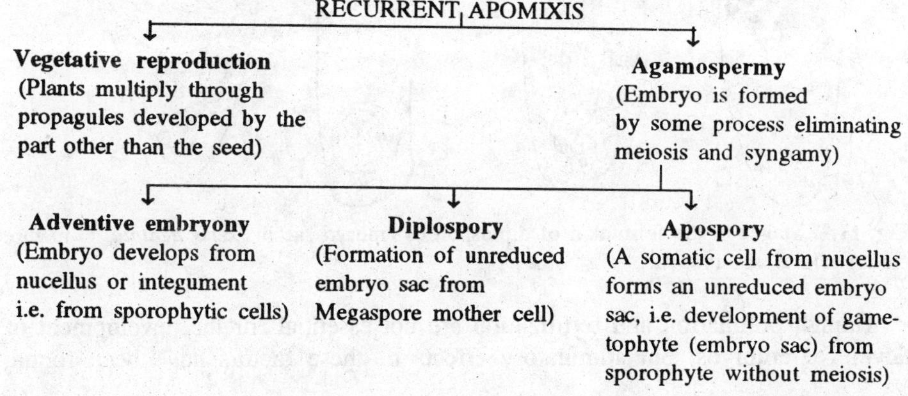

1. Vegetative Reproduction

Vegetative propagation includes reproduction by means of bulbils, bulbs, runners, suckers etc. These are formed by the sporophytic parts of the plants only and may arise either from vegetative parts or from floral regions, as the case may be.

2. Agamospermy

These plants retain seed formation for propagation. In addition to that agamospermy is found in them. The embryo is formed by some process without involving meiosis and syngamy. Three different types of agamospermy are found.

(i) **Adventive embryony.** The embryo is formed from the cells of nucellus or the integument, i.e. diploid tissue of parent sporophyte and not from the cells of gametophyte. Adventive embryony is sometimes called sporophytic budding. Gametophytic generation in such cases is completely eliminated. Sometimes, zygotic embryo may also develop in addition to adventive embryo. In such cases, the adventive embryos are distinguished from zygotic embryo because of their lateral position and absence of suspensor. Formation of adventive embryos has been reported in *Citrus, Euphorbia dulcis, Eugenia* sp., *Capparis frondosa, Mangifera indica, Hiptage madablota* etc.

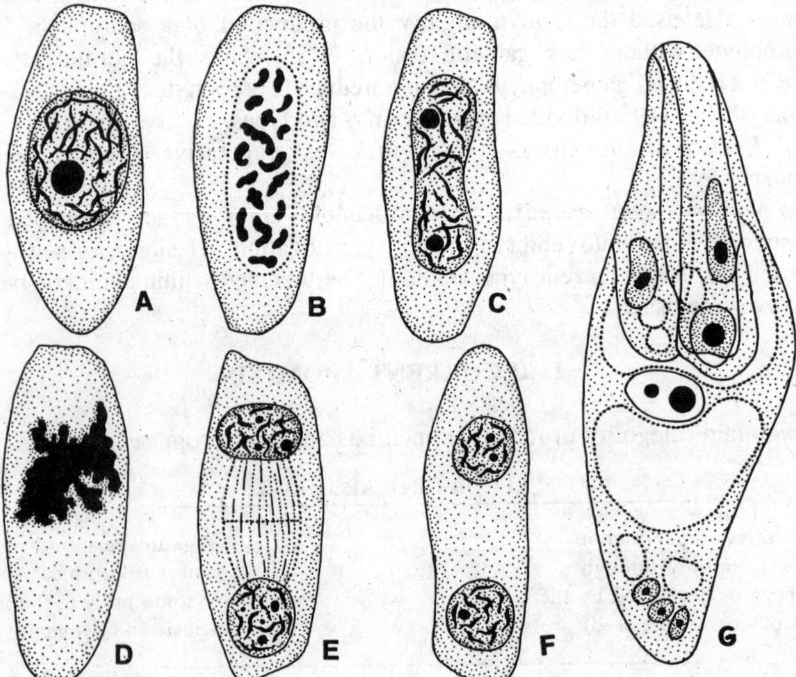

Fig. 11.1. Stages in the formation of diplosporous embryo sac in *Ixeris dentata*. G-mature eight-nucleate embryo sac.

Though pollination and fertilization are not essential for the development of adventive embryos, but stimulatory effects of these factors have been found.

Formation of endosperms has been recorded in all cases and these may originate either by triple fusion or without it. However, endosperm formation does not occur in *Opuntia aurantiaca*.

(ii) Diplospory (Generative apospory). When diploid embryo sac is formed from megaspore mother cell (MMC) without undergoing meiotic division, it is called diplospory. Diplospory may be divided into two types on basis of the type of divisions found in MMC.

In ***type first***, MMC undergoes a division which is semi-heterotypic type. There is no pairing, or some degree of pairing may be found and then a restitutional nucleus is formed. This nucleus remains diploid. It has been reported in triploid forms of *Ixeris dentata* and polyploid forms of *Taraxacum*. This nucleus divides by three mitotic divisions to form eight-nucleate embryo sac.

In ***type second***, the nucleus of MMC divides directly by mitotic divisions as reported in *Eupatorium glandulosum* and *Calamogrostis* species.

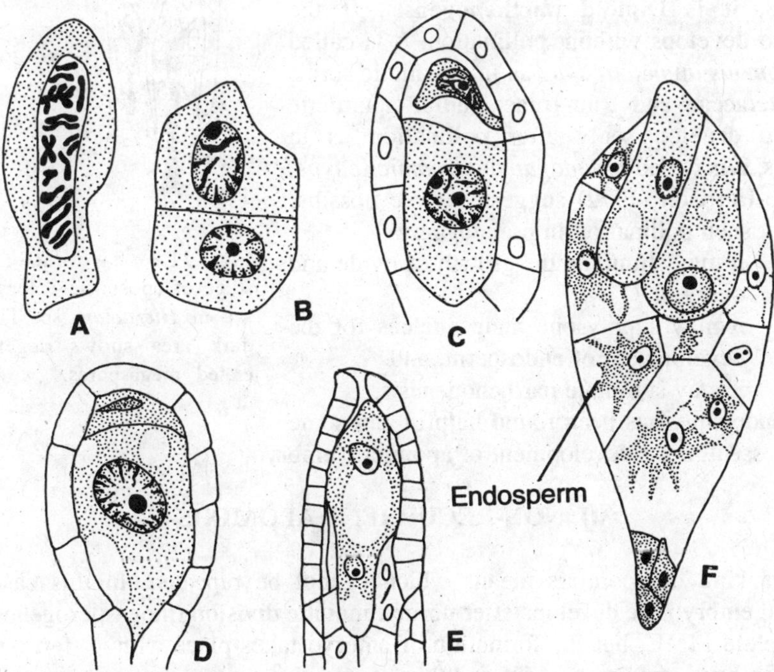

Fig. 11.2. Stages in the formation of diplosporous embryo sac in *Taraxacum albidum*. F-mature embryo sac in which development of endosperm has started.

(iii) Apospory. Among angiosperms, apospory was for the first time reported in *Hieracium* species by Rosenberg (1907). In this case, MMC divides by usual meiosis but all four megaspores gradually degenerate. Simultaneously, a nucellar cell becomes activated and develops into an embryo sac. This type of apospory is also called somatic apospory.

Apospory has been reported in some grasses, *Crepis, Cliftonia, Parthenium* and *Rubus.* In grasses, more than one embryo sac may develop in the same nucellus, e.g. in *Poa arctica,* the number of aposporic embryo sacs reach upto twelve. In aposporic forms of family Asteraceae, only one nucellar cell functions as the mother cell that gives rise to a normal, eight-nucleate embryo sac.

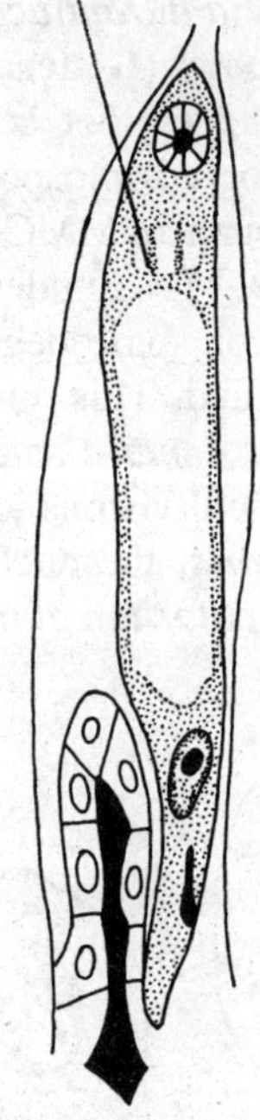

Fig. 11.3. Two-nucleate stage of aposporic embryo sac in *Hieracium* sp. The dark area shows degenerated megaspores.

The *aposporic* embryo-sacs are diploid. For the occurrence of normal sporophyte, their diploid egg cell must develop into embryo without fusing with male nucleus (fertilisation). Formation of embryo from an unfertilised egg is called **Parthenogenesis** and in this case, since the egg is diploid, it is **Diploid parthenogensis**. If the embryo develops without pollination, it is called *autonomous apomictic type* as in apomictic forms of Asteraceae and Rubiaceae, and if apomictic embryo develops only after pollination as in grasses, it is called *pseudogamous apomictic type*, Heslop-Harrison (1972) suggested three possible roles of such pollination in pseudogamy—

(i) it may stimulate the growth of ovule and ovary,
(ii) it may supply one male nucleus for the development of endosperm, and
(iii) it may stimulate parthenogenesis.

Pseudogamy has been found helpful and some times essential for development of apomictic embryo.

II. NON-RECURRENT APOMIXIS

This kind of apomixis means which cannot be repeated. In this case, the haploid embryo sac develops after normal meiotic division (megasporogenesis) of the diploid MMC, but the formation of embryo takes place without fertilization. Since embryo remains haploid and thus sterile, the process cannot be repeated in the next generation. The haploid embryo may develop from egg cell, antipodal cell or synergid cell. If embryo develops from egg cell, the phenomenon is called **haploid parthenogenesis** as in case of *Oenothera* and if it develops either from a synergid cell or an antipodal cell, the phenomenon is called **haploid apogamy** as in case of *Lilium.*

Maheshwari (1950) suggested that the egg cell remains unfertilized due to, (i) absence of pollen tube, (ii) male nuclei not discharged from pollen tube, (iii) feeble attraction between male and female nuclei, (iv) earlier disintegration of male gametes, or (v) the egg cell remains immature at the time of penetration by the male gamete.

GENERAL QUESTIONS

1. Describe the various aspects of apomixis.
2. Write short notes on—
   (i) Apomixis
   (ii) Parthenogenesis
   (iii) Adventive embryony
   (iv) Apospory
   (v) Apogamy

CHAPTER 12

Polyembryony

Maheshwari (1950) defined polyembryony as the occurrence of more than one embryo in a seed. Since, the additional embryos generally degenerate in the course of seed development, the development of more than one embryo is frequently not seen. However, polyembryony has been found as a regular feature in *Citrus* and *Mangifera*.

TYPES OF POLYEMBRYONY

Polyembryony was first reported by Leeuwenhoek (1719) in certain orange (*Citrus*) seeds. Later, a considerable study was made by several embryologists like Braun (1859), Ernst (1918), Schnarf (1929), Webber (1940), Gustafsson (1946) etc. Schnarf (1929) divided polyembryony into two types-

(1) ***True polyembryony***—in which there is a development of multiple embryos within or by projection into a single embryo sac.

(2) ***False polyembryony***—in which the multiple embryos are developed from different embryo sacs

Gustafsson (1946) proposed that *true polyembryony* is the development of embryos derived from one or several embryo sacs present in the same nucellus. The term *false polyembryony* should be restricted to those cases in which two or more nucelli, each with its embryo sac, fuse and the seed may appear single or double, depending upon degree of fusion.

Braun (1859) classified true polyembryony into four types—

(1) Cleavage of zygote or zygotic embryo.
(2) Formation of embryos from the cells of the embryo sac other than the egg.
(3) Development of more than one embryo sac within the same ovule.
(4) Activation of some initial sporophytic cells of the ovule.

The following types of polyembryony are discussed here.

1. Cleavage Polyembryony

When polyembryony arises by the cleavage of zygote or pro-embryo into two or more units, the condition is called cleavage polyembryony. Cleavage polyembryony is less common in angiosperms while it is commonly found in gymnosperms. Among angiosperms, it is very common in orchids. Rao (1965) reported that apical promeristem of the embryo in *Vanda* (orchid) divides into a number of primordia (3 to 9), each further organising into an embryo. Swamy (1943) reported it in *Eulophia epidendraea* arising in three different ways.

(i) Zygote divides irregularly to form a mass of cells, of which those lying towards the chalazal end grow simultaneously and give rise to multiple embryos (fig. 12.1A)

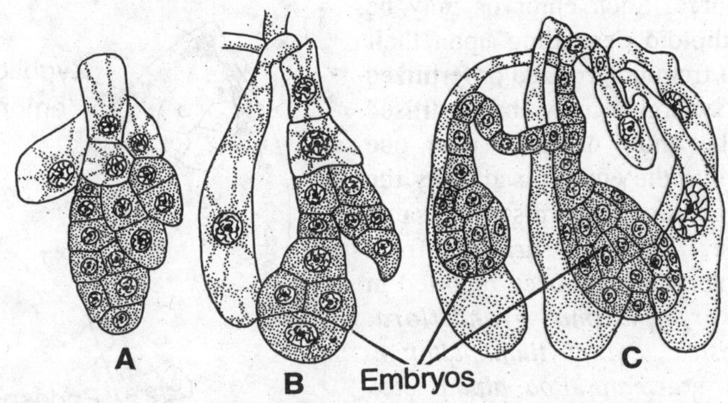

Fig. 12.1. Cleavage polyembryony in *Eulophia epidendraea*. A-three embryos arising by cleavage of the zygote, B-budding of the embryo and C-two embryos developing by splitting of a single embryo (*Redrawn* from Swamy, 1943).

(ii) The pro-embryo gives out small buds or outgrowths which may themselves function as embryos (fig. 12.1B).

(iii) The filamentous embryo becomes branched and each of the branches grows into an embryo. (fig. 12.1C).

Jeffery (1895) reported cleavage polyembryony in *Erythronium americanum*, Ernst (1910) in *Tulipa gesneriana*, Hall (1902) in *Limnocharis emarginata*, Whitehead and Chapman (1949) in *Cocos nucifera*, Rao (1965) in *Vanda* etc. Abnormal type of cleavage polyembryony has been reported in some cases. Cooper (1943) reported in *Nicotiana rustica* an ovule with two embryos of which the smaller had apparently arisen as an outgrowth from the apex of the primary embryo. Abnormal embryos arising from suspensor have been reported by Crete

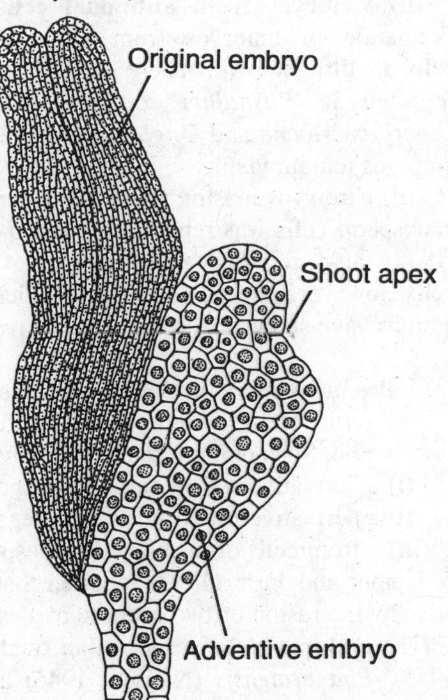

Fig. 12.2. Formation of adventive embryos from suspensor cell in *Eranthris*.

(1944) in *Lobelia syphlitica*, by Subramanyam (1946) in *Isotoma longiflora*, by Maheshwari and Nagi (1950) in *Dipteracanthus patulus*, by Ram (1959) in *Exocarpus* and by Haccius (1965) in *Eranthris hiemalis* (fig 12.2)

2. **Embryos from cells of embryo sac other than the egg**

Such embryos may arise either from the synergids or from the antipodal cells.

(i) **Embryos from synergid cells**. Embryos from the synergids are the most common source. Such embryos may be haploid or diploid depending upon their development from unfertilized or fertilized synergids. Synergids may be fertilised either by the entry of more than one pollen tube into the embryo sac or by the presence of additional sperms in the same pollen tube. Embryos arising from fertilized synergids have been reported in *Najas minor, Tellima grandiflora, Fragaria vesica, Peltiphyllum peltatum, Aristolochia bracteata, Poa alpina* etc., while from unfertilised synergids in *Argemone mexicana, Plantago lanceolata, Phaseolus vulgaris, Lolium* species etc.

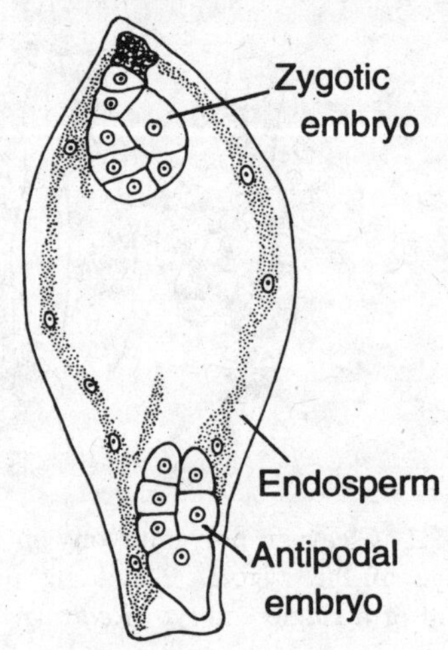

Fig. 12.3. Embryo sac of *Ulmus glabra* showing antipodal and zygotic embryos.

(ii) **Embryos from antipodal cells**. Formation of embryos from antipodal cells is little known. These have been reported in *Paspalum scrobiculatum, Ulmus americana* and *U. glabra* (fig. 12.3). It is not certain whether or not such embryos remain viable.

(iii) **Embryos arising from endosperm cells**. The formation of embryo from endosperm cells was reported by Treub (1898) in *Balanophora elongata*. Kajale (1942) also reported development of additional embryos from endosperm cells.However, on basis of recent studies, such reports were doubtful. Reinvestigations have shown clear cut negative results.

(3) **False polyembryony** (*More than one embryo sac in an ovule*)

The term refers to embryos arising from other embryo sacs found within the same ovule. Such embryo sacs in an ovule may arise from—

(i) derivatives of the same megaspore mother cell,

(ii) derivatives of two or more megaspore mother cells, or

(iii) from cells of nucellus (by apospory)

Kumar and Joshi (1942) reported that the polyembryonic condition may also arise by the fusion of two or more ovules, as in case of *Rhizophora mucronata*.

The multiple embryo sacs in an ovule have been reported in *Citrus* (Bacchi, 1943), *Poa pratensis* (Nielsen, 1946) and in *Casuarina equisetifolia* (Swamy, 1948). Twin embryos in a seed have been reported in many cases and these may be formed in different combinations, i.e. DIPLOID-DIPLOID in *Zea mays*, or

DIPLOID-TRIPLOID in *Triticum vulgare* and *Secale cereale* or HAPLOID-TRIPLOID in *Phleum* or HAPLOID-DIPLOID in *Zea mays.*

(4) Adventive polyembryony (*Embryos arising from tissue outside the embryo sac*)

Embryos arising from the tissue outside the embryo sac are called ***adventive embryos***. These tissues are maternal sporophytic tissues, such as nucellus or integuments. Adventive embryos are commonly observed in *Citrus* and *Mangifera.* The nucellar polyembryony is known in *Opuntia dillenii* and *Trillium undulatum.* Not all species of *Citrus* show polyembryony, but it is commonly met in *C. microcarpa, C. reticulata* and *C. unshiu.* In these cases, adventive embryos arise by the proliferation of the nucellar cells. The inception of nucellar embryos take place outside the embryo sac but these are gradually pushed into the embryo sac cavity where these differentiate into mature embryos. Kobayashi and coworkers (1981), Wilms and coworkers (1983) and Wakana and Uemoto (1987) have shown that in *Citrus,* the initiation of adventive embryos takes place all around the embryo sac and these are not affected by pollination, fertilization, developing zygotic embryo etc. However, lack or poorly developed endosperm result in poor development of adventive embryos. It has also been found that the presence of endosperm promotes development of adventive embryos at the micropylar end but represses their development towards the chalazal end.

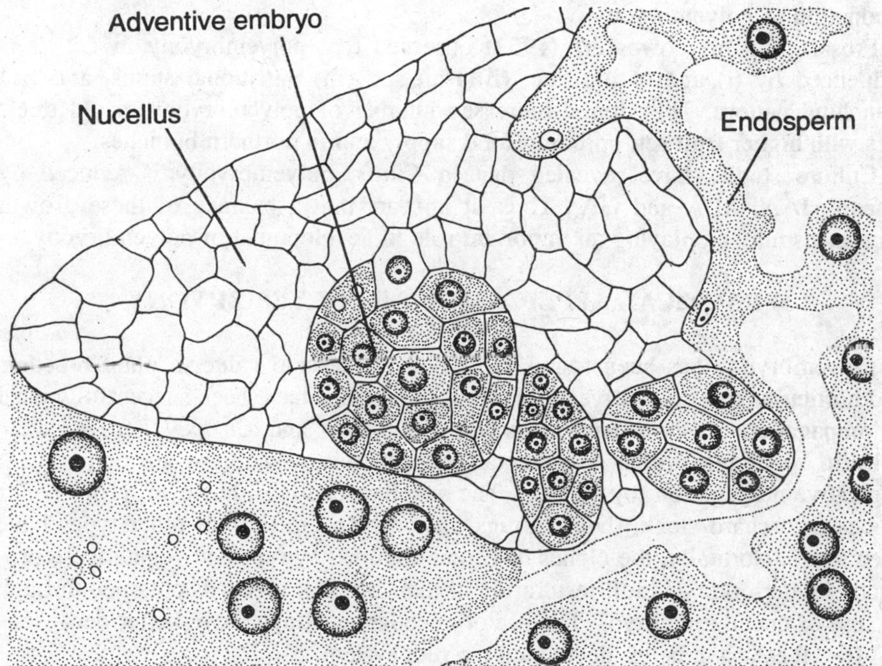

Fig. 12.4. Figure showing nucellar embryos growing into the embryo sac cavity in *Mangifera odorata* (Redrawn from Sachar and Chopra, 1957).

Naumova (1981) reported that nucellar embryony is commonly found in *Crassinucellate ovules*, e.g. in *Citrus* and *Opuntia* while integumentary embryony occurs in *Tenuinucellate ovules*, e.g. in *Euonymus*.

CAUSES OF POLEMBRYONY

Several theories have been proposed to explain the causes of polyembryony but none of them has been fully accepted. These theories relate with hybridisation, necrohormone, effect of recessive genes etc.

Haberlandt (1922) while working on *Oenothera* proposed the **necrohormone theory**. He believed that a hormone is considered to be released by the degenerating cells of nucellus which stimulates the development of adventive embryo. However, due to lack of experimental proofs, this theory was not found to be satisfactory.

Leroy (1946) proposed **recessive gene theory** taking the example of mango. He believed that the monoembryonate condition is due to dominant gene in mango, which is found in primary centre of its origin, like India. The polyembryonate condition arose due to recessive gene and is found in places of secondary origin, like China, Sudan, Phillipines etc. This theory is not fully accepted because some Indian varieties of mango like *Langra, Fazli, Bombai* etc. are commonly polyembryonate. Some scientists like Ernst (1918), Baker (1960) and Cesca (1961) believe **hybridization** to be the factor responsible for the production of polyembryony.

Frusato and his coworkers (1957) observed that polyembryony in *Citrus* is influenced by (i) age of the tree, (ii) fruit set, (iii) nutritional status, and (iv) branching pattern. They found increased number of polyembryony in old trees, ears with higher fruit set, optimum food supply and in northern branches.

Culture studies have revealed that in *Citrus*, polyembryony is reduced by ethanol, IAA, ABA and GA_3. Thus, it appears that a balance of these growth substances may be playing an important role in development of polyembryony.

PRACTICAL APPLICATION OF POLYEMBRYONY

Polyembryony has been found to be of considerable value to plant breeders and horticulturists. Adventive embryos from nucellus have been successfully used in propagation of some fruit trees like *Mangifera* and *Citrus*. These produce uniform seedlings similar to parental type. In this way, these are at par with vegetative propagation (by cutting). The nucellar seedlings, however, prove better clones or orchard stock, than cuttings upon which grafts from other types have been made. Normally, the clones of *Citrus* obtained by repeated vegetative reproduction deteriorate and full restoration of their original vigour is obtained through the use of nucellar embryos. The use of nucellar seedlings prove better because-

(i) these have better developed tap root system,

(ii) these show a restoration of vigour, and

(iii) these are free from disease. So far nucellar polyembryony is the only practical method to develop virus-free clones of polyembryonate varieties.

Ammirato (1989) suggested that before commercialisation of artificial seeds, some problems are to be solved, e.g. (i) occurrence of somaclonal variations in tissue culture, (ii) low rate of viability, (iii) induction of somatic embryogenesis, and (iv) insufficient knowledge about regulation of embryo development. Likewise, there can be several other problems related with the practical use of polyembryony.

GENERAL QUESTIONS

1. What is polyembryony? Give a brief account with suitable examples.
2. Describe the various types of polyembryony met in angiosperms.
3. Write short notes on-
   (i) Cleavage polyembryony
   (ii) Adventive polyembryony
   (iii) False polyembryony
   (iv) Practical application of polyembryony

CHAPTER 13

Seed

The ovule after fertilization develops to form the seed whereas the ovary develops into the fruit. A true seed may be defined *as a fertilized mature ovule*, which encloses the embryonic plant within, stored food in the form of cotyledons and a protective seed-coat. The term seed is sometimes used to include single-seeded dry fruits, such as caryopsis of cereals, cypsella of compositae and structure of vegetative propagation, e.g. bulbils, potato-tubers etc.

Seeds show considerable variation in size, shape, colour and seed surface. The size of the seeds may range from tiny dust particles, e.g. orchids, to large double-coconuts, e.g. *Lodoicea maldivica*. The tiny orchid seeds weigh about 20.33μ g, whereas a double-coconut (seed) weighs approximately 5-6 kg. The shape of the seeds may be spherical, ovoid, oblong, flattened or cylindrical and may possess a smooth, wrinkled, ribbed, furrowed, reticulate, hairy or tuberculate seed-surface.

In the life cycle of a plant, the seed is a dormant, temporary phase in which the metabolic life activities have been suspended. It is an intermediate stage which helps to tide over unfavourable climatic and nutritional conditions and on return of favourable conditions, germinates to form a new plant. Structurally a mature seed consists of an embryo (baby plant), endosperm (if present) and an enclosing structure (seed coat). The ovule after fertilization is stimulated to startling changes, forming ultimate structure.

Ovule	⟶	Seed
Funicle	⟶	Stalk
Integuments	⟶	Seed-coat
Nucellus	⟶	Perisperm
Egg cell	⟶	Embryo
Endosperm nucleus	⟶	Endosperm

STRUCTURE AND DEVELOPMENT OF SEED COAT

After fertilization, the integuments surrounding the ovule undergo significant changes to form the seed-coat. The seed-coat, however, may be formed from both the integuments or from one integument only. Ovules possessing two integuments are known as *bitegmic*, whereas those possessing one integument are known as *unitegmic*.

I. Bitegmic ovules

Members of Polypetalae and moncots possess bitegmic ovules i.e. each ovule has two integuments. In bitegmic ovules, two conditions of integument development are possible.

(i) *Both integuments contribute to the formation of seed-coat*

Gossypium herbaceum possesses a bitegmic ovule, in which both the integuments participate in the formation of the seed-coat. Prior to pollination, the integuments are differentiated into (1) *Outer integument*-composed of 4-6 layers of thin-walled cells, and (2) *Inner integument*-composed of 8-15 layers of cells.

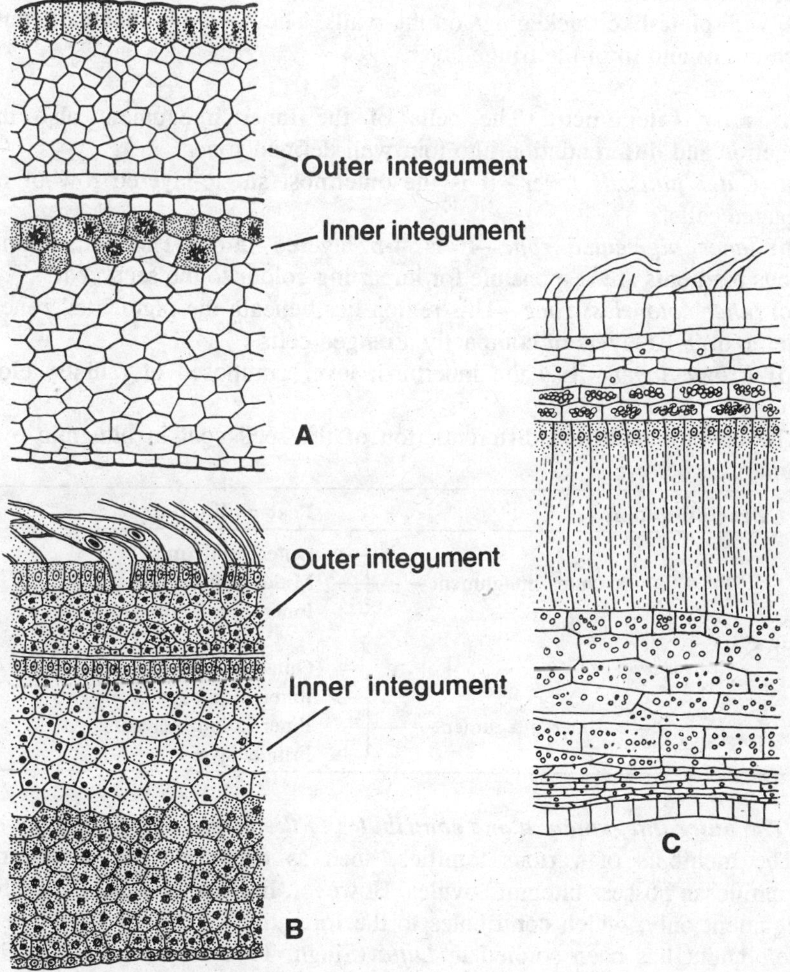

Fig. 13.1. L.S. of seed coat of *Gossypium herbaceum*, A-L.S. of a part of inner integument before fertilization, B-L.S. of portion of integument after 5-6 days of pollination and C-L.S. of a portion of mature seed coat.

After pollination, significant changes take place in both the integuments, giving rise to the complicated seed-coat.

1. Outer integument. In *G. herbaceum*, the outer integument develops extensively and gets differentiated into three zones—

(a) *Outer epidermis*—It consists of thick-walled, radially elongated cells and gives rise to the cotton fibres of commerce. These single-celled and thin-walled cells are popularly known as *lint hairs*. The commercially important fibres consist of cellulose, callose, some non-cellulosic polysaccharides and wax.

(b) *Middle zone*—It is 4-5 layered thick, comprising pigmented cells filled with starch and tannin.

(c) *Inner epidermis*—The innermost region is constituted of slightly elongated cells, with plate-like thickenings on the walls. They contain abundant compound starch grains and form the fringe layer.

2. Inner integument. The cells of the inner integument also undergo elongation and differentiation into four well-defined zones.

(a) *Outer palisade layer*—It is the outermost single-layered row of radially elongated cells.

(b) *Inner pigmented zone*—It is 4-5 layered thick. The cells, containing various pigments are responsible for imparting colour to the seed-coat.

(c) *Inner colourless zone*—This region lies beneath the pigmented zone and is made up of 9-10 layers of compactly arranged cells.

(d) *Fringe layer*—It is the innermost layer composed of radially elongated cells.

The development and differentiation of the seed coat in bitegmic ovules is summarised below:

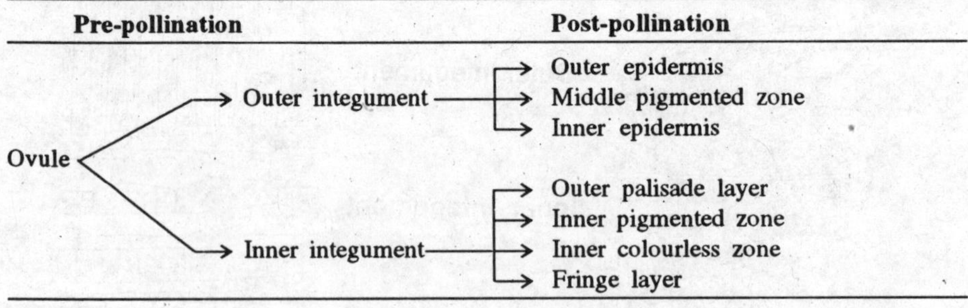

(ii) *The outer integument alone contributes to the formation of the seed-coat*

The members of various families, such as Cucurbitaceae, Cruciferae and Leguminosae possess bitegmic ovules. However, in these members, it is the outer integument only, which contributes to the formation of the seed-coat. Seed-coat development has been studied in *Luffa* (Singh, 1971). Prior to pollination, the bitegmic ovule has an outer 10-15 layered thick *outer integument* and 2-3 layered thin *inner integument*. Pollination brings about significant changes in the development and mature structure of the integument. The mature seed-coat is derived from the outer integument only, whereas the inner integument degenerates.

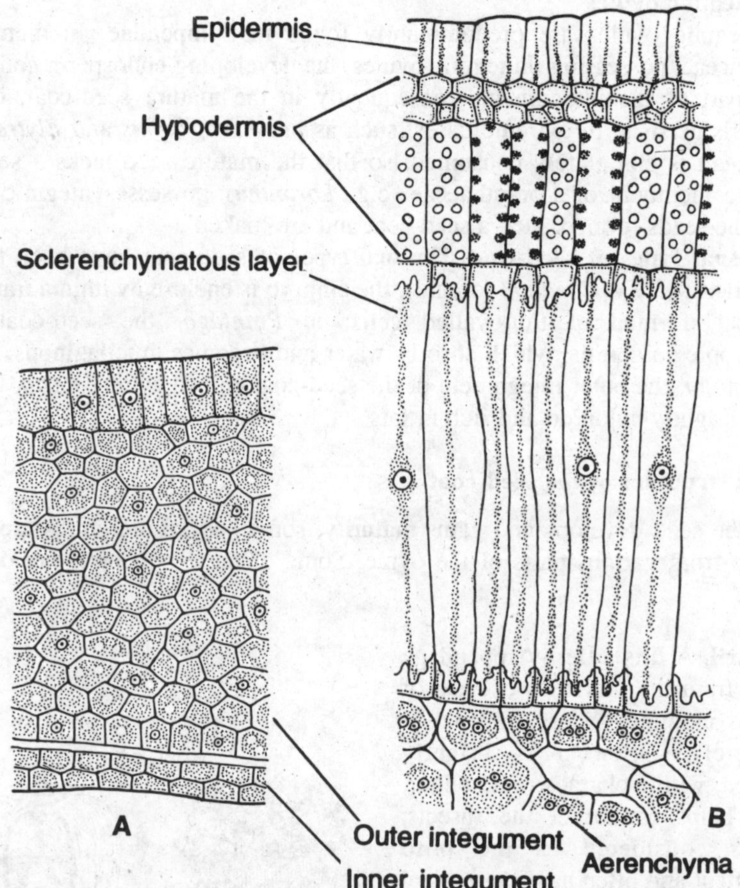

Fig. 13.2. Seed-coat development in *Luffa*. A-L.S. of integument before fertilization and B-L.S. of a portion of seed coat (after fertilization). (Redrawn from Singh, 1971)

The mature seed-coat of *Luffa* may be differentiated into five zones (fig 13.2).

(a) *Epidermis*—It is the outermost layer of cells possessing rod-like thickenings on their radial walls.

(b) *Hypodermis*—The multilayered hypodermis lies beneath the epidermis and is composed of thick-walled cells. The innermost layer of hypodermis comprises radially elongated cells.

(c) *Mechanical layer*—It is the middle region composed of sclerenchyma like-cells, which are elongated and form sclereids. The mechanical layer provides mechanical strength to the seed-coat.

(d) *Aerenchyma layer*—This region lies beneath the mechanical layer and is made up of spongy cells. These cells are more or less concentrated at the ends and margins of the seeds.

(e) *Chlorenchyma layer*—The innermost layer, known as chlorenchyma layer is constituted of 10-12 layers of tangentially elongated, pigment-containing cells.

II. Unitegmic ovules

Unitegmic ovules are predominantly found in Sympetaiae. In members of Acanthaceae, possessing unitegmic ovules, the developing endosperm consumes a major part of the integument, consequently in the mature seed-coat, only the epidermis persists. In extreme cases, such as in *Andrographis* and *Elytraria*, the integument is completely consumed, so that the mature seed lacks a seed-coat. Parasitic members of Loranthaceae, e.g. *Loranthus* possesses ategmic ovules, hence the seeds formed, lack a seed-coat and are 'naked'.

Unusual types of seed-coat. Unusual types of seed-coat have been found in some plants. In *Cypripedium* (orchid), the embryo is enclosed within a transparent seed-coat, formed of thin-walled cells; in *Plantago,* the seed-coat has a hygroscopic epidermis, which absorbs water and becomes mucilaginous, whereas in *Magnolia,* the outer integument of the seed-coat known as *sarcotesta,* becomes fleshy, brightly coloured and rich in fats.

Special structure of the seed-coat

As the seed develops and gains maturity, some special sturctures are seen to develop from various parts of the ovule. Some of these structures are described briefly.

1. Aril. Aril is a fleshy appendage arising from the funiculus or testa of the seed and more or less surrounds it completely. Often it is brightly coloured, with a characteristic odour, which helps to attract the insects. Aril is considered as the third integument and often has commercial importance, such as, the aril of nutmeg (*Myristica fragrans*) after drying, forms the 'mace' of commerce, the fleshy aril of litchi (*Litchi sinensis*) is the edible part of the fruit and in Passifloraceae, the aril contains a yellow-red pigment, which may be used as a dye.

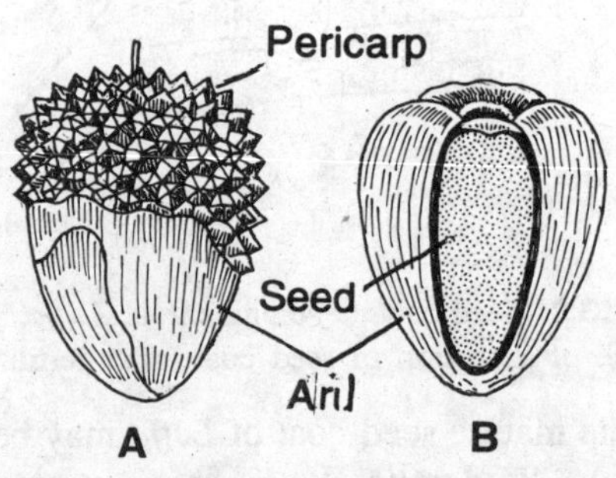

Fig. 13.3. Fruit of *Litchi*. A-entire fruit (part of pericarp removed to show aril) and B-L.S. of fruit.

2. Caruncle. In some members of Euphorbiaceae, e.g. *Ricinus*, a fleshy, whitish structure is present at the micropylar end of the seed. The structure known as *caruncle* is formed by hypertrophy of cells of the outer integument and which proliferates out through the micropyle in the form of a knob-like structure. The caruncle performs two important functions—

(i) ***Seed dispersal***—The caruncle is sweet in taste and is eaten by ants, which thus cause the dispersal of the seeds.

(ii) ***Water-absorbing function***—The caruncle is hygroscopic in nature and absorbs water, which is then passed on to the embryo.

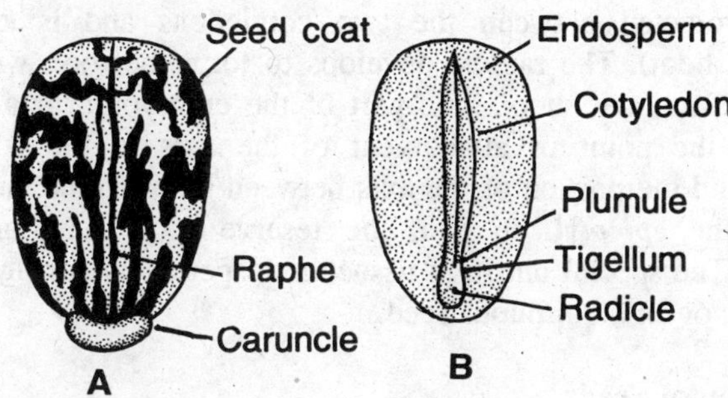

Fig. 13.4. Castor. A-a seed and B- L.S. of seed.

3. Operculum. A number of monocotyledonous families (Araceae, Lemnaceae, Musaceae, Marantaceae and Zingiberaceae) and some dicotyledonous families (Begoniaceae, Droseraceae, Melastomaceae and Nymphaeaceae) bear a plug-like structure in the micropylar part of the seed, known as *operculum*. The operculum may originate from the apex of the nucellus or endostome or exostome to form a dome-shaped structure. It consists of thick-walled cells, containing a bright orange-coloured substance. The operculum probably influences seed germination.

Functions of seed-coat

The seed-coat performs three important functions.

(i) **Protective**. The seed-coat encloses the seed and provides protection to the young baby plant within, from the mechanical injury and environmental changes.

(ii) **Nutritive**. The seed-coat may also be a source of nutrition to the developing embryo, e.g. the seed-coat of *Pisum sativum* contains starch and proteins which is supplied to the embryo during its development.

(iii) **Seed-dispersal**. The seed-coat of some seeds possess various structures which help in seed dispersal, e.g. seed hairs in *Gossypium*, wings in *Oroxylon* and air-filled cavities in *Enalus acoroides*.

STRUCTURE OF MATURE SEED

A mature seed may be differentiated into three distinct parts - (i) the embryonic plant, (ii) cotyledons–one or two in number and (iii) the covering seed-coat. The typical structure of a seed is illustrated by the garden pea (*Pisum sativum*). The mature fruit or pod contains a number of seeds attached to the fruit wall by a short stalk, the ***funiculus***. The funiculus abscisses on maturity, leaving a scar on the seed-coat, known as ***hilum***. Lying beneath the hilum is a small pore, the ***micropyle***, through which water is absorbed during germination. The hilum is continuous with a slightly elevated ridge on the seed-coat, known as ***raphe***. The seed is enclosed by a smooth seed-coat, commonly referred to as seed ***testa***. On removing the seed-coat, the embryo is exposed. It consists of two large, green and hemispherical ***cotyledons*** enclosing the embryonic axis. One end of the embryonic axis is seen to project out between the two cotyledons like a curved structure and is termed as ***radicle*** (embryonic root), whereas the other end of the axis lies protected between the two cotyledons and is known as ***plumule*** (embryonic shoot). The radicle develops to form the root, whereas the plumule develops to form the shoot. The part of the embryonic axis lying between the radicle and the point of attachment of the axis to the cotyledons is called ***hypocotyl***, and the portion of the axis between the plumule and the cotyledons is termed as the ***epicotyl***. In pea, the reserve food material is stored in the cotyledons. The special nutritive tissue, endosperm, is lacking, hence it is a non-endospermic or ex-albuminous seed.

Classification of Seeds

Seeds are classified into two groups, based on the number of cotyledons present.

1. **Dicotyledonous seed**—Seeds possess two (*di=two*) cotyledons, enclosing the embryonic plant. Dictoyledonous seeds may be further subdivided into two types, based on the presence or absence of endosperm.

(i) *Albuminous or Endospermic seeds*—in which the endosperm is present, e.g. Castor (fig. 13.4), custard-apple, four o'clock plant etc.

(ii) *Ex-albuminous or non-endospermic seeds*—in which the endosperm is absent, e.g. pea, gram, country bean, *Capsella* (fig. 13.5), mango, sunflower etc.

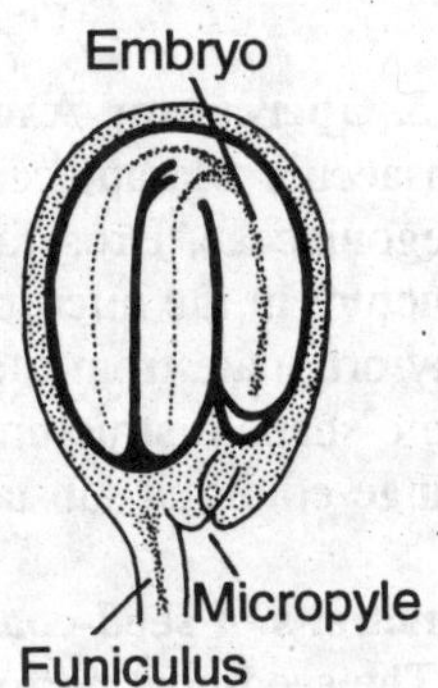

Fig. 13.5. L.S. of seed of *Capsella*.

2. **Monocotyledonous seeds**—Seeds possess one (*mono=one*) cotyledon only. Monocotyledonous seeds may be of two types depending upon the presence or absence of endosperm.

(i) *Albuminous or endospermic seeds*—in which the endosperm is present, e.g. rice, maize, onion, wheat, coconut (fig. 13.6), palms etc.

(ii) *Ex-albuminous or non-endospermic seeds*—in which the endosperm is absent, e.g. orchids, water plantain, arrowhead etc.

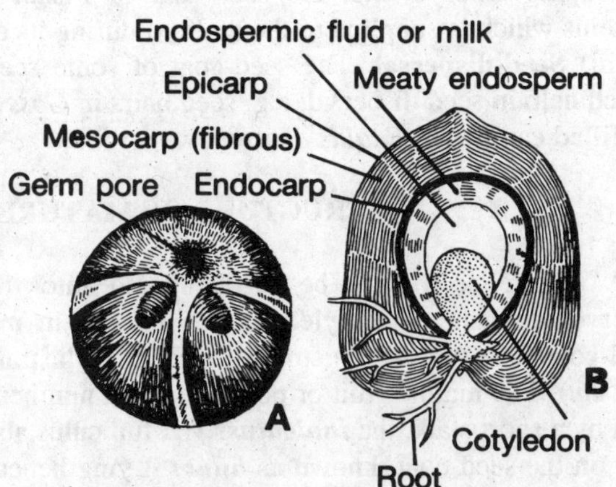

Fig. 13.6. Coconut. A-seed and B- L.S. of seed.

Structure of a dicot seed

(a) ***Albuminous or endospermic dicot seed.*** An endospermic dicot seed may be studied by examining the castor-seed. The castor seed is differentiated into three distinct parts (fig. 13.4).

1. **Seed-coat**. The castor-seed has a highly characteristic outer, hard, black and white designed seed-coat known as **testa**. At one end of the seed, towards the micropylar end, is present a knob-like, whitish outgrowth, known as **caruncle**. Beneath the caruncle and hidden by it, is present the **hilum** which marks the point of attachment of the seed to the stalk. At the base of the caruncle is present the **micropyle**. From the hilum runs a ridge, downwards along the length of the seed, known as **raphe**. Beneath the hard, outer testa is present a thin and membranous covering known as **tegmen**, which represents the inner seed-coat.

2. **Endosperm**. The seed-coat encloses a white, fleshy mass-the **endosperm**, which is visible only after removing the seed-coat. The endosperm is a storehouse of reserve food material and is specially rich in oil (castor oil).

3. **Embryo**. The endosperm encloses and completely surrounds the young embryo. The embryo consists of two, thin and papery **cotyledons** and the **embryonic axis.** The axis is divisible into the radicle, facing the micropylar end and the plumule, lying in between the cotyledons. The radicle germinates to form the root, whereas the plumule develops to form the shoot.

(b) ***Exalbuminous or non-endospermic dicot seed.*** Seeds lacking an endosperm are known as non-endospermic or exalbuminous seeds, e.g. pea, gram, *Capsella* (fig. 13.5), bean etc. They possess the structure of a typical dicot pea seed.

Structure of a monocot seed (fig. 13.7) **or Albuminous or endospermic seed**

The maize grain represents a typical endospermous monocot seed. It is a one-seeded fruit, in which the fruit wall is adherant to the seed wall and cannot be separated. On one side of the grain a whitish, suppressed area is distinct, which marks the position of the embryo within. A longitudinal section of the grain through this region reveals the following structures—

1. **Seed-coat**. The seed-coat is thin, single-layered and inseparably fused with the fruit wall, the pericarp.

2. **Endosperm**. Internally, the grain is distinctly divisible into two unequal portions, separated by a layer known as **epithelium**. The region lying above the epithelium represents the endosperm. It functions as a food storage tissue and is rich in starch, which may be tested by staining with iodine solution, whence the endosperm turns blue.

3. **Embryo**. The portion lying on the lower side of the epithelium represents the embryo. It is differentiated into—

(i) ***One single cotyledon***—which is shield-shaped, lying above and known as *scutellum* and

(ii) ***Embryonic axis***—which is differentiated into an upper, leaf-bearing **plumule** and a lower **radicle**. The plumule is further capped by a protective

sheath known as coleoptile, whereas the radicle is enclosed by the radicle-sheath known as coleorrhiza.

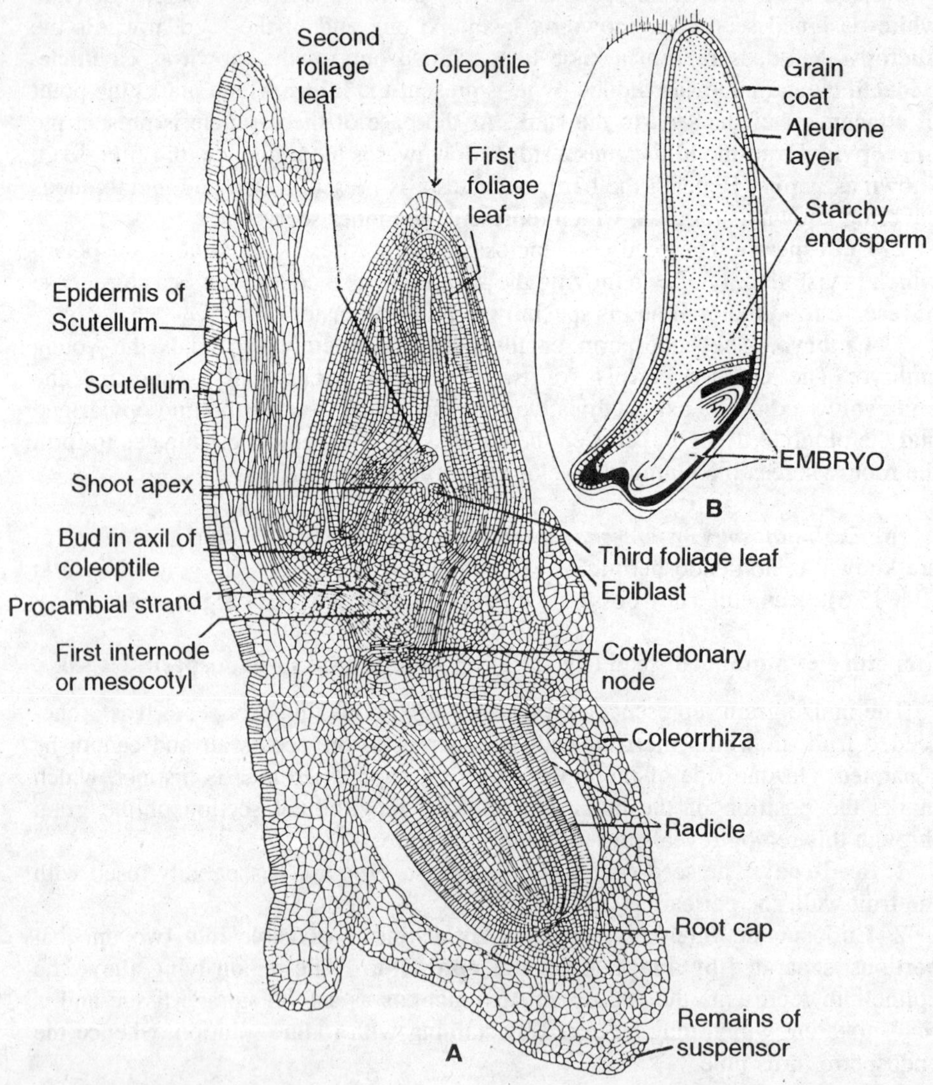

Fig. 13.7. L.S. of seed of *Triticum* (monocot seed) (diag.) and B-L.S. of embryo of *Triticum* showing details (Redrawn from Esau (1953) and Hayward (1938).

The separating sheath, i.e. epithelium, performs the important function of digesting and absorbing the food material, present in the endosperm and transferring it to the young baby plant.

Seeds of dicotyledons and monocotyledons are similar in their basic structures, but exhibit some significant differences.

Table 13.1 Differences between dicot seed and monocot seed (grain)

Dicot seed	*Monocot seed (grain)*
1. It is a true seed, developing from the ovule and formed inside the fruit.	1. The seed is known as the grain, and represents a single-seeded fruit called caryopsis.
2. The seed-coat is distinct from the fruit-coat or pericarp.	2. The seed-coat or testa is fused with the pericarp.
3. The seed-coat may be differentiated into an outer testa and an inner tegmen.	3. The seed-coat is uni-layered and insepar-able from the pericarp.
4. The seed encloses two cotyledons.	4. The grain has only one cotyledon, known as scutellum.
5. Seeds may or may not possess endosperm.	5. Majority of the monocot members possess endospermic grains.
6. The hilum, raphe and micro-pyle are distinctly present.	6. The hilum, raphe and micropyle are absent.
7. The two cotyledons enclose the embryonic axis in between them.	7. Internally, the grain is distinctly divided into two parts by the epithelium.
8. Endosperm, if present encloses the embryo.	8. Endosperm, usually present, lying above the embryo and separated from it by the epithelium.
9. The plumule and the radicle are unprotected.	9. The plumule is protected by a distinct sheath-the coleoptile whereas the radicle is protected by the coleorrhiza.
10. The hypocotyl and epicotyl are dis-tinct parts of the embryonic axis.	10. The hypocotyl and epicotyl are lacking.
11. Reserve food material is stored in the cotyledons and used during seed germination.	11. Reserve food material is stored in endosperm, which is passed on to the embryo, via the cotyledon.

Importance of seeds

Seeds are perhaps the most important and commercially valuable part of the plant and of immense utility to mankind. Some of their economic importance are enumerated below.

1. For the plant itself, especially the annuals, seeds are the only organs of perpetuation and continued existence. The seeds remain dormant and on return of favourable conditions, germinate to produce new plants.
2. Seeds or grains of cereals are rich in stored food such as cabohydrates, minerals, proteins, vitamins etc. which are stored in the endosperm. It is because of their high food value, that cereals constitute a major part of the human diet.
3. Cereal grains are dry and can be stored for long periods without deteriorating.

4. Seeds of *Gossypium* bear seed hairs, which constitute the cotton fibres of great economic importance.
5. Seeds are the source of various commodities, useful to man, such as edible oil (mustard, and groundnut), commercial oil (tung oil), beverages (coffee, cocoa), spices (cardamom) and medicine (nux-vomica, isabgul).

RESERVE FOOD MATERIALS OF THE SEED

The seed houses the baby plant and is entrusted with the dual function of providing protection and nutrition to the growing embryo. Within the seed, considerable amount of food is stored in an insoluble form within the cotyledons and endosperm. At the time of germination, the seed absorbs water, which converts the insoluble food into soluble form, so that it can be easily acted upon by the enzymes, e.g. *zymase, catalase, oxidase* etc. Enzymatic activity renders the food more absorbable and simple so that it may be conveniently utilized by the growing embryo for various functional activities, such as respiration, nutrition, cell formation, growth of protoplasm and development and differentiation of the embryonic parts. Food material in the seed is stored in the following forms-

(i) **Hemicellulose**. The endosperm of many palm seeds, e.g. betel-nut palm, date-palm, vegetable ivory-palm. etc. are rich in hemicellulose which are deposited on the thickened cell walls.

(ii) **Oils**. Oils are the liquid form of fats and are present to a greater or lesser extent in all the seeds. However, some seeds are extra-rich in oil and are known as oil-producing seeds, e.g. coconut (coconut oil), castor (castor oil), gingelly (gingelly oil), groundnut (groundnut oil), safflower (safflower oil), mustard (mustard oil) etc.

(iii) **Proteins**. Proteins are the structural and functional units of living organisms. As such all seeds contain proteins in varying degrees. However, seeds like pulses contain a higher percentage of proteins, e.g. Soya-bean contains the highest percentage of protein (30-50%) along with 11 important amino acids. During seed germination, the stored proteins are broken down by the action of proteolytic enzymes, into simple amino acids, which migrate to the embryo, where they are synthesized into new proteins.

(iv) **Starch**. The photosynthetic product, carbohydrate is stored in the form of insoluble starch. All seeds contain starch, but cereals like rice, wheat, maize, barley etc, are particularly rich in starch.

GENERAL QUESTIONS

1. Describe briefly the internal structure of a monocot and a dicot seed giving suitable examples.
2. Give a comprehensive account of structure of seed coat.
3. Write short notes on
   (i) Aril
   (ii) Caruncle
   (iii) Castor seed
   (iv) Reserve food materials stored in seeds
   (v) Importance of seeds.

CHAPTER 14

Embryology in Relation to Taxonomy

Taxonomy is the science dealing with the naming and classification of plants. The modern system of classification takes into consideration anatomical, embryological, palynological, cytological and even chemosystematic data at different levels. However, the significance of embryological studies as compared to anatomical and cytological studies have been lately appreciated. The eminent German embryologist, Schnarf (1931) was the first to highlight the important role of embryology in taxonomic groupings, which was later supported by Hofmeister and Strasburger. A significant aspect of the role of embryology in taxonomical studies is that often a single embryological character is sufficient to identify a family, such as, the occurrence of pseudo-embryo sac is characteristic of the family Podostemaceae, the formation of a single pollen grain from a pollen mother cell is characteristic of Cyperaceae family etc.

Embryological characteristics of use in taxonomic identification

A number of embryological features have been considered to be of great significance in evaluating taxonomic position and identification. Some of these are—

1. **Nature of division of microspore mother cell**. The pattern of division of the microspore mother cell is an important taxonomic feature, *e.g.* whether the divisions are successive or simultaneous.

2. **Anther structure**. The structure of the anther, the nature of the anther wall, number and arrangement of the chambers (locules) and the structure of the tapetum, whether amoeboid or glandular are taken into taxonomic consideration.

3. **Structure of pollen grain**. The morphological features of the pollen grain, such as the exine ornamentation, the number and position of the germ pores and the NPC system of the apertures are an important taxonomic consideration.

4. **Structure of the ovule**. Various structural aspects of the ovule have been considered for taxonomic identification, *e.g.* the number of integuments, the vascular supply to the ovule, crassinucellate or tenuinucellate type of ovule, presence of aril or caruncle etc.

5. **Development of the embryo sac**. The various developmental aspects of the embryo sac which find use in taxonomy are the position of the functional

megaspore, monosporic, bisporic or tetrasporic nature of the embryo sac, nuclear fusion during embryo sac development etc.

6. **Structure of the mature embryo sac**. The structure and form of the mature embryo sac, the persistence or absence of antipodal cells, formation of embryo sac haustoria etc. are of great help in taxonomic classification.

7. **Fertilization**. The path of pollen tube, its entry into the ovule, the time interval between pollination and fertilization have been considered for identification.

8. **Endosperm**. The nature and structure of the endosperm have proved useful in taxonomic identification. These are the cellular, nuclear or helobial nature of the endosperm, presence or absence of endosperm haustoria etc.

9. **Embryo**. The mature structure of the embryo, the presence or absence of suspensor etc. are sometimes considered for classification.

10. **Seed coat**. The development and organisation of the seed coat may prove useful in taxonomic identification.

SIGNIFICANCE OF EMBRYOLOGY IN TAXONOMY

Embryological studies and investigations have helped to evaluate the correct taxonomic position of a number of genera and families. Some of the important taxa and families are briefly discussed.

Taxo-embryological studies in various taxa

1. **Actinida**. Vijayaraghavan (1965) made embryological studies on *Actinida* and showed that the ovules possess incipient integumentary tapetum and are anatropous, unitegmic and tenuinucellate, the endosperm is cellular and the embryo development is of Solanad type. Based on these characters, *Actinida* has been removed from family Dilleniaceae (Bentham and Hooker, 1876) and placed in a newly created family Actinidiaceae, under the order Theales.

2. **Cassytha**. Embryological studies in *Cassytha* (family Lauraceae) have shown that it possesses certain distinctive characters, such as secretory type of anther tapetum, formation of multiple embryo sacs, cellular endosperm etc. which support its promotion to the status of a separate monogeneric family, Cassythaceae.

3. **Cercidiphyllum**. The genus *Cercidiphyllum* of family Hamamelidaceae exhibit certain embryological features, such as cellular nature of endosperm, extensive elongation of the embryo sac, broad and ill-defined furrow on the pollen grain etc. These features support the removal of *Cercidiphyllum* from the family Hamamelidaceae and assigning it to a separate family Cercidiphyllaceae.

4. **Daphniphyllum**. Hooker (1885) and Rendle (1938) included *Daphniphyllum* in Euphorbiaceae, however, the genus exhibits specific characters quite dissimilar to Euphorbiaceae. Engler and Prantl (1889) placed it in a separate family Daphniphyllaceae on the basis of two or three-layered anther wall, thickenings on radial and inner tangential walls of endothecium, bicarpellate ovary, parietal placentation, minute embryo etc.

5. **Exocarpus**. Gangnepain and Boureau (1947) proposed that *Exocarpus* should be removed from family Santalaceae and placed under Taxaceae (Gymno-

sperms), due to presence of "naked ovule" and pollen chamber. However, Ram (1959) studied the embryology of *Exocarpus* in detail and confirmed the inclusion of the genus under Santalaceae.

6. **Paeonia**. The classical system of classification assigned *Paeonia* to the tribe Paeonieae of family Ranunculaceae. However, embryological studies strongly support the removal of *Paeonia* from Ranunculaceae to a separate family Paeoniaceae. The embryological differences between *Paeonia* and Ranunculaceae are enumerated in Table 14.1.

Table 14.1. Embryological differences between *Paeonia* and Ranunculaceae

Embryological feature	*Paeonia*	*Ranunculaceae*
Microsporangium	Endothecium multilayered with 2-3 layered tapetum.	Endothecium and tapetum one-layered.
Microspore	Exine shows reticulate pits, generative cell large and elongate.	Exine is smooth, granular or papillate, small lenticular generative cell present.
Female archesporium	It is multicellular in which many cells function as megaspore mother cells.	It is uni or multicellular in which only one cell is the functional megaspore mother cell.
Embryogeny	It is of unique type. The diploid nucleus divides repeatedly to form a coenocytic structure in which the nuclei are arranged peripherally around a central vacuole. Soon wall formation starts and a number of the peripheral cells form the embryo initials, of which only one matures into adult embryo.	It is of Onagrad or Solanad type.
Seed	Possesses aril.	Does not possess aril.

7. **Pentaphragma**. The genus *Pentaphragma* was assigned by earlier botanists to the family Campanulaceae (Bentham and Hooker (1876) and Engler and Prantl (1889), but Shaw (1954) removed it from Campanulaceae and placed it in a new family Pentaphragmataceae, based on the presence of distinct embryological characters, such as an unicelled, uninucleate micropylar haustorium, a short suspensor and the presence of extramicropylar protusion of the embryo sac.

8. **Trapa**. The importance of embryological features in evaluating taxonomic position is well illustrated in the genus *Trapa*, which was previously placed in

family Onagraceae, but has now been included in the monogeneric family, Trapaceae (Hutchinson, 1959). Table 14.2 elucidates the differences between *Trapa* and family Onagraceae.

Table 14.2. Differences between *Trapa* and Onagraceae

Embryological feature	*Trapa*	*Onagraceae*
Microspore	Pyramidal in shape with 3 crests	Triangular and basin shaped
Ovary	Semi-inferior, biloculed, each locule having a single anatropous ovule	Inferior, usually trilocular with many ovules per locule
Embryo sac	It is of Polygonum type	It is of Oenothera type
Endosperm	Absent	Present
Embryo	Solanad type	Onagrad type
Suspensor	Well developed haustorial suspensor	It is short and inconspicuous
Seed	Possesses one reduced and one normal cotyledon	Both cotyledons equal in size

Taxo-embryological studies in families

1. **Canellaceae**. Wettstein (1935) and Vestal (1937) suggested that the family Canellaceae is closely related to the family Myristicaceae and Hutchinson (1959) placed the family Canellaceae in the order Magnoliales. Wilson (1960) proposed that Canellaceae show close affinities to the families Dilleniaceae, Eupeteleaceae, Eupomatiaceae of the order Ranales. However, Parameswaran (1962) on the basis of embryological studies confirmed the close relationship between Myristicaceae and Canellaceae.

2. **Cneoraceae**. The family Cneoraceae is represented by the genus *Cneorum*. Engler and Prantl (1897) placed the family in the order Geraniales, whereas Van Tieghem (1906) placed it in the order Rhamnales and Hutchinson (1959) placed the family in the order Celastrales. On the basis of embryological studies, Souges (1955) confirmed the inclusion of the family Cneoraceae in the order Rhamnales.

3. **Gentianaceae**. The family Gentianaceae was sub-divided into two sub-families *viz.* Menyanthoideae (marshy plants) and Gentianoidae (terrestrial plants) by Bentham and Hooker (1862) and Rendle (1889). The two sub-families, however, show distinct embryological traits, such as cellular endosperm and Asterad type of embryogeny in Menyanthoideae, whereas Gentianoideae exhibits nuclear endosperm and Solanad type of embryogeny. Based on these embryological

differences, Wettstein (1935) raised the two sub-families to two independent families and which was supported by Don (1938), Lindsey (1938) and later by Maheshwari Devi (1962).

4. **Nelsonioideae**. The taxonomic position of Nelsonioideae has been controversial as to whether it should be included in family Acanthaceae or Scrophulariaceae. However, embryological studies conducted by Johri and Singh (1959) and Mohan Ram and Masand (1963) confirmed the inclusion of Nelsonioideae in the family Acanthaceae. The confirmation of the taxonomic position of Nelsonioideae is based on the asymmetrical growth of the endosperm and the presence of a persistent and ruminate endosperm.

5. **Orobanchaceae**. The family Orobanchaceae has been related to the family Gesneriaceae or Scrophulariaceae by different workers. Tiagi (1963) suggested the alliance between Orobanchaceae and Scrophulariaceae, based on embryological similarities, such as the development of the male and female gametophytes, endosperm, embryo and the seed. He further proposed that the Orobanchaceae may have been derived from the Scrophulariaceae.

6. **Stackhousiaceae**. The family Stackhousiaceae was placed in the order Celastrales by Bentham and Hooker (1862–1867), whereas Engler (1897) placed the family in the order Sapindales, between Salvadoraceae and Staphyleaceae. Detailed embryological studies made by Narang (1953) on *Stackhousia linariaefolia* showed that the bitegminal and tenuinucellar nature of the ovule, the *Polygonum* type of embryo sac and the nuclear type of endosperm confirm the inclusion of the family in the order Celastrales.

7. **Turneraceae**. Hutchinson (1959) placed the family along with Losaceae in the order Losales, whereas Engler suggested its relation to Passifloraceae and included it in the order Violales. The embryological features of Turneraceae, such as anatropous, bitegminal and crassinucellate ovules, nuclear type of endosperm development, Onagrad type of embryogeny, arillate ovules etc. are found to be absent in Losaceae (Losales), but present in Passifloraceae (Violales). As such Turneraceae should be placed in order Violales, close to the Passifloraceae.

Families distinguished by their embryological characters

All families have one or two embryological features which are common to all, however, some families can be marked out and identified by their special embryological features. These are-

Family	Embryological feature
Cyperaceae	The microspore mother cell gives rise to only one pollen grain. The pollens are shed at the 3-celled stage.
Onagraceae	Embryo sac is of *Oenothera* type and is formed from the micropylar megaspore. Antipodals are absent.

Podostemaceae	Pseudo-embryo-sacs are formed. Pollen grains occur in pairs. Presence of prominent suspensor haustoria and absence of triple fusion (consequently endosperm is absent) are characteristic.
Proteaceae	The embryo-sac forms a tubular, coiled and coenocytic structure called "vermiform appendage."

Although embryological data have helped to evaluate the position of a number of genera and families, it is essential to take into consideration morphological, anatomical and cytological characteristics to evolve a modern system of classification.

GENERAL QUESTIONS

1. Give an account as to how embryological studies have helped to evaluate taxonomic position.
2. Discuss briefly the various embryological features useful in identifying taxonomic position.
3. Discuss briefly the significance of taxo-embryological studies in various taxa.
4. Write notes on-
   (i) *Daphniphyllum*
   (ii) *Paeonia*
   (iii) *Trapa*
   (iv) Gentianaceae
   (v) Turneraceae
5. Write an essay on "Embryology in relation to taxonomy."

CHAPTER 15

Experimental Embryology

Embryology originated as a descriptive science, however, since 1960 it gained a new dimension as an experimental science. Experimental embryology is concerned with two main objectives, *viz.* (i) to determine the various factors that control the different embryological processes, and (ii) to develop methods of control for modifying these processes, by an alteration in the environmental conditions of the entire plant or excised plant parts.

Significant progress in the field of experimental embryology has been made possible through the technique of "*in vitro* culture, which consists of isolating plants cells, tissues, organs and growing them in nutrient medium, under aseptic conditions, to give rise to plantlets. *In vitro* culture of plant parts is collectively known as *tissue culture.*The main types of tissue culture are:

1. Plant culture—Culture of seedlings or larger plants.

2. Embryo culture—Culture of isolated embryos.

3. Organ culture—Culture of isolated plant organs such as root tips, stem tips, leaf tips, flowers etc.

4. Nucellus and ovary culture—Culture of nucellus and ovary.

5. Cell or tissue suspension culture—Culture of suspensions of single isolated cell or group of cells together in liquid media.

6. Protoplast culture—Culture of mechanically or enzymatically isolated protoplasts.

In vitro culture technique comprises of three main aspects—

(i) ***Culture or nutrient medium***—The basic requirement for culturing all types of plant parts, is an appropriate nutrient enriched culture medium. A *culture medium* is defined as an artificially prepared nutrient medium in which excised plant tissue and organs can be grown *in vitro*.

(ii) ***Aseptic conditions***—It is extremely important to maintain complete aseptic conditions inside the culture dishes. The nutrient rich culture medium can support a luxuriant growth of microorgainsms, which would eventually check the growth of the plant tissue. Hence, the culture medium, culture dishes, the tissue to be cultured and accessory apparatuses should be thoroughly sterilized. Inoculation is performed under aseptic conditions in a laminar air flow cabinet, to prevent microbial contamination.

(iii) ***Aeration***—Proper aeration of the cultured tissue is necessary for maintaining healthy growth. Aeration is provided by shaking the flasks or dishes on an automatic shaker.

CULTURE MEDIUM

The study of culture medium is essential, since it is the base for all experimental procedures,

Every tissue or organ has its special requirements for optimal growth which need to be worked out. Most media contain a mixture of inorganic salts (to supply macro- and micro-elements), vitamins and sucrose. Such a medium is commonly called *basal medium.* To this may be added, for better results, some growth regulators such as auxins, cytokinins and gibberellins in various combinations or as natural plant extracts, such as coconut milk (liquid syncytium of endosperm), various fruit juices, casein hydrolysate and yeast extract. Natural plant extracts are not commonly used as these increase chances of contamination.

Technique of preparing basal medium

Philip White's medium is commonly used as basal medium for demonstration of tissue culture technique. It contains both macro- and micro-nutrients essential for growth. For preparing it, the following five stock solutions, each in one litre of distilled water are prepared.

	grams
Solution 1	
$Ca(NO_3)_2$	2.0
KNO_3	0.8
KCI	0.65
NaH_2PO_4	0.165
Solution 2	
$MgSO_4$	36.0
Solution 3	
$MnSO_4$	0.45
$ZnSO_4$	0.15
H_3BO_3	0.15
KI	0.075
[$CuSO_4$	0.002]
[Na_2MoO_4	0.021]
Solution 4 (must be freshly prepared)	
$Fe_2(SO_4)_3$	0.25
Solution 5	
Glycine	0.3
Thiamine	0.01
Pyridoxine	0.01
Nicotinic acid	0.05

During the experiment only small amount of stock solutions are used. In a one litre flask, 500 ml of distilied water is taken into which 100 ml of stock solution 1 and 10 ml each of stock solutions 2, 3 and 4 are added. After these have been thoroughly mixed by shaking, 20 g of sucrose is added. Distilled water is now added to make up the quantity to 1 litre. If necessary to solidify the medium, about 0.8 per cent agar is added. The pH of medium is adjusted around 5.8 (slightly acidic). The solution is then divided by pouring 50 ml amounts into smaller flasks, plugged with non-absorbant cotton wrapped in cheese cloth. These flasks are placed, in an autoclave for fifteen minutes at a pressure of fifteen pounds per square inch. It brings about sterilization of the medium. The flasks may be stored for a few days in a refrigerator until transfers can be made.

Though White's medium is preferably used, the other media have also been recommended. The composition of some nutrient media used for *in vitro* culture of plant tissues is given here. These may also be tried to demonstrate the technique of tissue culture.

Table 15.1. Composition of some nutrient media (basal media)

Constituents	*Murashige and Skoog's (M.S.) medium* $mg l^{-1}$	*Nitsch's medium* $mg l^{-1}$	*Nagata and Takebe's medium* $mg l^{-1}$
INORGANIC			
KNO_3	1900	950	950
NH_4NO_3	1650	720	850
$CaCl_2$	440	166	220
$MgSO_4.7H_2O$	370	185	1233
KH_2PO_4	170	68	680
$Mn\ SO_4.4H_2O$	22.3	25	22.3
H_3BO_3	6.2	10	6.2
$Zn\ SO_4.4H_2O$	8.6	10	8.6
$Na_2MoO_4.2H_2O$	0.25	0.25	0.25
$CuSO_4.5H_2O$	0.025	0.025	0.025
$CoCl_2.6H_2O$	0.025	—	—
$CoSO_4$	—	—	0.08
KI	0.83	—	0.88
$Fe\ SO_4.7H_2O$	27.8	27.8	27.8
Na_2EDTA	37.3	37.3	37.3
ORGANIC			
Niacin	0.5	5.0	—
Glycine	2.0	2.0	—
Pyridoxin HCl	0.5	0.25	—
Thiamine HCl	0.1	0.5	1
Folic acid	—	0.5	—
Biotin	—	0.05	—
Myoinositol	100	100	—
Sucrose	30,000	20,000	10,000
IAA	1.30	—	—
NAA	—	—	3
Kinetin	0.4-10	—	—
Benzylaminopurine	—	—	1

However, the use of Murashige and Skoog (M.S.) medium has also given promising results and several tissue culture labs are frequently using it. Nitsch's medium (1974) has been found to be quite successful for obtaining haploids of *Datura* and *Nicotiana* from isolated pollen culture.

It has now become possible to raise continuously growing cultures from most plant tissue (including those which would have never divided in nature) with the greater knowledge of the nutritional requirements of cultured tissues and the discovery of several growth promoting substances. In many cases, callus remains as an undifferentiated mass in the culture medium. In this reference, the work of Skoog and Miller stands as a milestone. They discovered the role played by auxin and cytokinin in the control of shoot and root formation or cellular differentiation. They showed that in the presence of cytokinin (0.2 mg/1 kinetin), a vigorously growing callus of tobacco pith tissue was formed, which with the supply of normal medium had shown poor growth. When it was transferred to the medium containing 3 mg/1 of IAA (auxin) and 0.2 mg/1 of kinetin (cytokinin), this callus differentiated only roots. If a portion of the same callus was planted on a medium with higher concentration of kinetin (1 mg/1) and lower concentration of IAA (0.03 mg/1), it differentiated only shoots.

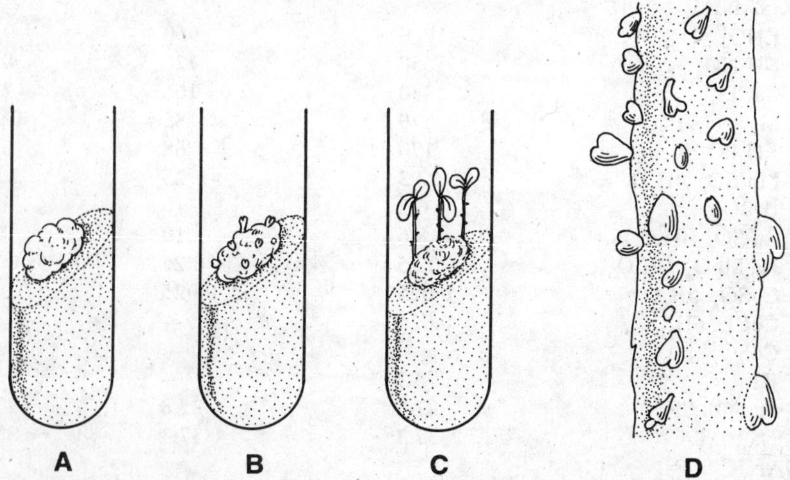

Fig. 15.1. Figures explaining procedure of development of embryoid in tissue culture of *Ranunculus*; A-callus, B-numerous embryoids in six week old cultures, C-plantlets developed from embryoids and D-an enlarged portion showing embryoids (redrawn from Konar and Nataraja, (1965).

The important findings of Steward and his coworkers (1964) with carrot culture is another classical example in the field of tissue culture and in demonstration of *cellular totipotency*. They grew small pieces of phloem of mature carrot root in a liquid medium supplemented with coconut milk, in special containers. By gently shaking the cultures, cell clusters and ultimately cells became free and floated in the medium. Some of these developed into rooting clumps. When these clumps were transferred to the tubes containing a semi-solid medium, they gave rise to whole plant that flowered and set seeds.

A precise procedure for developing tobacco plant from a single cell has been worked out by Vasil and Hildebrandt (1965). They were able to raise whole plant of tobacco by first growing isolated single cells into masses of tissue, in microchambers and later transferring them to flasks. Konar and Nataraja (1965) got success in embryoid formation in tissue culture of *Ranunculus* (fig. 15.1).

Guha and Maheshwari (1966) convincingly demonstrated the formation of embryoids of single cell origin, i.e. from pollen grains. In the process, the excised anther is cultured on a suitable medium. Some of the pollen inside the anther undergoes repeated divisions forming a multicellular tissue within the parent wall. Eventually, the pollen wall bursts releasing the callus which directly develops into an embryoid and germinates to form a haploid sporophyte (**androgenic haploid**).

Some specific examples related with experimental embryology are discussed here.

EMBRYO CULTURE

Hanning (1904) first successfully cultured embryos of *Raphanus* and *Cochleria* under aseptic conditions. A great achievement in the field of embryo culture was made by Laibach (1925) who successfully cultured the abortive embryos obtained from interspecific crosses of *Linum*. His technique of using cotton waddings on moist filter papers containing sucrose or lactose is now widely followed.

Later studies revealed that embryos have abilities to synthesise most of the substances required for their growth. That is why embryos can be successfully grown in relatively simple basal medium, containing only a few mineral salts and sucrose, during their autotrophic phase. But during heterotrophic phase, i.e. nymph stage, embryos obtain their nutrition from the endosperm, under natural condition, and for growing such embryos an elaborate nutrient medium is needed. The basal medium has to be supplemented with growth substances like coconut milk, yeast extract or casein hydrolysate etc.

Embryo culture studies have revealed that under natural conditions, the maternal tissues present in the surrounding of developing embryos play some very important role in directing the course of normal embryogenesis. Norstog (1965) noted that excised pre-globular embryos when cultured, fail to follow normal embryo developmental phases and it grows into unorganised fashion, which later formed many plantlets. Similar case was earlier reported by Johri and Bajaj (1962) while working on *Dendrophthae*.

The significant contributions made by various scientists in the development of embryo culture are summarized in Table 15.2.

Table 15.2. Important contributions in embryo culture

Scientist	*Year*	*Contribution*
E. Hanning	1904	Father of plant embryo culture, cultured embryos of *Raphanus sativus* and *Cochleria danica* and grew them into planlets,
M. Dubard, J.A. Urbain, G.D. Buckner and J.H. Kastle	1913-17	Studied the role of endosperm in the growth of embryos,
D.I. Andronescu	1919	Demonstrated that the embryo of Graminae deprived of the scutellum, grew weakly in the culture medium.
L. Knudson	1922	Germinated orchid embryos into plant lets by growing them on sugar-containing agar medium.
K. Dieterich	1924	Introduced the term 'precocious germination' for the malformation of seedlings, raised from the embryo of immature seed, *in vitro*.
F. Laibach	1925-29	Demonstrated the application of embryo culture technique in rearing viable seedlings from unsuccessful crosses.
C.D. LaRue	1936	Cultured a number of monocot, dicot and gymnospermous embryos.
J. van Overbeek	1941-42	Demonstrated that proembryos could be cultured in a highly enriched, coconut milk-containing medium.

Categories of Embryo culture

Embryo culture has been categorised into following types.

(i) ***Culture of proembryos or immature embryos***. The heart-shaped, globular embryo is known as proembryo. Culture of proembryo helps in understanding the differentiation process as well as the nutritional requirements of developing embryos.

(ii) ***Culture of intact seed containing undifferentiated embryo***. In orchids, the seeds contain morphologically undifferentiated embryos, lacking both radicle and plumule. As such, the entire seed is considered to be an embryo, and its culture is treated as embryo culture.

(iii) ***Culture of mature and intact seed embryo***. The culture of mature seed embryos helps to determine the diverse parameters of embryonic growth.

(iv) ***Culture of dissected embryo***. The mature embryo is cut up into small segments, which are then cultured separately. This helps to analyse the inter-relationship of different parts of the embryo.

(v) ***Culture of inviable or abortive embryo***. Interspecific or intergeneric breeding may give rise to abortive embryos, which in turn produce non-viable seeds. Normally, these seeds fail to germinate. However, *in vitro* culture of abortive embryos has helped to produce hybrid plants.

Technique for Embryo culture

The basic principle involved in embryo culture is the aseptic excision of the embryo from within the seed and its transfer to a suitable nutrient medium. The most important aspect of embryo culture is the selection of the culture medium, fortified by additional nutrients, to sustain continued growth of the embryo. The embryo culture of *Capsella bursa-pastoris* is taken as an exemplary protocol, executed by the following steps—

(i) The fruit or capsule in the required stage of development is surface sterilized for 5-10 minutes in 0.1% $HgCl_2$ and then washed repeatedly in distilled water.

(ii) The sterilized capsules are next placed in a depression slide, containing few drops of liquid medium, under a specially designed dissecting microscope.

(iii) With the help of the microscope, the fruit is carefully dissected to expose the ovules. The enclosed embryo is now freed by making a small incision in the ovule and applying slight pressure.

(iv) The excised embryos (about 6-8) are now transferred by micropipettes to petridishes, containing 25 ml of solidified nutrient medium. The petridishes are sealed with cellotape.

(v) The sealed cultures are now placed in a culture room at 25±1°C and exposed to 16 hours illumination by cool, white flourescent tube. After about four weeks interval, subcultures are made into fresh medium, where organogenesis followed by plantlet formation starts.

Work on embryo culture under *in vitro* conditions has been done more comprehensively as compared to ovary and ovule cultures. Besides the general technique of embryo culture described, new techniques have also been adopted. One of them is the *endosperm transplant technique*.

Endosperm transplant technique. It refers to culturing of excised embryos (or *ex ovule* embryos) on cultured endosperm. Kruse (1974) reported that procedure of implanting the immature hybrid embryos on cultured endosperm, increase 30-40 per cent chances of survival, in comparison to growing the embryos directly on culture medium. Since then, endosperm transplant technique has become very popular particularly for culturing immature hybrid embryos.

Williams (1980) suggested a little modification in this technique i.e., to insert the excised embryo into the dissected out endosperm of another parent or of closely related species and to culture both on an artificial medium. This technique succeeded in culturing interspecific hybrids of *Trifolium*.

Importance of Embryo Culture

In vitro embryo culture in relation to biological knowledge is of great significance in evaluating fundamental problems of plant development.

(i) *In vitro* embryo culture studies help to analyse the various parameters of embryonic growth.

(ii) It helps to ascertain the various physical and nutritional factors that regulate the growth of organ primordia of the seedling.

(iii) Changes in nutritional requirements of embryos at different stages of growth and differentiation can be studied.

(iv) The effect of the nutrient medium, various amino acids, vitamins, growth hormones etc. on the growth, development and differentiation of the embryo can be studied by *in vitro* culture technique.

Studies on Embryo Morphogenic Problems

In order to study, (i) the regeneration potentialities of various parts of the embryo and (ii) the interaction between different parts of embryo during germination, microsurgical experiments with embryos in artificial medium were carried out with different modifications. Lee (1955) probably for the first time cultured various segments of lupine embryos and found that only plumule-containing segments of embryo grew into full seedlings. Bajaj (1966) made comprehensive studies with *Dendrophthae falcata* and observed *in vitro* response of various segments of embryos, while growing in IAA and casein hydrolysate supplemented basal medium.

Further studies have revealed that the explanted radicular end formed only holdfast with haustoria, while segments consisting of (i) radicular end, with plumule and adjoining portions of cotyledons, (ii) plumular region and adjoining portions of cotyledons and (iii) plumular region with complete cotyledons, formed complete plantlets. The explant segment consisting of cotyledons only formed merely undifferentiated callus.

These findings indicated that cotyledons play some very important role in seedling development particularly in the growth of plumule. Such observations have been substantiated by several other workers in *Avena, Pennisetum, Hordeum* and *Zea mays,* where scutellum (the cotyledon) played that important role. Rangaswamy and Rangau (1971) extended their studies further and observed the effect of cotyledons on seedling formation in *Cassytha filiformis*. Their observations can be summarised as follows (Table 15.3).

Table 15.3. Response of segments of cotyledons on seedling formation in *Cassytha filiformis*

Cotyledon portion removed		*Growth response*
1. Plumular half of one cotyledon	—	Seedling well developed
2. Radicular half of one cotyledon	—	" " "
3. Plumular half of both cotyledons	—	Seedling well developed
4. One entire cotyledon	—	Growth of hypocotyl and radicle inhibited
5. Radicular half of one cotyledon and plumular half of other cotyledon	—	Growth of plumule inhibited
6. One entire cotyledon and a part of other cotyledon	—	Plumule growth quiescent

The observations on *Cassytha filiformis* revealed that the minimal segment of cotyledons necessary to permit normal differentiation of plumule is the radicular half of one cotyledon together with plumular halves of both the cotyledons or radicular halves of both the cotyledons.

The morphogenetic studies of various plant organs such as ovary, endosperm, nucellus or embryo needs very precise and sophisticated techniques and is therefore still in process. The coming decade is expected to resolve several hitherto unknown mysterical problems of morphogenesis.

Practical applications of Embryo culture

Embryo culture is being routinely used in some crop improvement programmes to obtain interspecific hybrids, propagation of orchids, over-coming of dormancy etc. A few of them are described here:

(1) **Recovery of distant hybrids**—In general, the distant or interspecific crosses often fail either due to inability of pollen to germinate and failure of pollen tube to grow or due to degeneration of endosperm. In such cases, embryo culture is used to recover hybrid plants. Some recent examples are recovery of hybrids from *Hordeum vulgare* X *Secale cereale, Triticum aestivum* X *Agropyron repens, Hordeum vulgare* X *Triticum aestivum* etc. One interesting example is of *Triticale* which is a hybrid plant of cross between *Triticum aestivum* X *Secale cereale*. The viable seeds are rarely produced in it due to presence of two dominant genes, the Kr_1 and Kr_2. The majority of seeds are poorly developed and small and show poor germination. The viability percentage obtained ranges between 5 to 10 per cent. The recovery of hybrid seedlings is much greater (50-70 per cent) when 10-14 day old embryos are removed and cultured in suitable medium.

(2) **Propagation of orchids**—Seeds of orchids are smallest among angiosperms and these lack any sort of stored food. The embryo of orchid is virtually naked and in many cases it may remain immature or underdeveloped. That is why propagation of orchids in nature is very slow, tedious and highly time-consuming. Since orchids are beautiful creations of nature and are largely valued for their ornamental and economic value, this problem is now being largely solved by using tissue culture technique. Young or mature orchid embryos are removed from seeds and placed on suitable nutrient media which develop into seedlings either directly or through callus formation.

(3) **Shortening of breeding cycle**—Seeds often take 15 to 20 days to mature after the full development of embryo. In some cases, maturity period is very long. By applying embryo culture technique, embryos are made to develop directly into seedlings and thus avoiding a long time of seed maturity.

(4) **Overcoming dormancy**—In most fruit trees, embryos require a period of after-ripening i.e. dormancy before germination. This period of dormancy can be effectively eliminated by embryo culture. Such successful experiments have been done in case of *Prunus, Taxus, Iris* etc.

(5) **Propagation of rare plants**—Embryo culture technique can be used in maintaining rare or endangered plants. For example, '*makapuno*' is a special delicious variety of coconut which is served during special ceremonial occasions in Philippines. This variety in fact contains soft, solid and fatty tissues in place of

liquid endosperm. De Guzman and Del Rosario (1974) using embryo culture technique succeeded in establishing sufficient number of *makapuno* plants.

(6) **Rapid seed viability test**—Branton (1961) suggested that germination of excised embryos, using tissue culture technique, is more reliable procedure to perform seed viability test, rather than adopting usual staining technique. Thus, it provides authoritative correct information of seed viability.

(7) **To study the mutagenic effect**—Embryo culture is a very useful technique to evaluate the mutagenic effect of seeds irradiated by X-rays and gamma rays. It also helps to ascertain the safety doses of irradiation.

(8) **Improvement of horticultural plants**—Embryo culture is an important research technique employed for the improvement of horticultural plants, both fruit trees and flowering plants.

(9) **Overcoming self-sterility of seeds**—Propagation by vegetative means is the natural method in some economically important plants like banana and kachoo. Seeds produced, fail to germinate due to natural self-sterility, which may be overcome by embryo culture.

SOMATIC EMBRYOGENESIS

The diploid zygote germinates to form a diploid, multicellular embryo. Such types of embryo formed as a result of the sexual process is known as *zygotic embryo*. Embryos may also be formed from the sporophytic tissue surrounding the embryo sac (*non-zygotic embryos*) or from an unfertilized egg (*parthenogenetic embryos*). But, the formation of embryo from cells outside the female gametophyte, has not been reported in nature. However, in plant tissue culture, totipotency of the somatic cell, has been established, which means that any cell of the plant has the genetic potential to produce the entire plant. J. Reinart, a German botanist in 1959, working on callus structures derived from carrot storage root tissues, found small, bipolar organised structures bearing cotyledons developing from the callus. This phenomenon was later termed as *somatic embryogenesis*, which is defined *as the developmental pathway of well-organised, small embryoids, resembling the zygotic embryo from the embryogenic potential somatic plant cell of the callus tissue, or cells of suspension culture*. The *embryoid* is a small well-organised structure comparable to the sexual embryo, which is produced in tissue culture of dividing embryogenic potential somatic cells.

Somatic embryogenesis may be induced by two methods.

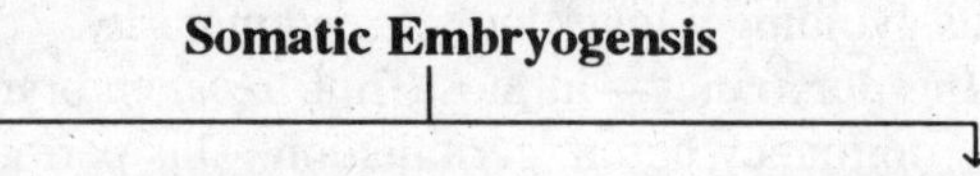

Direct Method

Embryoids are produced directly from proembryonic determined cells, without callus formation. e.g. somatic embryos have been formed from leaf mesophyll cells of *Dactylis glomerata* (orchard grass).

Indirect method

The cells of the explant undergo callus formation and embryoids develop from within the callus tissue, from induced embryogenic cells.

Somatic embryogensis is affected by a number of factors such as auxin concentration, gibberellins, reduced nitrogen, dissolved oxygen, potassium etc.

Importance of Somatic Embryogenesis

The importance and potential application of *in vitro* somatic embryogenesis are as follows—

(i) Somatic embryos arise from single cells and is of special significance in mutagenic studies.

(ii) In many cases, the somatic embryos formed are free from virus and other pathogens.

(iii) The mass production of somatic embryos in cell culture is considered to be an ideal propagation system.

(iv) The somatic embryo develops directly into a plantlet, hence a separate rooting phase (necessary in shoot culture) is eliminated.

(v) Somatic embryos, however, lack food reserves and the hybrid vigour found in zygotic embryos.

ARTIFICIAL SEEDS

In nature the fertilised ovule forms the true seed, which consists of a zygotic embryo enclosed by protective coats, developed from the integument. The concept of artificial seed has arisen from the phenomenon of somatic embryoids, which are formed from the somatic diploid cells, cultured *in vitro*. Somatic embryoids lack any protective coats and hence cannot be directly sown into field soil. So, tissue culturists tried to develop methods of encapsulating somatic embryoids by protective gel-like substance, so that they could survive field conditions. Thus the concept of artificial seed took birth. Artificial seeds were first successfully developed by Kitto and Janick (1982), who coated carrot embryoids with polyoxethylene. Redenbangh *et al.* discovered that hydrogels, such as sodium alginate could be used to encapsulate somatic embryoids to produce single embryoid artificial seed. Artificial seeds are living seed-like structures derived from somatic embryoids by encapsulating them by a hydrogel. They may be used as a substitute for natural seeds and function like true seeds when sown in soil.

Technique for making artificial seeds

Artificial seeds are made by the following technique, illustrated by the flow chart—

Establishment of callus culture ⟶ Induction of somatic embryogenesis in callus culture ⟶

Completion of embryo development ⟶ Maturation of somatic embryos ⟶ Encapsulation of somatic embryos ⟶

Test for embryoid to plant conversion ⟶ Green-house and field planting.

The last two steps are for testing the viability of the artificial seeds. Water soluble hydrogels have been found to be suitable for encapsulation of embryoids. Some useful hydrogels are sodium alginate, Gel-rite™, Carrageenin with Locust bean gum, sodium alginate with gelatin etc.

Importance of Artificial Seeds

Though the production cost of artificial seeds is considerably high, there are some advantages for its use.

(i) True seeds are obtained only after completion of the reproductive phase, but artificial seeds may be obtained at least once a month.

(ii) Production of true seeds is dependent upon the flowering season and the execution of the pollination and fertilization processes, whereas artificial seed is totally independent of these natural processes.

(iii) Germination of natural seeds is delayed due to dormant stage of the seeds, which in turn, delays the fruiting time. But, by using artificial seeds, which germinate immediately, the life cycle of a plant could be shortened.

(iv) Successful somatic embryogenesis and production of artificial seeds have been observed in number of crops, like *Daucus carota, Zea mays, Apium graveolens, Lactuca sativa, Brassica* sp., *Gossypium hirsutum, Medicago sativa* etc. In the near future, it may be possible to produce artificial seeds in almost any crop.

(v) Artificial seeds may be used for large scale monocultures and mixed genotype plantations as well as for studying the role of endosperm and seed coat formation.

ANTHER CULTURE

The main plant body of angiosperms is diploid or a sporophyte. But in nature, haploid sporophytes are also found. Blakeslee (1922) reported the first haploid plant in nature which was *Datura stramonium*. Soon, another haploid plant of *Nicotiana tabacum* was also discovered. Subsequently, haploid plants of *Triticum compactum, Crepis capilaris, Solanum nigrum, Oenothera rubricalyx, Oryza sativa, Zea mays, Hordeum vulgare* and *Gossypium barbadense* were also found in nature. The potential value of such plants for breeding and for studying their biology was soon realised.

In angiosperms, the gametophytic or haploid phase is extremely brief and represented by pollen grains or one or more cells in the embryo sac. The pollen grain possesses an unique genome, where every gene is present as a single copy. The discovery of totipotency of the plant cell is the basis of exploitation of this unique genetic unit of the pollen and the production of androgenic haploids.

The first successful production of androgenic haploid was done by Guha and Maheshwari (1964, 66). They found that the pollen grains of *Datura innoxia* by themselves have the potentially to develop into haploid plantlets. These may be called *pollen embryoids* or *androgenic haploids*. Such discovery proved useful to plant breeders, physiologists and biochemists for their research works.

Maheshwari and Guha's work stimulated the scientists of the world and soon the haploids were successfully raised by anther culture in *Nicotiana,* in France, by Bourgin and Nitsch (1967) and in *Oryza sativa,* in Japan, by Niizeki and Oono (1968). Now, a little over 247 species have been brought under anther culture technology.

Ontogeny of Haploid plants

The microspores at the time of first mitosis, or just after, are most suitable for the induction of pollen embryogenesis *in vitro*. Some success has also been achieved using anthers at tetrad stage or at binucleate stage. In solanaceous plants chances for embryoid formation are very high at Gl phase of cell cycle.

Androgenesis is defined as the *in vitro* development of haploid plants, originating from totipotent pollen grains through a series of cell division and differentiation. During culture practices, two basic categories are well recognised,

(i) haploids arise directly from the microspores, and
(ii) haploids arise indirectly i.e. through intervention of callus formation phase.

These two routes of producing pollen embryoids have their own merits and demerits.

The **direct pollen embryogenesis** has following advantages.

1. It is highly desirable for carrying *single copy genetic* information.
2. It is desirable for its genetic variability.
3. It is a less time-consuming process.

The **indirect pollen embryogenesis** also has its own advantages, such as:

1. It offers opportunity to use wide range of media and possibility of their manipulation for callus culture.
2. From it, the regenerants can be obtained in large numbers in a limited time.
3. There are possibilities of obtaining homozygous diploids directly from callus by manipulating the media.
4. Plants with different ploidy status may be obtained.
5. These render a handy cellular system for mutational, tranformational and biochemical studies.

Explaining these two categories, Rashid (1983) proposed the concept of *pollen dimorphism*. He stated that basically the pollen population are dimorphic, the embryogenic pollen (E-pollen) and non-embryogenic pollen (G-pollen). E-pollens are smaller, thin-walled, weakly staining and characterised with presence of attenuated cytoplasm, poor ribosomal population and condensed mitochondria. On the other hand G-pollens are larger in size and possess a thick exine, dense cytoplasm, rich ribosomal population and normal mitochondria. It is possible to separate two types of pollen grains on the basis of density gradient centrifugation.

Factors controlling development of haploid plants (pollen plants):—

The successful, controlled-formation of pollen plants of 247 species indicates that establishment of a callus is not an unsolved problem, rather it needs knowledge of technical details. Some of the key factors that affect the success of obtaining pollen plants are:

(i) Culture conditions and nutritional media (basal media),
(ii) Selection of anther donor plant,
(ii) Pollen age,
(iv) Genetic factors,
(v) Chilling pre-treatments etc.

The basal media proposed by White (1943), Murashige and Skoog (1962) and Nitsch (1969) are found most suitable with or without hormonal supplements and natural extracts. Natural extracts prepared from potato, sweet potato, sprouted wheat, tomato, lotus rhizome, coconut milk etc. have been successfully used. The age of anther, light photoperiod and temperature affect the embryogenic potential of anther, so these points must be considered during the production of pollen embryoids. Consideration of age of pollens is equally important as some pollens are extremely short lived, e.g. pollens of barley and rice have very short life. The pollens of *Gossypium* are viable for about 24 hours and those of brinjal and date palm, one to three days and one year, respectively. Recent researches point out the fact that cold pre-treatment (4-5°C for 2-7 days) of the excised anthers, flower buds and inflorescences increase the chances for successful pollen culture.

Culture technique

The basic steps of anther culture and production of androgenic plants or haploid plants are as follows-

1. The flower buds are excised from the parent plant at the appropriate time, i.e. the first mitotic division of the uninucleate microspore tetrad.
2. The excised buds are suface sterilized with alcohol, chlorine water or 0.1% $HgCl_2$ to remove adhering surface microbes.
3. The anthers are now transferred onto the culture medium, to which 1% agar has been added, to provide solid base. The nutrient medium commonly used are Nitsch medium, Murashige-Skoog medium etc.
4. After 4-5 days of transference, the anthers begin to show slight swelling, which indicates that cell division has been initiated in the connective tissue region. Rapid cell division leads to formation of pale-green callus tissue.
5. The pollen sac now swells up and ruptures lengthwise, along its upper margin, revealing a mass of pale-green pollen callus. However, the pollen callus grows at a slower rate as compared to connective callus.
6. The culture medium is now supplemented with high concentration of kinetin, which stimulates callus development. The ratio of kinetin/auxin percentage may be adjusted, according to whether root or shoot formation is desired.

The various steps of anther culture technique are illustrated in fig. 15.2.

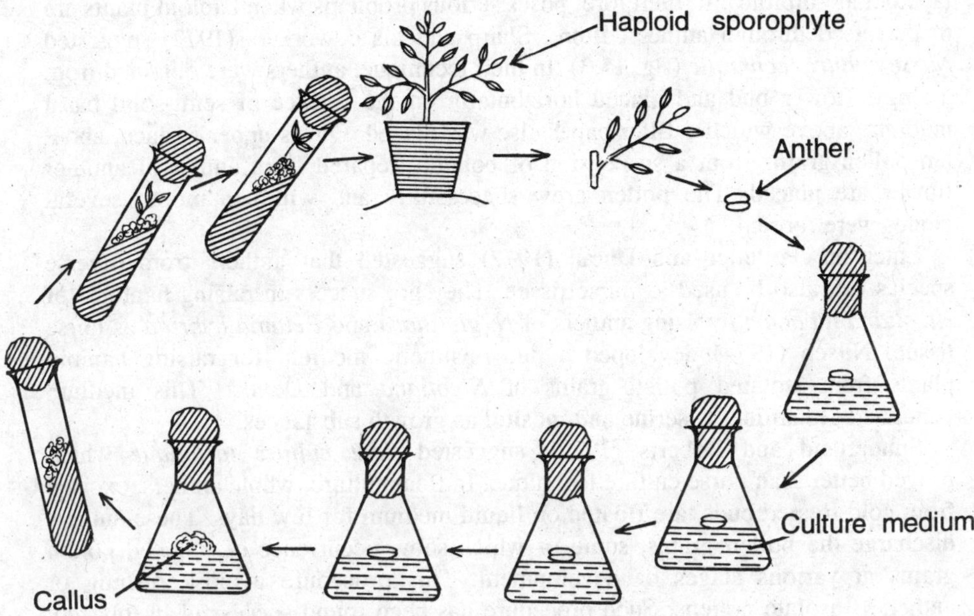

Fig. 15.2. Technique of anther culture.

In addition to the basic technique, several modified techniques have also been successfully practised. These are—

1. Isolated pollen culture. Anther is genetically heterogenous structure that contains both haploid and diploid cells (pollens are haploid and anther wall and

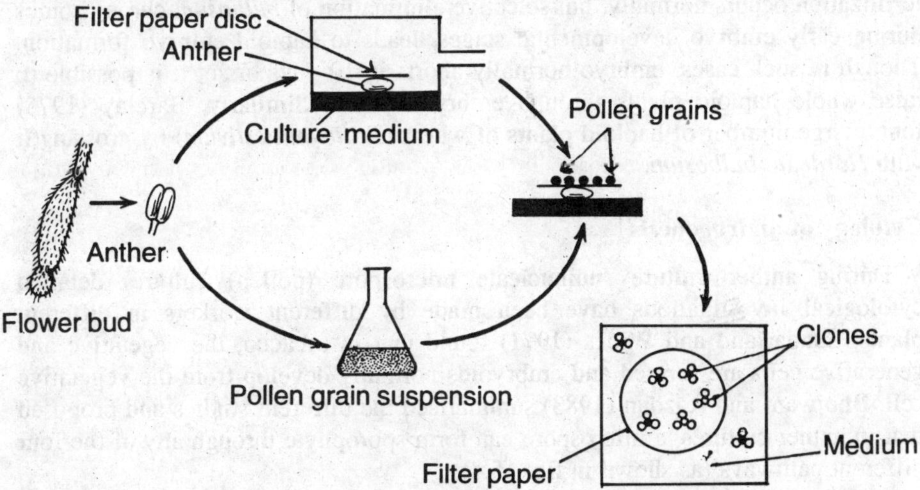

Fig. 15.3. Illustration of nurse culture technique for raising clones from isolated pollen grains of *Lycopersicon* (after Sharp *et al.*, 1972).

tapetum are diploid). It, therefore, poses serious problems when haploid plants are to be raised through anther culture. Sharp and his coworkers (1972) suggested *Nurse culture technique* (fig. 15.3). In their technique, anthers were obtained from a single flower bud and placed horizontally on the surface of semi-solid basal medium, above which a filter paper disc was placed. On its upper surface, about ten pollen grains from a suspension of pollen, prepared from anther of another flower are placed. The pollen grew successfully and within a month several clones were formed.

Later on, Pelletier and Duran (1972) suggested that anthers from diverse species can also be used as nurse tissue. They got success in raising haploids of *Nicotiana tabacum* by using anthers of *N. glutinosa* and *Petunia hybrida* as nurse tissue. Nitsch (1974) developed a pure synthetic medium for raising haploid plants from isolated pollen grains of *Nicotiana* and *Datura*. This medium contained glutamine, L-serine and inositol as growth substances.

Sunderland and Roberts (1979), suggested *float culture technique* which proved better than nurse culture technique. In float culture, whole anthers, excised from cold-treated buds, are floated on liquid medium for few days. These anthers discharge the pollen grains, some of which show occurrence of induced pollen grains at various stages of development. This procedure avoids crushing of anthers to isolate pollens. Such procedure has been found successful in tobacco, barley and datura.

2. Haploid production through interspecific hybridisation. Kasha and Kao (1970) suggested this new procedure to obtain pollen plants or haploid plants. They succeeded in isolating haploid plants of barley. A cross was made between tetraploid species of barley, *i.e. Hordeum vulgare* with *H. bulbosum* and di-haploid plants were formed. These di-haploid plants contain a haploid set of chromosome of *H. vulgare* and another of *H. bulbosum*. In these plants, double fertilization occurs normally, but selective elimination of *bulbosum* chromosomes during early embryo developmental stages, leads to haploid embryo formation. Though in such cases, embryo normally aborts in 10 days or so, it is possible to raise whole haploid plants through embryo culture. Similarly, Barclay (1975) found large number of haploid plants of wheat (*Triticum aestivum*) by crossing it with *Hordeum bulbosum*.

Cytology of androgenesis

During anther culture (uninucleate microspore (pollen) culture) detailed cytological investigations have been made by different workers in different plants. Sundarland and Wicks (1971) found that in tobacco, the vegetative and generative cells are formed and embryoids normally develop from the vegetative cell. Bhojwani and Razdan (1983) summarised the different studies and proposed that in anther cultures, a microspore can form sporophyte through any of the four different pathways (as shown in fig. 15.4).

Pathway 1. In *Datura innoxia*, microspores divide by an equal division and both daughter cells equally contribute towards the formation of sporophyte.

Pathway 2. In *Hordeum vulgare* and *Nicotiana tabacum*, the microspore divides by unequal divisions and the sporophyte arises from the vegetative cell alone.

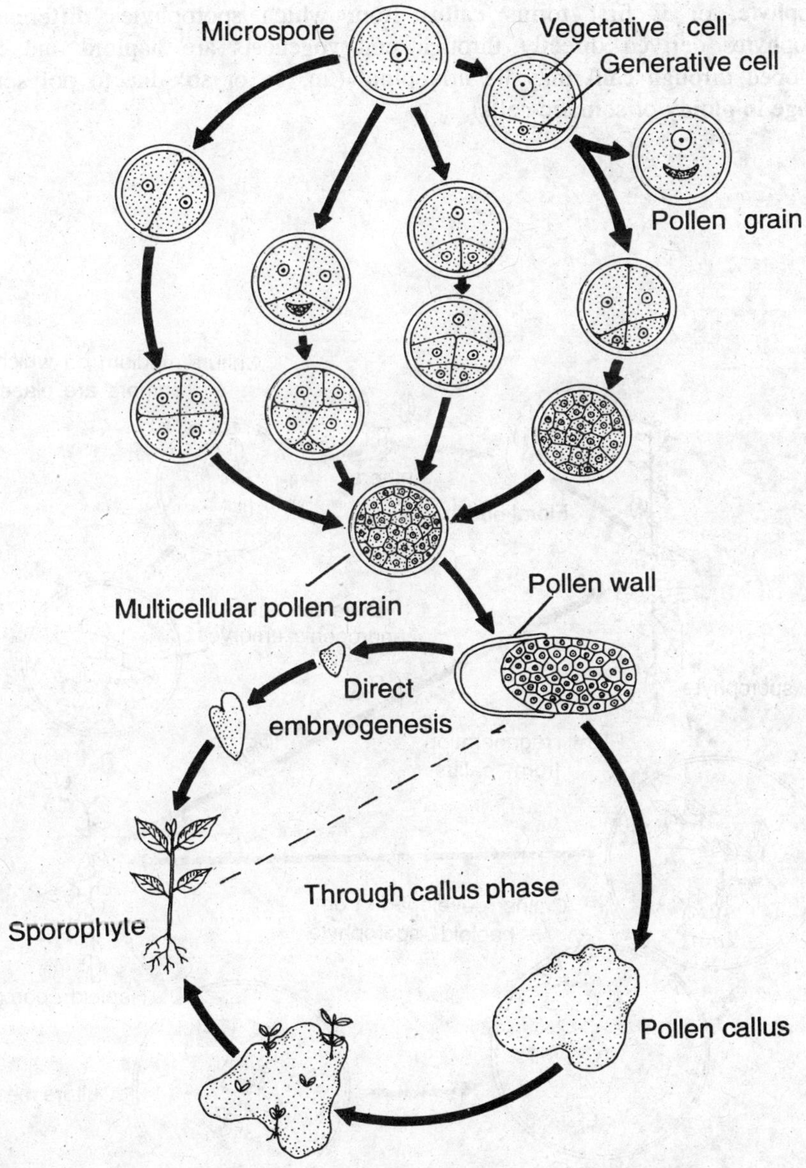

Fig. 15.4. Schematic details of anther cultures (Redrawn from Bhojwani and Razdan, 1983).

Pathway 3. In *Hyoscyamus niger*, the microspore divides by an unequal division and the sporophyte arises from the generative cell only.

Pathway 4. Microspore divides unequally and both vegetative and generative cells contribute towards formation of the sporophyte.

In the process, ultimately a multicellular mass is released by bursting of the pollen wall. This mass, either forming directly embryoids develop into a sporophyte, or it first forms callus from which sporophytes differentiate. Sporophytes derived directly through embryogenesis are haploid and those developed through calli may be mixoploid (2n, 3n or so) due to polysomaty (change in ploidy of somatic cells).

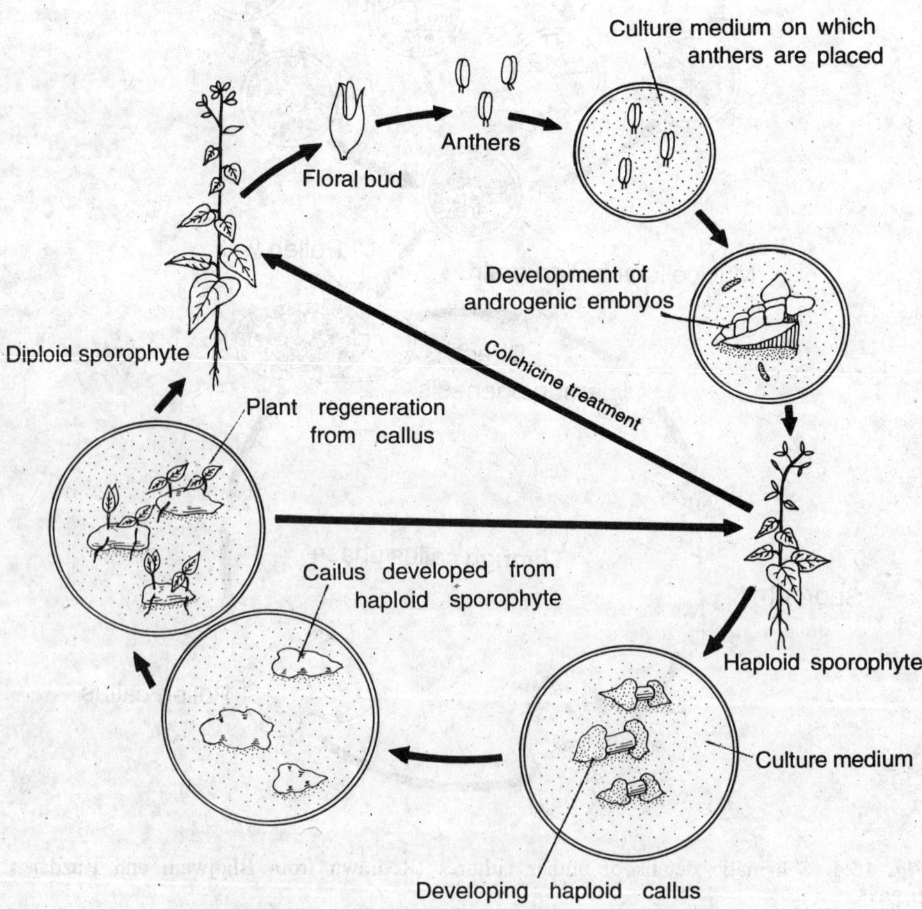

Fig. 15.5. Schematic explanation to obtain homozygous diploids from androgenic haploids (Redrawn from Bhojwani and Razdan, 1983).

Diploidisation of haploids

Androgenic haploids may grow upto flowering stage but these cannot show normal differentiation of viable gametes due to lack of meiosis and homologous chromosomes. It is now possible to produce homozygous diploids by certain techniques. The most common procedure has been use of colchicine for duplication of chromosomes. Another procedure is now possible, through callus, developed from haploid sporophyte. Some of the callus cells undergo endomitosis and produce diploid (sporophyte) cells. By transferring such calli to a suitable culture medium, fertile diploids can be obtained. The process is explained in fig. 15.5.

Application of anther culture techniques

The haploids are highly valued as these offer new possibilities for the investigation of several biological problems. Some important points are as follows:

1. The anther culture drastically reduces the time period in producing pure homozygous lines after diploidisation. Otherwise the traditional methods take 6 to 8 generation times and even then may not be absolutely pure. Nakamura and his co-workers (1971) successfully produced a high quality of tobacco line from haploids. Carlson (1973) using androgenesis and cell culture technique obtained tobacco line resistant to MSO (methionine sulfoximine). The MSO is similar in structure and function to the toxin produced by *Pseudomonas tabaci,* causing wild fire disease in tobacco;
2. The haploidy is generated rapidly by anther culture technique and these can be used in breeding programmes. Pollen plants greatly help in the detection of new characters in F_1 while the detection of new combination by conventional method would be possible after F_2 or F_3 generation;
3. Haploids have been used at still complex levels by geneticists. Dey and his coworkers (1973) have successfully used haploids of *Arabidiopsis thaliana* and *Lycopersicon esculentum* for the transfer and subsequent expression of three system of genes from the bacterium *Escherichia coli*. The genes involved are galactose, lactose and mutant suppressor gene which can correct amber (nonsense) codon in mRNA. Similarly, free cell culture and protoplasts of haploids provide useful materials for various kinds of genetic manipulations. Hu Han (1983) stated that anther culture is a rapid and efficient method for the creation of basic material of chromosome engineering;
4. Haploids are useful in detecting recessive mutations, as in diploid phase, recessive mutations do not express themselves in the presence of their dominant alleles on homologous chromosomes. Mutations induced in haploids can be easily detected, because haploids have single set of genes and there is no interference by dominant allele; and
5. Haploids with desirable mutations can be picked up and their chromosomes duplicated to get fertile diploids with desirable mutations in a single generation.

OVULE CULTURE

For the study of the behaviour of the zygote or very young zygote, the ovule culture is quite valuable as it provides vital information about their physiological requirements and growth mechanism. It was Prof. White (1932) who for the first time attempted to culture ovule of Snapdragon (*Antirrhinum majus*), but perfection in technique of ovule culture was achieved by an Indian scientist, Prof. Maheshwari (1958). Later, Chopra and Sabharwal (1963) successfully cultured ovule of *Gynandropsis* and *Impatiens* and Guha and Johri (1966) cultured ovule of *Allium cepa*. Maheshwari and his co-workers (1961) could raise viable seeds of poppy plant (*Papaver somniferum*) from the excised ovule. They found Nitsch's medium ideal for ovule culture without any supply of growth substance. Initially, the growth appeared slow in comparison to growth in nature (i.e. *in vivo*) but after a month or so, the growth became rapid and better than that found in nature.

The details, from the level of entry of pollen tube into the ovule and upto the development of embryo, has been successfully worked out in orchids, while grown in only ten per cent sucrose medium, by a lady scientist, P. Arnoldi (1960). At the same time, growing ovules *in vitro* under aseptic conditions is quite difficult. For example, Sachar and Kapoor (1959) failed to grow excised ovules in Nitsch's medium even after supplying growth substances such as kinetin, IAA and GA. In certain cases, natural juices of cucumbers and water melons stimulated the growth of ovule to the level of seeds under *in vitro* conditions, such as in case of berseem (*Trifolium repens*). Wakizuko and Nakajima (1974) found that five per cent cucumber juice was essential to raise fully developed seeds from the ovules of *Petunia hybrida*. Joshi and Johri (1972) using the extracts of ovules of same plant or supplementing White's basal medium with casein hydrolysate, achieved success in growing ovules of *Gossypium hirsutum* (cotton plant). Later, it was found that most ovules of *G. hirsutum* if cultured two days after anthesis (opening of the flower) showed the development of fibres on seed surface. It thus appears that growth requirements of young ovules are quite variable.

During recent years, ovule culture techniques are widely used in raising interspecific hybrids of economically important plants. These hybrids fail to grow in nature. Punir (1967) achieved success in growing embryos produced from hybridisation of *Gossypium arboreum* X *G. hirsutum*. Similarly Steward and Hsu (1978) could raise four different hybrids in the genus *Gossypium* which are otherwise unknown.

OVARY CULTURE

Ovary culture is quite valuable to horticulturists to acquire knowledge of fruit physiology, for improving the quality of fruits. The general usage of chemicals cannot ascertain that whether the chemical tested is exercising its effect in its original form or is getting modified in the plant system before actually reaching the ovary. To overcome such problem, ovary culture is very helpful because it allows the direct feeding of the ovary with the chemicals.

La Rue (1942) probably made the first successful attempt to raise aseptic cultures of excised flowers and upto certain extent the limited growth of ovaries. However, Nitsch (1951) carried out comprehensive studies and achieved success in growing excised ovaries of *Cucumis anguria, Lycopersicon esculentum, Nicotiana tabacum* and *Phaseolus vulgaris* on synthetic medium. He succeeded in growing ovaries of pollinated flowers of *Lycopersicon* which developed into mature fruits with viable seeds (fig. 15.6). But the fruits were relatively of smaller size than those produced in nature. Similar attempts have been successfully made in growing fruits from just pollinated ovaries of *Iberis amara, Hyoscyamus niger* and *Linaria maroccana.* Efforts have been made using Nitsch's or White's medium supplemented with different types of growth substances such as vitamins, hormones, amino acids etc. However, in case of *Anethum graveolens*, Johri and Sehgal (1966) using Nitsch's medium, supplemented with coconut milk, got success in producing larger size of fruits, in *in vitro,* than natural fruits.

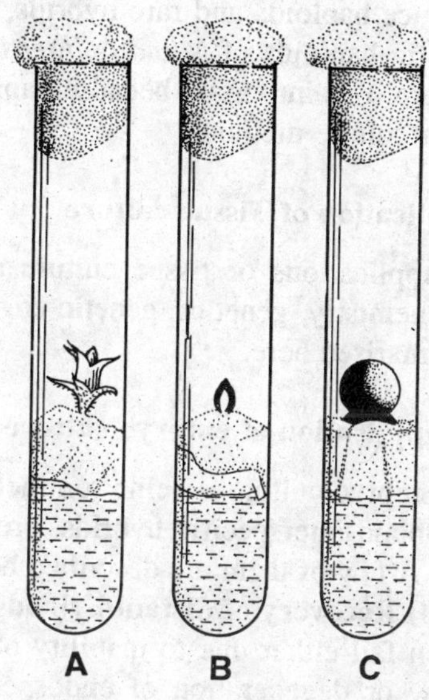

Fig. 15.6. Development of *Lycopersicon* ovaries on synthetic medium. A, B and C—three different stages of culture, C—small fruit developed *in vitro* (after Nitsch, 1951).

It is interesting to note that fertilized ovaries are cultured easily in mineral-salt, sugar-vitamin medium but those of unpollinated flowers cannot be grown in such medium. Nitsch (1951) obtained seedless fruits of tomato from unpollinated ovaries, while growing in basal medium supplemented with IAA, 2, 4-D or NAA (Naphthalene acetic acid). IBA (Indole butyric acid) has been successfully used for the same purpose in case of *Althaea rosea.* It appears thus that auxins can replace the pollination stimulus.

Studies on *in vitro* cultures of ovaries have also unravelled the mystery of the role of perianth lobes or calyx, and it is now proved that these structures do not function only as protective structures for the sex organs, but also play an important role in the development of fruits. This fact has been convincingly demonstrated in *Allium cepa, Althaea rosea, Hordeum vulgare, Hyoscyamus niger, Iberis amara* and *Lycopersicon esculentum.* In case of *L. esculentum,* Nitsch (1963) proposed that calyx supplies some nitrogenous substances essential for fruit growth.

Production of viable seeds from cultured ovaries is a big problem. In most cases of successful ovary cultures, non-viable seeds are formed. Sachar and Guha (1962), however, got some success in producing viable seeds while working on *Ranunculus sceleratus.*

While ovary culture studies have revealed a comprehensive knowledge of physiology of fruit development, it has been successfully used as an aid to produce haploids and rare hybrids. Mess and Wagner (1976) produced parthenogenetic haploids of *Mimitus lutens* by pollinating their exposed ovules in ovary culture. It is important because haploid in this plant could not be produced by any other culture methods.

Application of Tissue culture

Applications of tissue culture are enormous in the field of morphogenesis, biochemistry, genetics, genetic engineering and plant breeding. Some of them are summarised here.

1. Application of embryo culture

Embryo culture is being routinely used in some crop improvement programmes to obtain interspecific hybrids, propagation of orchids, overcoming of dormancy etc. A few of them are described here.

(i) Recovery of distant hybrids. In general, the distant or interspecific crosses often fail either due to inability of pollen to germinate, failure of pollen tube to grow or degeneration of endosperm. In such cases, embryo culture is used to recover hybrid plants. Some recent examples are recovery of hybrids from *Hordeum vulgare* X *Secale cereale. Triticum aestivum* X *Agropyron repens, Hordeum vulgare* X *Triticum aestivum* etc. One interesting example is of *Triticale* which is a hybrid plant of cross between *Triticum aestivum* X *Secale cereale*. The viable seeds are rarely produced in it due to presence of two dominant genes, the Kr_1 and Kr_2. The majority of seeds are poorly developed, small and show poor germination. The viability percentage obtained, ranges between 5 to 10 per cent. The recovery of hybrid seedlings is much greater (50-70 per cent) when 10-14 days old embryos are removed and cultured in a suitable medium.

(ii) Propagation of orchids. Seeds of orchids are smallest among angiosperms and these lack any stored food. The embryo of orchid is virtually naked and in many cases it may remain immatured or underdeveloped. That is why propagation of orchids in nature is very slow, tedious and highly time-consuming. Since orchids are beautiful creations of nature and are largely valued for their ornamental and economic value, this problem is now being largely solved by using tissue culture technique. Young or mature orchid embryos are removed from seeds and placed on suitable nutrient media. The embryos develop into seedlings either directly or through callus formation.

(iii) Shortening of breeding cycle. Seeds often take 15 to 20 days to mature after the full development of embryo. In some cases, maturity period is very long. By applying embryo culture, embryos are made to develop directly into seedlings and thus avoiding a long time of seed maturity.

(iv) Overcoming dormancy. In most fruit trees, embryos require a period of after-ripening i.e. dormancy before germination. This period of dormancy can be effectively eliminated by embryo culture. Such successful experiments have been done in case of *Prunus, Taxus, Iris* etc.

(v) Propagation of rare plants. Embryo culture technique can be used in maintaining rare or endangered plants. For example, '*makapuno*' is a special delicious variety of coconut which is served during special ceremonial occasions in Philippines. This variety infact contains soft, solid and fatty tissues in place of liquid endosperm. De Guzman and Del Rosario (1974) using embryo culture technique succeeded in establishing sufficient number of *makapuno* plants.

(vi) Rapid seed viability test. Branton (1961) suggested that germination of excised embryos, using tissue culture technique, is more reliable procedure to perform seed viability test, rather than adopting usual staining technique. Thus, it provides authoritative correct information of seed viability.

2. Application of meristem culture

Meristem culture is highly beneficial in crop improvement, particularly in clonal propagation, recovery of virus free stocks, germplasm exchange and germplasm conservation.

(i) Vegetative propagation. Meristem culture has been considerably used for vegetative propagation of many crop plants and fruit trees. It is basically used for obtaining clones and hence sometimes termed as *mericloning, i.e.* cloning through meristem. In culture, a single meristem produces many plantlets which may again be used in meristem culture. Thus a large number of clones can be obtained in a short time, particularly of vegetatively propagating crops. Success has been achieved in many fruit-bearing and timber trees.

A landmark success has been achieved in mass propagation of *Dioscorea floribunda*, using meristem culture at N.B.R.I., Lucknow. Chaturvedi (1979) has developed a *tissue bank* to maintain propagation stocks of high-yielding clones in his lab. Field cultivation of the *in vitro* raised plants of *D. floribunda* at CIMAP farm, Lucknow, is a wonderful example of tissue culture technique adopted by him. This plant is medicinally important for commercial production of steroidal drugs (wonder drug) and sex hormones and in preparing oral contraceptive pills.

(ii) Recovery of virus free stocks. Seeds from viral infected plants rarely contain virus. Therefore, sexual progenies are almost free from such infections except for new infections. But in case of asexually reproducing crops, viral infections spread rapidly, because vegetative propagules from infected plants contain virus particles. However, meristems and young tissues are generally free from the virus. Thus, meristem culture is useful to overcome such problem. Some valuable clones in potato and sugarcane have been freed from viral infections through meristems culture.

(iii) Germplasm exchange. The commonly used clones or propagules are bulky and large in size which create difficulty in exchange and quarantine problem. Seedlings in test tubes, obtained from apical meristems, would prove extremely useful in germplasm exchange of such crops. This application has great scope in future.

(iv) Germplasm conservation. The storage of roots and tubers requires large space and is expensive, as these lose viability rapidly. So germplasm conservation in such clonal crops creates many problems. Freeze preservation of meristems and cells in liquid nitrogen at –196°C has been suggested for long term preservation of their germplasm. However, this technique is in developing stage.

3. Application of anther or pollen culture

Anther or pollen cultures are useful in obtaining haploid plants which are greatly valued in genetic studies and in plant breeding programmes. From haploids, homozygous diploids can be produced in a short time. Haploids obtained from pollen grains are being used for improvement of tobacco in which disease-resistant and good varieties have been developed through this technique. Similar technique has been adopted in the improvement of rice and wheat in China. The famous Tanfeng 1 (meaning haploid-derived, high-yielding) has been released for commerical cultivation. Haploids have not been successfully produced in all cases and therefore wider application of pollen culture in crop improvement remains restricted.

4. Application of tissue and cell culture

The various applications of plant tissue and cell culture are summarised here as follows:

(i) Clonal propagation. Tissue culture is well suited for quick vegetative propagation of plants and specially suited to those cases, where regeneration of plants through this technique is easy. It has also been used for obtaining disease-free and virus-free plants.

(ii) Mutant isolation. Biochemical mutants are far more easily isolated from cell cultures than from whole plant population. This is also because millions of cells can be effectively screened for biochemical mutant cells, whereas screening of many plants would be impossible even. The biochemical mutants could be selected for several reasons, such as for disease resistance, improvement of nutritional quality, adaptation of plants to stress conditions and to increase the biosynthesis of plant products used for medicinal or industrial purposes.

(iii) Somatic hybridisation or parasexual hybridisation. Now it is possible to isolate protoplast of any plant species and to culture it, to produce callus. These can be subjected to hybridisation. Such parasexual hybridisation requires three steps—(i) isolation of somatic protoplast, (ii) fusion of isolated protoplasts and (iii) culture of the fused protoplasts in a nutrient medium. All these steps have been successfully carried out by tissue culture technique.

Hybrid plants have been obtained after fusion of protoplasts, *e.g.* from fusion of *Nicotiana glauca* protoplasts with those of *N. langsdorfii*. Though great success has been achieved with tobacco plants, it is doubtful that somatic hybridization would be of any practical value in crop improvement in near future. However, successful protoplasm fusion has been performed using barley and soybean, corn and soybean, potato and tomato, tobacco and soybean and sorghum and corn protoplast mixtures.

(iv) Genetic transformation (recombinant DNA technology). It has now become possible to transfer gene, by feeding cells with DNA, in case of eukaryotes. This raises the possibility of genetic modification of plant cells with the help of both homologous (from same species) and heterologous (from different species) DNA. The DNA containing plant viruses (such as cauliflower mosaic virus and potato leaf roll virus), plasmids (e.g. T_1 plasmid of

Agrobacterium) and transposons have been successfully used as carriers of genes for genetic modification of plant cells. Scientists are improving this technique for developing plants that are immune or resistant to disease, can extract nitrogen from the air, use sunlight more efficiently in photosynthesis and have foreign genes that will improve crop yields and produce more nutritious proteins.

(v) Germplasm conservation. Since cultured tissues may be frozen and stored at –196°C for long-term storage of germplasm, this would be of great value in conservation of germplasm of those crops, which normally do not produce seeds, but multiply through roots and tubers. This technique of freeze preservation, also called *Cryobiology*, of plant cells is still in the developing stage, but in future it may assume considerable importance in germplasm conservation. Tissue Bank of *Dioscorea floribunda* is being maintained at N.B.R.I., Lucknow, which furnishes an example of great success in tissue culture technology.

Since, tissue culture technology is highly sophisticated and requires both environmental and biological (contamination) control, its general application in plant improvement in near future may not be possible but above all, this modern branch is highly valuable.

PARTHENOCARPY

In normal conditions, the ovary, after pollination and fertilization develops to form the fruit and the fertilized ovules form the seed. However, in certain members of angiosperms, it has been observed that the ovary develops to form the fruit, without pollination or fertilization taking place, such as in banana, figs, tomato, pineapple, apple, guava, papaya etc. Noll (1902) coined the term **parthenocarpy** to this phenomenon of fruit development without pollination or any other stimulus. Later, the definition of parthenocarpy was modified by Nitsch (1965), who referred to it as the *formation of fruits without fertilization.*

A characteristic feature of all parthenocarpic fruits is the absence of seeds, which earned them the name "seedless fruits". However, the term is a misnomer, since in seedless fruits, the ovules may have been fertilized and later degenerated, as reported in some strains of *Vitis vinifera* var. *concord.* Also, seeded, parthenocrapic fruits formed by auxin treatement, has been reported in *Citrus* sp. pineapples, watermelons and grapes.

Classification of Parthenocarpy. Parthenocarpy may be of two types viz.

(i) ***Natural parthenocarpy.*** In a number of horticultural varieties of plants, such as bananas, cucumber, pineapples, tomatoes and figs, parthenocarpic fruits are known to develop naturally. The naturally developed parthenocarpic fruits may be grouped into two types—

(a) *Vegetative parthenocarpy*—the fruits are formed without pollination.

(b) *Stimulative parthenocarpy*—the ovary is stimulated to fruit formation only after pollination, which provides the necessary stimulus. Pollination is, however, not followed by fertilization. In some exceptional cases, fertilization also occurs, but the embryos (fertilized ovules) abort, hence the fruits lack seeds.

(ii) ***Induced parthenocarpy.*** Fruit formation is stimulated or induced by the application of growth promoting substances such as IAA, IBA, αNAA, phenylacetic acid and gibberellins.

Nitsch (1963) recognized three types of parthenocarpy, on the basis of the provided stimulus, namely (i) environmental, (ii) genetic and (iii) chemically induced.

(i) **Environmentally induced parthenocarpy.** Environmental conditions play a decisive role in the flowering and fruiting of a plant. It was observed that sudden environmental changes, such as low temperatures, frost or fog influenced the normal functioning of sex organs and induced parthenocarpic development of fruits. Lewis (1942) obtained parthenocarpic pears, by exposing their flowers to negative temperature for 3-19 hours. Increased parthenocarpic fruit formation was reported in *Capsicum* by transferring the plants from normal temperatures to 10°C—16°C, at the time of anthesis (Cochran, 1936) and in tomatoes under low temperatures and high light intensities (Osborne and Went,1953).

(ii) **Genetically induced parthenocarpy.** A number of horticultural plants exhibit seeded as well as parthenocarpic fruits, which arise due to mutations or hybridization. Genetically induced parthenocarpic fruits have been observed in *Cucurbita, Citrus, Eugenia, Musa, Punica* and *Vitis.* An interesting case of genetic parthenocarpy gave rise to the famous "naval oranges", which arose from a normal seeded variety of *Citrus*, through mutation in an axillary bud. This mutated axillary bud grew into a branch, bearing seedless fruits of oranges,

Hybridization technique may also give rise to parthenocarpic fruits, such as obtained by Wellington and Hawthorn (1929), who crossed "English forcing" variety of cucumber with "Arlington White spine" variety to produce parthenocarpic variety of cucumber.

(iii) **Chemically induced parthenocarpy.** The most successful method of inducing parthenocarpy is by the use of growth-promoting substances, such as auxins and gibberellins. Extensive research work has established that plants, normally bearing seeded fruits may be induced to bear parthenocarpic fruits by controlling pollination and applying chemicals to the pistil. Thiamann (1934) first suggested that many pollen grains possess growth promoting substances which provided the necessary stimulus for fruit formation. This "pollen hormone stimulus" could be substituted by the application of chemicals such as IAA, IBA, NAA, NOAA, 2, 4-D and 2, 4, 5-T. These growth-stimulating substances are mixed in a lanolin paste (0.5-1%) and smeared on the stigma, or as sprays. Auxins have been reported to produce parthenocarpic fruits in tomato, strawberry, black berry, figs, cucurbits, tobacco, pepper, guava, citrus and rose. Gibberellic acids have been reported to induce parthenocarpy in tomatoes, apples, grapes, figs and in rosaceous fruit trees.

It has been reported that the time of chemical application should be precise for the induction of parthenocarpy. Auxins are generally applied sometimes after anthesis (early application damages the flowers, resulting in seed abortion and fruit drop), whereas gibberellins should be applied as early as possible, preferably at anthesis.

Significance of Parthenocarpy. Parthenocarpic fruits are extremely popular among horticulturists, as these varieties of fruits are ideally preferred by food preservationists for preparing jams, jellies, sauces, fruit drinks etc. The edible, meaty part of the fruit may also be increased proportionately by the absence of seeds.

GENERAL QUESTIONS

1. What is experimental embryology? Discuss briefly.
2. Describe briefly about the embryo culture technique and its application.
3. Write a brief account of anther culture and its application.
4. Write short notes on—
   (i) Haploid plants
   (ii) Nurse culture technique
   (iii) Application of anther culture
   (iv) Application of haploid plants
   (v) Ovary culture
   (vi) Ovule culture
   (vii) Artificial seeds
   (viii) Parthenocarpy.
5. Explain the culture technique of pollen plants and the factors controlling its development.
6. Discuss briefly the application of tissue culture.
7. What are androgenic haploids? Discuss briefly.

CHAPTER 16

Palynology

Introduction

The beginning of the structural organisation of plants and its evolution is initiated and represented by the male reproductive unit, the *pollen grain*. Hence, the morphostructure of pollen grains assumes vital importance and forms the fundamental principle underlying the evolutionary process in plants, both of form and function.

Hyde and Williams (1944) used the term palynology for the study of external (morphological) features of mature pollen grains. Palynology term has been derived from *palynin* meaning "to scatter", since pollen grains and spores are often spread by the wind, and *logos* meaning study. At first, the branch failed to get any distinct recognition, but during recent past, it has now emerged as an important discipline of fundamental and applied botany. All over the world several laboratories are involved in studying this branch, and in India, National Botanical Research Institute, Lucknow (NBRI), Birbal Sahni Institute of Palaeobotany, Lucknow, Osmania University, Hyderabad and Bose Institute, Calcutta are important centres where studies on pollen of Indian flora are being carried out.

Information on the early studies in pollen morphology of Indian plants is restricted to occasional description contained in the various floras. Hooker (1872-97) recorded pollen characters of Apocyanaceae, Asclepiadaceae Leguminosae, Orchidaceae, Nymphaeaceae, Malvaceae and Passifloraceae and Duthie (1903-28) mentioned about pollens of Acanthaceae and Convolvulaceae.

With the start of 20th century, several Indian scientists such as Banerjee, Sayeeduddin, Nair, Sampat, Ramnathan, Mittre etc. studied pollens of various plants. Dr. P.K.K. Nair of N.B.R.I., Lucknow has established a modern and perfect lab of palynology where studies are made with the help of most sophisticated apparatuses such as SEM (scanning electron microscope).

Preparation of Pollen Slides

For studying their external morphology, pollen slides are prepared by acetolysis as suggested by Erdtman (1952). Some modifications have been suggested by Nair (1960) which may be summarised as follows.

The fresh or fixed polleniferous samples are collected in 70% alcohol and are kept undisturbed for an hour or so. These are then transferred to a plastic centrifuge tube and crushed by means of a glass rod. The dispersion is passed through a mesh (48 meshes/ cm), centrifuged, and the alcohol is decanted off. After adding 5 ml glacial acetic acid, it is again centrifuged. To the sediment is added 5 ml of acetoylsis mixture (one part conc. H_2SO_4 added drop by drop to nine parts of acetic anhydride). The tube is then kept at 70°C in a water bath and heated till boiling starts. Boiling is allowed for a minute or so. The waste acetolysis mixture is fully decanted off after centrifugation. Little amount of water is added in the tube and well shaken to form a homogenous mixture.

At this stage, the pollen grains with thick exines are subjected to chlorination, for which the dispersion in the tube is divided into two parts in the ratio 1:3. The tube containing 2/3 of the liquid is centrifuged, the supernatant water is decanted and then 5 ml. of glacial acetic acid and two drops of saturated sodium chlorate in water is added, followed by two drops of conc. HCl and the mixture is stirred. This process brings about bleaching of pollen grains. The dispersion is centrifuged and the waste chlorine liquid is decanted completely. It is now washed with distilled water and after centrifuging, the waste suspension is removed. These pollen grains are now chlorinated and are ready for study. For a large number of plants, the acetoylsed grains are mixed with alcohol-treated unacetolysed grains and are stained by safranin to ascertain the possible changes undergone by the process of acetolysis.

The prepared pollen grains are mounted in glycerine jelly, warmed, covered with a coverslip and sealed with wax. The slide is now ready for study.

POLLEN MORPHOLOGY

In general, characters of pollens are studied under the divisions—Polarity, Symmetry, Apertures, Size and Shape, Exine stratification and Exine *Palynogram.*

Polarity. Pollens are generally formed in tetrads from their mother cells. The tetrad stage is considered to be significant as far as the origin of certain morphological characters of exine are concerned. In a tetrad, the end of a pollen directed towards the centre of tetrad is called *proximal pole* while opposite is the *distal pole*. The distal pole faces peripheral side. The polar axis must always be perpendicular with the distal pole at the apex and the proximal pole at the base. The pollen grains are said to be *isopolar*, if two faces are similar, or *heteropolar*, if different. In heteropolar grains, one face has an aperture while the other has none (Fig. 16.1).

Symmetry. The pollens possess a symmetry, being bilateral or radial but asymmetric (with no symmetry at all) and some times filamentous pollen grains are also met with.

Apertures. The exine of pollen grain is provided with apertures which are thin, more or less distinctly delimited areas formed only of a hyaline membrane. Anonymous (1958) proposed some descriptive terms that may be applied to the type of aperture.

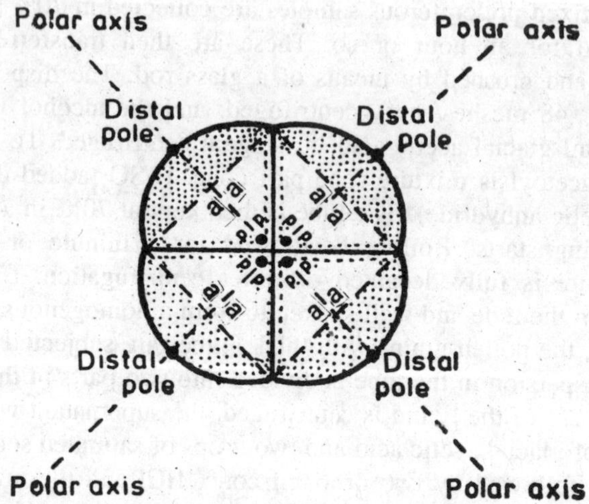

Fig. 16.1. Diagram of a pollen tetrad showing polar axis. (a-a-equatorial axis of equatorial plane and P-P-proximal pole of each pollen grain. (Redrawn from Walker, 1976)

The exine possesses apertures (called *aperturate)* or have no apertures (*inaperturate*). The apertures are either elongate (colpate) or circular (porate) in both of which the outer and inner surfaces are congruent (simple aperture). Some times these are not congruent and called *compound apertures*. In that case, the outer part (*Ectocolpium* or *Ectoporium*) is either elongate (*colporate*) or circular (*pororate*). In the colporate apertures, the inner surface is called *Endocolpium* which is a synonym of *Ora* or *Endoaperture*. It may either be circular (*Pororate*), longitudinally elongated (*Lolongate*) or laterally elongated (*Lalangate*).

In studying characteristics of apertures, their number (N), position (P) and character (C) are studied carefully. This may also be called **NPC** system. During recent years, term *treme* is preferred over *aperture* to classify different pollen grains. Pollen grains without aperture are called *atreme* (or N_0) but depending upon **number** of apertures, these are called *Monotreme* (N_1), *Ditreme* (N_2), *Tritreme* (N_3), *Tetratreme* (N_4), *Pentatreme* (N_5), *Hexatreme* (N_6) or so. Similarly at least seven categories have been recognised to refer **position** of aperture and these are called P_0 to P_6. If the aperture is on the proximal face, it is called *Catatreme* and if on distal face, *Anatreme*. The term *Zonotreme* is used to refer to the position of aperture on the equator and *Pantotreme* to refer to apertures more or less uniformly distributed all over the surface. The **character** groups are also recognised into seven types (C_0 to C_6). C_0 refers to query *i.e.* character of aperture is not known. C_1 refers to pollen with an aperture-like thin area called *Leptoma* and accordingly terms *Monolept* refers to pollen with one leptoma, *Catalept* to leptoma on proximal face and *Analept* to leptoma on distal face. In C_2 category pollen grains with 3-slit colpus *i.e. trichotomocolpate* are kept. The colpate, porate, colporate and pororate pollen grains are C_3, C_4, C_5 and C_6 types, respectively (fig. 16.2).

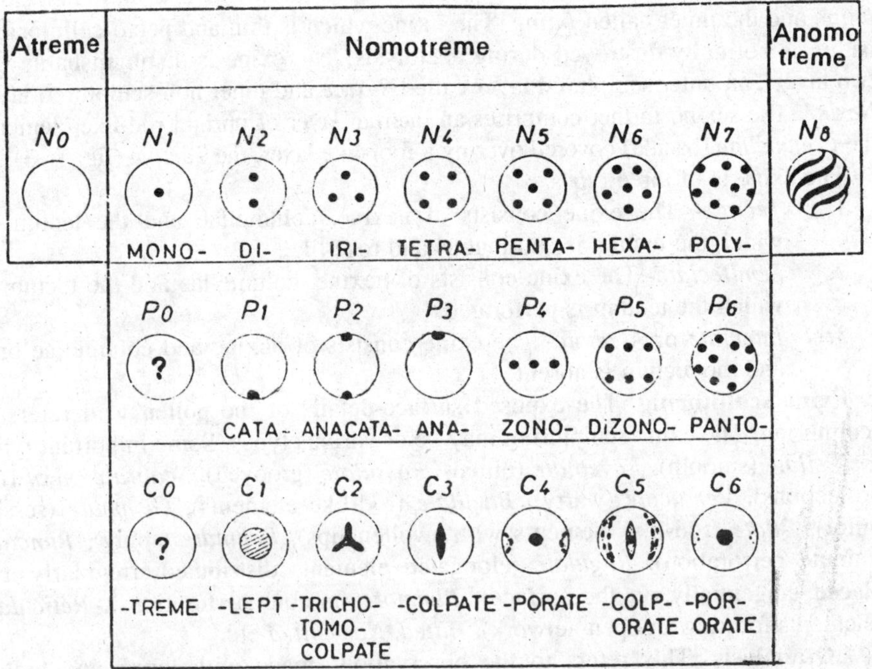

Fig. 16.2. The NPC System (Summary) (Redrawn from Erdtman (1969).

The margin of the aperture may be thin (*Tenuimarginate*) or thick (*Crassimarginate*). The outer aperture maybe sometimes placed on dome-like arc (called *Asphides*) protruding from the general suface of the grain, the *Aspidote*. The end of the colpus may be pointed or blunt, and free or united (*Syncolpate* and *Parasyncolpate*).

Size and shape. The size of pollen grains are quite variable which may range from 5 μm to 200 μm as in Cucurbitaceae. The shape of pollen may be flat called *Peroblate* or very elongate called *Perprolate* or any other standard type.

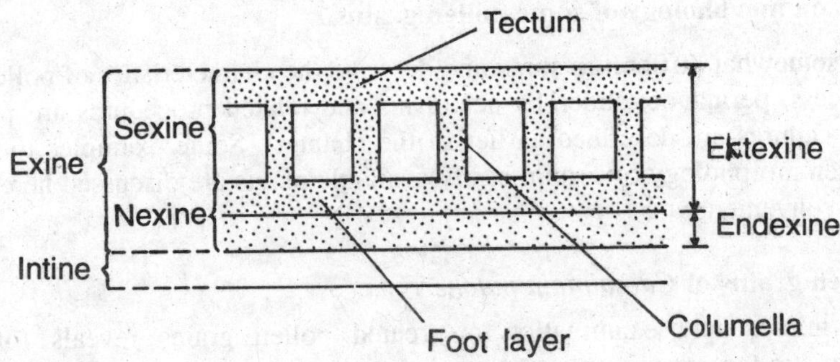

Fig. 16.3. Illustration of pollen wall.

Exine stratification. The pollen wall consists of two layers, the outer called *Exine* and the inner called *Intine*. The intine which is thin and pecto-cellulosic in nature is normally destroyed during acetolysis. The exine is distinguishable into two layers, an outer sculptured layer called *Sexine* and inner non-sculptured layer, *Nexine*. The sexine further comprises an internal layer of upright rod-like elements, the *Columellae* (bacula) covered over by a roof-like layer, the *Tectum* (fig. 16.3)

The exine is of three types:

1. *Tectate*- The exine consists of nexine, columellae and the tectum in which the tectum is continuous and roof-like.
2. *Semitectate*- The exine consists of nexine, columellae and the tectum in which the tectum is perforated.
3. *Intectate* or *Atectate*- The exine consists of nexine and columellae only and the tectum is absent.

Exine sculpturing. The exposed surface-details of the pollen wall refers to sculpturing. The exine sculpturing may be of various types. Some important types are *Psilate* (smooth), *Foveolate* (pitted), *Fossulate* (grooved), *Scabrate* (very fine projections), *Verrucate* (warty), *Baculate* (rod-like elements), *Gemmate* (sessile pillor), *Pilate* (rod-like elements with swollen tips), *Echinate* (spiny), *Punctate* (minute perforation), *Rugulate* (elongated elements distributed irregularly and placed tangentially on the surface), *Punctate* (minute perforations), *Reticulate* (elements forming an open network), *Pitted, Non-pitted* etc.

LO-analysis. This refers to the observations made with lower and higher planes of microscope adjustments. The sporoderm pattern, at a high plane of focus appear as bright islands separated by dark channels and on a lower adjustment, it shows dark islands separated by bright channels. The picture at high adjustments is just the reverse of what is visible at the lower plane of focus. (fig. 16.4).

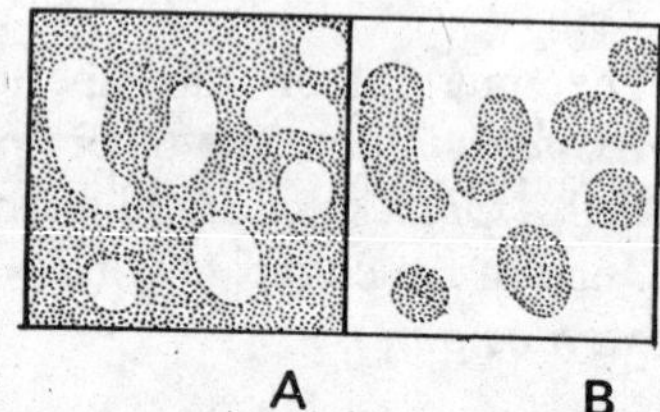

Fig. 16.4. Diagram to explain LO pattern (after Nair, 1970).

All these characters are taken into consideration to study the morphology of pollen which has been found to be highly helpful in solving the taxonomic riddles.

Studies on morphology of some pollen grains

It is somewhat difficult to study and describe the characteristics of pollens. For the process, permanent slides by acetolysis and related procedures are prepared (the procedures are described earlier in the chapter). Some examples to explain the pollen morphology of certain easily available plants are discussed here for the sake of convenience.

1. **Pollen grains** of *Caesalpinia pulcherrima*

The microscopic examination of treated pollen grains reveals following characteristic features:—

Shape of pollen grain is spheroidal; Size ranges in between 77 and 90μ; pollens 3-zoniaperturate; ends of pseudocolpi united at the poles producing a synpseudocolpate condition; Size of endoaperture 9.8 × 5.6 μ; Exine fairly thick 4 to 6 μ, thinner at pseudocolpium region narrowing into a thinner edge round the aperture; Synpseudocolpi very broad, formed by thinning of the exine, appearing as colpus margins; Ectine surface shows a gradation of patterns in the order areolate, regulate and reticulate on the mesocolpium but pilate pattern found in pseudocolpium.

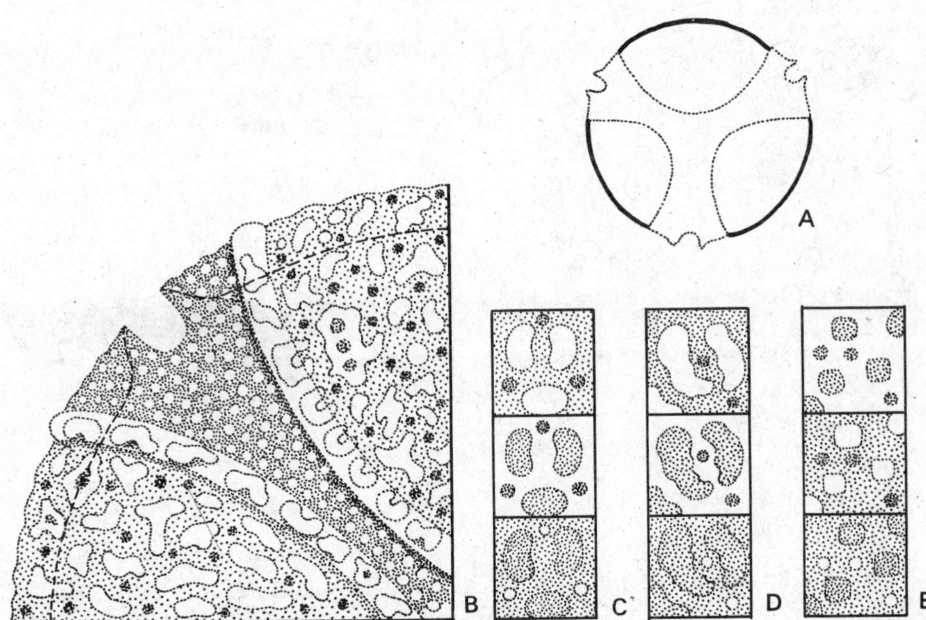

Fig. 16.5. Palynogram of *Caesalpinia pulcherrima*, A—main figure or general appearance of pollen in polar view, B—an enlarged position of polar view of pollen and C, D and E—patterns showing areolate, regulate and reticulate ornamentation. (Diag. redrawn from Nair).

2. **Pollen grains of *Althaea rosea***

The microscopic observations of treated pollen grains reveals following important structures (fig. 16.6).

Pollens are panporate and spheroidal; diameter ranging from 130-140 μ; Pores circular with average diameter 2μ; Interpolar distance ranges 7 μ or so; Exine thickness about 7μ; Exine spinate; Columellate layer (ectine) thinner than endine; Ends of all spines pointed; Basal cushion not formed; Spines dimorphic, one type being of longer height (about 20-21μ) and the other of shorter height (about 5-6μ).

3. **Pollen grains of *Carica papaya***

The detailed microscopic examination of pollen grains shows following characteristic features (fig. 16.7).

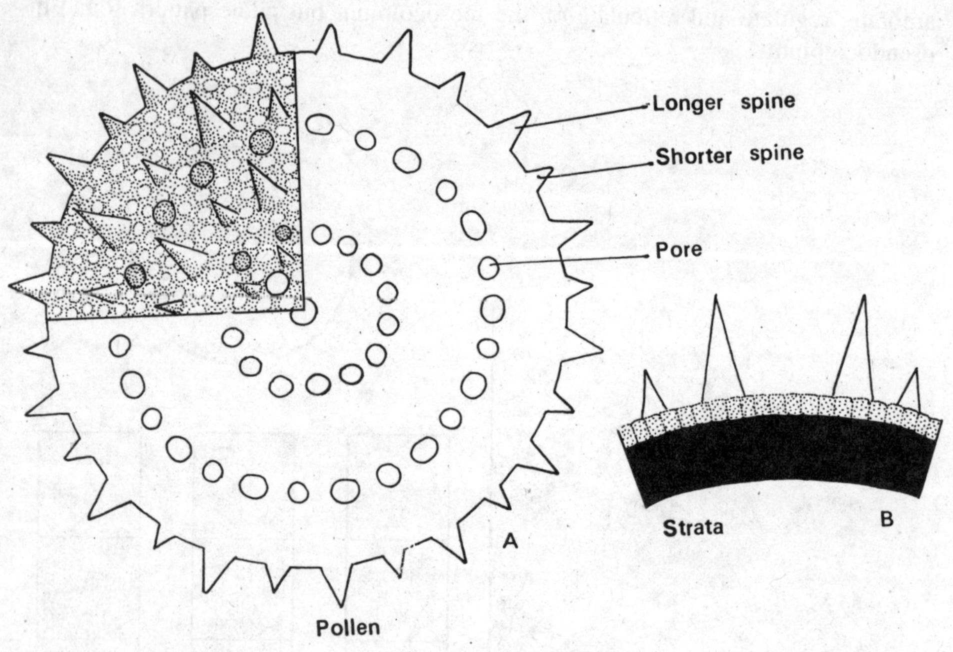

Fig. 16.6. Palynogram of *Althaea rosea*. A-main figure and B-details of exine strata.

Pollen grains 3-Zonicolporate, subprolate (range 32—35 × 28—32μ) or rarely spheroidal (28-35μ); Ectocolpium slightly constricted in the middle, margin slightly incrassate, ends slightly rounded or pointed; Endocolpium elongate (3 × 7μ) with lateral end slightly rounded; Exine 2μ thick; Ectine almost as thick as

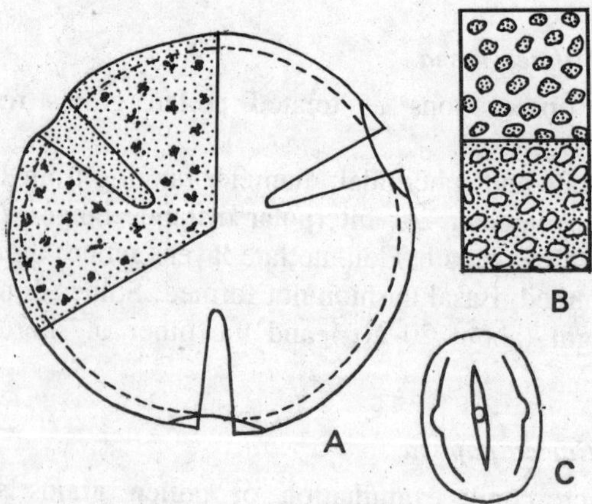

Fig. 16.7. Palynogram of *Carica papaya*. A-main figure in polar view, and B-details of exine ornamentation and C-details of zonicolpate pore.

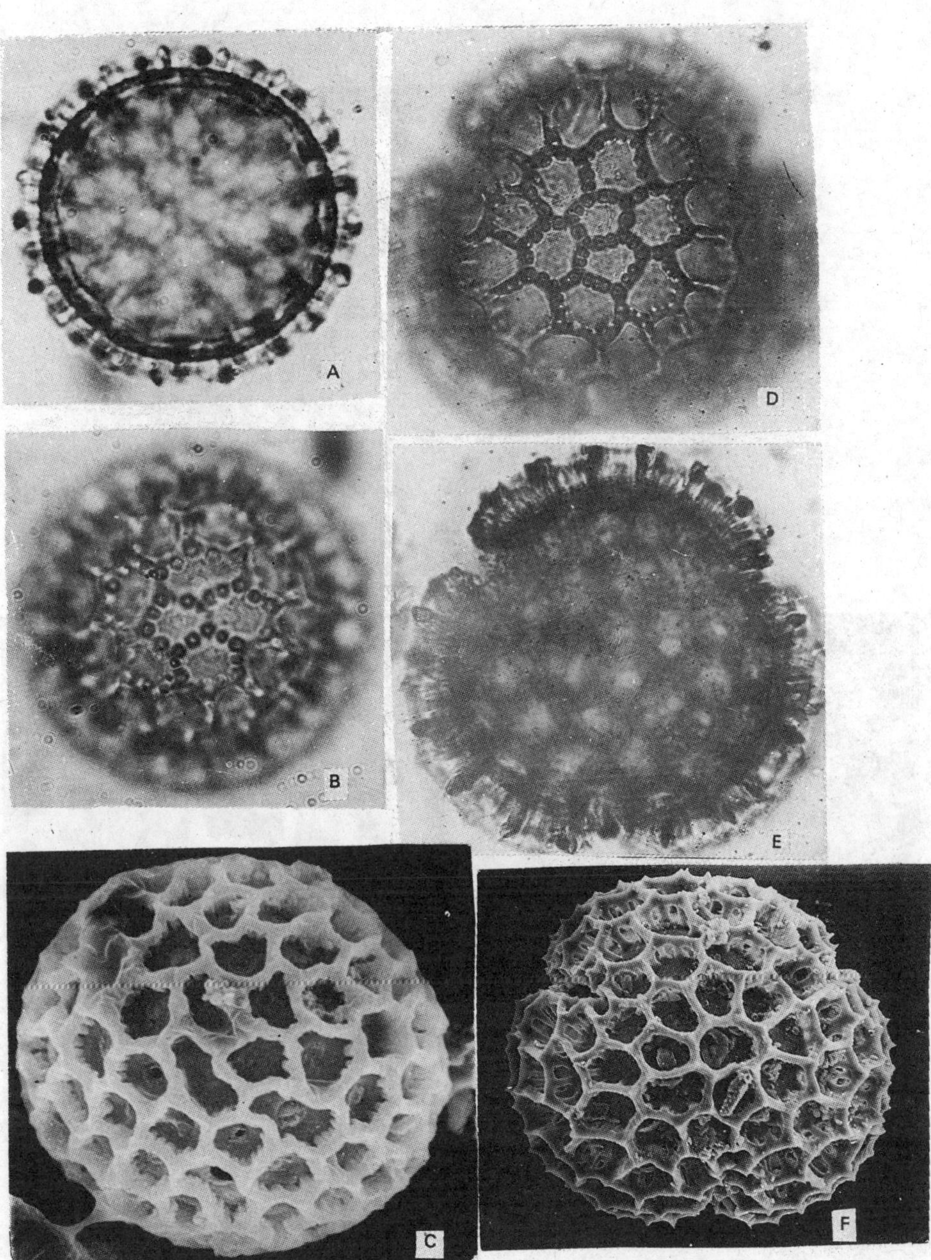

Fig. 16.8. Pollens. A-C-*Tribulus cistoides* and D-F-*Mansonia senegalensis*. A, B, D & E- LM and C & F- SEM photographs (C-panporate grains showing lophotriculate pattern and F-showing brochi with plane muri and spinules at corners).

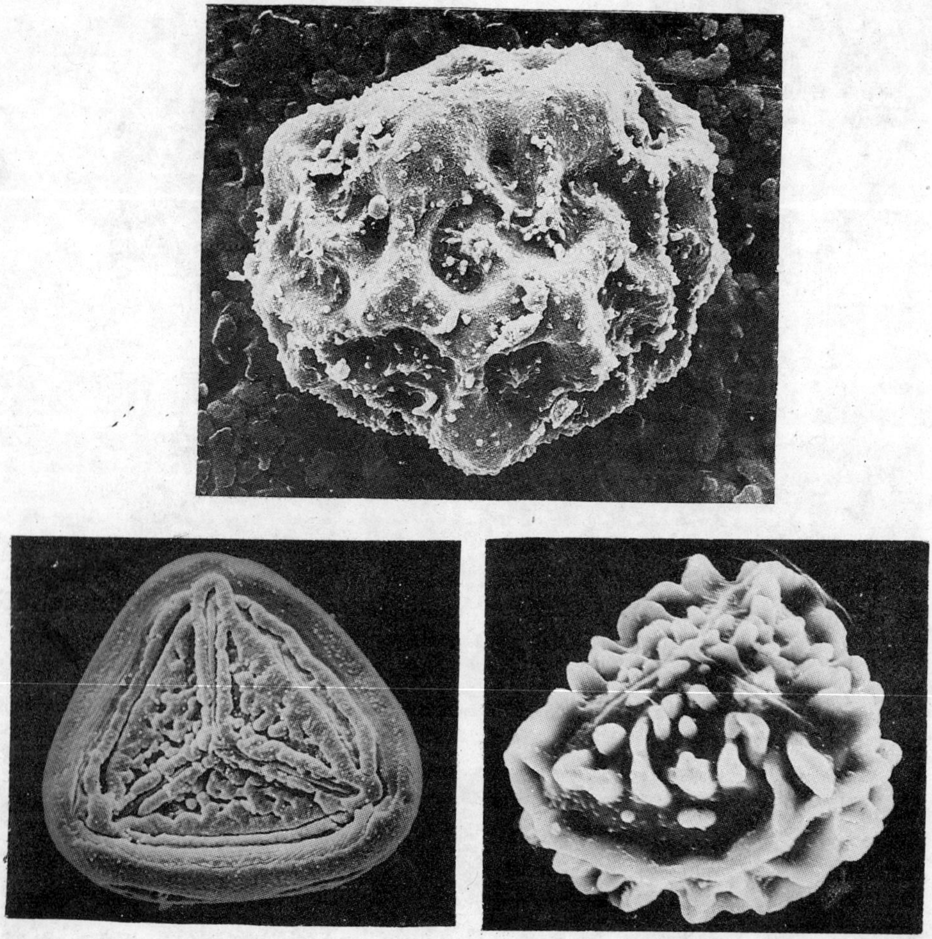

Fig. 16.9. SEM photographs of different pollens (photographs courtesy by Mrs. D. Yunus, N.B.R.I., Lucknow).

endine, with surface faintly reticulate with low muri; Columella not clearly discernible (since the pollen description refers to hermaphrodite flowers, the pollens of unisexual flowers may show variation).

APPLIED ASPECTS OF PALYNOLOGY

In general, palynology can be sub-divided into two broad categories which may be further divided as follows.

I. Basic palynology. It includes—

1. Pollen and spore morphology
2. Theoretical aspects of applied palynology

II. Applied palynology. It includes—

1. Palynotaxonomy. The study of pollen morphology which finds great use to plant taxonomists.

2. Geo or Palaeopalynology. The study of fossil pollen grains or spores (pollen-analytical investigation of Quaternary deposits; pollen and spore floras of-Pre-Quaternary deposits).

3. Melittopalynology. The study of everything about pollen grains found in honey.

4. Pharmacopalynology. The study of pollen grains in drugs, tablets etc.

5. Iatropalynology. The study of pollen grains etc. in connection with allergies.

6. Copropalynology. The study of pollen grains etc.in excrements.

7. Forensic palynology. The study of pollen grains etc. as an aid in criminology.

8. Aeropalynology. The study of pollen grains etc. found in atmosphere.

However, the applied value of pollen is considerably high which cannot be covered under these headings. Some of the important applied aspects are discussed here.

The ancient aspect

The importance of pollen in the improvement of plants was recognised as early as 717 B.C. by Assyrian artisans. The ancient Indian classics also contain some reference which records the refreshing pollen bath of Indian laburnum (*Cassia fistula*) given to Lord Vishnu by Lord Shiva, prior to giving a shower of Ganges water. The cosmetic chests of Egyptian queens are believed to have contained pollen powder for face skin cure. A *pollen cult* was developed by tribal American Indians who considered pollens of cattail (*Typha*) and corn as an emblem of peace, happiness and prosperity. The use of flowers in God worship even now may also be correlated with this fact.

As human food

The use of pollen as human food is not popular but has been reported amongst tribals. American Indians use pollens of *Typha* for preparing pollen soup or gruel. Similarly, a few tribes of New Zealand, India and Pakistan use *Typha* pollens as food. The pollens are mixed with water to form cakes and baked. The preparation is sweetish and light and similar to the taste of London ginger bread. The protein content of pollens varies from 7 to 11 per cent in pine to 35 per cent in date palm. The average fat content ranges from 1 per cent in birch to 17.5 per cent in black walnut. Bee collected pollens are said to be an excellent source of vitamin B and C as well as vitamin D and E.

In medicine

The medicinal importance of pollens and spores was long recognised. The *Lycopodium* spores (commercially popular as *Vegetable sulphur*) were used by ancient Arabian physicians to cure stomach disorders. Spores of *Adiantum philippines* are used in the treatment of leprosy and other skin diseases. The pollens of *Typha laxmanii* are astringent and styptic and that of *Cycas circinalis* narcotic. Flowers of several plants containing large amount of pollen find use in drug manufacture. However, direct effect of such pollen is not well established. There are several reports regarding use of certain pollen tablets, tonics and cream. The pollen tablets have been used in the treatment of prostatitis.

Pollen Allergy

The term 'allergy' was coined by Pirquet to indicate altered specific reactivity of all kinds, including both protective, immune responses and hypersensitivity. Allergy and hypersensitivity are frequently used interchangeably. During recent past it has been well documented that air-borne pollen grains and spores widely cause various allergic complaints such as hay-fever, eczema and asthma. Hay-fever has assumed enormous proportions in many parts of the world. The treatment involves use of pollens directly or in the form of pollen extracts.

The role of air-borne pollens in the etiology of respiratory allergy is well established. That is why aeropollen flora has been exhaustively studied all over the world. However, due attention has not been given in India to such diseases until very recently. During recent years, studies on this aspect have been conducted at Delhi, Lucknow, Jaipur, Poona, Calcutta, Kalyani, Vellore, Guwahati, Bhopal, Nagpur, Mysore etc.

Based on the distribution of pollen types in the atmosphere, usually three periods are recognised-

(i) Maximum pollen period from February to April,
(ii) Minimum pollen period from May to October, and
(iii) Moderate pollen period from November to January.

These three periods may show variation from place to place, e.g. Singh (1981) recognised only two principal pollen seasons in Shillong, – herbs and shrubs predominate from June to October and trees from February to April.

The important allergenic pollen producing plants found in India are-

1. **Herbs.** *Chenopodium, Amarantus, Cannabis, Phaseolus, Rumex, Parthenium, Plantago* (all dicots), *Agrostis, Cynodon, Digitaria, Eleusine, Eragrostis, Oryza, Panicum, Poa, Setaria, Zea, Sorghum* (all grasses, monocots), *Carex and Cyperus* (monocots).

2. **Shrubs.** *Datura, Ligistrum, Ricinus, Lantana, Bougainvillea* etc.

3. **Trees**. *Cassia, Eucalyptus, Prunus, Quercus, Prosopis, Albizzia, Citrus, Phoenix* etc.

Anand and his coworkers (1981) to test allergy reaction, conducted intradermal test with antigenic extracts of 17 different air-borne pollens on 172 cases of nasobronchial allergy in Bangalore. Their critical study revealed that antigenic pollen extracts of *Parthenium hysterophorus, Albizzia lebbeck, Cassia occidentalis, Artemesia scoparia, Ageratum conyzoides, Cassia siamea, Amarc* [illegible] *pinosus* and *Melia azadirachta* show 4+ reactions. Whereas 3+ and 4+ reactions were shown by *Argemone mexicana, Azadirachta indica, Cenchrus ciliaris, Chenopodium murale, Cynodon dactylon, Holoptelia integrifolia* and *Eucalyptus citriodora; Albizzia lebbeck* showed 65 positive reactions, while 56 were shown by *Cassia occidentalis,* 42 by *Parthenium hysterophorus,* 34 by *Cassia siamea* and 20 each by *Eucalyptus citriodora* and *Argemone mexicana.*

Some recent researches on the biochemistry of pollen allergy has revealed that, there are specific sites of antigen in the pollen wall, both in exine and intine. The exinous fraction hides the antigen within its interstices which is forced out when the pollen comes in contact with the human body. However, there is still much detailed work to be done regarding this problem.

Medicine value

Bee collected pollens mixed with coffee, milk, honey, potatoes and chocolates are prescribed to cure many disorders in children. As discussed earlier, these are also used to beautify the skin. These are specially suitable to the skin because they contain a well balanced content that a skin needs such as Vitamins B_1, B_2, C and E and different types of fats, sugars, proteins, amino acids, phosphates etc.

Agriculture

During recent times the significance of pollen in agriculture has been well recognised. Infact among scientific agricultural practices, pollen based agricultural researches are being programmed. The selection of pollen with the desired potential from desired individuals of any one crop population is considered as an essential prerequisite. Storage of pollen is as important as storage of seeds, in maintaining plant genetic resources. At many research institutes, the haploid plants are successfully used for the maintenance of the genetic stock of pollen plants. It is claimed by Chinese and Japanese scientists that haploid plants of rice (*Oryza sativa*) developed from pollens have been transferred from test tubes to field conditions. It is beyond doubt a great achievement.

Beside these scientific advancements, every one knows that successful pollination ensures abundant fruit set and increased seed production. For this, at many places, bee hives are installed in orchards to use honey bee as pollinators. In some scientifically advanced countries, stored pollens are disseminated over crop fields by means of sprayers drawn by tractors, or by breaking *Pollen bombs* or by hand, on days with moderately high winds.

Stored pollens are sometimes transported to distant places for use in breeding. For example, cold-resistant sugarcane pollens were shipped from Washington to

Palmira, Columbia (U.S.A.) for crossing the commercial varieties to increase the sugar cane yield and quality.

Pollen irradiation and breeding

It is being increasingly realised, that pollens are ideal material for experimentally controlling the genetic behaviour of plants, to produce more and more useful hybrids. Pollen irradiation is successfully used to meet the problems of intraspecific or interspecific incompatability, by inducing desired mutational changes in the pollen genome. Similarly induction of male sterility becomes a boon in crossing self-pollinated plants, like grapes. Both these points increase the enormous utility of palynology.

Weed Control

Aeropalynological studies of aeropollens have opened a new field of controlling the weed and this has been practised successfully in England. They suggested that weed control can be effectively planned and practised from the information gained from atmospheric pollens.

Pollen in honey

Honey and its related products are rich in pollens, in which these become partial source of vitamins, minerals and amino acids. Pollen grains collected by bees and added to the *bee bread* are indispensible to the growth of larvae and existence of *Queen bees.* It is interesting to note that nearly 25 to 30 kg. of pollens are required by a fair size colony of bees every year.

Study of honey-pollens provides information about the plants from which the nectar and pollens have been gathered. Such studies can also help in detecting adulteration in honey, and then it becomes easy to check the undesired traders. Pollens of undesired plants can be minimised by removing those plants from nearby areas. Similarly, a trained melittopalynologist can easily distinguish between honey from two different localities and can tell the period of honey production. Sharma and Nair (1965) stated that if honey contains 50 per cent or more pollens of one plant, it should be called **Unifloral honey** otherwise, it should be referred to as **Multifloral honey**. Nair and Chaturvedi (1974) reported that in India, *Eugenia, Nephelium, Sapindus, Putranjiva, Citrus* and *Brassica* are common nectar-yielding plants while *Holoptelea, Alnus* and *Borassus* and other plants are common pollen-yielding plants used by honey bees. Such a study can be helpful in selecting the area for establishing apiary gardens. Furthermore, the name of honey can be assigned on the basis of pollen contained in them which would ameliorate the select utility of honey.

Palaeobotany and Geology

Study of pollen fossils found in lignites, coal and oil-bearing sediments has been found to be of special significance in economic geology and in oil exploration. In oil exploration, it is of primary importance to study pollens and microfossils of the offshore sedimentary deposits. The quantitative analysis of various fossil-forms, from bore-hole sediments, preludes the ancient shore lines, as is evident from what is called **isobotanical lines.** Similarly, the category of fossil flora provides conclusive support to the correlation of coal and lignite seams in evaluating their grade and quality.

Pollen industry

An effort to develop pollen industries has been made considering the importance of pollen in human health and welfare. For the treatment of hay-fever, pollen grains are collected and sold on commercial scale. Though it is in initial phase, it is increasingly gaining importance. In U.S.A. one can find sale of pollen grains of *Quercus* and *Timothy*. Sugar-coated pollen tablets are sold under the trade names such as *Cernitin, Cernitol, Cerniton, Polloton* and *Pollison.* In Europe, bee collected pollens are sold at very high price by famous apiarists.

Pharmacognosy, smuggling and criminology

Studies of pollens provide the most reliable method in checking the adulteration of drugs prepared from floral parts. In Indian pharmacopoeia, nearly 100 such floral drugs are listed. Their detailed palynological knowledge can be used in minimising drug adulteration.

Application of palynology also finds place in checking smuggling and criminal activities. A few years back, smuggling of wheat bags in England was reported. The government could find its source only with the help of a palynologists, who studied pollen present in bags and found certain pollen of *Grevillea*, which is largely grown in Australia. The source of the contraband goods was thus established. Similarly, pollen grains recovered from mud present on a criminal's clothes, provide authoritative information on his movements, leading to the detection of crime.

All these examples offer a new hope for application of palynology for human advantage, which is fast emerging as an established science. Scientists are finding great many uses of pollen and studies on palynology.

POLLEN ALLERGY

The term '*allergy*' was coined by Pirquit to indicate altered specific reactivity of all kinds, including both protective, immune reponses and also hypersensitivity. Allergy and hypersensitivity are frequently used interchangeably. Allergy is also defined as altered and accelerated reaction of an individual to a second or subsequent exposures to a substance, to which his body has already become sensitized. During recent past, it has been well established that air-borne pollen grains and spores widely cause various allergic complaints such as hay-fever, eczema and asthma. Some important diseases are discussed here.

SOME IMPORTANT DISEASES

Hay fever

Hay fever is also called *allergic rhinitis* and its characteristic symptoms include profuse watery nasal discharge with sneezing, frequently accompanied by redness, irritated and watery eyes and headache. Such allergy is caused by aeroallergens that are produced by the following important plants.

GYMNOSPERMS—*Ginkgo biloba, Juniperus mexicana, Pinus contorta, Thuja plicata* etc.

ANGIOSPERMS—**Dicotyledons**—*Acer, Amarantus, Ambrosia, Xanthium, Chrysanthemum, Cannabis, Carica papaya, Casuarina, Chenopodium, Acacia, Eucalyptus, Rumex, Ranunculus* etc.

Monocotyledons—*Phoenix, Juncus, Luzula, Avena, Cynodon, Poa, Secale, Zea, Sorghum, Typha* etc.

Majority of the plants that induces allergic rhinitis are wind pollinated. Sometimes insect or bird pollinated plants also cause the disease. *Ambrosia* pollens are most allergenic and have high potentiality in inducing hay fever.

Anaphylactic shock

Anaphylaxis is the most severe of all allergic reactions. It is characterised by sudden vasomotor collapse, leading to shock and paroxysmal bronchio-constriction, and if immediate treatment is not given, it may result in death. Most cases of this category are associated with serum therapy, penicillin therapy and insect stings. Some recent studies have revealed that allergens are shared by the members of family of which *Ambrosia* and *Chrysanthemum* are a part.

Asthma

Asthma or bronchial asthma is considered as a form of allergy which has a chronic course much longer than allergic rhinitis. **Extrinsic asthma** is found among children and young adults which can be caused by pollens, spores, feathers and animal danders. Pollen from *Pinus contorta* has been known to cause bronchial asthma. **Intrinsic asthma** is generally found in older age and its likely agents can be bacteria or viruses. Recent studies reveal that chlorogenic acid is a strong allergen causing asthmatic diseases and is wide spread in plants. It is found in large quantity in coffee beans and castor beans.

Some of the common plants producing allergenic pollens, causing asthma are—*Acer, Populus, Ambrosia, Cynodon, Morus, Acacia, Quercus, Astragalus, Amarantus, Fraxinus, Olea* etc.

Dermatitis

Dermatitis refers to inflammatory reacton of the skin. It may be of two types, the *irritant dermatitis* and *photodermatitis*. Higher plants that provoke irritant dermatitis are *Narcissus, Urtica, Laportea, Urera, Ranunculus, Ananas* etc. and photodermatitis is caused by *Anona, Plumeria, Brassica, Ananas, Carica, Acacia, Jatropha, Allium, Ranunculus, Solanum, Capsicum* etc.

Dermatitis is mainly caused because of hypersensitivity developed due to certain chemicals present in various plant parts. Such chemicals may be found in pollens too. Some of the chemicals causing dermatitis are listed in table 16.1

Table 16.1. Some important chemicals causing dermatitis and their source.

Primary component (Chemical)		*Found in Plants*
Hydrocarbons	— Myrcene	*Humulus lupulus*
	Limonene	*Carum carvi, Citrus* spp.
	Phellandrene	*Mentha spicata*
		Cinnamomum zeylanicum
Alcohols	— Geraniol	*Pelargonium* sp.
	Linalool	*Coriandrum sativum*
	Citronellol	*Eucalyptus citriodora*
	Santalol	*Santalum album*
Ketones	— Camphor	*Cinnamomum camphora*
	Carvone	*Mentha spicata*
	Thujone	*Artemesia puligeoides*
Phenols	— Anethole	*Foeniculum vulgare*
	Eugenol	*Eugenia caryophyllus*
Quinones	— Primin	*Primula obconica*
Oleoresins	— Uroshiol	*Anacardium occidentale*
		Mangifera indica
	Capsaicin	*Capsicum frutescens*
	Asafoetida	*Ferula asafoetida*
Oleoresins	— found in pollen	*Acer pseudoplatanus*
		Ambrosia
		Fraxinus
		Erigeron
		Iva
		Mangifera
		Populus
		Xanthium

Many other plants have been found to cause contact dermatitis but their offending compounds have not been identified. *Rauvolfia, Thevetia, Calotropis, Rhoeo, Croton, Euphorbia, Cynodon, Magnolia, Anemone, Clematis, Delphinium, Datura, Sapindus* etc. are examples of such plants.

POLLEN ALLERGENS

During some recent years, a considerable amount of work on pollen allergy has been done in India by workers like Shivpuri (1969), Baruah (1969), Tilak (1980), Reddi (1974), Appana (1980), Shukla (1980) etc. Some of these workers have prepared pollen calenders of various regions of the country. Shivpuri (1973) reported, that about ten per cent population suffers from one or other forms of allergy. Studies have revealed that there are two peak periods of air-borne pollens, one during September to November and other during February to April. These two peaks show closer relationship with the incidence of various disorders.

The wind-borne pollens of tree origin are common allergens and most of them belong to *Quercus, Carya, Ulmus, Acer, Acacia* etc. However, allergenic pollens are quite frequently produced by weeds, may be herbs or shrubs. The most important family involved in producing allergenic pollens is the grass family, now called Poaceae (Graminae) which includes important plants like *Agrosti, Anthoxanthum, Dactylis, Digitaria, Phleum* etc. Next to Poaceae is the weed family Asteraceae (Compositae) which includes plants like *Ambrosia, Iva, Xanthium, Parthenium, Artemesia* etc. *Ambrosia* is now known to have the most potent allergens. In India, *Parthenium* perhaps tops the list among plants producing allergenic pollens.

The chemical analyses of pollen wall have revealed that some proteins present below the apertures in the intine or in the exine, are water-soluble, and when in contact with the mucosa of the skin of man, get released within a few seconds. These have been found causing allergy. Some of these proteins have been isolated recently. E is the major allergenic protein of *Ambrosia* while others are Ra, Ra 3 and Ra 5. Among the grasses, allergens identified are A and B for *Phleum pratense* and *alpha, beta* and *gamma* allergens in *Lolium perenne*. However, specific proteins in majority remains to be characterised.

CONTROL MEASURES

There are two important procedures to control pollen allergy, the first is prevention and the second is cure or prophylactic measures. For prevention, it is important to recognise plants releasing allergenic pollens, so that these should be avoided from growing or eliminated from that area. It is quite important for the sensitive patients to move away from the location of the offending allergens. The process of filtering of aeroallergens by appropriate air conditioning can also be adopted for preventing allergenic pollens.

The prophylactic measures involve injection of increasing amounts of the allergen, to develop a higher level of locking antibodies, before exposure to normal environment. During recent years, the acute cases are also treated by administration of antihistamines, ephedrine and corticosteroids. Infact, these provide symptomatic relief.

GENERAL QUESTIONS

1. Write an essay on applied aspects of palynology.
2. Write notes on— (i) Pollen industry, (ii) Pollen allergy, (ii) Melittopalynology:
3. Write short notes on—
   (i) NPC system, (ii) Branches of palynology.
4. How is the knowledge of palynology helpful to mankind? Describe.
5. Give an account of the pollen morphology.
7. What is allergy? Discuss the role of pollen in causing allergy to man.
8. What are the various kinds of pollen allergies? Discuss about source of allergenic pollens, nature of allergens and some control measures.
9. Write short notes on—
   (i) Hay fever (ii) Asthma (iii) Dermatitis (iv) Allergenic pollens.
10. What is pollen allergy? Describe a brief account of pollen allergy.
11. Mention some important pollen allergy symptoms and name the pollens responsible for it. Suggest the effective treatement.

Selected Readings

Archibald, E.E.A. 1939. The development of the ovule and seed of jointed cactus (*Opuntia aurantiaca* Lindley). *South African J. Sci.* **36:** 195-211.

Bajaj, Y.P.S. 1978. Regeneration of haploid toabcco plants from isolated pollen grains in drop culture *Ind. Exp. Biol.* **16:** 407-409.

Bhojwani, S.S. and S.P. Bhatnagar, 1974. *The Embryology of Anglosperms.* Vikas Publishing House Pvt. Ltd., New Delhi, India.

Blakeslee, A.F. and S. Satina, 1944. New hybrids from incompatible crosses in *Datura* through culture of excised embryos on malt media. *Science* **99:** 331-334.

Boke, N.H. 1949. Development of the stamens and carpels in *Vinca rosea* L. *Amer. J. Bot.* **34:** 535-547.

Bourgin, J.P. and J.P. Nitsch 1967. Obtention de *Nicotiana* haploides a partir etamines cultivees *in vitro. Ann. Physiol. Veg.* **9:** 377-382.

Campbell, D.H. 1899a. Die Entwicklung des Embryo sac von *Peperomia pellucida. Ber. deutsch. bot. Gesell.* **17:** 452-453.

Campbell, D.H. 1899b. A peculiar embryo sac in *Peperomia pellucida. Ann. Bot.* **13:** 626.

Campbell, D.H. 1901. The embryo sac of *Peperomia pellucida Ann. Bot.* **15:** 103-117.

Chopra, R.N. 1958. *In vitro* culture of ovaries of *Althaea rosea* Cav. In Maheshwari P. ed. Proc. Seminar mod. Dev. Plant Physiol. Univ. Delhi pp. 87-89.

Chopra, R.N. 1962. Effect of some growth substances and calyx in fruit and seed development of *Althaea rosea* Cav. In *Plant Embryology*-A Symposium C.S.I.R. New Delhi, pp. 170-181.

Davis, G.L. 1966. *Systematic Embryology of Angiosperms*, New York, John Wiley and Sons, 1966.

Eames, A.J. 1961. *Morophology of the Anglospersm*, Tata MC Graw-Hill Publishing Co. Ltdd., Bombay-New Delhi.

Echlin, P. 1968. *Pollen. Scient. Am.* **218**(4): 80-90.

Erdtman, G. 1964. Palynology. In *Vistas in Botany* Vol. 4. (Ed. Twill, W.B.) London: 23-54.

Esau, K. 1960. *Anatomy of Seed Plants.* pp. 376, Wiley Eastern Ltd. New Delhi.

Eunus, A.M. 1949. Contributions to the embryology of the Liliaceae (*Gloriosa superba*). Proc. 36th Indian Sci Cong. Allahabad, Bot Sect. No. **25.**

Fagerlind, F. 1938. Embryosack Von *Plumbagella and Plumbago. ärkiv for Bot.* **29B:** 1-8.

Fahn. A. 1989. *Plant Anatomy*, 3rd ed. Pergamon Press, London.

Farr, C.H. 1918. Cell division by furrowing in *Magnolia. Amer. J. Bot.* **5:** 379-395.

Ferguson, M.C. 1927 A cytological and genetical study of *Petunia.* I. *Bull. Torrey Bot. club* **54:** 657-664.

Focke, W.O. 1881. Die Pflanzenmischlinge, ein Beitrag zur Biologie der Gewachse, Berlin.

Fossard, R.A. 1969. De. Development and histochemistry of the endothecium in the anthers of *in vitro* grown *Chenopodium album* L. *Bot. Gaz.* **130:** 10-22.

Gemmel, A.R. 1969. *Developmental Plant Anatomy.* William Clowes and Sons Ltd., London.

Guha. S. and B.M. Johri. 1966. *In vitro* development of ovary and ovule of *Allium cepa* L. Phytomorphology **16:** 353-364.

Guha, S. and S.C. Maheshwari. 1964. *In vitro* production of embryos from anthers of *Datura. Nature* (London) 204-497.

Guha. S. and S.C. Maheshwari. 1966. Cell division and differentiation of embryo in the pollen grains of *Datura in vitro. Nautre* (London) **212**: 97-98.

Guha, S. and S.C. Maheshwari, 1967. Development of embryoids from pollen grains of *Datura in vitro*. Phytomorphology, **17**: 454-461.

Gupta, B.L. 1935. Studies in the development of the pollen grain and embryo sac of *Wolffia arrhiza. Curr. Sci.* (India). **4**: 104-105.

Hofmeister, W. 1859. Neue Beitrage zur Kenntnis der Embryobildung der Phanerogamen. I. Dikotyledonen mit ursprunglich einzelligem, nur durch. Zelltheilung Wachesendem Endosperm. *Abh. Konigl. Sachs. Gesell. Wiss.* 1959, 535-672.

Jensen, W.A. 1968. Cotton embryogenesis: the zygote, Planta **79**: 346-366.

Jensen, W.A. 1973. Fertilization in flowering plants. *Bio. Sci.* **23**: 21-27.

Johansen, D.A. 1945. A critical survey of the present status of plant embryology. *Bot. Rev.* **11**: 87-107.

Johnson. D.S. 1900. On the endosperm and embryo of *Peperomia pellucida. Bot. Gaz.* **80**: 1-11.

Johri, B.M. 1936. The life history of *Butomopsis lanceolata Kunth. Proc. Indian Acad. Sci.* Sect. B **4**: 139-162.

La Rue, C.D. 1936. The growth of plant embryos in culture. *Bull. Torrey Bot. Club.* **63**: 365-382.

La Rue, C.D. 1936. The rooting of flowers in sterile cultures. *Bull. Torrey Bot. Club.* **69**: 332-382.

Leopold, A.C. 1966. *Plant Growth and Development*, pp. 466. Mc Graw-Hill Book Company, New York and Maidenhead.

Maheshwari, P. 1950. *An Introduction to the Embryology of Angiosperms.* Mc Graw-Hill Book Company, New York.

McLean, R.C. 1956. and W.R. Ivimey-Cook, 1956. *Text Book of Theoretical Botany.* Vol. 2. Longmans; London.

Nair, P.K.K. 1970. *Pollen Morphology of Angiosperms.* Vikas Pub. House Pvt. Ltd. New Delhi.

Poddubnaya-Arnoldi, V.A. 1959. Study of fertilization in the living material of some angiosperms. *Phytomorph.* **10**: 185-198.

Rangaswamy, N.S. 1961. Experimental studies on female reproductive structures of *Citrus microcarpa* Bunge. *Phytomorph* **11**: 109-127.

Sattler, R.A. 1974. A new approach to gynoeical morphology, *Phytomorph.* **24**: 22-35.

Stanley, R.G. and H.F. Linskens, 1974. *Pollen*: Biology, Biochemistry, and Management. Springer, N.Y.

Strasburger, E. 1879. "*Die Angiospermen and die Gymnospermen*" Jen.

Sunderland, N. 1978. Strategies in the improvement of yields in anther culture. In: *Proc. Plant Tissue Culture*, Peking, pp. 65-86.

Sunderland, N. and F.M. Wicks. 1971. Embryoid formation in pollen grains of *Nicotiana tabacum. J. Exp. Bot.* **22**: 213-226.

Sunita. N. 1937. On the mature pollen grain in Angiosperm. *Bot. Mag.* (Tokyo) **51**: 424 530.

Swamy, B.G.L. 1943. Gametogenesis and embryogeny of *Eulophea epidendraea* Fischr. *Proc. Natl, Inst. Sci.* India **9**: 59-65.

Swamy, B.G.L. 1948. A contribution to the life history of *Casuarina*. *Proc. Amer. Acad. Arts. Sci.* **77**: 1-32.

Walker, J.W. and J.J. Skvaria. 1975. Primitively columellaless pollen: a new concept in the evolutionary morphology of angiosperms. *Science* **187**: 445-447.

Webber, H.J. 1900. Xenia, or the immediate effect of pollen in maize. U.S. Dept. Agr. Div. Veg. Path and Physiol. Bul 22.

White, P.R. 1932. Plant tissue culure: A preliminary report of results obtained in the culturing of certain plant meristems. Arch. Exp. Zellforch **12**: 602-620.

White, P.R. 1963. *The Cultiviation of Animal and Plant Cells*. 2nd edition Ronald Press, New York.

Glossary of Terms

Abaxial, the surface directed away (outside) from the axis.

Abscission, the detachment or separation of a part.

Abscission layer, a layer at the base of leaf or fruit stalk in which the parenchyma cells become separated from one another, through the dissolution of middle lamella.

Abscission zone, that zone containing the tissue which brings about the abscission of organ.

Accessory or Subsidiary cell, an epidermal cell surrounding the guard cells of stomata and which differs from other normal epidermal cells.

Acropetal, proceeding towards the apex (i.e. youngest towards apex and oldest towards base), e.g. leaves on a branch.

Actinostele, a protostele with star-shaped arrangement of xylem.

Adaxial, directed towards axis.

Adnation, attachment of one organ with another by its whole length.

Adventitious organ, an organ developed in an unusual position.

Aerenchyma, a parenchyma tissue characterised by the presence of large intercellular spaces, as found in cortex of certain hydrophytes.

Aerophyte, a plant growing attached on the aerial part of other plant, such as an epiphyte.

Aggregate fruit, a fruit developing from apocarpous gynoecium.

Albuminous cell, also called *Strasburger cell*, certain cells found in the phloem parenchyma of Gymnosperms which are connected morphologically and physiologically to the sieve elements.

Alburnum, see *sap wood*, also called splint wood.

Aleurone grain, a grain of reserve proteins.

Alkaloid, a basic nitrogenous organic substance produced in certain plants. They have a physiological effect on animals, e.g. morphine, atropine, nicotine etc.

Amphicribal vascular bundle, a concentric vascular bundle in which xylem remains surrounded by phloem, as seen in some ferns.

Amphiphloic siphonostele, a siphonostele with phloem on outer and inner side of xylem (i.e. presence of outer and inner phloem in a siphonostele).

Amphivasal vascular bundle, a concentric vascular bundle in which phloem remains surrounded by xylem, as in *Dracaena*.

Amyloplast, a leucoplast plastid concerned with the storage of starch.

Androecium, a collective term used for all the stamens of a flower.

Androgynophore, stalk-like structure found in between petals and androecium (androphore) and between androecium and gynoecium (gynophore) to elevate the androecium and gynoecium.

Anisocytic type (stomata), a stomata characterised by three subsidiary cells, of which one is small and other two are large, as found in Cruciferae and Solanaceae.

Anomocytic type (stomata), a stomata that remains surrounded by subsidiary cells which are indistinguishable from other epidermal cells, as found in Ranunculaceae and Papaveraceae.
Anther, the fertile part of stamen.
Anthesis, the time of maturation of the male and female organs of the flower.
Anticlinal (division), perpendicular to the surface (at right angle to the surface of apex).
Antipodals, cells of an embryo sac present towards chalazal end.
Aperture, (of pollen grain), a depressed area of germ pore.
Apical meristem, the growing point at the tip of root and shoot.
Apocarpy, a condition in a flower where carpels are free.
Apomixis, a phenomenon to refer reproduction without sexual process
Apposition, the formation of successive layers in growth of a cell wall.
Areole, (in mesophyll), the smallest area of mesophyll delimited by veins, and (in Cactaceae), the small outgrowths on the stem bearing spines.
Aril, a fleshy colourd covering on the seed, which arises as an outgrowth of the funicle or base of the ovule.
Articulate, jointed.
Astrosclereid, a branched sclereid.
Atactostele, a stele characterised with scattered vascular bundles, as found in stem of monocotyledons.
Autochory, refers to seed dispersal by a self-dispersal mechanism.
Axial organ, the root, stem, inflorescence of flower axis without their appendages.
Axillary, growing in axils as buds.

Bark, a collective term for all the tissues outside the vascular cambium.
Basipetal, proceeding towards base, youngest towards base and oldest towards apex.
Bast, a vague term, sometimes used to mean phloem.
Bicollateral, (vascular bundle), having two sides similar, i.e. a vascular bundle with phloem on both sides of xylem, as in Cucurbitaceae.
Bifacial, (leaf) also called dorsiventral leaf, a leaf in which palisade parenchyma is found on one side and spongy parenchyma on the other side.
Bilateral, having two sides symmetrical about an axis.
Bordered pits (found in tracheids and vessels), a form of pit with overarching border of secondary cell wall.
Brachysclereid or stone cell, a small isodiametric sclereid.
Bud, a rudimentary shoot.
Bulliform cell, an enlarged epidermal cell, commonly found in the leaves of Graminae.
Bundle sheath, a layer(s) of cells surrounding a vascular bundle, which may be parenchymatous or sclerenchymatous.

Callose, a polysaccharide (β-1,3-glucan) found in sieve areas (phloem), wall of pollen tubes, wall of fungal cells etc.
Callus, (i) corky secondary tissue developed by woody plants over a wound, (ii) amorphous mass of tissue developing in tissue culture.
Calyptrogen, a term used out of histogen theory, for a meristematic tissue that gives rise to root cap.
Cambium, a lateral meristem that gives rise to secondary tissue.
Carpophore, (i) a raised part of receptacle which bears stamens and carpels, (ii) the forked stalk from which mericarps are attached.
Caruncle, an arilloid outgrowth in seed near the micropyle (as in seed of castor).
Caryopsis, a fruit in which pericarp and testa are fused.

Casparian strip, a band-like thickening of lignin or suberin found characteristically in endodermal cells. (named after German botanist R. Caspary).

Chalaza, a region in an ovule where integument and nucellus connect with the funiculus.

Chlorenchyma, a chloroplast containing parenchyma.

Cisterna, a flattened sac-like structure of ER.

Coenocyte, multinucleate, aseptate cells.

Cohesion, fusion of organs of the same nature.

Coleoptile, a plumule sheath found in monocot seeds.

Coleorrhiza, a sheath surrounding radicle found in monocot seeds.

Collateral, (vascular bundle) bundles with xylem and phloem in same radius.

Collenchyma, a living tissue, the walls of which are usually unequally thickened, supporting the tissue of young organs.

Colleter, a multicellular hair which secretes sticky substance.

Companion cell, a narrow parenchymatous cell associated with sieve tube. It arises from the same mother cell as the sieve tube member and maintains physiological connection with it. These are found only in angiosperms.

Complementary cell, a loose tissue formed by the phellogen of lenticel; the cells may or may not have suberised cell wall.

Concentric, (vascular bundle), the V.B. with one tissue surrounding the other, i.e. amphicribal and amphivasal.

Conjunctive tissue, (1) parenchyma formed between specialised tissue, e.g. between vascular strands, (2) the parenchyma present between the secondary vascular bundles in monocots.

Connective, the region between the two lobes of an anther.

Cork, a layer of suberised, dead cells, cut off by phellogen on the outside.

Cork cambium, also called phellogen of periderm.

Cork cell, (i) a cell of cork, (ii) a small suberised epidermal cell found in members of Graminae.

Corpus, (of tunica corpus theory), the core of the apical meristem (within tunica), cells of which divides in various planes and increase the volume of the shoot apex.

Cortex, the tissue present between stele and epidermis of the stem.

Cotyledon, the first leaf of the embryo.

Crassulae, (also called bars of Sanio), the transversely oriented thickening between the pits found in Gymnosperms.

Cristae, the infolding of mitochondrial inner membrane.

Cuticle, a layer of cutin (a lipid derivative) found on the outer wall of the epidermal cells which is protective and impermeable to water.

Cutinization, the deposition of cutin in the cell walls.

Cystolith, a calcium carbonate deposition on the specific outgrowth of epidermal cells (as in *Ficus*).

Cytokinesis, a division of cytoplasm during cell division (distinct from nuclear division).

Dermatogen, the outer meristematic layer (of histogen theory) that gives rise to epidermis in plants.

Diacytic type, (stomata) in which stomata remains enclosed by a pair of subsidiary cells whose common wall is at right angles to the long axis of guard cells.

Diakinesis, a stage of prophase I of meiosis.

Dictyosome, a unit of Golgi bodies.

Dictyostele, a kind of siphonostele, characterised by presence of a large number of superimposed leaf gaps and appear broken up into meristeles in a cross section (as in fern rhizome).

Diffuse porous (wood), a secondary wood in which vessels are almost uniformally distributed (as found in tropical trees).
Diplotene, a stage of prophase I of meiosis.
Dorsiventral leaf, a bifacial leaf (as found in most dicots).
Druse, a compound crystal (containing ergastic substances such as calcium oxalate).
Duramen, also called heartwood, a part of secondary wood.

Ectoplast, the protoplasmic film or outer part of protoplast.
Elaioplast, an oil-storing leucoplast.
Elaiosome, an outgrowth from the surface of seed, containing large oil storing cells. These are helpful in seed dispersal by attracting ants.
Endarch (xylem), a type of primary xylem in which protoxylem is towards central axis (i.e. maturation of xylem is centrifugal).
Endocarp, the inner most layer of pericarp.
Endodermis, the inner most layer of cortex that surrounds stele.
Endogenous, developing from internal tissues (as the origin of lateral root from the pericycle).
Endoplasmic reticulum (ER), a kind of cell organelle found in eukaryotic cell. It consists of cisternae, tubules and vesicles.
Endosperm, a nutrient tissue formed within embryo sac in angiosperms.
Endothecium, a wall layer found in anther below the epidermis.
Epiblast, a small outgrowth found opposite the scutellum in the embryo of some members of Graminae.
Epiblema, the outer layer of root tissue (the piliferous layer).
Epicotyl, the stem of an embryo (or seedling) above the cotyledons.
Epidermis, the outermost layer of primary tissue of plant.
Epipetalous stamen, a stamen adnated to a petal.
Epithelium, layer of cells lining schizogenously formed secretory canals and cavities.
Ergastic matter, the non-protoplasmic (non-living) products of various metabolic processes.
Ergastoplasm, the endoplasmic reticuluum.
Essential oils, the volatile oils with characteristic odour.
Eumeristem, the meristem composed of isodiametric thin-walled cells.
Eustele, the most advanced type of stele, the vascular tissue of which forms a hollow cylinder comprising collateral or bicollateral vascular bundles.
Exalbuminous seeds, the non-endospermic seeds.
Exarch (xylem), a primary xylem in which protoxylem lies towards outside (i.e. maturation is centripetal).
Excentric, one sided.
Exine, the outer wall of mature pollen grain.
Exocarp, the outer most layer of pericarp (fruit wall).
Exodermis, a specialised more or less thickened (with suberin) layer below the piliferous layer (a type of hypodermis).
Exogenous, developing from external tissues.
Extra- cellular, found outside the cell.
Extra- stelar,found outside the stele.
Extra- xylary, occurring outside the xylem.

Fascicular, situated in a bundle of vascular tissue as a part of it.
Fibre, an elongated sclerenchymatous cell with tapering (pointed) ends.
Fibre, libriform, a fibre of the secondary xylem with thick walls and simple pits.

Fibre sclereid, a sclerenchyma cell, structurally intermediate between fibre and sclereid. It is found in phloem but non-conductive in nature.

Fibre tracheid, a sclerenchyma cell found in secondary xylem, which is structurally intermediate between tracheid.

Foraminate, having round pores.

Funiculus, the stalk of ovule by which it remains attached to the placenta.

Fusiform, elongated and tapering towards each end.

Gametophyte, an individual of the haploid generation in the life cycle.

Geitonogamy, the cross pollination between two flowers on the same plant.

Gland, an organ secreting a specific substance.

Glandular, with glands.

Golgi apparatus, a collective term for all Golgi bodies in a cell.

Golgi body, a cell organelle (also called dictyosome) comprising cisternae, tubules and vesicles.

Gonophyll, (gonophyll theory), a sterile leaf together with an ovule bearing branch, from which carpel has been derived.

Grana, a structure found in choroplasts of higher plants.

Growth ring, a distinguishable region in the secondary xylem which is formed during a single growth season.

Guard cells, a pair of specialised parenchyma cells bounding a stoma.

Gummosis, a condition which is expressed by the formation of gum.

Guttation, the secretion of water in liquid form from the plant.

Gynoecium, refers to all the carpels of a single flower.

Gynophore, an extended stalk between stamens and carpels, which elevates gynoecium.

Halophyte, a plant growing in saline soils.

Haplostele, a protostele in which the xylem core is more or less circular in cross section.

Hard wood, a general term used to refer wood of dicotyledons (angiosperms).

Heart wood (the duramen), the inner layers of secondary wood which is darker in colour and has lost the ability of conduction. No living cells are found in this region.

Hilum, (1) a small granule at the centre of a starch grain, (2) the scar on the testa of a seed where it was attached to the funicle.

Histogen, apical meristem showing three distinct zones, the *dermatogen*, the *periblem* and the *plerome* (as proposed by Hanstein).

Hydathode, a structure found in leaves through which water is secreted in the form of liquid.

Hypoblast, a term used for the suspensor of the mature monocot embryo.

Hypodermis, a layer of one or more cells thick, or thickened cells lying just below the epidermis.

Hypophysis, one of the cells of the embryo found in its early stage of development.

Hypsophyll, an inflorescence bract, (a leaf found on the top of the shoot which is structurally different from other foliage leaves).

Idioblast, a plant cell rich in ergastic substances and which differs from the surrounding cells.

Integument, an envelope surrounding the nucellus (in ovule).

Internode, the part of the stem found in between two nodes.

Interfasicular, situated between the vascular bundles.

Interxylary, within and surrounded by xylem tissue.

Intrafasicular, within a vascular bundle.

Intraxylary, on the innerside of xylem relative to the axis of the plant.

Intussusception, the growth in cell wall as a result of the interpolation of new wall material within the wall already formed.

Isobilateral, (1) (a leaf), the unifacial leaf as found in most monocots; (2) a form of bilateral symmetry where a structure is divisible in two planes at right angles.

Karyokinesis, the process of nuclear division.

Karyolymph, the nuclear sap.

Knot, a portion of branch, the wood of which has become heartwood and which is embedded in the developing wood of the stem, from which the branch arose.

Lacuna, (1) an intercellular space, (2) an interruption in the vascular tissue of the central cylinder.

Latewood, also called summer wood, the wood formed in the later part of the growth layer, it is denser and comprises small cells.

Laticifer, a cell or group of cells containing latex.

Leaf gap, a gap region in vascular cylinder which appears during the departure of a leaf trace and it consists of parenchyma.

Leaf trace, vascular tissue from the stem to the leaf.

Lenticel, the minute pore in the periderm of woody stems meant for gaseous exchange.

Leucoplast, a colourless plastid, it contains no pigment.

Libriform fibre, a xylem fibre with thick walls and simple pits.

Lignin, a polymer containing phenolic derivatives of phenylpropane.

Lithocyst, a cell containing a cystolith.

Lysigenous, the formation of space as a result of disintegration of cells.

Lysosome, a cell organelle rich in hydrolytic enzymes.

Maceration, a process of separating individual cells of a tissue by disintegration of middle lamella.

Macrosclereid, a some what elongated sclereid (columnar).

Malphighian cell, the macrosclereid found in seed coat of legume seeds and represents the epidermis.

Mantle, the outer layers of shoot apex showing regularly arranged cells.

Marginal initial, the meristematic cell along the margin of the growing lamina of leaf.

Marginal meristem, the meristem found along the margin and brings about growth of margin.

Mass meristem, group of meristematic tissues which divide in various planes.

Massula, a group of fused pollen grains, which participate in the formation of pollinium.

Medulla, a pith, ground tissue in the centre of a stem or root.

Meristele, one of the bundles of dictyostele.

Meristem, a tissue which produces cells.

Mesarch (xylem), xylem having metaxylem developing in all directions from the protoxylem (as in ferns)

Mesocarp, the middle layer of the pericarp.

Mesocotyl, refers to internode between the scutellar node and the coleoptile (as in grasses).

Mesophyll, the (inner) ground tissue of a leaf blade.

Metaphloem, part of primary phloem that differentiates after phloem.

Metaxylem, part of primary xylem which differentiates after the protoxylem.

Micelle, a unit of cellulose in which molecules are arranged parallel to one another, found in cell wall.

Micropyle, a sub-microscopic thread-like structure found in the cell wall.

Micropyle, an opening at the free end of ovule (in the integuments).
Microspórocyte, the cell that gives rise to microspore.
Mitochondrion, an organelle associated with aerobic respiration in eukaryotes.
Morphogenesis, development of form or structure in ontogeny or in regeneration.
Multiple epidermis, an epidermis with two or more cell layers.
Myrosin cell, a cell containing the myrosin enzyme (found in Cruciferae).

Nectary, a sugary fluid-secreting multicellular glandular structure.
Nexine, the inner layer of exine.
Node, a place on stem where one or more leaves are attached.
Nucellus, the tissue within ovule in which female gametophyte develops.
Nuptial nectary, the intrafloral nectary.

Obturator, an outgrowth of the placenta or stylar canal which brings the transmitting tissue closer to the micropyle.
Occlusion, the process by which wounds in stem are healed by the growth of callus.
Ontogeny, the process of development of tissue, organ or an organism towards maturation.
Open bundle, a vascular bundle in which cambium is present.
Osteosclereid, a bone-shaped sclereid.

Paracytic type (stomata), in which one or more subsidiary cells remain parallel with the long axis of the guard cells.
Parenchyma, a simple thin walled tissue with intercellular spaces.
Parenchyma, apotracheal, axial parenchyma of secondary xylem which remain distributed, independent of vessels.
Parenchyma, axial, parenchyma of vertical system of secondary xylem.
Parenchyma, paratracheal, axial parenchyma of secondary xylem associated with vessels.
Parthenocarpy, the production of fruit without fertilisation (i.e. seedless fruit).
Passage cells, the thin-walled cells of exodermis or endodermis which permit passage of solution.
Pelltate hair, a hair with discoid plate of cells borne on stalk.
Perforation plate, the perforated part of the cell wall of vessel.
Periblem, a layer of histogen which forms the cortex.
Pericarp, the fruit wall.
Periclinal, parallel to the surface.
Pericycle, the outer zone of stele present in between endodermis and the conducting tissue.
Periderm, the secondary protective tissue layers derived from the phellogen. It consists of *phellem, phellogen* and *phelloderm*.
Perisperm, a nutritive tissue in some seeds, derived from the nucellus of the ovule.
Phellem, *see* cork.
Phelloderm, the secondary suberised cortex formed on the inner side of cork cambium.
Phellogen, the cork cambium, the secondary lateral meristem which produces the phellem and phelloderm.
Phelloid cell, an unsuberised cork cell.
Phloem, the principal food conducting tissue of the vascular plants.
Phloem, interxylary, secondary phloem which occurs in between the xylem.
Phloem, intraxylary, also called internal phloem, phloem lying between the xylem and the centre of the stem (pith).

Phloem ray, a vascular ray present in phloem region.

Phragmoplast, a fibrous structure which appears during mitotic telophase between the two daughter nuclei and within which the cell plate is formed. It comprises microtubule.

Phyllome, a collective term to refer all types of leaves.

Phylogeny, the evolutionary history of an organism or taxa.

Piliferous cell, the trichoblast.

Piliferous layer, the young superficial layer of tissue of roots producing the root hairs, when present.

Pit, a depression in the cell wall with secondary thickening. In such area, only middle lamella and primary wall are present.

Pit membrane, the middle lamella and primary wall enclosing the pit cavity on the outerside.

Placenta, the region of the attachment of the ovules to the carpel.

Placentation, the position of placenta in the ovary.

Plasma membrane, the membrane on the outer surface of cytoplasm.

Plasmodesmata, a thin cytoplasmic strand which passes through a pore in the cell wall, and which connects protoplasts of two cells.

Plastid, a small self-propagating body in the cytoplasm of plant cell.

Plastochron, the intermediate time between the commencement of two successive and repetitive phenomena, e.g. between the initiation of two successive leaf primordia.

Plate meristem, a plate-like meristem, the cells of which divide anticlinally.

Plectostele, a type of protostele in which xylem is arranged in longitudinal plates.

Plerome, a zone of histogen that brings about formation of stelar system and associated ground tissue.

Plumule, the shoot apex of the embryo.

Pneumatode, a group of cells in the velamen with dense spiral thickening.

Pollinium, the entire pollen grain mass of a single pollen sac, when these adhere together to form one mass.

Polyderm, a special type of protective tissue comprising alternating bands of suberised and non-suberised parenchyma cells.

Polyembryony, the presence of more than one embryo in an ovule.

Porous wood, the secondary wood having vessels or pores.

Primary body, part of the plant that arises from the apical meristems and their derivatives before the secondary growth.

Primordium, an organ or organised group of cells in the earliest stage of differentiation.

Procambium, a primary meristem which undergoes differentiation to form the primary vascular tissues.

Proembryo, the embryo in the earliest stage of development.

Promeristem, the initials and their immediate derivatives in the apical meristem.

Prosenchyma, tissue of thick-walled elongated cells with overlapping tapering ends.

Protoderm, the meristem of the epidermis.

Protophloem, the first formed elements of the primary phloem.

Protostele, the simplest type of stele which comprises a solid core of xylem surrounded by the phloem.

Protoxylem, the first formed elements of the primary xylem.

Provascular bundle, a procambium strand.

Proximal, the part nearest to the axis.

Quiescent centre, a zone of inactive cells in the root apex of angiosperms.

Raphe, a ridge along the seed formed by that part of the funiculus which was fused to the ovule.

Raphide, a needle-shaped crystal of calcium oxalate found in bundles in certain plants.

Ray initials, cells in the vascular cambium that give rise to vascular cambium.

Reaction wood, the wood which develops under stress conditions, such as formed in parts of leaning stem.

Residual meristem, least differentiated part of the apical meristem.

Rhytidome, the outer bark consisting of periderm and tissues isolated by it.

Rib meristem, a zone in shoot apex.

Ring porous, that part of wood in which vessels of earlier wood are distinctly larger than those of late wood, as found in temperate trees.

Sap wood, also called **alburnum**, the outer and that portion of wood in living plants which contains living cells and reserve materials and in which conduction of water occurs.

Scar tissue, consisting of necrosed cells injured by the wounding and subjacent cells impregnated with protective substances.

Schizogenous, the manner of development of intercellular spaces by the separation of cells along their middle lamellae.

Schizo-lysigenous, the manner of development of intercellular spaces by both cell separation along their middle lamellae and cell disintegration.

Sclereid, a sclerenchymatous cell of various shape and which is not much elongated and often contains many pits.

Sclerenchyma, a simple tissue consisting of fibres and sclereids and non-living contents.

Scutellum, the specialised cotyledon of the embryo of monocots (Graminae).

Secondary cortex, also called phelloderm, the cells of periderm inner to the phellogen.

Secondary meristem, a meristem that develops from parenchyma (living permanent tissue).

Secondary tissue, the tissue which is derived from lateral meristem.

Separation layer, that layer in the abscission zone, the cells of which disintegrate (or separate) and cause the abscission of leaf or fruit.

Sexine, the outer layer of exine.

Sieve area, an area in the wall of sieve element which appears as a depression and contain pores commonly lined with callose.

Siphonostele, a stele in which pith is present in the central portion.

Soft wood, a general term to refer secondary wood of conifers (gymnosperms).

Solenostele, an amphiphloic siphonostele in which successive leaf gaps are widely separated.

Spring wood, the early wood (or first formed wood) of the annual ring.

Stele, that portion of plant axis which consists of vascular system and its associated ground tissue.

Stereome, the supporting tissue, i.e. sclerenchyma and collenchyma.

Stomatal crypt, a depression in leaf in which stomata are present.

Subsidiary cell, *see* accessory cell.

Succulent, fleshy and juicy.

Supporting tissue or mechanical tissue, the tissue with thick-walled cells which provide mechanical strength to the plant body.

Suspensor, that part of embryo which connects the main part of embryo to the basal cell.

Synergids, the cells present on either side of egg cell in a mature embryo.

Tapetum, the inner most layer of pollen sac wall.

Tegmen, the portion of seed coat produced by the inner integument.
Tension wood, the reaction wood produced in angiosperms.
Testa, the seed coat.
Tetrarch, the primary xylem of root with four protoxylem strands.
Tonoplast, the cytoplasmic membrane surrounding the vacuole.
Torus, the thickened central portion of the pit membrane in a bordered pit.
Trabecula, a rod-like structure or a cell running across a cavity.
Trachea, also called vessel.
Tracheary element, a xylem element which takes part in water transport. It consists of tracheid and vessel member.
Tracheid, a tracheary element of xylem which is not perforated.
Transfer cell, a parenchyma cell with wall ingrowth to increase the surface of plasma membrane and it is specialised in short distance movement of solute.
Trilacunar node, a node with three leaf gaps associated with one leaf.
Tunica, the outer or peripheral layer or layers of apical shoot meristem in which the cells divide only anticlinally.
Tylosis, an outgrowth of parenchyma cells in the lumen of tracheid or vessel.
Tylosoid, a proliferation of an epithelial cell into an intercellular cavity such as resin or gum duct, or of a parenchyma cell into a sieve element.

Unifacial leaf, a leaf having similar structure on both sides, as in a monocot leaf.
Unilacunar node, a node with one leaf gap related to one leaf.
Uniseriate ray, vascular ray which is only one cell wide.

Vacuole, a cell sap containing cavity in cytoplasm externally bounded by tonoplast.
Vascular bundle, a strand of conducting tissue comprises xylem and phloem.
Vein, the conducting strand of a leaf.
Velamen, the multiseriate spongy epidermis found in roots of certain orchid roots.
Venation, the arrangement of veins in a leaf.
Vessel, a series of vessel members joined end to end by their perforated end walls.

Wood, the secondary xylem tissue.
Wound gum, the gum formed as a result of some injury.

Xeromorphic, having the structure found in a typical xerophyte.
Xylem, a complex tissue concerned with conduction of water.
Xylotomy, the anatomy of xylem.

Index

Part One

PLANT ANATOMY

Part Two

Embryology and Palynology